Structural Health Monitoring of Composite Structures Using Fiber Optic Methods

Devices, Circuits, and Systems

Series Editor
Krzysztof Iniewski
Emerging Technologies CMOS Inc.
Vancouver, British Columbia, Canada

PUBLISHED TITLES:

Analog Electronics for Radiation Detection
Renato Turchetta

Atomic Nanoscale Technology in the Nuclear Industry
Taeho Woo

Biological and Medical Sensor Technologies
Krzysztof Iniewski

Building Sensor Networks: From Design to Applications
Ioanis Nikolaidis and Krzysztof Iniewski

Cell and Material Interface: Advances in Tissue Engineering, Biosensor, Implant, and Imaging Technologies
Nihal Engin Vrana

Circuits and Systems for Security and Privacy
Farhana Sheikh and Leonel Sousa

Circuits at the Nanoscale: Communications, Imaging, and Sensing
Krzysztof Iniewski

CMOS: Front-End Electronics for Radiation Sensors
Angelo Rivetti

CMOS Time-Mode Circuits and Systems: Fundamentals and Applications
Fei Yuan

Design of 3D Integrated Circuits and Systems
Rohit Sharma

Electrical Solitons: Theory, Design, and Applications
David Ricketts and Donhee Ham

Electronics for Radiation Detection
Krzysztof Iniewski

Electrostatic Discharge Protection: Advances and Applications
Juin J. Liou

Embedded and Networking Systems: Design, Software, and Implementation
Gul N. Khan and Krzysztof Iniewski

PUBLISHED TITLES:

Energy Harvesting with Functional Materials and Microsystems
Madhu Bhaskaran, Sharath Sriram, and Krzysztof Iniewski

Gallium Nitride (GaN): Physics, Devices, and Technology
Farid Medjdoub

**Graphene, Carbon Nanotubes, and Nanostuctures:
Techniques and Applications**
James E. Morris and Krzysztof Iniewski

High-Speed Devices and Circuits with THz Applications
Jung Han Choi

High-Speed Photonics Interconnects
Lukas Chrostowski and Krzysztof Iniewski

**High Frequency Communication and Sensing:
Traveling-Wave Techniques**
Ahmet Tekin and Ahmed Emira

Integrated Microsystems: Electronics, Photonics, and Biotechnology
Krzysztof Iniewski

Integrated Power Devices and TCAD Simulation
Yue Fu, Zhanming Li, Wai Tung Ng, and Johnny K.O. Sin

Internet Networks: Wired, Wireless, and Optical Technologies
Krzysztof Iniewski

Ionizing Radiation Effects in Electronics: From Memories to Imagers
Marta Bagatin and Simone Gerardin

Labs on Chip: Principles, Design, and Technology
Eugenio Iannone

Laser-Based Optical Detection of Explosives
Paul M. Pellegrino, Ellen L. Holthoff, and Mikella E. Farrell

Low Power Emerging Wireless Technologies
Reza Mahmoudi and Krzysztof Iniewski

Medical Imaging: Technology and Applications
Troy Farncombe and Krzysztof Iniewski

Metallic Spintronic Devices
Xiaobin Wang

MEMS: Fundamental Technology and Applications
Vikas Choudhary and Krzysztof Iniewski

Micro- and Nanoelectronics: Emerging Device Challenges and Solutions
Tomasz Brozek

Microfluidics and Nanotechnology: Biosensing to the Single Molecule Limit
Eric Lagally

PUBLISHED TITLES:

MIMO Power Line Communications: Narrow and Broadband Standards, EMC, and Advanced Processing
Lars Torsten Berger, Andreas Schwager, Pascal Pagani, and Daniel Schneider

Mixed-Signal Circuits
Thomas Noulis

Mobile Point-of-Care Monitors and Diagnostic Device Design
Walter Karlen

Multisensor Data Fusion: From Algorithm and Architecture Design to Applications
Hassen Fourati

Nano-Semiconductors: Devices and Technology
Krzysztof Iniewski

Nanoelectronic Device Applications Handbook
James E. Morris and Krzysztof Iniewski

Nanomaterials: A Guide to Fabrication and Applications
Sivashankar Krishnamoorthy

Nanopatterning and Nanoscale Devices for Biological Applications
Šeila Selimović

Nanoplasmonics: Advanced Device Applications
James W. M. Chon and Krzysztof Iniewski

Nanoscale Semiconductor Memories: Technology and Applications
Santosh K. Kurinec and Krzysztof Iniewski

Novel Advances in Microsystems Technologies and Their Applications
Laurent A. Francis and Krzysztof Iniewski

Optical, Acoustic, Magnetic, and Mechanical Sensor Technologies
Krzysztof Iniewski

Optical Fiber Sensors: Advanced Techniques and Applications
Ginu Rajan

Optical Imaging Devices: New Technologies and Applications
Ajit Khosla and Dongsoo Kim

Organic Solar Cells: Materials, Devices, Interfaces, and Modeling
Qiquan Qiao

Physical Design for 3D Integrated Circuits
Aida Todri-Sanial and Chuan Seng Tan

Radiation Detectors for Medical Imaging
Jan S. Iwanczyk

Radiation Effects in Semiconductors
Krzysztof Iniewski

PUBLISHED TITLES:

Reconfigurable Logic: Architecture, Tools, and Applications
Pierre-Emmanuel Gaillardon

Semiconductor Radiation Detection Systems
Krzysztof Iniewski

Smart Grids: Clouds, Communications, Open Source, and Automation
David Bakken

Smart Sensors for Industrial Applications
Krzysztof Iniewski

Soft Errors: From Particles to Circuits
Jean-Luc Autran and Daniela Munteanu

Solid-State Radiation Detectors: Technology and Applications
Salah Awadalla

Structural Health Monitoring of Composite Structures Using Fiber Optic Methods
Ginu Rajan and Gangadhara Prusty

Technologies for Smart Sensors and Sensor Fusion
Kevin Yallup and Krzysztof Iniewski

Telecommunication Networks
Eugenio Iannone

Testing for Small-Delay Defects in Nanoscale CMOS Integrated Circuits
Sandeep K. Goel and Krishnendu Chakrabarty

Tunable RF Components and Circuits: Applications in Mobile Handsets
Jeffrey L. Hilbert

VLSI: Circuits for Emerging Applications
Tomasz Wojcicki

Wireless Medical Systems and Algorithms: Design and Applications
Pietro Salvo and Miguel Hernandez-Silveira

Wireless Technologies: Circuits, Systems, and Devices
Krzysztof Iniewski

Wireless Transceiver Circuits: System Perspectives and Design Aspects
Woogeun Rhee

FORTHCOMING TITLES:

Advances in Imaging and Sensing
Shuo Tang and Daryoosh Saeedkia

High Performance CMOS Range Imaging: Device Technology and Systems Considerations
Andreas Süss

FORTHCOMING TITLES:

Introduction to Smart eHealth and eCare Technologies
Sari Merilampi, Krzysztof Iniewski, and Andrew Sirkka

Magnetic Sensors: Technologies and Applications
Laurent A. Francis and Kirill Poletkin

MRI: Physics, Image Reconstruction, and Analysis
Angshul Majumdar and Rabab Ward

Multisensor Attitude Estimation: Fundamental Concepts and Applications
Hassen Fourati and Djamel Eddine Chouaib Belkhiat

Nanoelectronics: Devices, Circuits, and Systems
Nikos Konofaos

Power Management Integrated Circuits and Technologies
Mona M. Hella and Patrick Mercier

Radio Frequency Integrated Circuit Design
Sebastian Magierowski

Semiconductor Devices in Harsh Conditions
Kirsten Weide-Zaage and Malgorzata Chrzanowska-Jeske

Structural Health Monitoring of Composite Structures Using Fiber Optic Methods

Edited by
Ginu Rajan
The University of Wollongong, Australia
B. Gangadhara Prusty
UNSW Australia

Krzysztof Iniewski MANAGING EDITOR
CMOS Emerging Technologies Research Inc.
Vancouver, British Columbia, Canada

CRC Press
Taylor & Francis Group
Boca Raton London New York

CRC Press is an imprint of the
Taylor & Francis Group, an **informa** business

CRC Press
Taylor & Francis Group
6000 Broken Sound Parkway NW, Suite 300
Boca Raton, FL 33487-2742

First issued in paperback 2020

ISBN-13: 978-1-4987-3317-5 (hbk)
ISBN-13: 978-0-367-73693-4 (pbk)

Library of Congress Cataloging-in-Publication Data

Names: Rajan, Ginu, editor. | Prusty, B. Gangadhara, editor.
Title: Structural health monitoring of composite structures using fiber optic methods / editor, Ginu Rajan and B. Gangadhara Prusty.
Description: Boca Raton : Taylor & Francis, a CRC title, part of the Taylor & Francis imprint, a member of the Taylor & Francis Group, the academic division of T&F Informa, plc, [2016] | Series: Devices, circuits, and systems
Identifiers: LCCN 2016006391 | ISBN 9781498733175 (hard back)
Subjects: LCSH: Structural health monitoring. | Composite materials--Testing. | Structural analysis (Engineering) | Fiber optics--Industrial applications.
Classification: LCC TA656.6 .S7745 2016 | DDC 624.1/8--dc23
LC record available at https://lccn.loc.gov/2016006391

Visit the Taylor & Francis Web site at
http://www.taylorandfrancis.com

and the CRC Press Web site at
http://www.crcpress.com

Contents

Foreword xi

Editors xiii

Contributors xv

Chapter 1   Introduction to Composite Materials and Smart Structures 1

B. Gangadhara Prusty, Ebrahim Oromiehie, and Ginu Rajan

Chapter 2   Structural Health Monitoring Methods for Composite Materials 21

*Wiesław M. Ostachowicz, Tomasz Wandowski, and
Paweł H. Malinowski*

Chapter 3   Introduction to Optical Fiber Sensors 41

Dipankar Sengupta and Ginu Rajan

Chapter 4   Structural Health Monitoring of Composite Materials Using
Fiber Bragg Gratings 69

*Christophe Caucheteur, Damien Kinet, Nicolas Lammens,
Thomas Geernaert, Francis Berghmans, Geert Luyckx,
Joris Degrieck, and Patrice Mégret*

Chapter 5   Structural Health Monitoring of Composite Materials Using
Distributed Fiber-Optic Sensors 105

Hideaki Murayama

Chapter 6   Importance of Strain Transfer Effects for Embedded Multiaxial
Strain Sensing and Optical Fiber Coating Optimization 157

*Nicolas Lammens, Geert Luyckx, Thomas Geernaert, Damien
Kinet, Francis Berghmans, Christophe Caucheteur, and
Joris Degrieck*

Chapter 7   Monitoring Process Parameters Using Optical Fiber Sensors in
Composite Production 201

Edmon Chehura

Chapter 8   FBG-Based Structural Performance Monitoring and Safety
Evaluation of a Composite Arch Bridge..269

Xiao-Wei Ye, Tan Liu, Chuan-Zhi Dong, and Bin Chen

Chapter 9   Smart Composite Textiles with Embedded Optical Fibers..............299

Tamer Hamouda and Abdel-Fattah M. Seyam

Chapter 10 Smart Aerospace Composite Structures...339

Gayan Kahandawa and Jayantha Ananda Epaarachchi

Chapter 11 Advances in FRP-Based Smart Components and Structures 371

Zhi Zhou, Yung William Sasy Chan, and Jinping Ou

Chapter 12 Fiber-Optic Structural Health Monitoring Systems for Marine
Composite Structures 415

Raju and B. Gangadhara Prusty

Index..481

Foreword

Since the development and deployment of optical fiber for communications in the 1970s, the application of optical fibers as sensors for a wide variety of measurands has been the subject of extensive research and has resulted in the development of a wide variety of sensor types for a vast range of areas, from the macro scale, for example, structural sensing for bridges and other civil structures, to the nanoscale, for biological and chemical sensing.

Compared with conventional electronic sensors, optical fiber sensors offer useful advantages such as high sensitivity, compact size, light weight, immunity to electromagnetic field interference, the capability to work in high-temperature environments, and the potential for remote operation. Optical fiber sensors can be classified as either point sensors, sensing a measurand at a single point, or distributed sensors, in which case sensing takes place in a continuum over distances that can range from centimeters to tens of kilometers.

Composite material structures are widely used in the aerospace, marine, aviation, transport, and civil engineering industries. They offer high strength coupled to light weight by comparison with traditional metal and other alternatives. However, as with all structures, composite structures are frequently subjected to external perturbations and varying environmental conditions, which may cause the structures to suffer from fatigue damage and/or failures. Thus, real-time structural health monitoring is needed in many applications, with the expectation that the structural health monitoring will be possible at all times during the working life of the structure. The goal of structural health monitoring is to detect, identify, locate, and assess the defects that may affect the safety or performance of a structure.

Sensors that are commonly employed for structural health monitoring are resistance strain gages, fiber optic sensors, piezoelectric sensors, eddy current sensors, and micro-electromechanical systems sensors. Traditional nondestructive evaluation techniques such as ultrasonic inspection, acoustography, low-frequency methods, radiographic inspection, shearography, acousto-ultrasonics, and thermography are effective for structural health monitoring of composite materials and structures, but are difficult or impossible to use in an operational structure due to the size and weight of the systems involved. Fiber-optic sensors, on the other hand, are very suitable candidates for structural health monitoring in composite materials during operation, since they are capable of achieving the goals of diagnostics as well as condition monitoring and can be embedded into such structures. Over the last decade and more, fiber-optic sensors embedded in composite structures have been shown to be capable of monitoring stress/strain, temperature, the composite cure process, vibration, thermal expansion, humidity, delamination, and cracking. The potential of optical fiber sensors for condition monitoring of composites has been a major foundation of the emergence of so-called smart composite structures, where the sensors act in many ways as the equivalent of a human nervous system for the material and structures.

This textbook provides a comprehensive overview of structural health monitoring for composite structures using fiber-optic methods. Drawing on the extensive experience of a wide range of distinguished contributors, this textbook covers the variety of topics relevant to structural health monitoring in composite materials. The book opens with a chapter covering the useful prerequisite of fabrication methods in advanced composite structures. The topic of structural health monitoring methods for composite materials in a general sense is then considered to put the use of fiber-optic sensors into a larger perspective. This is followed by an overview of fiber-optic sensors, covering a variety of sensor types and operating principles. In the chapter that follows, more specific coverage of fiber Bragg grating sensors is provided. Fiber Bragg gratings are one of the most common sensor types deployed in composite structures for structural health monitoring by point sensing, offering high sensitivity to strain and a mature, well-understood technology base. While multiple fiber Bragg gratings can be used for sensing at a number of points in a structure, for true distributed sensing, sensors such as polarimetric sensors and sensors based on scattering in fibers are used, and thus, a chapter is dedicated to such sensors. In the next chapter, fiber-optic acoustic emission sensors are addressed. Such sensors are a promising approach to detecting critical precursors to structural failure, such as crack growth and delamination. The complex topic of strain transfer within a composite structure is considered in the chapter that follows, reflecting the real-world importance of a good understanding of how sensors interact with the host material when embedded in composite materials.

Given the rapid rise in the use and applications of composite structures, composite fabrication and production is a topic of great importance. Live monitoring of the production process is increasingly the norm for complex and high-cost composite parts, and a chapter is dedicated to the use of optical fiber sensors for monitoring process parameters such as cure rate and temperature.

The chapters that follow cover a range of very topical and useful application areas for optical fiber sensors. Composite bridges are a novel approach to one of the most common civil engineering structures in use, and a chapter is thus dedicated to their monitoring and evaluation. In the chapter that follows, smart composite textiles with embedded optical fibers are considered. Textiles with embedded sensors are proving to be an attractive development with applications from large scales, in monitoring soil movement, to smaller scales, for sensing in smart clothing for medical applications. The chapter that follows considers the applications of sensors in the aerospace industry, reflecting the massive growth that has taken place in the last decade in the deployment of composites in aircraft to reduce weight and thus lower fuel usage. The subsequent chapter then describes recent advances in sensing in smart fiber-reinforced plastic and fiber-reinforced polymers. Finally, the closing chapter considers the case study of SHM for curved composite structures.

Through a combination of a wide variety of related topics in SHM for composite materials, this textbook is an invaluable reference for those who work with composite materials and structures. The material is suitable for scientists and engineers working in the field as well as undergraduate and graduate students studying or carrying out research in mechanical, mechatronic, structural, biomedical, and electrical engineering.

Professor Gerald Farrell,
Director, Photonics Research Centre
Dublin Institute of Technology, Ireland

Editors

Ginu Rajan is a vice-chancellor's research fellow/lecturer at the University of Wollongong, Australia, and is also a visiting fellow at the University of New South Wales, Australia. He received his BSc degree in Physics from the University of Kerala and MSc degree in Applied Physics from Mahatma Gandhi University, Kerala, India, in 2000 and 2002, respectively. He worked as a researcher at the Indian Institute of Astrophysics during the period 2003–2005. He subsequently undertook research in the area of optical fiber sensors, in which he gained a PhD from Dublin Institute of Technology, Ireland, in 2009.

During 2009–2012, Dr. Rajan worked as a project manager at the Photonics Research Centre of Dublin Institute of Technology in collaboration with the Warsaw University of Technology, Poland, and from 2012–2014 as a research fellow/lecturer at the University of New South Wales. He has published over 120 articles in journals, at conferences, and as book chapters, and two patents are also granted to him. He has also given invited talks at conferences and is a technical program committee member of several conferences in the smart structures and photonics area. He is currently a reviewer for more than 23 scientific journals and also a reviewer for grant applications of the Portugal Science Foundation and Australian Research Council. He is the editor of the book *Optical Fiber Sensors: Advanced Techniques and Applications* and is also an editorial board member of the *Scientific World Journal*. His research and teaching interest includes optical fiber sensors and their applications in biomedical engineering, fiber Bragg grating interrogation systems, photonic crystal fiber sensors, polymer fiber sensors, smart structures, and physics of photonic devices. He can be reached at ginu@uow.edu.au or ginurajan@gmail.com.

B. Gangadhara Prusty is a professor for the University of New South Wales Mechanical and Manufacturing Engineering and leads the School's Advanced Structures and Materials group. He is also the deputy director of the Centre for Sustainable Materials Research Technology at the University of New South Wales.

His research strength is in the mechanics of composites at nano, micro, and macro scales, embodied with the latest analysis and modeling techniques blended with material characterization.

Gangadhara has already contributed to a number of fundamental developments in the field of mechanics of composite materials and structures, such as the novel finite element formulation for stiffened structures, *in situ* monitoring of robotic composite manufacturing, hierarchical multiscale submodeling approach for the onset theory, efficient modeling of barely visible impact damage in post-buckled structures, robust design optimization for layups for shape-adaptive composite propellers, and mitigating creep and cracking in thermoplastic composite welding.

Professor Prusty has led a number of major internationally collaborative projects, such as Systems for Crashworthiness and Robust Optimisation for Imperfection Sensitive Composite Launcher Structures at the University of New South Wales, through external funding. His research is closely aligned with the emerging the

University of New South Wales research strength of next-generation materials and technologies.

Krzysztof (Kris) Iniewski manages R&D at Redlen Technologies Inc., a start-up company in Vancouver, Canada. Redlen's revolutionary production process for advanced semiconductor materials enables a new generation of more accurate, all-digital, radiation-based imaging solutions. Kris is also a founder of ET CMOS Inc. (www.etcmos.com), an organization of high-tech events covering communications, microsystems, optoelectronics, and sensors. In his career, Dr. Iniewski has held numerous faculty and management positions at University of Toronto, University of Alberta, Simon Fraser University, and PMC-Sierra Inc. He has published over 100 research papers in international journals and conferences. He holds 18 international patents granted in the USA, Canada, France, Germany, and Japan. He is a frequent invited speaker and has consulted for multiple organizations internationally. He has written and edited several books for CRC, Cambridge University Press, IEEE, Wiley, McGraw Hill, Artech House, and Springer. His personal goal is to contribute to healthy living and sustainability through innovative engineering solutions. In his leisure time, Kris can be found hiking, sailing, skiing, or biking in beautiful British Columbia. He can be reached at kris.iniewski@gmail.com.

Contributors

Francis Berghmans
Brussels Photonics Team
Department of Applied Physics and
  Photonics
University of Brussels
Brussels, Belgium

Christophe Caucheteur
Electromagnetism and Telecom
  Department
University of Mons
Mons, Belgium

Yung William Sasy Chan
School of Civil Engineering
Dalian University of Technology
Dalian, China

Edmon Chehura
Department of Engineering Photonics
Cranfield University
Cranfield, United Kingdom

Bin Chen
Department of Civil Engineering
Zhejiang University
Hangzhou, China

Joris Degrieck
Department of Materials Science and
  Engineering
Gent University
Gent, Belgium

Chuan-Zhi Dong
Department of Civil Engineering
Zhejiang University
Hangzhou, China

Jayantha Ananda Epaarachchi
School of Mechanical and Electrical
  Engineering
University of Southern Queensland
Toowoomba, Australia

Thomas Geernaert
Brussels Photonics Team
Department of Applied Physics and
  Photonics
University of Brussels
Brussels, Belgium

Tamer Hamouda
Textile Research Division
National Research Centre
Dokki, Egypt

Gayan Kahandawa
School of Engineering and Information
  Technology
Federation University
Ballarat, Australia

Damien Kinet
Electromagnetism and Telecom
  Department
University of Mons
Mons, Belgium

Nicolas Lammens
Department of Materials Science and
  Engineering
Gent University
Gent, Belgium

Tan Liu
Department of Civil Engineering
Zhejiang University
Hangzhou, China

Geert Luyckx
Department of Materials Science and
   Engineering
Gent University
Gent, Belgium

Paweł H. Malinowski
Institute of Fluid–Flow Machinery
Polish Academy of Sciences
Gdańsk, Poland

Patrice Mégret
Electromagnetism and Telecom
   Department
University of Mons
Mons, Belgium

Hideaki Murayama
Department of Systems Innovation
School of Engineering
University of Tokyo
Tokyo, Japan

Ebrahim Oromiehie
School of Mechanical and
   Manufacturing Engineering
University of New South Wales
Sydney, Australia

Wiesław M. Ostachowicz
Institute of Fluid–Flow Machinery
Polish Academy of Sciences
Gdańsk, Poland

and

Faculty of Automotive and Construction
   Machinery
Warsaw University of Technology
Warsaw, Poland

Jinping Ou
School of Civil Engineering
Harbin Institute of Technology
Harbin, China

and

School of Civil Engineering
Dalian University of Technology
Dalian, China

Raju
BMS College of Engineering
Bangalore, India

Dipankar Sengupta
École de technologie supérieure
Université du Québec
Montreal, Canada

Abdel-Fattah M. Seyam
College of Textiles
North Carolina State University
Raleigh, NC

Tomasz Wandowski
Institute of Fluid–Flow Machinery
Polish Academy of Sciences
Gdańsk, Poland

Xiao-Wei Ye
Department of Civil Engineering
Zhejiang University
Hangzhou, China

Zhi Zhou
School of Civil Engineering
Dalian University of Technology
Dalian, China

1 Introduction to Composite Materials and Smart Structures

B. Gangadhara Prusty, Ebrahim Oromiehie, and Ginu Rajan

CONTENTS

1.1 Introduction 1
1.2 Composite Materials: Structures and Types 2
    1.2.1 Reinforcements 3
    1.2.2 Matrix 4
1.3 Composite Manufacturing Methods and Technologies 6
    1.3.1 Prepreg Layup Method 6
    1.3.2 Wet Layup Method 7
    1.3.3 Spray Layup Method 8
    1.3.4 Filament Winding Method 9
    1.3.5 Pultrusion Method 10
    1.3.6 Resin Transfer Molding (RTM)/Structural Reaction Injection
           Molding (SRIM) Method 11
    1.3.7 ATP/AFP Method 11
    1.3.8 Design of Composite Structures 12
1.4 Smart Structures and Structural Health Monitoring 13
1.5 Demand for the SHM in Composite Structures 14
1.6 SHM System for Composite Structures 16
References 17

1.1 INTRODUCTION

Composite structures are widely used in the aerospace, marine, aviation, transport, sport/leisure, and civil engineering industries [1]. In the past decade, advanced composite materials have been widely used in a variety of load-bearing structures such as rotor blades, aircraft fuselage, and wing skins [2]. For example, 52% by weight of the new Dreamliner and 25% by weight of the new Airbus A380 are made from composite materials. The unique properties of composite materials such as their high strength-to-weight ratio, high creep resistance, high tensile strength at elevated temperatures, and high toughness have been attracting increasing interest in numerous applications in different industries such as the automotive and aerospace industries [1,3]. It should be noted that the annual growth in the composite industry is approximately 17%, mainly backed by four industrial areas: aerospace/defense, wind turbine, automotive, and sports [4]. One of the main factors that drives industry to use composite materials is the benefit of lower weight, resulting in improved energy efficiency and less CO_2 emissions. This increasing use of composite materials in safety-critical applications such as aviation brings with it a need to establish inspection and/or monitoring regimes to ensure structural integrity and safe operation throughout service life. This can make the cost of ownership high, with downtime being an important issue. Therefore, any structural health monitoring (SHM) methodologies that could increase the inspection intervals for a structure and indicate damage before costly failures occur would be very advantageous financially and, in many cases, could increase safety levels. This chapter provides a brief introduction to composite materials and smart structures. Additionally, the need for SHM for composite structures is also discussed.

1.2   COMPOSITE MATERIALS: STRUCTURES AND TYPES

A composite material is a macroscopic combination of two or more constituent materials with significantly different physical or chemical properties. When these materials are combined, a new material with characteristics different from the individual components is produced. The classifications of composite based on reinforcement and matrix materials are presented in Figures 1.1 and 1.2, respectively.

Among the composite materials presented in Figure 1.1, laminated and sandwich panels are being increasingly used for structural applications [9,10]. Laminated

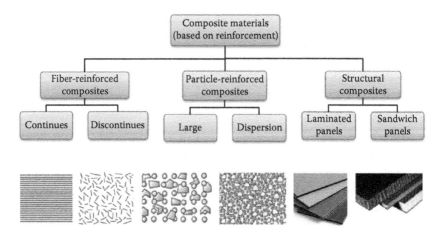

FIGURE 1.1   Classification of composite materials based on reinforcement [5–8]. (From City College of San Francisco. *Fiber Reinforced Composites*. Available from: http://fog. ccsf.cc.ca.us/~wkaufmyn/ENGN45/Course%20Handouts/14_CompositeMaterials/03_Fiber-reinforcedComposites.html; Khurshid, Z., et al., *Materials*, 8(2): 717–731, 2015; IndiaMart. *Indimart*. 2016. Available from: http://dir.indiamart.com/chennai/aluminium-composite-panels.html; Thomasnet. *Protech Composites*. 2016. Available from: http://www.thomasnet. com/productsearch/item/30192759-15241-1003-1003/protech-composites/custom-carbon-fiber-sandwich-panels/ [5–8].

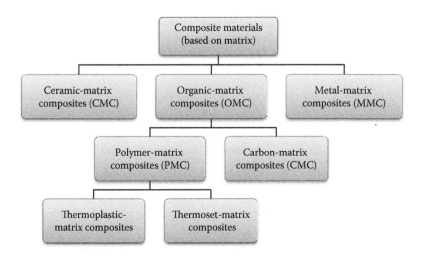

FIGURE 1.2 Classification of composite materials based on matrix.

composites are formed by bonding two or more layers (also called ply or lamina) together. The individual layer is made up of long, continuous high-modulus fibers as reinforcements surrounded by weaker matrix material.

1.2.1 REINFORCEMENTS

The fibers act as reinforcement that mainly carries the load. The role of the matrix is to distribute the fibers and transmit the load to the fibers. Typical fiber materials are carbon, glass, aramid, boron, and silicon carbide. The matrix materials are usually polymers such as epoxies and polyimides [11]. Boron fibers were the first commercially available advanced fibers. They are extremely stiff and strong and have a relatively large diameter, typically of the order of 200 μm. Their higher density and much higher cost limit their usage. Ceramic fibers, such as silicon carbide fibers and aluminum oxide fibers, were developed for use in high temperature applications such as for engine components.

Glass fiber in its various forms is perhaps the most popular synthetic reinforcement due to its low costs and easy usage. High stiffness and strength combined with low density and intermediate costs have made carbon fibers second only to glass fibers in use. However, there are other expensive fibers available in the market. Two of these are the carbon fiber and the graphite fiber that were originally developed for application in the aerospace industry but are now being used in other applications, such as in sports/leisure equipment. Their key advantage is the relatively high stiffness-to-weight and high strength-to-weight ratios. Carbon fibers vary in strength and stiffness with the processing variables, so that different grades are available such as high tensile modulus with lower strength or intermediate-tensile modulus with higher strength (tensile modulus from 207 up to 1000 GPa and tensile strength up to 4 GPa).

Regardless of the material, reinforcements are available in forms to serve a wide range of processes and end-product requirements. Materials supplied as reinforcement include roving, milled fiber, chopped strands, continuous, chopped, or thermoformable mat. Multidirectional reinforcement is obtained by weaving, knitting, or braiding continuous fibers into fabric or other sheet goods, unidirectional fabrics, and rovings.

1.2.2 MATRIX

Polymer-matrix materials are generally divided into two categories: thermosets and thermoplastics (Figure 1.2). An assessment of thermosets and thermoplastics based on their processing advantages is presented in Table 1.1 [12], and their mechanical properties are presented in Table 1.2 [13].

Thermoset matrices are the most common, being low cost and solvent resistant. The curing process in the thermosets is an irreversible chemical process. The most common type of thermoset is epoxy resin that possesses superior performance and is relatively low cost. Thermoplastic-based matrices can be softened by heating to an elevated temperature. Then, to form a thermoplastic composite material, the softened matrix is mixed with the reinforcement fiber, a process that is reversible. The selection of resin and its properties based on its real-life application is presented in Table 1.3 [14] for easy reference.

TABLE 1.1
Subjective Assessment of Thermoset (TS) and Thermoplastic (TP) Processing

Aspect	Advantage	Reason
Prepreg Formation		
Viscosity	TS	Lower
Solvents	TS	Greater choice
Hand	TS	More flexible
Tack	TS	Prepolymer variable
Storage	TP	Not reactive
Quality control	TP	Fewer variables
Composite Fabrication		
Layup	TS	Ease of handling
Degassing	TP	Fewer volatiles
Temperature change	TP	Fewer
Maximum temperature	TS	Lower
Pressure changes	TP	Fewer
Maximum pressure	TS	Lower
Cycle time	TP	Lower
Post-cure cycle	TP	Not required
Repair	TP	Remelt
Post forming	TP	Remelt

Source: Muzzy, J.D., and A.O. Kays. *Polymer Composites*, 5(3): 169–172, 2004.

TABLE 1.2

Comparisons of Thermoplastic and Thermoset Matrix (Resins)

Resin Family	Continuous Service Temperature (C/F)	Cure Time (Min)	Tensile Strength (ksi)	Tensile Modulus (Msi)
Thermoset Resins				
Phenolic (PH)	170/340	60+	6.9	0.55
Epoxy (E)	180/350	60–240	9.7	0.34
Cyanate ester (CE)	180/350	60–180	7.4–13.2	0.38–0.45
Bismaleimide (BMI)	230/450	120–240+	10.5	3.9
Polyimide (PI)	370/700	120+	16.6	—
Thermoplastic Resins				
Polycarbonate (PC)	120/250	<20	9.4	0.33
Polyphenylene sulfide (PPS)	240/464	<20	13.5	0.5
Polyetherimide (PEI)	200/390	<20	6	0.8
Polyetheretherketone (PEEK)	250/480	<20	14–33	0.58–3.4

Source: Composites world. http://www.compositesworld.com/articles/the-outlook-for-thermoplastics -in-aerospace-composites-2014–2023.

TABLE 1.3

Thermoplastic and Thermosetting Plastics Applications

Plastic Name	Products	Applications	Properties
(a) Thermoplastics			
Polyamide (Nylon)		Bearings, gear wheels, casings for power tools, hinges for small cupboards, curtain rail fittings and clothing	Creamy colour, tough, fairly hard, resists wear, self-lubricating, good resistance to chemicals and machines
Polymethyl methacrylate (Acrylic)		Signs, covers of storage boxes, aircraft canopies and windows, covers for car lights, wash basins and baths	Stiff, hard but scratches easily, durable, brittle in small sections, good electrical insulator, machines and polishes well
Polypropylene		Medical equipment, laboratory equipment, containers with built-in hinges, "plastic" seats, string, rope, kitchen equipment	Light, hard but scratches easily, tough, good resistance to chemicals, resists work fatigue
Polystyrene		Toys, especially model kits, packaging, "plastic" boxes and containers	Light, hard, stiff, transparent, brittle, with good water resistance

(*Continued*)

TABLE 1.3 (CONTINUED)
Thermoplastic and Thermosetting Plastics Applications

Plastic Name	Products	Applications	Properties
Low-density polythene (LDPE)		Packaging, especially bottles, toys, packaging film and bags	Tough, good resistance to chemicals, flexible, fairly soft, good electrical insulator
High-density polythene (HDPE)		Plastic bottles, tubing, household equipment	Hard, stiff, able to be sterilised
(b) Thermosets			
Epoxy resin		Casting and encapsulation, adhesives, bonding of other materials	Good electrical insulator, hard, brittle unless reinforced, resists chemicals well
Melamine formaldehyde		Laminates for work surfaces, electrical insulation, tableware	Stiff, hard, strong, resists some chemicals and stains
Polyester resin		Casting and encapsulation, bonding of other materials	Stiff, hard, brittle unless laminated, good electrical insulator, resists chemicals well
Urea formaldehyde		Electrical fittings, handles and control knobs, adhesives	Stiff, hard, strong, brittle, good electrical insulator

Source: Stephens Plastic Mouldings. http://www.stephensinjectionmoulding.co.uk/thermoplastics/.

1.3 COMPOSITE MANUFACTURING METHODS AND TECHNOLOGIES

Composite components and structures are manufactured using a wide variety of manufacturing methods. The ideal processing for a particular structure will depend on the chosen fiber and matrix type, processing volume, quality required, and the form of the component.

1.3.1 Prepreg Layup Method

The pre-impregnated (prepreg) layup method is one of the most labor-intensive manufacturing methods. It gives quite a large degree of flexibility with fiber orientation and layup sequence and is capable of producing very complex and high-strength components. The process is largely employed in the aerospace industry because of the strength of the components it can produce. In this process, tows or fabrics

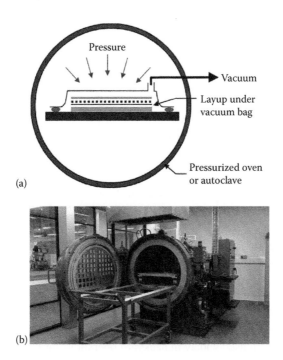

FIGURE 1.3 (a) Autoclave molding and (b) industrial-grade autoclave.

are impregnated with controlled quantities of resin before being placed on a mold. Prepreg layup is typically used to make high-quality components for the aerospace industry. This process results in particularly consistent components and structures. Because of this, prepreg techniques are often associated with sophisticated resin systems that require curing in autoclaves (Figure 1.3a and b) under conditions of high temperature and pressure.

In this process, first all the prepreg tapes or fibers are laid in an open mold in the desired sequence and orientation. Peel ply fabric is applied on top of the stacked plies prior to the vacuum bag preparations being made. The vacuum bagged part is placed in the autoclave for the required curing process, applying heat and pressure simultaneously. To control the fiber volume fraction, the prepreg sheets will generally contain excess amounts of resin. On heating, the resin will liquefy and through modifying the heating and pressure cycles, the excess resin can flow out of the component and into the bleeder material. Once the resin is cured, the part can be removed from the mold. Overall, the process is expensive, highly skilled, and labor intensive.

1.3.2 WET LAYUP METHOD

In the wet layup process, resin is applied to the mold and then reinforcement fibers are placed on top. A roller is used to impregnate the fibers and resin, as shown in Figure 1.4. Wet layup is widely used to make large structures such as the hulls of

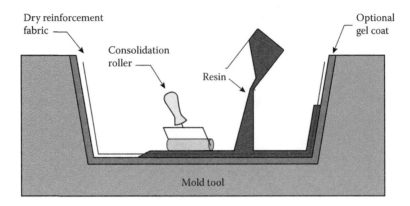

FIGURE 1.4 Wet layup process. (From Performance proven worldwide. http://www.pactinc. com/capabilities/wet-lay-uphand-lay-up-method/.)

small ships. The process was historically the most dominant manufacturing method largely due to the low initial setup and tooling costs and the low level of skill required to manufacture components [15].

The process is still used widely in backyard operations for prototyping purposes. However, the heavily labor-intensive process of applying each layer and "wetting-out" the fiber reinforcement has quite a few limitations. The parts produced are highly inconsistent in terms of weight, fiber volume fraction, and strength. The "bucket and brush" approach introduces a much higher risk of human error. Often, components can have areas where the fabric has not been wetted out correctly while other areas have excess resin. Vacuum bagging is commonly used with this method to minimize these variations and remove some of the excess resin typically involved with the wet layup procedure.

1.3.3 SPRAY LAYUP METHOD

This procedure is widely accepted as being the method of choice for producing a large quantity of components. The process is used in the marine and automotive sectors to produce components such as boat hulls and body panels. The spray layup process can be highly automated and, through the use of robotic equipment, can be fairly repeatable.

It is very similar to the wet layup. The only difference is in the method of applying resin and fibers to the mold. The process involves the application of the resin and fiber reinforcement through a specialized spray gun (Figure 1.5). The gun is designed to chop a roll of fiber, known as *fiber roving*, into smaller fiber pieces. The gun then atomizes the resin, combines the chopped fiber with atomized resin, and applies the combined mixture to the mold. The thickness is built up until the desired level is achieved and the part is allowed to cure.

The process often leads to a very resin-rich component. The high proportion of resin, combined with short fibers, means that the components produced are generally of fairly low strength. The process is, therefore, typically not considered for structural

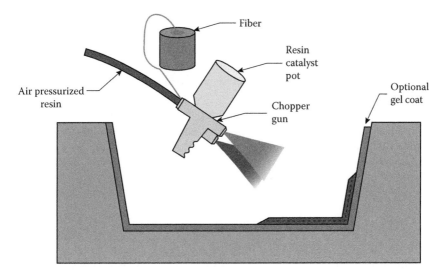

FIGURE 1.5 Spray layup process. (From Net Composites. http://www.netcomposites.com/ guide-tools/guide/manufacturing/spray-lay-up /.)

components. Additional reinforcement such as unidirectional tape and core materials can be applied during the process to increase strengths in certain areas [15].

1.3.4 FILAMENT WINDING METHOD

In this method, tows are wet-out with resin before being wound around a rotating mandrel. This process is used for cylindrical and spherical components such as pipes and pressure vessels (Figure 1.6). Winding is inherently automated, so it allows for

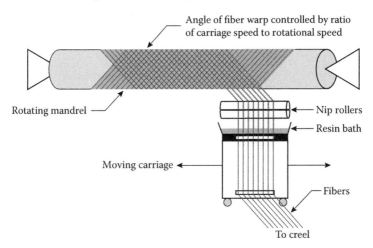

FIGURE 1.6 Filament winding. (From J & J Mechanic. http://www.jjmechanic.com/ process/f_winding.htm.)

the cheap manufacture of consistent components. However, the range of component geometries amenable to this method is limited to tubular parts and it is not suitable for making open structures. However, given a machine with enough degrees of freedom, it can be adopted to produce slightly more complicated shapes. Also, it is difficult to obtain uniform fiber distribution and resin content throughout the thickness of the laminate [15].

The filament winding process is highly repeatable and, therefore, can be quite cost effective. This method is able to produce relatively high-strength components by maintaining fiber volume fractions of 60:40 (fiber to matrix). This process has an added advantage of providing increased torsional strength to the finished beam as a result of the axially placed fibers.

1.3.5 PULTRUSION METHOD

This process is a completely automated composite manufacturing process. Typically, it is used to produce fiber-reinforced sections of various shapes and sizes. This process is widely available and is currently used mainly in the civil engineering fields. This method produces constant cross-section composite components. Tows or fabric are wetted out in resin and then fed through a heated die to cure (Figure 1.7). In this process, resin can also be directly injected into the die.

Pultrusion is the most efficient composite manufacturing process and has the highest level of productivity in comparison with other methods. The limitation, however, is that fibers generally are orientated only lengthways in the beam element [15].

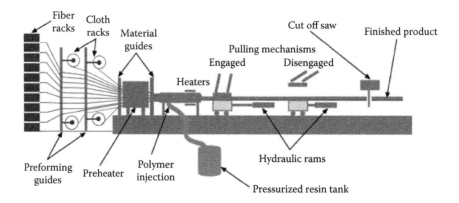

FIGURE 1.7 Pultrusion process. (From Net Composites. http://www.netcomposites.com/guide-tools/guide/manufacturing/spray-lay-up/.)

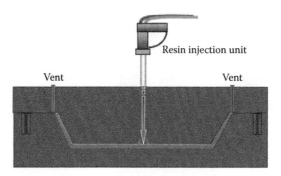

FIGURE 1.8 Resin transfer molding process. (From Moulded Fiber Glass Companies. http://www.moldedfiberglass.com/processes/processes/closed-molding-processes/ resin-transfer-molding.)

1.3.6 Resin Transfer Molding (RTM)/Structural Reaction Injection Molding (SRIM) Method

Resin transfer molding (RTM), also known as liquid transfer molding, is a relatively new manufacturing method for composite materials. The process is currently used in the aerospace and automotive industries and has the capability of producing high strength and complex components, while still maintaining high production volumes and dimensional tolerance [15].

This process is remarkably similar to the injection molding technique used for thermoplastics and involves infusing the resin into the closed mold. The difference, however, is the ability to use continuous fiber reinforcement called a *preform*. This preform is placed within the mold prior to infusing the resin (Figure 1.8).

Here, dry fibers are built up into intermediate preforms using tows and fabric held together by a thermoplastic binder. One or more preforms are then placed into a closed mold, after which resin is injected and cured to form a fully shaped component of high quality and consistency. The in-mold cycle time for RTM is approximately several minutes, while the time for structural reaction injection molding (SRIM) is measured in seconds. As fibers are manipulated in a dry state, these processes provide unmatched design flexibility. RTM produces good-quality components efficiently but incurs high initial costs for tooling and development. As a result, there is often a crossover point between prepreg layup and RTM for the manufacture of high-quality components, such as spinners for aero engines. At a lower level, SRIM is used for the manufacture of automotive parts that have a lower volume fraction of fibers.

1.3.7 ATP/AFP Method

ATP/AFP are the technologies that have revolutionized the production of composite structures for the aerospace industry in the last decades [2]. The advantages of these methods include higher accuracy of fiber placement; automatic debulking; no need

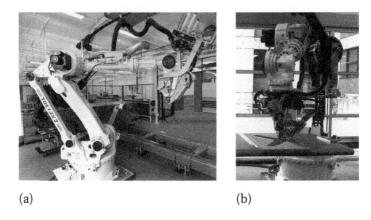

(a)   (b)

FIGURE 1.9 Automated Tape Placement (ATP) robot with thermoset head (a) and thermo-plastic head (b) (*Photos courtesy: Automated Composites Laboratory, UNSW Australia*).

for postprocessing, such as curing in oven or autoclave [20]; and reduction of material and labor cost [21]. Because of these capabilities, these techniques have been adopted by big manufacturers, such as Boeing and Airbus, and are used extensively for making highly precise components, such as wing skins, nose cones, and fuselages [22]. Besides that, several studies are being conducted to investigate utilizing these methods for making more complicated parts [23–25]. ATP/AFP have opened new markets, a wider range of applications as well as new levels of performance for advanced composite materials, and are entering into other industries, such as wind energy for making turbine blades [26–28]. The ATP/AFP are defined as additive manufacturing or inverse machining processes, as the composite part is built up by adding material [29–31]. The ATP machine consists of a placement head and a robotic arm which are computer controlled (Figure 1.9).

Several manufacturing stages, including layup, curing, and bonding, are merged together in a layup head that increases the productivity of this method [33]. With the running of the layup, the placement head brings the prepreg tape surfaces together under heat and pressure. Then, while the polymer matrix of plies is in direct contact, bonding is formed by defusing the polymer chains via thermal vibrations. The cooling process is carried out at room temperature [21]. The pressure is applied by squeezing the roller to force the air out of the composite structure and a hot gas torch or laser can be used as a heat source. Consequently, the quality of the final laminate depends on a number of factors including temperature, pressure, and layup speed, which should not exceed the tolerance. Due to better compaction, the ATP method has shown marked improvement in achieving high-quality composite laminate. However, producing defect-free composite laminate using this method needs appropriate attention.

1.3.8 Design of Composite Structures

For a composite manufacturer's requirement, it is important to select the appropriate fiber, matrix, reinforcement form, and manufacturing method taking into account

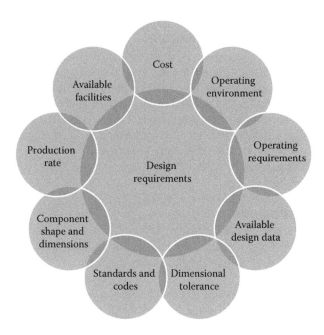

FIGURE 1.10 Composite design requirements.

the end user's application for the design of composite structures. The common design requirements for a designer are depicted in Figure 1.10 for easy reference.

1.4 SMART STRUCTURES AND STRUCTURAL HEALTH MONITORING

The technological advances of the late-twentieth century led to the use of structures with sensors and actuation systems. Figure 1.11 shows how structures and sensing and actuation systems can join together to create a new class of structures. Smart material

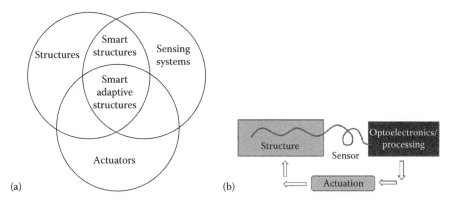

FIGURE 1.11 (a) Smart structures and smart adaptive structures and (b) implementation of structural health monitoring.

in principle should react and change in response to external conditions similar in many ways to the abilities of the human body. In order to realize this capability, a smart material needs to incorporate three components: sensors, a processing unit, and actuators. The system should be able to detect the damage, characterize it (recognize, localize, and quantify), and report it, providing important input for structure managers. The data resulting from a monitoring program can be used to optimize the operation, maintenance, repair, and replacement of the structure based on reliable and objective data. This kind of system is called a structural health monitoring system. A more detailed description of SHM is given in Chapter 2.

The important factors to consider while implementing an SHM system include

- Structural behavior and performance
- The expected loads; the design principles
- The maintenance requirements
- The available systems or devices for structural assessment
- The emerging technologies suitable for use with SHM

Smart materials and structures are widely used in many different applications to develop intelligent products in the aerospace, IT, automobile, space, and military industries. Some examples of smart materials and structures include composite materials with surface-attached/embedded sensors, electrorheological (ER) materials, and magnetorheological (MR) materials. Composite materials embedded with fiber-optic sensors (FOS) have been recognized as one of the prominent enabling technologies for smart materials and structures. The rapid increase in the interest in composite materials embedded with FOS has been driven by numerous applications, such as intelligent composite manufacturing/processing, and safety-related areas in aircrafts.

1.5   DEMAND FOR THE SHM IN COMPOSITE STRUCTURES

Even though the designed structural part is optimized for fail-safe performance, there is a possibility of damage during operation in extreme environmental conditions and mechanical failure of the structure due to external perturbations [34] and this necessitates the requirement for nondestructive SHM techniques throughout the lifetime of the composite structural part.

The demand for SHM in composites is, in the first instance, driven by the increased use of composite materials. The widespread use of carbon-reinforced fiber plastics (CRP) and glass fiber–reinforced plastic (GRP) composite-based structural parts has significantly increased in application areas such as aircraft, sport vehicles, wind turbines, and infrastructure constructions, because of the inherent advantages of composite materials. For example, in aircraft the advantages are well known: lighter weight for the aircraft, reduced requirements for maintenance, and increased passenger comfort. The use of composite materials in aircraft is growing, as shown in Figure 1.12a. Aircraft that utilize a high proportion of lightweight composite materials consume less fuel and result in a high-cost benefit that yields energy savings of up to 18%. Some existing commercial aircraft models such as the Boeing 787 and Airbus

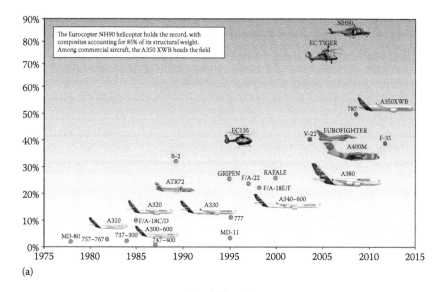

(a)

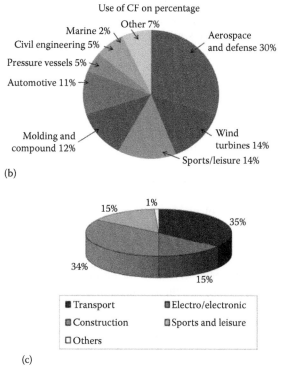

(b)

(c)

FIGURE 1.12 (a) Use of composite aerospace applications in the last 40 years, (b) use of CFP composites by industry, and (c) GRP production in Europe for different application industries. ([a,b] From Elmar, W., and J. Bernhard. Composites market report. The European GRP market [AVK], The global CRP market [CCeV], 2014. [c] Source data from Hexcel Corp. Aerostrategy.)

A350 are made up of about 50% composite materials by weight. Market research conducted in 2013 by Bernhard Jahn and Dr. Elmar Witten indicates that there will be a growing demand for such aircraft in the coming years [35]. Figure 1.12b and 1.12c shows the overall use of CFP composites by different industries and GRP production in different European industries, respectively.

The driving force behind the need for SHM in composite structures is the necessity to monitor composite parts during operation, for safety, and early detection of failure. During a typical 20-year service life, composite structures such as wind turbine blades, helicopter blades, and aircraft parts are subjected to static and dynamic lift, drag, and inertial loads over a wide range of temperatures and often severe environmental conditions. With the increasing demand for utilizing composite components for different applications, advanced manufacturing techniques have been introduced and developed during the last two decades for mass production purposes. Therefore, greater accuracy in producing defect-free components is required. Consequently, the composite industry has increasingly focused on utilizing nondestructive techniques for SHM over the service life of composite components.

Among several available nondestructive sensing techniques, optical fiber sensors have achieved wide acceptance due to their attractive properties such as small size, immunity to electromagnetic interference, and low cost [1]. Optical fiber sensors embedded in various structures are very useful for strain/temperature monitoring applications in extreme environmental conditions. For example, structural deformations due to delaminations and debonds can be monitored and avoided by implementing smart composite structures with embedded fiber-optic sensors. Also, the formation of ice on the wings of an aircraft can be detected by implementing smart solutions by embedded fiber-optic sensors within the wings' skin. Monitoring of the wing bend or loading due to ice accumulation combined with temperature data can improve the efficiency of the aircraft's deicing systems.

1.6 SHM SYSTEM FOR COMPOSITE STRUCTURES

Currently, different types of nondestructive methods such as ultrasonic and thermography are being used for the health monitoring of composite materials [35–38]. However, these techniques are more effective for laboratory conditions. Due to the size and weight of devices [39], there are some limitations to utilizing these techniques in the operational service or for on-line monitoring of manufacturing processes. In addition to these traditional methods, other techniques such as piezoelectric, resistance strain gauges, optical fiber sensor, eddy current, and microelectromechanical system sensors can also be used for SHM [40]. Among them, optical fiber sensors are considered suitable for composite materials because of their low cost, small size, light weight, wide bandwidth, possibility of remote operation, and electromagnetic interference immunity [41]. They can also be embedded into composite structure and can behave like a nervous system for determining defects as well as the condition of a structure [42,43]. Some of the applications of these sensors for tidal blades and aircraft wing skins are shown in Figure 1.13.

The focus of this book, therefore, is optical fiber–based SHM systems. In Chapter 2, an overview of various SHM systems is presented. The optical fiber sensors and

(a)

(b)

FIGURE 1.13 Structural health monitoring in (a) tidal blades (from Epsilon optics. http://www.epsilonoptics.com/tidal-energy-generation.html.) and (b) wing skins (from Black, S., High-performance composites [structural health monitoring: composites get smart], in *Composites World*. September 9, 2008 [45]).

their implementation for health monitoring both during and after the manufacturing of composite structures are discussed in the rest of the chapters. In addition, some demonstrated examples using fiber optics are also presented.

REFERENCES

1. Garg, D., M. Zikry, and G. Anderson. Current and potential future research activities in adaptive structures: An ARO perspective. *Smart Material Structures*, **25**: 539–544, 2001.
2. Landry, A. The needs for automation in composite design and manufacturing. In *The Second International Symposium on Automated Composites Manufacturing*. April 23–24, Montreal, Canada: Concordia University, 2015.
3. Balageas, D. *An Introduction to Structural Health Monitoring*. New York: Wiley Online Library, 2006.
4. Zhu, Y.-K., G.-Y. Tian, R.-S. Lu, and H. Zhang. A review of optical NDT technologies. *Sensors*, **11**(8): 7773–7798, 2011.
5. City College of San Francisco. Fiber reinforced composites. Available from: http://fog.ccsf.cc.ca.us/~wkaufmyn/ENGN45/Course%20Handouts/14_CompositeMaterials/03_Fiber-reinforcedComposites.html.
6. Khurshid, Z., M. Zafar, S. Qasim, S. Shahab, M. Naseem, and A. AbuReqaiba. Advances in nanotechnology for restorative dentistry. *Materials*, **8**(2): 717–731, 2015.
7. IndiaMart. *Indimart*. 2016. Available from: http://dir.indiamart.com/chennai/aluminium-composite-panels.html.
8. Thomasnet. Protech composites. 2016. Available from: http://www.thomasnet.com/productsearch/item/30192759-15241-1003-1003/protech-composites/custom-carbon-fiber-sandwich-panels/.
9. Chung, D.D.L. *Composite Materials: Science and Applications*. 2nd edn., p. 371, Berlin, Germany: Springer, 2010.
10. Bannister, M. Challenges for composites into the next millennium: Reinforcement perspective. *Composites Part A: Applied Science and Manufacturing*, **32**(7): 901–910, 2001.

11. Tong, L., A.P. Mouritz, and M.K. Bannister. Chapter 1: Introduction. In L. Tong, A.P. Mouritz, and M.K. Bannister, eds., *3D Fibre Reinforced Polymer Composites.* pp. 1–12, Oxford, UK: Elsevier Science, 2002.

12. Muzzy, J.D., and A.O. Kays. Thermoplastic vs. thermosetting structural composites. *Polymer Composites*, **5**(3): 169–172, 2004.

13. Red, C., *The Outlook for Thermoplastics in Aerospace Composites*, in *Composite World*, 9 January, 2014.

14. Stephens Plastic Mouldings. Common thermoplastics & thermosetting and their uses. Available from: http://www.stephensinjectionmoulding.co.uk/thermoplastics/.

15. Mazumdar, S.K., *Composites Manufacturing: Materials, Product, and Process Engineering*. Boca Raton, FL: CRC Press, 2002.

16. Performance Proven Worldwide. Wet lay-up process. Available from: http://www. pactinc.com/capabilities/wet-lay-uphand-lay-up-method/.

17. Net composites. Manufacturing. http://www.netcomposites.com/guide-tools/guide/ manufacturing/spray-lay-up/.

18. J & J Mechanic. Filament winding. Available from: http://www.jjmechanic.com/ process/f_winding.htm.

19. Molded FiberGlass Companies. Resin transfer molding (RTM) process. 2016. Available from: http://www.moldedfiberglass.com/processes/processes/closed-molding-processes/ resin-transfer-molding.

20. Béland, S. *High Performance Thermoplastic Resins and Their Composites*. Park Ridge, NJ: Noyes Data, 1990.

21. August, Z., G. Ostrander, J. Michasiow, and D. Hauber. Recent development in automated fibre placement of thermoplastic composites. *SAMPE*, **50**(2): 30–37, 2014.

22. Quddus, M., A.D. Brux, B. Larregain, and Y. Galarneau. A study to determine the influence of robotic lamination programs on the precision of automated fiber placement through statistical comparisons. In *The Second International Symposium on Automated Composites Manufacturing*. April 23–24, Montreal, Canada: Concordia University, 2015.

23. Jonsson, M., D. Eklund, A. Järneteg, M. Åkermo, and J. Sjölander. Automated manufacturing of an integrated pre-preg structure. In *The Second International Symposium on Automated Composites Manufacturing*. April 23–24, Montreal, Canada: Concordia University, 2015.

24. Di Francesco, M., P. Giddings, and K. Potter. On the layup of convex corners using automated fibre placement: An evaluation method. In *The Second International Symposium on Automated Composites Manufacturing*. April 23–24, Montreal, Canada: Concordia University, 2015.

25. Girdauskaite, L., S. Krzywinski, C. Schmidt-Eisenlohr, and M. Krings. Introduction of automated 3D vacuum buildup in the composite manufacturing chain. In *The Second International Symposium on Automated Composites Manufacturing*. April 23–24, Montreal, Canada: Concordia University, 2015.

26. MIKROSAM. Innovative composites manufacturing solutions. 2016. Available from: http://www.mikrosam.com/new/article/en/automated-fiber-placement-the-complete-sys tem/?gclid=CK334JSyoMgCFYIHvAodf4kF3w.

27. Akbarzadeh, A.H., M. Arian Nik, and D. Pasini. Structural analysis and optimization of moderately-thick fiber steered plates with embedded manufacturing defects. In *The Second International Symposium on Automated Composites Manufacturing*. April 23–24, Montreal, Canada: Concordia University, 2015.

28. Barile, M., L. Lecce, M. Esposito, G.M. D'Amato, and A. Sportelli. Advanced AFP applications on multifunctional composites. In *The Second International Symposium on Automated Composites Manufacturing*. April 23–24, Montreal, Canada: Concordia University, 2015.

29. Gutowski, T., *Advanced Composites Manufacturing*. New York: Wiley, 1997.

30. August, Z., G. Ostrander, and D. Hauber. Additive manufacturing with high performance thermoplastic composites. In *The Second International Symposium on Automated Composites Manufacturing.* April 23–24, Montreal, Canada: Concordia University, 2015.

31. Werner, D. Multi-material-head novel laser-assisted tape, prepreg and dry-fiber placement system. In *The Second International Symposium on Automated Composites Manufacturing.* April 23–24, Montreal, Canada: Concordia University, 2015.

32. Wu, K.C., B.K. Stewart, and R.A. Martin. ISAAC advanced composites research capability. In *The Second International Symposium on Automated Composites Manufacturing.* April 23–24, Montreal, Canada: Concordia University, 2015.

33. Jian, C., Q. Lei, Y. Shenfang, S. Lihua, L. PeiPei, and L. Dong. Structural health monitoring for composite materials. In *Composites and Their Applications*, p. 438. India: InTech, 2012.

34. Elmar, W., and J. Bernhard. Composites market report. The European GRP market (AVK), The global CRP market (CCeV), 2014.

35. Maldague, X.P. *Theory and Practice of Infrared Technology for Nondestructive Testing.* 1st edn., New York: John Wiley, 2001.

36. Shokrieh, M.M., and A.R. Ghanei Mohammadi. Non-destructive testing (NDT) techniques in the measurement of residual stresses in composite materials: An overview. In M.M. Shokrieh, ed., *Residual Stresses in Composite Materials.* pp. 58–75, Cambridge, MA: Woodhead, 2014.

37. De Angelis, G., M. Meo, D.P. Almond, S.G. Pickering, and S.L. Angioni. A new technique to detect defect size and depth in composite structures using digital shearography and unconstrained optimization. *NDT & E International*, **45**(1): 91–96, 2012.

38. Amenabar, I., A. Mendikute, A. López-Arraiza, M. Lizaranzu, and J. Aurrekoetxea. Comparison and analysis of non-destructive testing techniques suitable for delamination inspection in wind turbine blades. *Composites Part B: Engineering*, **42**(5): 1298–1305, 2011.

39. Yong-Kai, Z., G.-Y. Tian, R.-S. Lu, and H. Zhang. A review of optical NDT technologies. *Sensors*, **11**(12): 7773–7798, 2011.

40. Cai, J., L. Qiu, S. Yuan, L. Shi, P. Liu, and D. Liang. *Structural Health Monitoring for Composite Materials.* Cambridge, MA: InTech, 2012.

41. Méndez, A., and T. Graver. Overview of fiber optic sensors for NDT applications. In *IV NDT Panamerican Conference.* Buenos Aires, Argentina, 2007.

42. Gros, X.E. Current and future trends in non-destructive testing of composite materials. *Annales de Chimie Science des Matériaux*, **25**(7): 539–544, 2000.

43. Oromiehie, E., G. Prusty, and G. Rjan. Thermal sensitivity and relaxation of carbon fiber foam sandwich composites with fiber optic sensor. *Sandwich Structures and Materials*, in press.

44. Epsilon Optics. Tidal energy generation. Available from: http://www.epsilonoptics.com/tidal-energy-generation.html.

45. Black, S., *High-Performance Composites (Structural Health Monitoring: Composites Get Smart)*, in *Composite World.* 2008: United States.

2 Structural Health Monitoring Methods for Composite Materials

Wiesław M. Ostachowicz, Tomasz Wandowski, and Paweł H. Malinowski

CONTENTS

2.1 Introduction 21
2.2 Elastic Waves-Based Method 22
2.3 Acoustic Emission Method 26
2.4 Electromechanical Impedance Method 28
2.5 Vibration-Based Method 30
2.6 Other Methods 33
2.7 Summary 34
References 35

2.1 INTRODUCTION

Composite materials have a lot of advantages that make them very attractive in application to many industrial branches such as aerospace, automotive, maritime, and wind energy. Their most important advantages are low weight, high mechanical strength, high stiffness, and vibration damping ability. However, there are also many problems with the exploitation of composite materials due to their common disadvantages. Most important is their susceptibility to initiation and growth of damage in the internal structure in the form of delamination. This type of damage is located between layers of composite material and is initiated by impact. Another type of damage is matrix cracking. These two types of damage can be hidden in the internal structure and may not be visible on the surface of composites. Because of this, a need existed to develop methods to detect and localize these defects. Initially, classical nondestructive testing (NDT) methods were utilized in order to assess the state of composite structures. Later, many researchers focused their attention on the development of structural health monitoring (SHM) systems and methods that could be utilized for composite structures. In 1993, Rytter, in his PhD thesis [1], formulated four levels of SHM:

1. Damage detection
2. Damage localization

3. Quantification of damage severity
4. Prediction of remaining life of structure

In the work of Worden et al. [2], a set of fundamental axioms related to SHM was defined. First of all, each material has some internal defect and flaw. Moreover, defects cannot be detected directly by sensors because sensors do not measure damage but, rather, signals related to some physical properties that can be connected to damage initiation and growth (temperature, pressure, charge, voltage, etc.).

SHM systems can be divided generally into passive and active systems [3]. In an active system, SHM actuators and sensors are utilized. Actuators generate diagnostic signals when the sensors receive these signals in various locations of the structure. In a passive system, only the sensing ability, which is only listening for signals generated by growing damage, is utilized [3]. In SHM systems, many different methods are utilized in order to assess the state of a structure. The following sections describe the methods developed by researchers from universities, research institutes, and industries.

2.2   ELASTIC WAVES-BASED METHOD

Elastic waves that propagate in materials can be effectively employed for SHM purposes. This is due to the fact that elastic waves scatter at materials homogeneously, in particular at damage. This phenomenon has been widely used as one of the NDT methods that is based on elastic ultrasonic waves. However, the purpose of the SHM system is to interrogate large surfaces; therefore, elastic waves that propagate along the structural element and not through the thickness are preferred. In general, in unlimited three-dimensional solids, one observes only longitudinal and transverse (shear) waves. Classification is made by the direction of the vibration in relation to the elastic wave propagation path. In the case of longitudinal waves, the vibration and propagation directions are identical. The transverse waves vibration is perpendicular to the direction of wave propagation. If the surface of a three-dimensional solid is limited, two new types of elastic waves appear: Rayleigh and Love waves. Rayleigh waves propagate on the surface of solids where the thickness is many times larger than the length of the propagating waves. The wave motion trajectories are in a circular form. The characteristic feature of Rayleigh waves is the amplitude decay as one goes deeper into the material. Love waves have a more complex displacement field than Rayleigh waves. Similar to Rayleigh, though, Love wave amplitude decays with increasing depth. If one considers plates or shells, the transverse (shear) waves can be split into two types: shear horizontal (SH) waves and shear vertical (SV) waves. In the case of SH waves, the oscillations are horizontal; whereas, in the case of SV waves, the oscillations occur in a vertical direction. As a result of internal reflections of SV longitudinal waves, new forms of waves are created: Lamb waves. These waves are named after Horace Lamb, who described them in 1917 [4,1]. These waves exist as symmetrical and antisymmetrical modes. The number of modes is infinite but the number of propagating modes depends on the product of excitation frequency and the thickness of the material. Both symmetrical and antisymmetrical modes of Lamb waves are dispersive, which means that their velocity of propagation changes with frequency. Lamb waves are guided by the curvature of the element in which they propagate. Horace Lamb described them only in isotropic materials. Taking this into consideration, the more general name for these kinds of waves is guided waves.

Most commonly these waves are employed for SHM purposes using piezoelectric sensors, often named PZT (lead zirconate titanate) sensors, indicating the common material used for their manufacture. The direct and converse piezoelectric effect allows the use of these sensors for both sensing and excitation of propagating waves. A wave excited in an aluminum plate is depicted in Figure 2.1. One can notice a clear circular wave front propagating out from the centrally located PZT sensor. In the case of a composite material, the wave front pattern is more complex (Figure 2.2). First, one can observe two waves propagating: the faster S_0 mode and the slower A_0 mode. Second, the wave front of the slower one looks like a flattened circle, whereas the faster one is distorted even more, so that it looks similar to an ellipse. The reason for such wave front behavior is the anisotropy of the material. In this particular case, the composite material was unidirectional with horizontally oriented glass fibers. Obviously, the circular-like wave front is also possible in the case of composite materials for a more isotropic-like orientation of the layers.

The proper placement of PZT sensors is an extensive topic of research. The sensors are placed in various manners in order to ensure good damage detection abilities. They are placed in radar-like forms as concentrated arrays in order to interrogate the area around them [5–9] or in a form of a distributed array [10,11]. An example of a part of a structural element with a uniform sensor network and also a

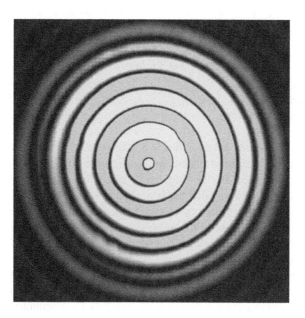

FIGURE 2.1 Elastic wave propagating in 1 mm thick aluminum plate; excitation with a piezoelectric sensor at 100 kHz; response measured with laser-scanning vibrometer on the sample surface.

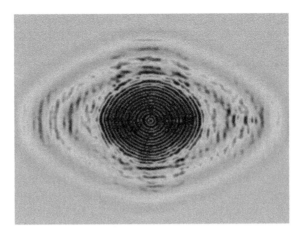

FIGURE 2.2 Elastic wave propagating in five-layer unidirectional glass fiber–reinforced polymer plate; excitation with a piezoelectric sensor at 100 kHz; response measured with laser scanning-vibrometer on the sample surface.

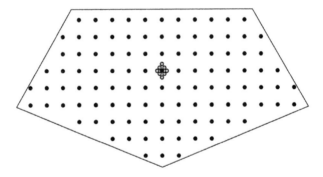

FIGURE 2.3 Comparison of a uniform sensor network and a concentrated array placed on an arbitrarily chosen shape.

concentrated array is depicted in Figure 2.3. The type of sensor array should be in agreement with the method of elastic wave generation and sensing. Two methods of elastic wave generation and sensing can be distinguished: pitch catch and pulse echo. In the case of the pitch-catch method, elastic waves are excited by one sensor and then propagate through the potentially damaged area; next they are sensed by the second sensor. In the case of the pulse-echo method, elastic waves propagating in the structure reflect from damage, and these reflections are sensed in the second (or the same) sensor. The phased array method is a special case of the pulse-echo method that uses elastic wave interference phenomena. As a result of firing sensors in a concentrated network with a specially adjusted time delay, a wave front is created that propagates in a chosen direction (Figure 2.4). Each approach needs a dedicated signal processing algorithm tailored for the sensor array.

The PZT sensors are not the only effective tools used for the elastic wave–based method. A commonly used device is the laser vibrometer, also named a *laser Doppler*

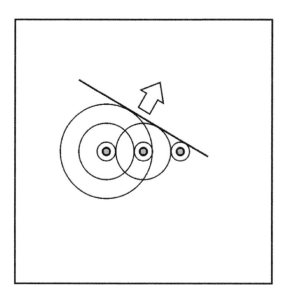

FIGURE 2.4 Phased array concept; excitation of wave with predefined time delays results in a plane wave front propagating in a desired direction.

vibrometer (LDV). An LDV uses a laser beam in order to register the vibration velocity at a measuring spot. Thanks to the use of three laser heads, one can obtain the vibration characteristic at the measurement spot in three orthogonal directions. Moreover, the scanning capability of the laser heads (scanning laser Doppler vibrometer [SLDV]) allows for rapid scanning of the investigated object surface. Obviously, the idea of an SHM system with sensors integrated within a structure cannot be realized using an LDV because one needs to use one of three bulky laser heads for measurements. However, the device is still very useful for SHM research in order to obtain a deeper insight into wave propagation phenomena and to design and test sensor arrays that can be made later with surface mounted or embedded sensors [5,6]. A laser source has been proven to be a very effective means of Lamb wave excitation [12]. In previous works, the laser excitation was combined with an air-coupled sensor in order to obtain a full noncontact method of measurement for damage detection [13]. By connecting the laser vibrometer with an additional laser source, one obtains a laser ultrasonic (LUS) device that can be utilized in traditional NDT/nondestructive evaluation (NDE) inspection and elastic-guided waves excitation and sensing. This was carried out for a carbon fiber–reinforced polymer (CFRP) aircraft wing in the work of Park et al. [14]. A promising technique for elastic wave sensing can be based on fiber Bragg grating (FBG) sensors. These sensors can be either surface mounted or embedded into composite material. In the work of Pereira et al. [15], the authors showed the embedded FBG response related to growing cracks. Miesen et al. [16] showed the impact-excited wave-sensing ability of these sensors.

The other branch of research dedicated to elastic wave–based SHM is focused on nonlinear approaches. The excitation of two waves with different frequencies (f_1 and f_2) in the structure causes higher harmonics ($2f_1$, $2f_2$, …) as well as modulations

$(f_2 - f_1, f_{21} + f_1)$ to be visible in the spectra. However, the two frequencies should be carefully selected. To solve this issue the broadband excitation can be used with a pulse laser. This has been proposed together with laser nonlinear wave modulation spectroscopy (LNWMS) [17]. LNWMS was successfully applied to a composite glass fiber–reinforced polymer (GFRP) 10 kW wind turbine blade.

The experimental research in the area of elastic wave propagation has been supported by various numerical modeling techniques. In the spectral methods, Fourier series or approximating polynomials possessing special properties are used for solving the problem. Such polynomials are usually orthogonal Chebyshev polynomials or very-high-order Lobatto polynomials. The finite element method is employed to solve various numerical problems, and it has also been successfully applied to wave propagation modeling. The spectral finite element method is a combination of both these methods, taking advantage of higher order polynomials and discretization of the space, giving very promising results in the analysis of elastic wave propagation phenomena [18].

It should be noted that the damage sensitivity of the elastic wave can be altered due to external (environmental) factors unrelated to the inspected structure. Factors such as temperature, loads, or absorbed moisture change the registered signals [19,20]. This should be taken into account when designing an elastic wave–based system.

2.3   ACOUSTIC EMISSION METHOD

The acoustic emission (AE) method is a passive approach based on wave propagation. Sudden internal stress redistribution caused by changes in the internal structure initiates wave propagation. Most of the sources of AE are damage related (crack initiation and growth, crack opening, fiber breakage, etc.) and that makes the method promising for SHM applications [21]. In the AE method, the structure needs to be externally loaded in order to cause the wave emission. Although the method is called acoustic, the frequency range of the waves can be really broad, up to the megahertz region. The elastic wave can also be initiated during impact on the material surface. The impact can either initiate damage or not in the materials. Low-velocity impact forces are not high enough to cause fiber break in fiber-reinforced composites but can result in internal delamination or debonding [22]. Both AE and impact detection can be, in general, called a problem of acoustic source localization. In order to localize the acoustic source, one needs a network of sensors (Figure 2.5). As the research of Markmiller [22] showed, it is possible to identify the impact force as well as its location on a surface of carbon-stiffened panel using a network of piezoelectric disc sensors [22]. The optimal sensor network for this task was determined using a genetic algorithm indicating that a regular network gives worse POD (probability of detection). The piezoelectric discs have an omnidirectional sensing characteristic similar to commercial AE sensors available on the market [23]. But there are also attempts at using directional sensors. In the work of Matt and di Scalea [24], piezoelectric rosettes that exploit macro-fiber composite (MFC) rectangular patches were used. These patches were comprised piezoelectric fibers, resulting in directional wave-sensing characteristics (Figure 2.6). This allowed for the determination of the direction of an incoming wave independently of the wave velocity in the material.

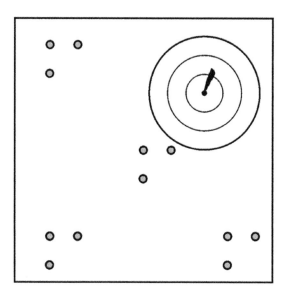

FIGURE 2.5 AE source in upper right corner causes elastic waves to propagate; the response is gathered by the sensors distributed on the inspected surface. These types of sensors and consideration of their optimal placement is an ongoing field of research.

FIGURE 2.6 Macro-fiber composite rosette used for acoustic emission source localization.

The procedure is based on the computation of the wave strain principal angle, and knowledge of the wave speed in contrast to TOF (time of flight) based procedures is not required. It was observed that two rosettes do not cope with an AE source that is located along the line connecting the two rosette positions. Using a third rosette removes this localization problem. Moreover, wave attenuation plays an important role in the AE source localization so, for example, better results were obtained for a composite plate than a sandwich panel [24].

In the task of AE source (impact event) localization, sophisticated signal process-ing techniques are employed. In the research reported by Ciampa et al. [25], the authors realized the AE source localization and determined the velocity of elastic waves in plate-like composite structures. Differences of the stress waves were measured by six surface-attached piezoelectric sensors. The arrival time of the Lamb wave was determined by the peak magnitude of the wavelet scalogram. The coordinates of the impact and the group velocity were obtained by solving a set of nonlinear equations. The proposed algorithm does not require *a priori* knowledge of the wave velocity profile, mechanical properties, stacking sequence, or thickness of the structure. The active methods based on elastic wave beamforming techniques are similarly employed for AE source sensing. Instead of delaying the excitation of waves (see Figure 2.4), the time delays of the wave sensed by the phased array elements indicate the direction of wave arrival. However, the wave velocity profile is needed as an input. To overcome this drawback, a technique using only six receiving sensors was proposed [26]. It does not need the direction dependent velocity profile or the solving of a nonlinear equations system. Moreover, it should be underlined that the strain rosette technique does not work for anisotropic plates. In an anisotropic plate, the wave propagation direction does not necessarily coincide with the principal strain directions measured by the strain rosette technique [26].

Summarizing, six sensors are enough for AE source localization with unknown material properties and wave velocity profile. This number reduces to four for isotropic plates. Following the work of Kundu [26], it can be said that the AE localization problem in plate- and shell-type structures has been essentially solved. Researchers need to focus on more sophisticated structures in which the wave's propagation path does not follow a straight line from the acoustic source to the receiver.

2.4   ELECTROMECHANICAL IMPEDANCE METHOD

The electromechanical impedance (EMI) method is based on the measurement of electrical parameters of the PZT sensor bonded to a host structure (Figure 2.7). Due to the electromechanical coupling of a PZT sensor with a structure, mechanical resonances of the structure can be seen in the electrical characteristics of the PZT sensor. Although this method is called electromechanical impedance, rather than impedance, very often it is such parameters as admittance, resistance, conductance, or susceptance that are measured.

Liang is cited in most papers as a precursor of the EMI method. In one paper [27], Liang presented an analytical model for electromechanical impedance. However, Soh [28] cited a paper by Crawley [29] as the first related to electromechanical impedance. It should be underlined that in this work a one-dimensional model was developed [29]. This model consisted of a PZT sensor in the form of a thin spring simulating the structure. However, this model was investigated only in the frame of static and it neglected the influence of important parameters from the point of view of dynamics: inertia and damping. Soh and Annamdas also conducted a lot of work applying this method to metallic structures. For example, Annamdas investigated the influence of load effect on the EMI method [30,31]; Soh investigated the influence of bonding layer thickness on EMI measurements [32]; and they both conducted a detailed review of the EMI method [33]. Lots of work related to the application of the EMI method for metallic structures has been published [34]. In one work [35],

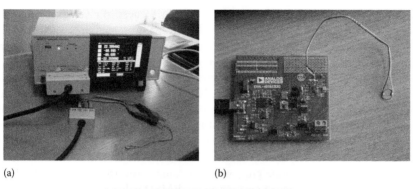

(a)   (b)

(c)

FIGURE 2.7 EMI method equipment: (a) electrical-impedance analyzer used for EMI method, (b) low-cost impedance analyzer manufactured by Analog Devices (AD5933), and (c) PZT sensor bonded to a CFRP sample for EMI measurements.

Giurgiutiu et al. utilized the multiphysic finite element method in order to model the electromechanical coupling for a free PZT sensor and for a sensor bonded to the host structure made out of a glass fiber–reinforced composite. The authors investigated the problem of sensor debonding as well as the influence of delamination of composite structure on the electromechanical impedance signatures. The problem of an influence of sensor debonding on real and imaginary parts of impedance and admittance was investigated in the work of Wandowski et al. [36]. In this case, the sensor was bonded to a GFRP sample. Results obtained for the EMI method were compared with results obtained for terahertz spectroscopy. The last mentioned method allowed for the visualization of the distribution of the bonding layer between the piezoelectric sensor and the GFRP sample.

In large composite structures, the sensitivity and range of the EMI method is reduced due to strong damping properties. Na and Lee [37] proposed a solution that enhances the sensitivity of this method for damage detection. In their presented approach, the resonant frequency range (below 80 kHz) was utilized for the analysis. This allowed for the possibility of obtaining a large sensing area. The idea behind this approach was to create resonant peaks in the part of the characteristic without it. This was realized by placing the PZT sensor on the metallic parts that had low resonant frequencies in the lower range. The same authors proposed a solution using complex geometry that avoids problems with the brittleness of PZT material during its attachment to the composite structure [38]. In the presented approach, the PZT sensor was attached to the steel wire. In the work of Cherrier et al. [39], a damage localization map-creating algorithm was proposed. This map was created based on the measurements of the EMI spectrum combined with an inverse distance weighting interpolation. This approach was tested in the case of a one-dimensional aircraft composite stiffener and thin composite panel. In both cases, impact damage was created. In the work of Schwankl et al. [40], the EMI method was also utilized for damage detection in a composite panel with stiffeners. The authors investigated the different sizes of the defect and distance from the sensor. The root-mean-square deviation (RMSD) index value grew with the damage severity. However, the influence of distance from damage to sensor on the RMSD value was ambiguous for the investigated frequency range (10–55 kHz). The damage detection process was conducted based on the measure of the real part of admittance.

In the work of Lim et al. [41], the kernel principal component analysis was proposed in order to improve damage sensitivity under varying operational conditions (temperature and load). Varying operational conditions were stated as a source of changing the impedance signature. This method was tested in the case of a metallic fitting lug that joined a composite wing to the fuselage. During tests, damage was simulated by bolt loosening.

In a paper by Sun et al. [42], the EMI method was utilized for debonding detection in a reinforced concrete (RC) beam strengthened with a fiber-reinforced polymer (FRP). Experimental results were compared with results from two-dimensional EMI models. Due to the high frequency measurement, the numerical model was based on the spectral element method. The numerical model consisted of a PZT sensor bonded to the FRP strengthened beam.

In the work of Malinowski et al. [43], the EMI method was used for the assessment of adhesive bonds in CFRP samples. Different scenarios of weak bonds were investigated: chemical contamination (silicon-based release agent), moisture, and improper curing temperatures. In this approach, conductance spectra were investigated. It should be mentioned that the authors utilized very high frequencies, around 3.8 MHz. This approach was based on thickness mode. Two parameters were considered during the analysis: RMSD and frequency shift of frequency peak for thickness mode. The proposed method was able to separate results for free sensors, bonded sensors on referential samples, and bonded sensors on samples with weak bonds. The EMI method was also utilized for moisture-level detection in CFRP samples [44]. The influence of moisture was seen as a frequency shift of the resonant peak, similar to the temperature influence. It needs to be mentioned that the sensitivity of this method to temperature changes is its main drawback and needs to be considered during utilization [41,45,46].

2.5 VIBRATION-BASED METHOD

The next method that can be used in SHM of composite materials is the vibration-based method (VBM). This method utilizes the vibration response of a structure in order to identify its properties and to detect possible damage. Damage can be a source of changes in stiffness, mass, damping, or boundary condition. These changes will then be sources of changes in the dynamic parameters of the structure (resonant frequencies, mode shapes, and modal damping) [47]. In the works of Farrar and Jauregui [47,48], the results of experimental and numerical studies were presented. The object of investigation was the Interstate 40 Bridge. Authors compared the different damage detection approaches for VBM: damage index, curvature mode shape, change in flexibility, and change in uniform load surface curvature. The VBM is continuously being improved and, as a result, many new approaches for metallic and composite structures (measurement techniques and techniques and algorithm for signal analysis) can be found. In Yam et al. [49], an integrated method for damage detection based on the frequency response function of structure and the combination of wavelet transform with artificial neural networks was proposed. The aim of a neural network is to identify damage location and severity. Studies were performed on a polyvinyl chloride (PVC) panel with introduced crack.

In the VBM, very often accelerometers are used for the purpose of registering the frequency response function or identifying displacement mode shapes. However, in Yam et al. [49] and Hamey et al. [50], smart materials in the form of PZT sensors were used. They were utilized for sensing and actuation. In the latter work [50], curvature modes of the investigated composite beam were acquired.

In the work of Park et al. [51], the frequency response function was measured by utilizing an MFC piezoelectric sensor. The presented approach was based on an active sensing system. The piezoelectric actuator was utilized in order to perform excitation of the structure where sensors registered frequency responses. Measurements were made for the structure at referential and damaged states. The proposed approach was suitable for delamination detection in a quasi-isotropic carbon-based composite plate.

Noncontact measurement methods can also be utilized in the VBM. One of these noncontact methods is electronic speckle pattern interferometry (ESPI). This method utilizes laser beams that illuminate the reflective surface of a structure. As a result, a speckle pattern is formed by a focusing lens and is recorded by video camera. This method can be utilized for vibration mode shape measurements [52]. In the work of Wang and Hwang [53], an amplitude fluctuation (AF) ESPI method was presented for the measurement of vibration of a composite plate with a crack. This method was based on video signal subtraction. However, in this approach a reference image is no longer needed for a stress-free state but, rather, for a vibrating state. This method allows for the achievement of doubled and stable fringes without additional optical equipment. This method was also utilized in Hong et al. [54] for the vibration measurement of symmetrical and asymmetrical panels manufactured from carbon fiber–reinforced plastic. It should be mentioned that this is an NDT method rather than an SHM technique.

Nowadays, many researchers utilize SLDV as a noncontact method for vibration measurements (Figure 2.8). Although this approach is not applicable to SHM, like the ESPI method, it is very useful in laboratory conditions for vibration analysis (frequency response function and mode shapes), especially of complex structures. This approach can also be applied to investigations related to a better understanding of damage influence on structural vibration in complex structures. Moreover,

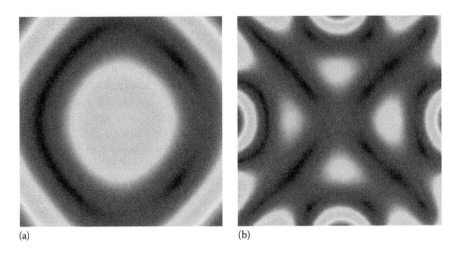

(a) (b)

FIGURE 2.8 Chosen mode shapes (a) and (b) of CFRP plate measured by scanning laser vibrometer.

the noncontact method does not need sensors, such as, for example, accelerometers. Such an approach eliminates change of mass distribution (sensors) and also damping properties (cables). In the work of Herman et al. [55], the SLDV method was used for vibration modal analysis of a composite T-stiffened panel. The main aim of the research was the detection of damage in the form of delamination and porosity based on mode shape curvature. In Bai et al. [56], an interesting approach, termed a scale waveform dimension analysis of two-dimensional mode shapes, was proposed. The method consisted of two parts: decomposition of mode shapes into scale mode shapes and waveform dimension analysis of scale mode shapes. The approach presented in this paper allowed for the identification of delamination in a composite plate with great accuracy while additionally taking into account noisy conditions.

The utilization of strain gauges instead of accelerometers allows for the identification of strain-based mode shapes [57,58]. Strain-based mode shapes can also be utilized for damage detection. In the work of Marques dos Santos et al. [58], vibration-based strains are measured in order to detect the damage in the SHM system. In that paper, two experimental objects were tested. The first one was a carbon fiber composite plate equipped with strain gauge rosettes. During this research, excitation with the impact form was utilized. It was concluded that strain mode shapes are more sensitive to damage than conventional displacement mode shapes. The second one was the vertical stabilizer of an F–16 aircraft. In this case, two shakers had been used for effective excitation. One of the wings was instrumented with strain gauges and accelerometers. The aim of this research was to compare and show the differences in displacement and strain mode shapes of the wing. It was shown that strain distribution cannot always be understood based on displacement mode shapes.

In Garcia et al. [59], a multivariate statistical procedure for damage identification based on time domain structural response was utilized. Delamination was simulated by the use of Teflon inserts between the layers of the composite plate. Different sizes and locations of damage were investigated. The authors compared the results coming

FIGURE 2.9   Surface-mounted FBG strain sensor in stainless steel carrier.

from numerical simulation with experimental results obtaining good agreement. In this approach, strain gauges were also utilized in order to register strain response.

In the work of Frieden et al. [60], FBG sensors (Figure 2.9) were used in order to measure dynamic strains for modal analysis. Authors investigated the influence of impact damage on change of eigenfrequencies of a carbon fiber–reinforced plate. The authors concluded that fiber-optic FBG sensors have sufficient dynamics for identification of natural frequencies in the case of the investigated plate. Moreover, fiber-optic strain sensors can replace traditional accelerometers, strain gauges, or laser vibrometers. The proposed method can be used for *in situ* strain measurements in SHM systems based on eigenfrequency monitoring. Similarly, in the work of Oman et al. [61], FBG sensors were used in order to measure dynamic strains but with a more complex composite structure with impact damage. The investigated structure was a composite plate with stiffeners manufactured in a single layup and curing process. FBG sensors were bonded to the stiffener's surface as well as embedded in the composite material during its manufacture.

2.6   OTHER METHODS

The previous sections introduced the methods used in SHM techniques for composite structures. However, not all techniques could be classified under the four chosen sections. In this section other methods are described.

FBG sensors can also be utilized for measurements of static or quasi-static strains in order to assess damage of composite structures. In the work of Minakuchi et al. [62], FBG sensors were used in order to measure strain in the cross section of an L-shaped composite corner part. The aim was to detect the life cycle–induced strain change. The nonaxisymmetric strain measured by FBG sensors was a good indicator of spring-in distortion in the manufacturing process. Furthermore, this approach is suitable for through-thickness failure that may occur during operation. An interesting approach was proposed in a paper by Pinto et al. [63] where the authors used a method of strain measurement based on carbon fibers and carbon nanotubes. The authors presented a comparative study of two types of strain sensors: in the first case, carbon fibers in a composite part working as a sensor and, in the second case, nonstructural carbon nanotube (CNT) film applied as spray deposition working as a sensor. Strains were extracted based on measurements of resistance. The carbon fiber–based solution showed better sensitivity to strain change and damage sensing than the CNT solution. But the CNT-based solution can be applied for glass fiber composites. This method is very promising for SHM systems dedicated to composite structures.

Comparative vacuum monitoring (CVM) is a nondestructive technique that can be utilized in SHM for crack initiation detection or crack propagation monitoring. This method is based on measurements of pressure difference between pipes with low vacuum and pipes with atmospheric pressure. Where there is no damage, the vacuum pressure remains stable. Due to crack initiation, the pipe breaks and pressure levels change due to the airflow through the connection of two pipes with different pressures [64].

One of the methods that is hard to classify is the nonlinear vibro-acoustic method. It employs both vibration and ultrasonic wave–based methods [65]. The weak high frequency (HF) ultrasonic wave is modulated by a stronger low-frequency (LF) modal/vibration excitation. Researchers use these two frequencies simultaneously. In the work of Pieczonka et al. [65], it was used for a sandwich panel with a foam core and impact damage. When the panel was undamaged, the amplitude of sidebands did not increase with the LF excitation amplitude. There were sidebands caused by intrinsic nonlinearities (e.g., material nonlinearities). In contrast, when the panel was damaged, the amplitude of at least one sideband increased with the LF excitation level and the intensity of modulation increased with the severity of damage. The damage-related modulations can be separated from vibro-acoustic modulations caused by other nonlinear factors. The same method was used for a carbon/epoxy unidirectional prepreg sample with impact-induced delamination [66].

Another hard-to-classify method is based on electrical resistance measurements. Resistive ladder sensors serve to detect and identify the linear sizes of cracks [67]. The sensor's structure and manufacturing process are similar to that of a foil strain gauge. There are several conductive paths connected in a parallel manner. In using this type of sensor, one needs to know the inspected structure because it has to be placed in a hot spot where a crack can potentially develop and grow. When a fatigue crack propagates under the sensor, it is causing a local deformation, and it gradually tears the foil of the sensor, causing a break of the conductive path [67]. As a result, the measured electrical resistance changes due to the lower number of paths that have been "cut" by the crack. In the work of Kurnyta et al. [67], the method was successfully used for crack propagation tracking during a full-scale fatigue test (FSFT) of a military aircraft.

2.7   SUMMARY

This chapter focused on the methods and techniques oriented toward SHM of structures manufactured from composite materials. These are mainly CFRP and GFRP. Fiber-reinforced composites are utilized in many branches of industry: aerospace, automotive, maritime, and renewable energy. The main feature that makes composite materials so attractive is their strength-to-weight ratio. These materials are light and simultaneously their strength is very high, but they are very sensitive to mechanical impacts or thermal and chemical degradation. As a consequence, delamination that is not visible during the conventional visual inspection can initiate in a composite structure. Such an internal delamination can further grow until it reaches a critical size that is dangerous for structural integrity. Because of this, the concept of SHM was raised.

In SHM, different methods can be utilized. In this chapter, methods based on elastic waves, electromechanical impedance, acoustic emission, and vibration methods were investigated. The problem of sensor placement was also discussed as was the experimental investigations of these methods on simple and complex composite elements (with stiffeners, honeycomb core, etc.). The numerical modeling aspects of the phenomena related to those methods were addressed, as was the research conducted on the use of PZT sensors, fiber-optic sensors, and laser sensing. The signal processing approach dedicated to each method is crucial, allowing for extraction of damage-related features from the gathered signals. Investigated damage comes in the form of delamination, thermal degradation, moisture, and chemical contamination. Some of the presented methods are also suitable for assessing the performance of bonded joints. The influence of external factors (such as temperature and load) on investigated methods was also addressed. The characteristics of each method were summarized by a critical analysis, giving the disadvantages that need to be addressed in future research.

REFERENCES

1. A. Rytter, Vibrational based inspection of civil engineering structures. Department of Building Technology and Structural Engineering, Aalborg University, Fracture and Dynamics; No. 44, Vol. R9314, PhD thesis, **1993**.
2. K. Worden, C. R. Farrar, G. Manson, G. Park, The fundamental axioms of structural health monitoring, *Philosophical Transactions of the Royal Society A*, 463 (2082), 1639–1664, **2007**.
3. W. J. Staszewski, S. Mahzan, R. Traynor, Health monitoring of aerospace composite structures: Active and passive approach, *Composites Science and Technology*, 69 (11–12), 1678–1685, **2009**.
4. H. Lamb. On waves in an elastic plate, *Proceedings of the Royal Society of London A*, 95, 114–128, **1917**.
5. L. Ambrozinski, T. Stepinski, T. Uhl, Efficient tool for designing 2D phased arrays in lamb waves imaging of isotropic structures, *Journal of Intelligent Material Systems and Structures*, 26 (17), 2283–2294, **2014**.
6. T. Wandowski, P. H. Malinowski, W. M. Ostachowicz, Circular sensing networks for guided waves based structural health monitoring, *Mechanical Systems and Signal Processing*, 66–67, 248–267, **2016**.
7. A. S. Purekar, D. J. Pines, Damage detection in thin composite laminates using piezoelectric phased sensor arrays and guided Lamb wave interrogation, *Journal of Intelligent Material Systems and Structures*, 21 (10), 995–1010, **2010**.
8. Y. Sun, S. Ji, Analysis, realization and experiment of Lamb wave phased arrays for damage detection and imaging in carbon composite structures, *Journal of Vibroengineering*, 17 (1), 188–202, **2015**.
9. Y. Shenfang, Z. Yongteng, Q. Lei, W. Zhiling, Two-dimensional near-field multiple signal classification algorithm-based impact localization, *Journal of Intelligent Material Systems and Structures*, 26 (4), 400–413, **2015**.
10. J. S. Hall, P. Mc Keon, L. Satyanarayan, J. E. Michaels, N. F. Declercq, Y. H. Berthelot, Minimum variance guided wave imaging in a quasi-isotropic composite plate, *Smart Materials and Structures*, 20, 025013, 8, **2011**.

11. N. Salowitz, Z. Guo, S. Roy, R. Nardari, Y.–H. Li, S.–J. Kim, F. Kopsaftopoulos, F.–K. Chang, Recent advancements and vision toward stretchable bio-inspired networks for intelligent structures, *Structural Health Monitoring*, 13, 609–620, **2014**.
12. H. Kim, K. Jhang, M. Shin, J. Kim, Application of the laser generated focused–Lamb wave for non-contact imaging of defects in plate, *Ultrasonics*, 44, 1265–1268, **2006**.
13. H. Kim, K. Jhang, M. Shin, J. Kim, A noncontact NDE method using a laser generated focused–Lamb wave with enhanced defect-detection ability and spatial resolution, *NDTE International*, 39, 312–319, **2006**.
14. B. Park, Y.–K. An, H. Sohn, Visualization of hidden delamination and debonding in composites through noncontact laser ultrasonic scanning, *Composites Science and Technology*, 100, 10–18, **2014**.
15. G. Pereira, L. P. Mikkelsen, M. McGugan, Embedded fibre Bragg grating sensor response model: Crack growing detection in fibre reinforced plastic materials, *Journal of Physics: Conference Series*, 628, 012115, **2015**.
16. N. Miesen, Y. Mizutania, R. M. Groves, J. Sinke, R. Benedictus, Lamb wave detection in prepreg composite materials with fibre Bragg grating sensors, *Proceedings of SPIE*, 7981, **2011**.
17. P. Liu, H. Sohn, B. Park, Baseline-free damage visualization using noncontact laser nonlinear ultrasonics and state space geometrical changes, *Smart Materials and Structures*, 24 (6), **2015**.
18. W. Ostachowicz, P. Kudela, M. Krawczuk, A. Zak, Spectral finite element method, in *Guided Waves in Structures for SHM: The Time-Domain Spectral Element Method*, Wiley, Chichester, UK, **2012**.
19. K. J. Schubert, C. Brauner, A. S. Herrmann, Non-damage-related influences on Lamb wave-based structural health monitoring of carbon fiber-reinforced plastic structures, *Structural Health Monitoring: An International Journal*, 13 (2), 158–176, **2014**.
20. K. J. Schubert, A. S. Herrmann, On the influence of moisture absorption on Lamb wave propagation and measurements in viscoelastic CFRP using surface applied piezoelectric sensors, *Composite Structures*, 94, 3635–3643, **2012**.
21. M. Huang, L. Jiang, P. K. Liaw, C. R. Brooks, R. Seeley, D. L. Klarstrom, Using acoustic emission in fatigue and fracture materials research, *Journal of the Minerals, Metals & Materials Society*, 50 (11), **1998**. http://www.tms.org/pubs/journals/jom/9811/huang/huang-9811.html.
22. J. F. C. Markmiller, F. K. Chang, Sensor network optimization for a passive sensing impact detection technique, *Structural Health Monitoring: An International Journal*, 9, 25–39, **2010**.
23. T. Kundu, H. Nakatani, N. Takeda, Acoustic source localization in anisotropic plates, *Ultrasonics*, 52, 740–746, **2012**.
24. H. Matt, F. Lanza di Scalea, Macro-fiber composite piezoelectric rosettes for acoustic source location in complex structures, *Smart Materials and Structures*, 16 (4), 1489–1499, **2007**.
25. F. Ciampa, M. Meo, E. Barbieri, Impact localization in composite structures of arbitrary cross section, *Structural Health Monitoring: An International Journal*, 11, 643–655, **2012**.
26. T. Kundu, Acoustic source localization, *Ultrasonics*, 54, 25–38, **2014**.
27. C. Liang, F. P. Sun, C. A. Rogers, An impedance method for dynamic analysis of active material systems, in *Proceedings of 34th Structures, Structural Dynamics and Materials Conference*, La Jolla, CA, 3587–3599, **1993**.
28. C.-K. Soh, Y. Yang, S. Bhalla, *Smart Materials in Structural Health Monitoring, Control and Biomechanics*, Springer, Berlin, Germany, **2012**.
29. E. F. Crawley and J. De Luis, Use of piezoelectric actuators as elements of intelligent structures, *AIAA Journal*, 25 (10), 1373–1385, **1987**.

30. V. G. M. Annamdas, Y. Yang, C. K. Soh, Influence of loading on the electromechanical admittance of piezoceramic transducers, *Smart Materials and Structures*, 16 (5), 1888–1897, **2007**.

31. V. G. M. Annamdas, J. H. L. Pang, K. Zhou, B. Song, Efficiency of electromechanical impedance for load and damage assessment along the thickness of lead zirconate titanate transducers in structural monitoring, *Journal of Intelligent Material Systems and Structures*, 24 (16), 2008–2022, **2013**.

32. A. V. G. Madhav, C. K. Soh, An electromechanical impedance model of a piezoceramic transducer-structure in the presence of thick adhesive bonding, *Smart Materials and Structures*, 16, 673, **2007**.

33. V. G. M. Annamdas, C. K. Soh, Application of electromechanical impedance technique for engineering structures: Review and future issues, *Journal of Intelligent Material Systems and Structures*, 21 (1), 41–59, **2010**.

34. V. Giurgiutiu, *Structural Health Monitoring with Piezoelectric Wafer Active Sensors*, Elsevier Academic Press, the Netherlands, **2008**.

35. M. Gresil, L. Yu, V. Giurgiutiu, M. Sutton, Predictive modeling of electromechanical impedance spectroscopy for composite materials, *Structural Health Monitoring*, 11, 671, **2012**.

36. T. Wandowski, J. Moll, P. Malinowski, S. Opoka, W. Ostachowicz, Assessment of piezoelectric sensor adhesive bonding, *Journal of Physics: Conference Series*, 628, 012114, **2015**.

37. S. Na, H. K. Lee, Resonant frequency range utilized electro-mechanical impedance method for damage detection performance enhancement on composite structures, *Composite Structures*, 94, 2383–2389, **2012**.

38. S. Na, H. K. Lee, Steel wire electromechanical impedance method using a piezoelectric material for composite structures with complex surfaces, *Composite Structures*, 98, 79–84, **2013**.

39. O. Cherrier, P. Selva, V. Budinger, F. Lachaud, J. Morlier, Damage localization map using electromechanical impedance spectrums and inverse distance weighting interpolation: Experimental validation on thin composite structures, *Structural Health Monitoring*, 12 (4), 311–324, **2013**.

40. M. Schwankl, Z. S. Khodaei, M. H. Aliabadi, C. Weimer, Electro-mechanical impedance technique for structural health monitoring of composite panels, *Key Engineering Materials*, 525–526, 569–572, **2013**.

41. H. J. Lim, M. K. Kim, H. Sohn, C. Y. Park, Impedance based damage detection under varying temperature and loading conditions, *NDT & E International*, 44, 740–750, **2011**.

42. R. Sun, E. Sevillano, R. Perera, Debonding detection of FRP strengthened concrete beams by using impedance measurements and an ensemble PSO adaptive spectral model, *Composite Structures*, 125, 374–387, **2015**.

43. P. Malinowski, T. Wandowski, W. Ostachowicz, The use of electromechanical impedance conductance signatures for detection of weak adhesive bonds of carbon fibre–reinforced polymer, *Structural Health Monitoring: An International Journal*, 14 (4), 332–334, **2015**.

44. T. Wandowski, P. Malinowski, L. Skarbek, W. Ostachowicz, Moisture detection in carbon fiber reinforced polymer composites using electromechanical impedance technique, *Proceedings of the Institution of Mechanical Engineers, Part C: Journal of Mechanical Engineering Sciences*, 230 (2), 331–336, **2016**.

45. G. Park, K. Kabeya, H. Cudney, D. Inman, Impedance-based structural health monitoring for temperature varying applications, *JSME International Journal Series A*, 42, 249–258, **1999**.

46. F. G. Baptista, D. E. Budoya, V. A. Almeida, J. A. Ulson, An experimental study on the effect of temperature on piezoelectric sensors for impedance-based structural health monitoring, *Sensors*, 14, 1208–1227, **2014**.

47. C. R. Farrar, D. A. Jauregui, Comparative study of damage identification algorithms applied to a bridge: I. Experiment, *Smart Materials and Structures*, 7 (5), 704–719, **1998**.

48. C. R. Farrar, D. A. Jauregui, Comparative study of damage identification algorithms applied to a bridge: II. Numerical study, *Smart Materials and Structures*, 7 (5), 720–731, **1998**.

49. L. H. Yam, Y. J. Yan, J. S. Jiang, Vibration-based damage detection for composite structures using wavelet transform and neural network identification, *Composite Structures*, 60, 403–412, **2003**.

50. C. S. Hamey, W. Lestari, P. Qiao, G. Song, Experimental damage identification of carbon/epoxy composite beams using curvature mode shapes, *Structural Health Monitoring*, 3 (4), 333–353, **2004**.

51. G. Park, A. C. Rutherford, J. R. Wait, B. Nadler, C. R. Farrar, T. N. Claytor, High-frequency response functions for composite plate monitoring with ultrasonic validation LA-UR-04-7688, *AIAA Journal*, 43 (11), 2431–2437, **2005**.

52. K.-S. Kim, H.-C. Jung, K.-S. Kang, Y.-J. Kang, Y.-H. Cha, W.-G. Jung, Experimental analysis of vibration modes of plates using ESPI, KSME, *International Journal*, 13 (10), 677–686, **1999**.

53. W.-C. Wang, C.-H. Hwang, Experimental analysis of vibration characteristics of an edge-cracked composite plate by ESPI method, *International Journal of Fracture*, 91, 311–321, **1998**.

54. K. M. Hong, Y. J. Kang, N. K. Park, W. J. Ryu, A study on the vibration characteristics of symmetry and asymmetry laminated composite materials by using ESPI, *Key Engineering Materials*, 353–358, 2366–2370, **2007**.

55. A. P. Herman, A. C. Orifici, A. P. Mouritz, Vibration modal analysis of defects in composite T–stiffened panels, *Composite Structures*, 104, 34–42, **2013**.

56. R. B. Bai, M. Radzienski, M. S. Cao, W. Ostachowicz, Z. Su, Non-baseline identification of delamination in plates using wavelet-aided fractal analysis of two-dimensional mode shapes, *Journal of Intelligent Material Systems and Structures*, 26 (17), 2338–2350, **2015**.

57. T. Kranjc, J. Slavic, M. Boltezar, The mass normalization of the displacement and strain mode shapes in a strain experimental modal analysis using the mass-change strategy, *Journal of Sound and Vibration*, 332, 6968–6981, **2013**.

58. F. L. Marques dos Santos, B. Peeters, J. Lau, W. Desmet, L. C. S. Goes, The use of strain gauges in vibration-based damage detection, *Journal of Physics: Conference Series*, 628, 012119, **2015**.

59. D. Garcia, R. Palazzetti, I. Trendafilova, C. Fiorini, A. Zucchelli, Vibration-based delamination diagnosis and modelling for composite laminate plates, *Composite Structures*, 130, 155–162, **2015**.

60. J. Frieden, J. Cugnoni, J. Botsis, T. Gmur, Vibration-based characterization of impact induced delamination in composite plates using embedded FBG sensors and numerical modeling, *Composites: Part B*, 42, 607–613, **2011**.

61. K. Oman, B. Van Hoel, K. Aly, K. Peters, G. Van Steenberge, N. Stan, S. Schultz, Instrumentation of integrally stiffened composite panel with fiber Bragg grating sensors for vibration measurements, *Smart Materials and Structures*, 24, 085031, **2015**.

62. S. Minakuchi, T. Umehara, K. Takagaki, Y. Ito, N. Takeda, Life cycle monitoring and advanced quality assurance of L–shaped composite corner part using embedded fiber-optic sensor, *Composites: Part A*, 48, 153–161, **2013**.

63. B. Pinto, S. Kern, J. J. Ku-Herrera, J. Yasui, V. La Saponara, K. J. Loh, A comparative study of a self strain–monitoring carbon nanotube film and carbon fibers under flexural loading by electrical resistance changes, *Journal of Physics: Conference Series*, 628, 012098, **2015**.

64. M. Wishaw, Comparative vacuum monitoring: A new method of *in-situ* real time crack detection and monitoring, in *10th Asia-Pacific Conference on Non-Destructive Testing*, Structural Monitoring Systems, Brisbane, Australia, 1–9, **2001**.

65. L. Pieczonka, P. Ukowski, A. Klepka, W. J. Staszewski, T. Uhl, F. Aymerich, Impact damage detection in light composite sandwich panels using piezo-based nonlinear vibro-acoustic modulations, *Smart Materials and Structures*, 23, 105021, **2014**.

66. A. Klepka, L. Pieczonka, W. J. Staszewski, F. Aymerich, Impact damage detection in laminated composites by non-linear vibro-acoustic wave modulations, *Composites Part B: Engineering*, 65, 99–108, **2014**.

67. A. Kurnyta, M. Dziendzikowski, K. Dragan, A. Leski, Approach to health monitoring of an aircraft structure with resistive ladder sensors during full scale fatigue test, The e-Journal of Nondestructive Testing, 20 (2), in *Proceedings of 7th European Workshop on Structural Health Monitoring*, Nantes, France, **2014**.

3 Introduction to Optical Fiber Sensors

Dipankar Sengupta and Ginu Rajan

CONTENTS

3.1 Introduction 41
3.2 Optical Fiber Sensors: Concept 42
3.3 Measurement Capabilities and Advantages of Optical Fiber Sensors........... 43
    3.3.1 Types of Optical Fiber Sensors.. 44
        3.3.1.1 Intrinsic and Extrinsic Sensors 44
        3.3.1.2 Point Sensor and Distributed Sensor 44
    3.3.2 Modulation Schemes.. 45
        3.3.2.1 Intensity-Modulated Sensors 45
        3.3.2.2 Phase-Modulated Fiber-Optic Sensors 50
        3.3.2.3 Wavelength-Modulated Sensors.. 54
        3.3.2.4 Fluorescent-Based Fiber Sensors 57
        3.3.2.5 Polarimetric Fiber Sensors.. 58
    3.3.3 Fiber-Optic Scattering-Based Distributed Fiber Sensors 59
    3.3.4 Surface Plasmon Resonance Sensors 61
    3.3.5 Photonic Crystal Fiber Sensors.. 61
    3.3.6 Optical Microfiber Sensors... 62
3.4 Optical Fiber Sensor Instrumentation 62
3.5 Optical Fiber Sensors: Industry Applications and Market 63
3.6 Conclusion 65
References.. 66

3.1 INTRODUCTION

Sensing has become a key enabling technology in many areas, from entertainment technology to health, transport, and many industrial technologies. In many such advanced applications, where miniaturization, sensitivity, and remote measurement are vital, optical fiber–based sensing techniques can provide novel solutions. As a result of this, optical fiber sensing (OFS) technology has been developed as a powerful and rich technology that is currently being implemented in a wide variety of applications [1,2].

Research work on OFS began in the 1960s, but it is the development of modern low-loss optical fibers that has enabled the transition from the experimental stage to practical applications. Some of the first sensing experiments using low-loss optical fibers [3–6] were demonstrated during the early 1970s. The field of OFS has continued to progress and has developed enormously since that time. For instance, distributed fiber-optic sensors have now been installed in bridges and dams to monitor the performance of and structural damage to these facilities. Optical fiber sensors are used to monitor the conditions within oil wells and pipelines, railways, wings of airplanes, and wind turbines. Decades of research has led to the development of accurate optical fiber measuring instruments, including gyroscopes, hydrophones, temperature probes, and chemical monitors. As the OFS field is complementary to optical communication, it has benefited from advances in optical fibers and optoelectronic instrumentation. With the rapid development of optical communication networks, the cost of fiber sensors has significantly reduced because of the commercially available key components, such as light sources and photodetectors.

It is anticipated that research on OFS technology will continue to grow and the technology will become widespread. Several new types of fiber sensors are progressively emerging, and some of the fiber sensing technologies are in a mature state and have been commercially produced and used in different applications. As lightwave technology is becoming the new enabling technology behind many advanced high-end applications, optical fiber sensors can play an integral part in those developments.

The aims of this chapter are to introduce the basic principles of OFS, to provide a brief overview of some of the different types of fiber sensors developed over the years, and to form the basis for the rest of the chapters in this book.

3.2   OPTICAL FIBER SENSORS: CONCEPT

For a given process or operation, a system needs to perform a given function, which requires the condition of one or multiple parameters to be qualified and/or quantified. Therefore, these parameters must be measured and processed so that the system can carry out its function. To facilitate the process, a sensor can be used, which is a device that provides this information by measuring any physical/chemical/biological measurand of interest (pressure, temperature, stress, pH, etc.) and converting it into an alternative form of energy (e.g., an electrical or optical signal) that can be subsequently processed, transmitted, and correlated with the measurand of interest.

The optical fiber alternative to traditional sensors, *optical fiber sensors*, can be defined as a means through which a physical, chemical, or biological measurand interacts with light guided through an optical fiber, or guided to an interaction region by an optical fiber, to produce a modulated optical signal with information related to the measurement parameter. The basic concept of a fiber-optic sensing system is demonstrated in Figure 3.1, where the fiber (guided light) interacts with an external parameter and carries the modulated light signal from the source to the detector. The input measurand information can be extracted from this modulated optical signal. Depending on the type of fiber sensor and its operating principle, the sensor system can operate either in transmission mode or in reflection mode, which is elaborated in later sections of this chapter.

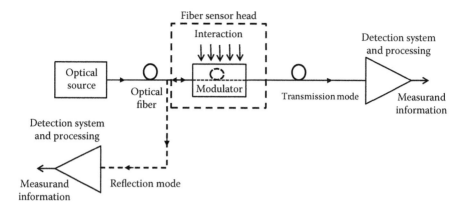

FIGURE 3.1 Basic concept of an optical fiber sensor.

3.3 MEASUREMENT CAPABILITIES AND ADVANTAGES OF OPTICAL FIBER SENSORS

One of the biggest advantages of fiber sensors is that they have the capability to measure a wide range of parameters based on the end user requirement, once a proper fiber sensor type is used. Some (not an exhaustive list) of the measurement capabilities of fiber sensors are

- Strain, pressure, force
- Rotation, acceleration
- Electric and magnetic fields
- Acoustics and vibration
- Temperature, humidity
- Liquid level
- pH and viscosity
- Bio-sensing: single molecules, chemicals, and biological elements such as DNA, single viruses, and bacteria

Given the advancement in the development of optical fibers and instrumentation, the development and measurement capability of optical fiber sensors have also increased, and they are currently being employed in many structural health monitoring and biomedical applications [7–9]. Optical fiber sensors share all the benefits of optical fibers and are excellent candidates for monitoring environmental/external changes. Some of the advantages of fiber sensors over conventional electronic sensors are

- Light weight
- Passive/low power
- Resistance to electromagnetic interference
- High sensitivity and bandwidth
- Environmental ruggedness

- Complementarity to telecom/optoelectronics
- Multiplexing capability
- Multifunctional sensing possibility
- Easy integration into a wide variety of structures
- Robustness, greater resistance to harsh environments

Though optical fiber sensors have many advantages compared with their electrical counterparts, there are also some concerns over this technology. Major drawbacks include cost, long-term stability, less efficient transduction mechanism, and complexity in its interrogation systems. Another disadvantage of the fiber sensing system is the unfamiliarity of the end user in dealing with the system. As more research, commercialization, and standardization efforts are ongoing, it is expected that some of the current disadvantages may become extinct in the near future.

3.3.1 TYPES OF OPTICAL FIBER SENSORS

3.3.1.1 Intrinsic and Extrinsic Sensors

As the optical fiber sensors operate by modifying one or more properties of light passing through the fiber, they can be broadly classified as extrinsic or intrinsic. In the extrinsic sensor types, the optical fiber is used as a means to carry light to an external sensing system, while in the intrinsic type, the light does not have to leave the optical fiber to perform the sensing function, as shown in Figure 3.2a. The intrinsic fiber sensor types are more attractive and widely researched, as this scheme has many advantages compared with the extrinsic type, such as their in-fiber nature and the flexibility in the design of the fiber sensor head (Figure 3.2b).

3.3.1.2 Point Sensor and Distributed Sensor

Another classification of fiber optic sensors is based on whether they allow measurements of the physical parameter of interest locally, in which case they are called *point sensors*, or provide the value of the physical parameter over a distance and as a function of position along the fiber-distributed sensors. Point transducers have a finite length and therefore provide the value of the physical parameter averaged over a given volume of space corresponding to the length of the point transducer. If multiple point measurements are required, a number of point sensors need to be used, which in many cases requires installing multiple lead-in/lead-out fibers,

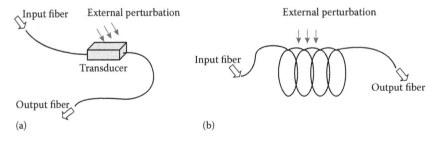

FIGURE 3.2 Optical fiber (a) extrinsic sensor and (b) intrinsic sensor.

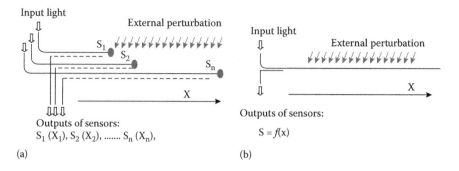

FIGURE 3.3 Fiber optic (a) point sensor and (b) distributed sensor.

arrays of detectors, and so on. The number of point sensors, which defines the spatial resolution of the system, is limited by the cost, and therefore becomes insufficient for many practical applications, as shown in Figure 3.3a.

The distributed fiber-optic sensing approach has no real equivalent among other types of sensors and has become an important differentiator of fiber-optic sensing. Distributed sensors are most often intrinsic, and the sensing principles and practical implementation are significantly different from those used for point sensing. Since only one fiber needs to be installed to sense the physical parameter of interest at many points, as shown in Figure 3.3b, it is possible to share the optical source and detector and often eliminate the other equipment required for multipoint sensing using point sensors. This results in much lower cost per point measurement and also much higher weight and space efficiency of the distributed sensing, making this technique the most powerful monitoring option in many applications, especially for structural monitoring of civil and aerospace structures.

3.3.2 MODULATION SCHEMES

Furthermore, depending on the light property that is modified, optical fiber sensors can be mainly classified into four main categories:

- Intensity-modulated sensors
- Phase-modulated (interferometric) sensors
- Polarization-modulated (polarimetric) sensors
- Wavelength-modulated (spectrometric) sensors

Optical fiber sensors can also be further classified based on their spatial positioning as well as on their measurement capabilities. A summary of different classifications of OFS is shown in Table 3.1.

3.3.2.1 Intensity-Modulated Sensors

Intensity-modulated sensors are one of the earliest types and perhaps the simplest type of optical fiber sensor. In the intensity modulation scheme, optical signal is transmitted through optical fibers, and then its intensity is modulated by various

TABLE 3.1

Different Classifications of Optical Fiber Sensors

Classification Based on the Working Principle	Classification Based on the Spatial Positioning	Classification Based on the Measurement Parameters
Intensity-modulated sensors Detection through light power	**Point sensors** Discrete points, different channels for each sensor/measurand	**Physical sensors** Temperature, strain, pressure, and so on
Phase-modulated (interferometric) sensors Detection using the phase of the light beam	**Distributed** Measurand can be determined along a path, surface, or volume	**Chemical sensors** pH content, gas sensors, spectroscopic study, and so on
Polarimetric sensors Detection of changes in the state of polarization of the light	**Quasi-distributed** Variable measured at discrete points along an optical link	**Bio-sensors** DNA, blood flow, glucose sensor, and so on
Spectrometric sensors Detection of changes in the wavelength change of the light	**Integrated** Measurand is integrated along an optical link giving a single-value output	

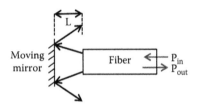

FIGURE 3.4 Simple extrinsic intensity-modulated sensor.

means, such as by fiber bending, reflectance, or changing the medium through which the light is transmitted. The main advantages of intensity-modulated sensors are ease of fabrication, simple detection system and signal processing requirements, and low price-performance. A simple example of a reflection-type intensity-modulated fiber optic sensor using a single fiber that can be used to measure distance, pressure, and so on is shown in Figure 3.4.

Compared with the single-fiber type, two-fiber-type reflection sensors are often used as proximity sensors in a variety of applications. In this case, one fiber is used as the input fiber, and a second fiber is used to collect the reflected light from a reflecting surface. The schematic of a two-fiber configuration and its typical intensity response is shown in Figure 3.5. The modulation function of a two-fiber reflection-type sensor is given as [10]

$$M_s = \frac{\Phi_{RS}}{\Phi_E} = \frac{S_R \cos^4 \theta}{\pi (2h \text{NA})^2} \quad (3.1)$$

where:

    Φ_E is the radiant flux from the emitting fiber

    Φ_{RS} is the radiant flux intercepted by the end face of the receiving fiber

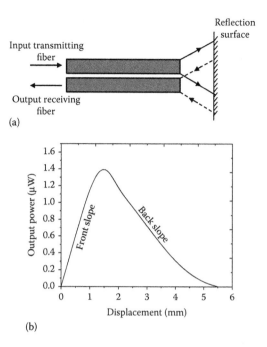

FIGURE 3.5 (a) Two-fiber reflection-type sensor; (b) its typical characteristic response.

S_R is the area of the end face of the receiving fiber
θ   is the angle between the axis of the receiving fiber and the line connecting the centers of the receiving fiber and the emitting fiber (virtual image)
NA   is the numerical aperture of the fiber

In contrast to other sensing principles, intensity modulation can be obtained by using a simple arrangement; however, optical fiber bending, coupling misalignments, source power fluctuation, and so on can cause signal attenuation and signal intensity instability, which leads to a less reliable sensor system. One solution to this problem is intensity referencing, in which a fraction of the input light (obtained using a fiber splitter) is used for monitoring the input power fluctuation. An example of such a configuration is shown in Figure 3.6 [11], where the power ratio of the modulated signal and reference signal is used.

Other types of intensity-modulated fiber-optic sensors include evanescent wave sensors, microbend, and macrobend sensors.

3.3.2.1.1   Evanescent Field-Based Fiber Sensor
The evanescent field–based fiber sensor is one of the most useful and commonly used intensity-modulated fiber-optic sensors. When light propagates along the optical fiber with an incidence angle at the core/cladding interface greater than the critical angle, the beam is totally reflected back into core of the fiber (Figure 3.7).

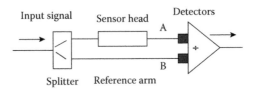

FIGURE 3.6 An example of ratiometric power measurement method for intensity referencing.

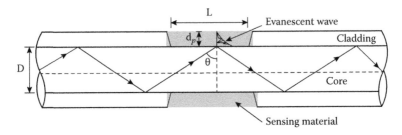

FIGURE 3.7 Cross-sectional view of an evanescence field sensor.

Interesting enough, light is not instantaneously reflected when it reaches the interface. Rather, the superposition of the incident and reflected beams results in the formation of a standing electromagnetic wave. The maximum of the electric field amplitude of the standing wave is located at the interface, but the field decays exponentially in the direction of an outward normal to the interface. The decaying field is called an *evanescent field*. The electric field amplitude, E, at a distance x along the normal is given by

$$E = E_0 \exp\left(-\frac{x}{d_p}\right) \tag{3.2}$$

where:
    E_0 is the electric field at the interface
    d_p is the penetration depth

The magnitude of the penetration depth is given by

$$d_p = \frac{\lambda}{2\pi n_1 \sqrt{\sin^2\theta - \left(\dfrac{n_2}{n_1}\right)^2}} \tag{3.3}$$

where:
    λ is the wavelength of the light source
    θ is the angle of incidence of the light at the core/cladding interface
n_1 and n_2 are the refractive indices (RI) of the core and cladding, respectively

Figure 3.7 is a cross-sectional view of a fiber cut at the long axis, and provides a graphical representation of the penetration depth of the evanescent field, the angle of incidence, and a hypothetical ray of light propagating along the fiber. The magnitude of the evanescent field depends on the cladding thickness. The increase in evanescent field forms the basis for increased sensitivity toward the surrounding medium. Evanescent wave absorption sensors are typically used to monitor the concentration of an absorbing fluid, humidity, chemical sensing, and so on [11].

In single-mode (SM) and multimode (MM) fibers, the effects associated with bending found their application in two early type of intensity-modulated point sensors popularly known as *microbend* and *macrobend* fiber sensors. The terms *micro* and *macro* in this context refer to the scale of the fiber bends rather than to the scale of the volume within which the parameter of interest is measured. The concepts of micro- and macrobending of fiber have certain similarities as well as differences.

3.3.2.1.2 Microbending

Microbending can be achieved by mechanical perturbation of the fiber and causes a redistribution of light power among the many modes in the fiber. When fiber is bent, consequently the guided mode of SM fiber couples to cladding modes, and the highest guided mode of MM fiber couples to the first radiated mode due to the deformations of the refractive index (RI) distribution or the geometry of the fiber. Coupling to cladding modes or radiated modes causes power to be lost from the guided core mode, and the power traveling in the cladding modes will be rapidly dissipated because of the high losses in these modes. The loss of power also depends on the period of the tooth spacing Λ on the moving plate (Figure 3.8).

For the step index fibers, this expression can be approximated as

$$\Lambda = \frac{\pi a}{\sqrt{\Delta}} \tag{3.4}$$

where:
    a is the fiber core radius
    Δ is the normalized index difference between core and clad

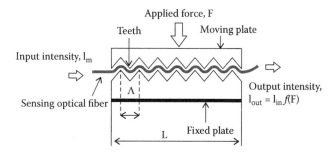

FIGURE 3.8 Structure of microbend force sensor.

The step index fiber exhibits a threshold response and the graded index fiber a resonant response. For a graded index fiber, it is shown that to achieve maximum microbend sensitivity, the spatial period of the deformer must match the expression $\Lambda = \sqrt{2}\pi a/\sqrt{\Delta}$. Typically, the spatial period is within the range of 0.2–2 mm. Although not as sensitive as interferometric sensors, microbenders have good sensitivity to small displacements in the order 10^{-10} m/$\sqrt{\text{Hz}}$ and a very large dynamic range, which is one of the major advantages of these sensors [12].

3.3.2.1.3 Macrobend Sensors

Macrobend sensors are another type of intensity-modulated, intrinsic fiber-optic sensors, for which a single-mode fiber (SMF) is usually used, and it is bent at relatively large diameters; typically, the bend radius is in the order of a few centimeters. Power loss in the core of an SMF due to a uniform bend consists of two components: a pure bend loss and a transition loss. The pure bend loss results from the loss of guidance at the outer portion of the evanescent field of the fundamental mode [13]. This loss of guidance is due to the phase velocity of the outer part of the evanescent field becoming equal to the speed of light in the cladding. The smaller the bend radius, the greater the fraction of the evanescent field affected, and hence, the greater the percentage of light lost at the bend. The transition loss arises from the coupling of light from the fundamental mode to leaky core modes whenever there is a change in curvature of the fiber axis, for example, from straight to curved or vice versa. Based on macrobending loss, physical parameters such as temperature and displacement can be measured. Figure 3.9 shows a displacement sensor using high–bend loss fiber.

3.3.2.2 Phase-Modulated Fiber-Optic Sensors

Phase-modulated sensors, also known as interferometric sensors, are based on the phase difference of coherent light traveling along two different paths, either in

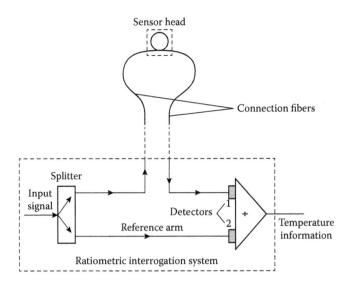

FIGURE 3.9 Schematic diagram of macrobend fiber temperature sensor system.

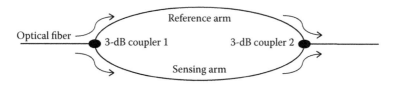

FIGURE 3.10 Mach–Zehnder sensor configuration.

the same fiber or in different fibers. These sensors are often considered as high-sensitivity sensors due to their capability to respond to small changes in the external measurands. The common interferometric fiber sensors include fiber-optic Mach–Zehnder, Michelson, Fabry–Perot, and Sagnac interferometers.

The fiber-optic *Mach–Zehnder interferometric sensor* (MZI) has two independent arms, the reference arm and the sensing arm, as illustrated in Figure 3.10. Incident light is split into two arms by a fiber coupler and then recombined by another fiber coupler.

The recombined light has an interference component according to the optical path difference (OPD) between the two arms. For sensing applications, the reference arm is kept isolated from external variation, and only the sensing arm is exposed to the variation. Then, the variation in the sensing arm induced by factors such as temperature, strain, and RI changes the OPD of the MZI, which can be easily detected by analyzing the variation in the interference signal. With the advent of the long-period fiber grating (LPFG), this scheme has been rapidly replaced by the scheme of the in-line waveguide interferometer (Figure 3.11).

This in-line type of MZI has the same physical length in both the reference arm and the sensing arm, but has different optical path lengths due to the modal dispersion; the cladding mode beam has a lower effective index than the core mode beam. The reported [46] RI sensitivity achieved is 1.8×10^{-6} [14]. Fiber tapering has been applied to the separated region between two LPFGs to improve the sensitivity [15]. Various effective in-line MZI configurations have also been developed using a lateral offset between the fibers [16], different core sizes [17], tapering a fiber at two points along the fiber [18], using a double-cladding fiber [19], micro-cavities [20], and a twin-core fiber [21].

Fiber-optic sensors based on Michelson interferometers (MIs) are quite similar to MZIs. However, each beam is reflected by the mirror at the end of each arm in MI configuration (Figure 3.12a).

MI sensors are operated under reflection modes and are compact and handy in practical uses and installation. Multiplexing capability with parallel connection of

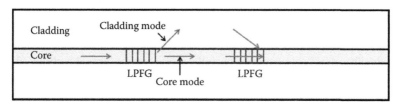

FIGURE 3.11 Inline MZI configuration using LPFG.

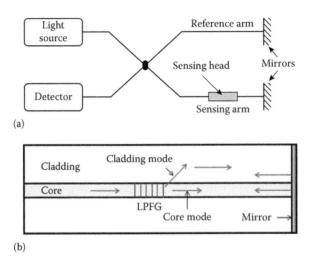

(a)

(b)

FIGURE 3.12 (a) Fiber-optic Michelson interferometer (MI) sensor; (b) in-line MI sensor configuration.

several sensors is another beneficial point of MIs. However, it is essential to adjust the fiber length difference between the reference arm and the sensing arm of an MI within the coherence length of the light source [22,23]. An in-line configuration of MI is also possible, as shown in Figure 3.12b. In this configuration, a part of the core mode beam is coupled to the cladding mode(s), which is reflected along with the uncoupled core mode beam by the common reflector at the end of the fiber. MI-based sensors are especially useful for measurements of the temperature and RI of liquid specimens.

Fiber-optic Fabry–Perot interferometers (FPIs) offer the advantages of high resolution, compact structure, and immunity to electromagnetic interference. They play important roles in a large number of sensing applications, such as RI, strain, temperature, and pressure. In FPIs, sensor interference occurs due to the multiple superpositions of both reflected and transmitted beams at two parallel surfaces, sometimes called *etalon* [24]. The basic configuration is different from MZI and MI. Fiber-optic FPI can be simply formed by intentionally building up reflectors either inside or outside the fibers, and based on this, they are classified into two categories, extrinsic and intrinsic FPI, as shown in Figure 3.13a and b, respectively. Extrinsic FPI is useful for obtaining a high-finesse interference signal, is easy to fabricate, and does not need any high-cost equipment. However, the low coupling efficiency and need for careful alignment and packaging are drawbacks compared with intrinsic FPI.

On the other hand, the intrinsic FPI fiber sensors have reflecting components within the fiber itself, for example, when the reflectors are formed within a fiber by any means, as in Figure 3.13b. The local cavity of the intrinsic FPI can be formed by a range of methods, such as micromachining [25], fiber Bragg gratings (FBGs) [26], chemical etching [27], and thin film deposition [28]. However, they still have the problem of using high-cost fabrication equipment for the cavity formation.

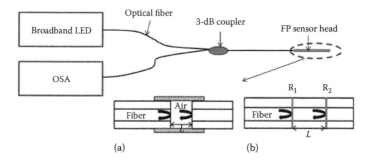

FIGURE 3.13 (a) Extrinsic FP interferometric sensor head; (b) intrinsic FP interferometric sensor head.

The reflection or transmission spectrum of an FPI can be described as the wavelength-dependent intensity modulation of the input light spectrum, which is mainly caused by the optical phase difference between two reflected or transmitted beams. The phase difference of the FPI is simply given as

$$\delta_{PPI} = \frac{2\pi}{\lambda} n2L \tag{3.5}$$

where:
λ is the wavelength of incident light
n is the RI of the cavity material or cavity mode
L   is the physical length of the cavity

When perturbation is introduced to the sensor, the phase difference is influenced by the variation in the OPD of the interferometer. For example, longitudinal strain applied to the FPI sensor changes the physical length of the cavity and/or the RI of the cavity material, which results in phase variation. By measuring the shift of the wavelength spectrum of an FPI, the strain applied on it can be quantitatively obtained.

3.3.2.2.1 Fiber-Optic Sagnac Interferometric (SI) Sensors

Fiber-optic Sagnac interferometric (SI) sensors are simple in structure, easy to fabricate, and environmentally robust. An SI consists of an optical fiber loop, in which the input light is split into two directions by a 3 dB fiber coupler, and the two counterpropagating beams are combined again at the same coupler (Figure 3.14). Unlike other fiber-optic interferometers, the OPD is determined by the polarization-dependent propagating speed of the mode guided along the loop. To maximize the polarization-dependent feature of SIs, birefringent fibers are typically used in sensing parts. The polarizations are adjusted by a polarization controller (PC) attached at the beginning of the sensing fiber. The signal at the output port of the fiber coupler is governed by the interference between the beams polarized along the slow axis and the fast axis. The phase of the interference is simply given as

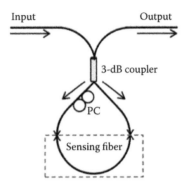

FIGURE 3.14 Schematic of sensor based on a Sagnac interferometer.

$$\delta_{sl} = \frac{2\pi}{\lambda} BL, \ B = |n_f - n_s|$$

(3.6)

where:

B is the birefringent coefficient of the sensing fiber
L is the length of the sensing fiber
n_f and n_s are the effective indices of the fast and slow modes, respectively

In general, high birefringent fibers (HBFs) or polarization-maintaining fibers (PMFs) are chosen as the sensing fibers to acquire a high phase sensitivity. The SI has been principally used to measure rotation and is a replacement for ring laser gyros and mechanical gyros [29]. It may also be employed to measure time-varying effects, such as acoustics and vibration, and slowly varying phenomena, such as strain.

3.3.2.3 Wavelength-Modulated Sensors

Wavelength-modulated fiber sensors exhibit a change in the propagating optical wavelength (spectral modulation) when interacted with by an external perturbation. Some common wavelength-based fiber sensors include Bragg grating sensors, fluorescence sensors, and black body sensors.

Arguably, one of the most widely used wavelength-based sensors is the Bragg grating sensor. Originating with the discovery of the photosensitivity of germanium-doped silica, FBG sensing technology has attracted widespread attention and has been the subject of continuous and rapid development since FBGs were used for the first time for sensing purposes in 1989. An elementary FBG comprises a short section of an SMF in which the core RI is modulated periodically using an intense optical interference pattern, typically at UV wavelengths. The change in the core RI is between 10^{-5} and 10^{-3}, and the length of a Bragg grating is usually around 10 mm. A schematic of an FBG is shown in Figure 3.15. The periodic index modulated structure enables the light to be coupled from the forward-propagating core mode into a backward-propagating core mode, generating a reflection response.

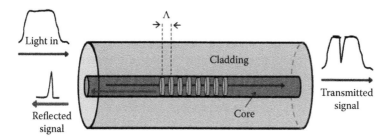

FIGURE 3.15 Structure of FBG.

The light reflected by the periodic variations of the RI has a central wavelength [30]

$$\lambda_B = 2n_{eff}\Lambda \tag{3.7}$$

where:

Λ is the grating periodicity

n_{eff} is the effective RI of the waveguide mode

The basic principle of operation of any FBG-based sensor is that any local strain or temperature change alters the effective index of core refraction and the grating period, followed by changes in wavelength of the reflected light. The corresponding Bragg wavelength shifts are typically of the order of 1 pm/µε and 10 pm/°C.

Monitoring of the changes in the reflected wavelength can be achieved by employing edge filters, tunable narrowband filters, tunable lasers, or charge-coupled device (CCD) spectrometers. In comparison with other fiber-optic sensors, FBGs have a unique advantage in that they encode the wavelength, which is an absolute parameter and which does not suffer from disturbances in the light path and optical source power fluctuations.

This wavelength-encoding nature of FBGs facilitates wavelength-division multiplexing (WDM) of multiple sensors; FBG sensors can be particularly useful when gratings with different periods are arranged along an optical fiber. Each of the reflected signals will have a unique wavelength and can be easily monitored, thus achieving multiplexing of the outputs of multiple sensors using a single fiber. Thus,

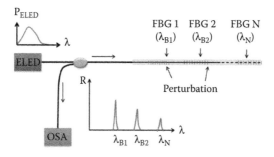

FIGURE 3.16 Quasi-distributed FBG sensor. (OSA, optical spectrum analyzer; ELED, edge-emitting light emitting diode.)

strain, temperature, or pressure can be measured at many locations, making FBGs a kind of typical quasi-distributed sensor (Figure 3.16). FBG sensors are preferred in many civil engineering applications and have been successfully employed in many full-scale structures requiring multiple-point sensing distributed over a long range.

Most of the research on FBG sensors has focused on the use of these devices to provide quasi-distributed point sensing of strain or temperature. FBGs have also been used as point sensors to measure pressure, magnetic field, high voltage, liquid level, RI of liquid, and acoustic emission levels, and to monitor the curing process of composites. One disadvantage of the FBG technology in real engineering applications is associated with its cross-sensitivity effect, which causes simultaneous sensitivity of Bragg wavelength shift to both strain and temperature. Therefore, some methods to discriminate both measurements are usually needed.

Another type of grating, LPFG, couples light from a guided mode into forward-propagating cladding modes, where it is lost due to absorption and scattering. The coupling from the guided mode to cladding modes is wavelength dependent, so we can obtain a spectrally selective loss. It is an optical fiber structure with the properties periodically varying along the fiber, such that the conditions for the interaction of several co-propagating modes are satisfied (Figure 3.17).

The period of such a structure is of the order of a fraction of a millimeter. In contrast to the fiber Bragg gratings, LPFGs couple co-propagating modes with close propagation constants; therefore, the period of such a grating can considerably exceed the wavelength of radiation propagating in the fiber. Because the period of an LPFG is much larger than the wavelength, LPFGs are relatively simple to manufacture. Since LPFGs couple co-propagating modes, their resonances can only be observed in transmission spectra. The transmission spectrum has dips at the wavelengths corresponding to resonances with various cladding modes (in an SMF). Depending on

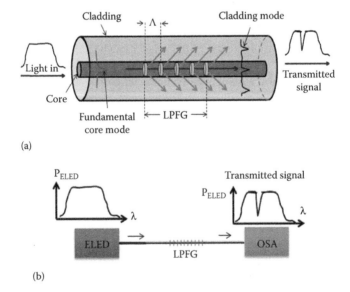

(a)

(b)

FIGURE 3.17   (a) Structure of LPFG; (b) schematic experimental setup for LPFG.

the symmetry of the perturbation that is used to write the LPFG, modes of different symmetries may be coupled. For instance, cylindrically symmetric gratings couple symmetric LP_{0m} modes of the fiber. Microbend gratings, which are antisymmetric with respect to the fiber axis, create a resonance between the core mode and the asymmetric LP_{1m} modes of the core and the cladding.

LPFGs with periods ranging from several hundred microns to several millimeters couple incident light guided by a fundamental mode in the core to different forward-propagating cladding modes of high diffraction order, m, in an optical fiber, which decay rapidly because of the radiation from scattering losses. The coupling of the light into the cladding region generates a set of resonant bands centered at wavelength λ_m in the transmission spectrum of the fiber. The resonance wavelengths, λ_m, of an attenuation band are solutions of the phase-matched conditions

$$\lambda_m = \left[n_{co}^{eff} - n_{cl,m}^{eff} \right] \Lambda = \delta n_{eff} \Lambda \qquad (3.8)$$

where:

Λ is the period of the grating

n_{co}^{eff} is the effective RI of the fundamental core mode at the wavelength of λ_m, which also depends on the core RI and cladding RI

In addition, $n_{(cl,m)}^{e} ff$ is the effective RI of the mth radial cladding mode ($m = 2, 3, 4, \ldots$) at the wavelength λ_m, which is also a function of cladding RI and in particular the RI of the surrounding medium, n_s. It is noted that both indices depend on the temperature and the strain experienced by the fiber. The spectral properties of individual cladding modes are determined by the fiber structure and may be observed through their associated attenuation bands. When the RI of the surrounding medium changes, $n_{cl,m}^{eff}$ also changes, and a wavelength shift can be obtained in the transmission spectrum. An LPFG can be very sensitive to changes in temperature and deformations due to fiber imperfections, loading, and bending, which also produce a noticeable wavelength shift in loss peaks. Therefore, to precisely measure variations in RI changes, temperature changes and deformations must be compensated or avoided.

LPFGs have a wide variety of applications, including band-rejection filters, gain-flattening filters, and sensors. Various gratings with complex structures have been designed: gratings combining several LPFGs, LPFGs with superstructures, chirped gratings, and gratings with apodization. Various LPFG-based devices have been developed: filters, sensors, fiber dispersion compensators, and so on [31].

3.3.2.4   Fluorescent-Based Fiber Sensors

Fluorescent-based fiber sensors are widely used for physical and chemical sensing for measurands such as temperature, humidity, and viscosity. Among the different configurations used, two of the most common are the end tip sensor and the blackbody cavity types [32]. In the case of a blackbody sensor, a blackbody cavity is placed at the end of an optical fiber [33], and as the temperature of the cavity rises, it starts to glow and act as a light source. Using suitable detectors and narrow band filters, the profile of the blackbody curve can be determined, which provides the measurand information. Due to the capability to measure temperature to within a few degrees centigrade under intense RF fields, sensors of this type have been developed as optical fiber thermometers and have been successfully commercialized.

3.3.2.5 Polarimetric Fiber Sensors

Polarimetric fiber sensors are quite well known. As the light wave propagates along the optical fiber, its state of polarization changes because of the difference in the phase velocity of the two polarization components in a birefringent fiber. The polarization properties of light propagating though an optical fiber can be affected by stress, strain, pressure, and temperature acting on it, and in a fiber polarimetric sensor, the change in state of polarization is measured and is used to retrieve the sensing parameter [34]. A symmetric deformation effect or temperature variation in an SMF influences the propagation constant (β) for every mode because of the changes in the fiber length (L) and the RIs of the core and the cladding. Under the influence of longitudinal strain (ε) or temperature (T), for SMF polarimetric sensors, the change in the phase difference can be written as [35]

$$\frac{\delta(\Delta\Phi)}{\delta X} = \Delta\beta \frac{\partial L}{\partial X} + L \frac{\partial(\Delta\beta)}{\partial X} \tag{3.9}$$

where X stands for temperature, pressure, or strain.

A fiber-optic polarimetric sensor that operates in intensity domain using a polarizer–analyzer arrangement is shown in Figure 3.18. In a typical polarimetric sensor, linearly polarized light is launched at 45° to the principal axes of a birefringent fiber such that both the polarization modes can be equally excited. The polarization state at the output is converted to intensity by using a polarizer analyzer oriented at 90° to the input polarization state. For such polarimetric sensors, the change in the output intensity at a wavelength λ due to externally applied perturbation can be described by the formula

$$I_s(\lambda) = \frac{I_0}{2}\left[1 + \cos(\Delta\Phi)\right] \tag{3.10}$$

Therefore, a change in polarization can be observed as a change in intensity, and by correlating the output change in intensity with the measurand, a polarimetric fiber sensor can be effectively used as a sensor for a variety of applications. The phase difference between the two orthogonal polarizations can also be extracted

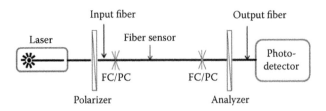

FIGURE 3.18   Working principle of a typical polarimetric fiber sensor.

using an experimental setup consisting of a tunable laser source and a polarimeter/polarization control system.

For polarimetric fiber sensors, typically high birefringent optical fibers are used, such as panda fiber, bow-tie fiber, elliptical core, or polarization-maintaining photonic crystal fibers. Due to the high sensitivity of polarimetric fiber sensors to external parameters such as strain and temperature, cross-sensitivity is often a major problem with these types of sensor. The capability of polarimetric fiber sensors to measure strain, temperature, and pressure is demonstrated in a variety of applications, such as in structural health monitoring [35–38]. Fiber polarimetric sensors are also used for current and voltage measurements, in which a range of polarization-based effects (optical activity, Faraday Effect, electro-optic effect) are used. A detailed description of polarimetric fiber sensors and their applications can be found in Chapter 11.

3.3.3 FIBER-OPTIC SCATTERING-BASED DISTRIBUTED FIBER SENSORS

Distributed fiber-optic sensors are capable of providing a continuous measurand profile over the length of the optical fiber and thus are most suitable for large structural applications such as bridges, buildings, and pipelines. Three main physical effects are used for distributed sensing: Rayleigh scattering, Raman scattering, and Brillouin scattering [39]. When a monochromatic light of frequency f_0 at $\lambda_0 \sim 1550$ nm propagates through an optical fiber, there is always Rayleigh, Brillouin, and Raman scattering within the fiber. The spectrum of these scatterings is shown in Figure 3.19. The components whose frequency is beyond f_0 correspond to anti-Stokes, while those below f_0 correspond to Stokes.

The Rayleigh scattering effect can be used for both strain and temperature monitoring. It is based on the shifts in the local Rayleigh backscatter pattern, which is dependent on the strain and the temperature. Thus, the strain measurements must be compensated for temperature. The main characteristics of this system are high resolution of measured parameters and good spatial resolution (typically a few meters),

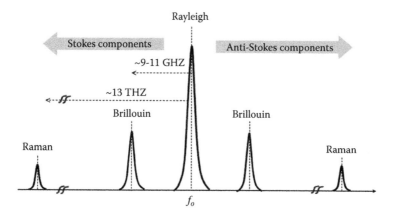

FIGURE 3.19 Schematic spectrum of Rayleigh, Brillouin, and Raman scattering in optical fibers.

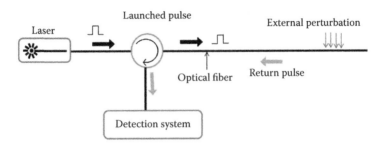

FIGURE 3.20 Schematic experimental setup for distributed sensing.

but the maximal length of the sensor is limited to a relatively short distance (70 m). This is due to the low-level signal arising as a result of low nonlinear coefficients for the silica that constitutes the fiber, resulting in quite a long system response time, resulting from the necessity to integrate small signals received over many pulses. Thus, this system is suitable for monitoring localized strain changes.

Brillouin scattering is the result of the interaction between optical and sound waves in optical fibers. Thermally excited acoustic waves produce a periodic modulation of the RI. Brillouin scattering occurs when light propagating in the fiber is diffracted backward by the moving grating, giving rise to a frequency-shifted component by a phenomenon similar to the Doppler shift. This process is called *spontaneous Brillouin scattering*. A schematic experimental setup for sensing is shown in Figure 3.20. The main challenge in using spontaneous Brillouin scattering for sensing applications resides in the extremely low level of the detected signal. This requires sophisticated signal processing and relatively long integration times.

Acoustic waves can also be generated by injecting into the fiber two counter-propagating waves with a frequency difference equal to the Brillouin shift. Through electrostriction, these two waves will give rise to a traveling acoustic wave that reinforces the phonon population. This process is called *stimulated Brillouin amplification*. Systems based on the stimulated Brillouin amplification have the advantage of working with a relatively stronger signal but face another challenge. To produce a meaningful signal, the two counterpropagating waves must maintain an extremely stable frequency difference. This usually requires the synchronization of two laser sources, which must inject the two signals at the opposite ends of the fiber under test.

Brillouin scattering sensor systems are able to measure strain or temperature variations of fiber with length up to 50 km, with spatial resolution down in the meter range. For temperature measurements, the Brillouin sensor is a strong competitor to systems based on Raman scattering, while for strain measurements, it has practically no rivals.

Raman scattering is the result of a nonlinear interaction within the light traveling in silica fiber. When an intense light signal is launched into the fiber, two frequency-shifted components called Raman Stokes and Raman anti-Stokes will appear in the backscattered spectrum. The relative intensity of these two components depends only on the local temperature of the fiber. If the light signal is pulsed, and the backscattered intensity is recorded as a function of the round trip time, it becomes possible to obtain a temperature profile along the fiber. The insensitivity of this parameter to strain is an advantage compared with Rayleigh and Brillouin-based temperature monitoring, since no particular packaging of the sensor is needed to keep the sensing fiber strain free. Systems based on Raman scattering have been commercialized by SMARTEC (Switzerland), Sensornet, and Sensa (UK). Typically, a temperature resolution of the order of 0.1°C and a spatial resolution of 1 m over a measurement range up to 8 km are obtained for MMFs. Since the leakage of pipelines, dams, and so on often changes the thermal properties of surrounding soil, besides the temperature monitoring, Raman-based systems are used for leakage monitoring of large structures.

Distributed fiber sensors are often categorized based on the interrogation method and the physical effect underpinning the operating principle: (1) optical time domain reflectometry (OTDR) and optical frequency domain reflectometry (OFDR), both based on Rayleigh scattering; (2) Raman OTDR (ROTDR) and Raman OFDR (ROFDR), both based on Raman scattering; and (3) Brillouin OTDR (BOTDR) and Brillouin OFDR (BOFDR), both based on Brillouin scattering.

3.3.4   Surface Plasmon Resonance Sensors

Surface plasmon resonance (SPR) refers to the excitation of surface plasmon polaritons (SPPs), which are electromagnetic waves coupled with free electron density oscillations on the surface between a metal and a dielectric medium (or air). SPPs or surface plasmon waves (SPWs) propagate along the interface of the metal and the dielectric material. For nanoscaled metallic structures such as metallic nanoparticles, or nanorods, a light of the appropriate wavelength can be used to excite the localized oscillation of charges, a process that is referred to as *localized SPR*.

When the energy as well as the momentum of both the incident light and the SPW match, a resonance occurs that results in a sharp dip in the reflected light intensity. The resonance condition depends on the angle of incidence, the wavelength of the light beam, and the dielectric function of both the metal and the dielectric. In an SPR-based fiber-optic sensor, all the guided rays are launched, and hence, the angle of incidence is not varied. In this case, the coupling of the evanescent field with surface plasmons strongly depends on wavelength, fiber parameters, probe geometry, and the metal layer properties.

The SPR-based fiber-optic sensor has a large number of applications for quantitative detection of chemical and biological species. These include food quality, medical diagnostics, and environmental monitoring. The sensing principle is based on detecting the change in RI of the medium around the metallic coating [40].

3.3.5   Photonic Crystal Fiber Sensors

Since the first demonstration of microstructured fiber (MOF) in 1996 (photonic crystal fiber), it has caused enormous attention and excitement in the photonics community. The unique features of the fiber, such as light guidance, dispersion properties, endlessly single-mode nature, higher birefringence, and enhanced nonlinear effects have led to MOFs being applied in many fields, such as in optical communications,

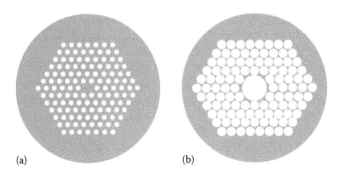

(a)   (b)

FIGURE 3.21   (a) Solid-core PCF and (b) hollow-core PCF. White: air; light gray: silica.

nonlinear optics, and sensing. Unlike conventional fibers, MOFs can be optimized for a large range of applications by tailoring the size, number, and geometry of the air holes that form the confining microstructure around the core region (Figure 3.21a and b). This led to great interest among researchers in exploiting the advantages and using MOF-based sensors in a variety of applications. In a short span of time since the first introduction of the MOF, a variety of microstructured fibers have been reported, ranging from silica to polymer, for applications ranging from environmental sciences to medicine, industry, and astronomy [41]. The fascinating physics and applications of MOFs are still of great interest among researchers.

3.3.6 OPTICAL MICROFIBER SENSORS

Another rapidly growing fiber sensor type is the optical microfiber-based sensor. Optical microfibers exhibit many desirable characteristics, such as large evanescent field, strong optical confinement, bend insensitivity, configurability, compactness, robustness, and feasibility of extremely high-Q resonators [42,43]. The resulting sensors have numerous advantages over their standard optical fiber counterparts, including high responsivity, fast response, selectiveness, nonintrusiveness, and small size. With sensing areas spanning acceleration, acoustic, bend/curvature, current, displacement, electric/magnetic field, roughness, rotation, and temperature, the future of optical microfiber technology looks exceptionally promising (Figure 3.22).

3.4 OPTICAL FIBER SENSOR INSTRUMENTATION

Though sensor heads based on optical fibers are relatively cheap, the optical and electronic instrumentation required to interrogate the sensor is often expensive and complicated. Based on the type of fiber sensor used, the instrumentation requirements for interrogating the sensor will vary. The basic characteristics of any fiber-optic sensing instrumentation are its suitability for the type of fiber sensor and sufficient sensitivity and resolution with an adequate measurement range.

The simplest instrumentation requirement is for the intensity-based systems, where a simple light source (light-emitting diode [LED], semiconductor laser, etc.)

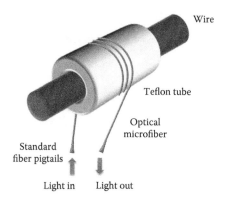

FIGURE 3.22   An optical microfiber temperature sensor.

and an optical receiver with or without an amplifier would be sufficient in most applications. However, based on the stability and accuracy required, further modifications might be necessary, such as ratiometric measurement (Figure 3.6). For spectrometric fiber sensors, devices that can measure the spectral change are required, such as an optical spectrum analyzer or an optical spectrometer. One of the most common types of widely used spectrometric fiber sensor are FBG sensors, and for their interrogation, FBG interrogation systems are widely used. The operating principles of these interrogation units vary, and a range of techniques that can be used to interrogate an FBG are reported.

For polarimetric fiber-based sensing systems, a polarimeter, a polarization controller, and so on are most often used to improve the accuracy and reliability of the system.

In general, optical fiber sensor instrumentation plays a vital part in the overall development of the fiber sensing area, and over the years we have seen a tremendous increase in new developments in the instrumentation area, supported by the growth of the optical fiber networks area, which shares a common interest when it comes to instrumentation requirements. It should also be noted that the instrumentation requirements for different fiber sensor types are sometimes unique to these sensor types, and as a result, overall system development is required to realize a complete fiber sensing unit. Some examples of OFS instrumentation are shown in Figure 3.23 [44,45].

3.5   OPTICAL FIBER SENSORS: INDUSTRY APPLICATIONS AND MARKET

Over the years, optical fiber sensors have grown rapidly in quality and functionality and have found widespread use in many areas, such as defense, aerospace, automotive, manufacturing, biomedical, and various other industries. Rapid development of sensor devices and instrumentation over the years has made OFS an attractive alternative to conventional sensors for measuring a range of parameters, such as pressure, temperature, strain, vibration, acceleration, rotation, magnetic field, electric field,

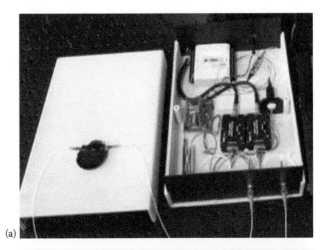

FIGURE 3.23 (a) An OEM interrogation system for a fiber temperature sensor. (From Rajan et al., *Sensors and Actuators A: Physical*, 163, 88–95, 2010.) (b) A portable FBG interrogator. (From http://www.smartfibres.com/FBG-interrogators.)

viscosity, and a wide range of other measurands. Improvements in performance and standardization, together with decrease in price, have helped OFS to make an impact on the industry.

Among the different types of optical fiber sensors, FBGs, fiber gyroscopes, and distributed fiber sensors are the most widely used and popular. High-speed and reliable interrogation systems for data acquisition, processing and interpretation have also been developed for these devices. They also meet application-specific demands that are not efficiently satisfied by conventional electronic sensors. Examples of such applications include the use of fiber-optic sensors for pressure and high-temperature monitoring in the oil and gas exploration industry, which is a major market for distributed fiber sensors.

Fiber-optic sensors can contribute to improving safety by providing an early warning of failure in critical structures and components, for example in the aeronautical and civil engineering sectors. Therefore, structural health monitoring (SHM) is emerging as a promising market for optical fiber sensors. As the growth of production of composite parts has increased, the composite/aerospace industry has increasingly focused on damage/failure-free composite structures and non-destructive techniques for SHM over their lifetime. Compared with conventional nondestructive sensing techniques, optical fiber sensors have achieved wide acceptance due to their attractive properties, such as small size and capability to be embedded within the host structure. Optical fiber sensors embedded in various structures are very useful for strain and temperature monitoring applications in extreme environmental conditions. For example, issues such as bend loading, and icing on wind turbine and helicopter blades, can be monitored and avoided by implementing smart composite structures with embedded optical fiber sensors. Such smart structures can enhance the safety of advanced machines, structures, and devices.

The medical and biomedical industry also provides many challenging instrumentation requirements for which optical fiber sensing can provide novel and reliable solutions. One such challenge is the need to monitor within radioactive medical environments and in magnetic resonance imaging (MRI) systems, where electrical sensors cannot operate. Another is the need for *in vivo* sensing during keyhole surgical procedures and catheter procedures, where space is a premium and where a miniature fiber sensor is the largest intrusion that can be tolerated. Recent success with fiber sensors on the biomedical front includes commercial use of fiber-based catheters and miniature fiber temperature and pressure transducers for medical applications.

The acceptance of optical fiber sensors among different industries/applications is expected to experience rapid growth during the coming years due to a number of initiatives being taken by market participants [46]. Another driving force behind the growth of OFS is that a number of research initiatives and efforts to achieve standardization are also ongoing and being promoted by various research groups and industry leaders. However, more initiatives are required, such as ensuring the reliability, longevity, and accuracy of fiber sensor products to achieve a wider acceptance of the technology across various industrial sectors. According to analysis by a market research firm (BCC Research), [47] the global fiber-optic sensor market reached almost a billion dollars in 2010, and is expected to reach 2.2 billion dollars by 2018 [47].

3.6 CONCLUSION

This chapter provided a brief summary of the basic broad principles of different types of optical fiber sensors. Descriptions of fiber sensors based on different classifications, such as modulation methods, spatial positionings, and measurement parameters, were also discussed. Some of the emerging optical fiber sensors were also introduced, and finally, the fiber-optic instrumentation and market trends were discussed. With the emergence of a variety of new types of fibers and new configurations, OFS and instrumentation is still growing and is expected to further widen the scope of its applications in the near future.

REFERENCES

1. Clushaw, B., Kersey, A., Fiber-optic sensing: A historical perspective, *Journal of Lightwave Technology*, Vol. 26, No. 9, pp. 1064–1078, 2008.
2. Bogue, R., Fibre optic sensors: A review of today's applications, *Sensor Review*, Vol. 31, No. 4, pp. 304–309, 2011.
3. Bucaro, J. A., Dardy, H. D., Carome, E. F., Fibre optic hydrophone, *Journal of the Acoustical Society of America*, Vol. 52, p. 1302, 1977.
4. Culshaw, B. B., Davies, D. E. N., Kingsley, S. A., Acoustic sensitivity of optical fiber waveguides, *Electronics Letters*, Vol. 13, pp. 760–761, 1977.
5. Vali, V., Shorthill, R. W., Fibre ring interferometer, *Applied Optics*, Vol. 15, No. 5, pp. 1099–1100, 1976.
6. Butter, C. D., Hocker, G. E., Fiber optics strain gauge, *Applied Optics*, Vol. 17, p. 2867, 1978.
7. Mishra, V., Singh, N., Tiwari, U., Kapur, P., Fiber grating sensors in medicine: Current and emerging applications, *Sensors and Actuators A: Physical*, Vol. 167, pp. 279–290, 2011.
8. Rajan, G., Ramakrishnan, M., Lesiak, P., Semenova, Y., Boczkowska, A., Woliński, T., Farrell, G., Composite materials with embedded photonic crystal fiber interferometric sensors, *Sensors and Actuators A: Physical*, Vol. 182, pp. 57–67, 2012.
9. Lopez-Higuera, J. M., Cobo, L. R., Incera, A. Q., Cobo, A., Fiber optic sensors in structural health monitoring, *Journal of Lightwave Technology*, Vol. 29, No. 4, pp. 587–608, 2011.
10. Zhao, Z., Lau, W. S., Choi, A. C. K., Shan, Y. Y., Modulation functions of the reflective optical fiber sensor for specular and diffuse reflection, *Optical Engineering*, Vol 33, No. 9, pp. 2986–2991, 1994.
11. Angela, L., Mohana Shankar, P., Mutharasan, R., A review of fiber-optic biosensors, *Sensors and Actuators B*, 125 (2007) 688–703.
12. John, W. B., Historical review of microbend fiber-optic sensors, *Journal of Lightwave Technology*, Vol. 13, No. 7, pp. 1193–1199, 1995.
13. Rajan, G., Semenova, Y., Farrell, G., An all-fiber temperature sensor based on a macro-bend singlemode fiber loop, *Electronics Letters*, Vol. 44, pp. 1123–1124, 2008.
14. Allsop, T., Reeves, R., Webb, D. J., Bennion, I., A high sensitivity refractometer based upon a long period grating Mach-Zehnder interferometer, *Review of Scientific Instruments*, Vol. 73, pp. 1702–1705, 2002.
15. Ding, J.-F., Zhang, A. P., Shao, L.-Y., Yan, J.-H., He, S., Fiber-taper seeded long-period grating pair as a highly sensitive refractive-index sensor, *IEEE Photonics Technology Letters*, Vol. 17, pp. 1247–1249, 2005.
16. Choi, H. Y., Kim, M. J., Lee, B. H., All-fiber Mach-Zehnder type interferometers formed in photonic crystal fiber, *Optics Express*, Vol. 15, pp. 5711–5720, 2007.
17. Ngyuen, L. V., Hwang, D., Moon, S., Moon, D. S., Chung, Y. J., High temperature fiber sensor with high sensitivity based on core diameter mismatch, *Optics Express,* Vol. 16, pp. 11369–11375, 2008.
18. Tian, Z., Yam, S. S.-H., Barnes, J., Bock, W., Greig, P., Fraser, J. M., Loock, H.-P., Oleschuk, R. D., Refractive index sensing with Mach-Zehnder interferometer based on concatenating two single-mode fiber tapers, *IEEE Photonics Technology Letters*, Vol. 20, pp. 626–628, 2008.
19. Pang, F., Liu, H., Guo, H., Liu, Y., Zeng, X., Chen, N., Chen, Z., Wang, T., In-fiber Mach-Zehnder interferometer based on double cladding fibers for refractive index sensor, *IEEE Sensors Journal*, Vol. 11, pp. 2395–2400, 2011.
20. Jiang, L., Yang, J., Wang, S., Li, B., Wang, M., Fiber Mach-Zehnder interferometer based on microcavities for high-temperature sensing with high sensitivity, *Optics Letters*, Vol. 36, pp. 3753–3755, 2011.

21. Frazao, O., Silva, S. F. O., Viegas, J., Baptista, J. M., Santos, J. L., Kobelke, J., Schuster, K., All fiber Mach-Zehnder interferometer based on suspended twin-core fiber, *IEEE Photonics Technology Letters*, Vol. 22, pp. 1300–1302, 2010.

22. Yuan, L.-B., Zhou, L. M., Wu, J. S., Fiber optic temperature sensor with duplex Michelson interferometric technique, *Sensors Actuators A*, Vol. 86, pp. 2–7, 2000.

23. Kashyap, R., Nayar, B., An all single-mode fiber Michelson interferometer sensor, *Journal of Lightwave Technology*, Vol. LT-1, pp. 619–624, 1983.

24. Byeong Ha Lee, Young Ho Kim, Kwan Seob Park, Joo Beom Eom, Myoung Jin Kim, Byung Sup Rho, Hae Young Choi, Interferometric fiber optic sensors, *Sensors*, Vol. 12, pp. 2467–2486, 2012.

25. Ran, J., Rao, Y., Zhang, J., Liu, Z., Xu, B., A miniature fiber-optic refractive-index sensor based on laser-machined Fabry–Perot interferometer tip, *Journal of Lightwave Technology*, Vol. 27, pp. 5426–5429, 2009.

26. Wan, X., Taylor, H. F., Intrinsic fiber Fabry–Perot temperature sensor with fiber Bragg grating mirrors, *Optics Letters*, Vol. 27, pp. 1388–1390, 2002.

27. Zhang, Y., Chen, X., Wang, Y., Cooper, K. L., Wang, A., Microgap multicavity Fabry–Pérot biosensor, *Journal of Lightwave Technology*, Vol. 25, pp. 1797–1804, 2007.

28. Zhao, J. R., Huang, X. G., He, W. X., Chen, J. H, High-resolution and temperature-insensitive fiber optic refractive index sensor based on Fresnel reflection modulated by Fabry–Perot interference, *Journal of Lightwave Technology*, Vol. 28, pp. 2799–2803, 2010.

29. Lefevre, H., *The Fiber Optic Gyroscope*, Artech, Norwood, MA, 1993.

30. Othonos, A., Kalli, K., *Fiber Bragg Gratings: Fundamentals and Applications in Telecommunications and Sensing*, Artech House, Norwood, MA, 1999.

31. James, S. W., Tatam, R. P., Optical fibre long-period grating sensors: characteristics and application, *Measurement Science and Technology*, Vol. 14, pp. R49–R61, 2003.

32. Schwab, S. D., Levy, R. L., In-service characterization of composite matrices with an embedded fluorescence optrode sensor, *Proceedings of SPIE*, Vol. 1170, p. 230, 1989.

33. Grattan, K. T. V., Selli, R. K., Palmer, A. W., A miniature fluorescence referenced glass absorption thermometer, *Proceedings of 4th International Conference Optical Fiber Sensors*, Tokyo, p. 315, 1986.

34. Wolinski, T. R., Polarimetric optical fibers and sensors, *Progress in Optics*, Vol. 40, pp. 1–75, 2000.

35. Wolinski, T. R., Lesiak, P., Domanski, A. W., Polarimetric optical fiber sensors of a new generation for industrial applications, *Bulletin of the Polish Academy of Sciences, Technical Sciences*, Vol. 56, pp. 125–132, 2008.

36. Murukeshan, V. M., Chan, P. Y., Seng, O. L., Asundi, A., On-line health monitoring of smart composite structures using fibre polarimetric sensor, *Smart Materials and Structures*, Vol. 8, pp. 544–548, 1999.

37. Ma, J., Asundi, A., Structural health monitoring using a fibre optic polarimetric sensor and a fibre optic curvature sensor-static and dynamic test, *Smart Materials and Structures*, Vol. 10, pp. 181–188, 2001.

38. Ramakrishnan, M., Rajan, G., Semenova, Y., Boczkowska, A., Domański, A., Wolinski, T., Farrell, G., Measurement of thermal elongation induced strain of a composite material using a polarization maintaining photonic crystal fibre sensor, *Sensors and Actuators A: Physical*, Vol. 190, pp. 44–51, 2013.

39. Xiaoyi Bao, Liang Chen, Recent progress in distributed fiber optic sensors, *Sensors*, Vol. 12, pp. 8601–8639, 2012.

40. Gupta, B. D., Verma, R. K., Surface plasmon resonance based fiber optic sensors: Principle probe designs and some applications, *Journal of Sensors*, Vol. 1, pp. 1–12, 2009.

41. Pinto, A. M. R., Lopez-Amo, M., Photonic crystal fibers for sensing applications, *Journal of Sensors*, Vol. 2012, p. 21, 2012.

42. Chen, G. Y., Ding, M., Newson, T. P., Brambilla, G., A review of micro fiber and nano fiber based optical sensors, *The Open Optics Journal*, Vol. 7, No. 1, pp. 32–57, 2013.

43. Lou, J., Wang, Y., Tong, L., Microfiber optical sensors: A review, *Sensors*, Vol. 14, pp. 5823–5844, 2014.

44. Rajan, G., Semenova, Y., Mathew, J., Farrell, G., Experimental analysis and demonstration of a low cost temperature sensor for engineering applications, *Sensors and Actuators A: Physical*, Vol. 163, pp. 88–95, 2010.

45. Smart Fibres, FBG interrogators. http://www.smartfibres.com/FBG-interrogators.

46. Frost and Sullivan Research Service, Fiber optic sensors (technical insights), Frost & Sullivan Research Service, Mountain View, CA, 2010.

47. Remote Sensing Technologies and Global Markets, BCC Research, Wellesley, MA, 2016. http://www.bccresearch.com/market-research/instrumentation-and-sensors/remote-sensing-technologies-markets-report-ias022e.html.

4 Structural Health Monitoring of Composite Materials Using Fiber Bragg Gratings

Christophe Caucheteur, Damien Kinet,
Nicolas Lammens, Thomas Geernaert,
Francis Berghmans, Geert Luyckx,
Joris Degrieck, and Patrice Mégret

CONTENTS

4.1 Introduction 69
4.2 Fundamentals on FBGs 70
    4.2.1 Basic Principle of Uniform FBGs...70
    4.2.2 Temperature Sensitivity of Uniform FBGs...72
    4.2.3 Axial Strain Sensitivity of Uniform FBGs 73
    4.2.4 Pressure Sensitivity of Uniform FBGs 76
    4.2.5 Transverse Strain Sensitivity of Uniform FBGs.....................................77
    4.2.6 Other Kinds of Fiber Gratings of Interest to Be Embedded into Composite Materials...78
4.3 Demodulation Techniques Used to Retrieve the Bragg Wavelength Shift 81
    4.3.1 Techniques Based on Wavelength Detection 81
    4.3.2 Reflectometric Demodulation Techniques..82
4.4 Discrimination between Temperature and Strain Effects 83
    4.4.1 FBG Pairs with One of Them Embedded into a Glass Capillary 85
    4.4.2 FBG Pair Comprising Type I and Type IA FBGs 86
    4.4.3 Weakly Tilted FBG...86
    4.4.4 Comparison of Performances 87
4.5 Residual Strain Sensing during the Manufacturing Process 88
4.6 Conclusion 95
References 95

4.1 INTRODUCTION

Nowadays, newly produced materials are increasingly smart in the sense that they embed sensors and/or metrological tools able to provide real-time information about the internal stress state. As such, they intrinsically ensure predictive maintenance. This feature is highly appreciated, as it can mitigate the risk of failure due to a sudden breakage, on the one hand, and decrease the maintenance costs, on the other hand. Of course, smart composite materials are also deployed. Considering the composition and geometry of composite materials, optical fiber sensors are very well suited to being embedded into composite materials without altering their mechanical performance.

Optical fiber sensors most often exploit telecommunication-grade single-mode optical fibers made of a thick silica core approximately 8 µm thick, doped with germanium oxide, surrounded by a 125 µm thick cladding in pure silica. The minimum attenuation (approximately 0.2 dB/km) occurs around 1550 nm. Therefore, this wavelength range is mainly privileged for sensing purposes. Optical fiber sensors present decisive benefits [1]. In terms of reliability, they are passive devices and are characterized by a long lifetime (more than 20 years). They are stable over time (no calibration required) and do not suffer from corrosion. Cables and connectors are telecommunication grade. In terms of performances, they are compatible with multiplexing, offering (quasi-) distributed measurement capabilities. Hence, several tens of sensing points can be cascaded along a single optical fiber. The use of light ensures their immunity against electromagnetic radiation and their insensitivity to radio-frequency interference. Also, there is no risk of high-voltage discharge, and they are explosion safe. Finally, their light weight and small dimensions allow them to be easily embedded.

Among the different optical fiber configurations, fiber Bragg gratings (FBGs) photo-inscribed in the core of an optical fiber have attracted considerable attention for insertion and use in composite materials [2]. They are obtained through a refractive index modulation of the fiber core along the fiber axis and behave as selective mirrors in wavelength. They are intrinsically sensitive to temperature, pressure, and axial strain and yield a wavelength-encoded response, which can be straightforwardly recorded and processed [3].

In practice, as they have to correctly sustain the fabrication process of composite materials, their integration and their subsequent proper operation require some know-how, as issues at different levels have to be solved. In this aspect, review papers that deal with strain measurements in composite laminates with FBG sensors have been recently published [4,5]. This chapter will start with some basic information about the operating principle of FBG sensors. After an overview of the demodulation techniques available to process their amplitude spectrum, the focus will be on two specific challenges arising from their use: the discrimination between temperature and strain effects, on the one hand, and the residual strain sensing during the manufacturing process, on the other hand. Each time, the main solutions will be reviewed, while their relative performances will be compared. Other concerns related to the distortion of the composite materials in the surroundings of the optical fibers and arising from multiaxial strain sensing will not be addressed here. Interested readers are invited to consult references [4,6,7] for details about these aspects.

4.2 FUNDAMENTALS ON FBGS

4.2.1 BASIC PRINCIPLE OF UNIFORM FBGS

A uniform FBG is a periodic and permanent modification of the core refractive index value along the optical fiber axis [2,8]. This modification is usually obtained

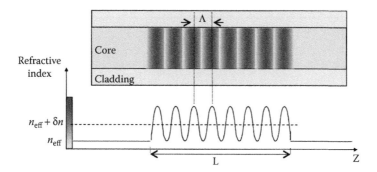

FIGURE 4.1 Sketch of an FBG and its physical parameters.

by transversally exposing the core of a photosensitive optical fiber to an intense interference pattern of ultraviolet light at a wavelength around 240 nm. Indeed, due to the presence of germanium oxide dopants inside the core, an optical fiber is photosensitive (i.e., it benefits from the property of permanently changing its refractive index when exposed to light) in a wavelength band centered around 240 nm. For this reason, continuous-wave frequency-doubled argon-ion lasers emitting at 244 nm or pulsed excimer lasers emitting at 248 nm are most often used to manufacture FBGs. Bright interference fringes locally increase the mean refractive index of the core. Mass production of identical gratings can be quite straightforwardly achieved with the phase mask technique. A phase mask is a diffractive element made in a pure silica substrate that is optimized to maximize the diffraction in the first (+1 and −1) diffraction orders.

An interference pattern is therefore produced in the fiber core when the optical fiber is located in close proximity behind the mask. An FBG is defined by several physical parameters, as sketched in Figure 4.1. The grating length L is the optical fiber length along which the refractive index modulation is realized. The periodicity and the amplitude of the refractive index modulation are labeled Λ and δn, respectively.

The order of magnitude of these parameters typically varies between 200 and 1000 nm for Λ, from a few millimeters to a few tens of centimeters for L, and from 10^{-5} to 10^{-3} for δn. Such a perturbation induces light coupling between two counter-propagating core modes. This mode coupling occurs for some wavelengths around the Bragg wavelength, defined by

$$\lambda_B = 2n_{eff}\Lambda \tag{4.1}$$

where n_{eff} is the effective refractive index of the core mode at the Bragg wavelength. A uniform FBG acts as a selective mirror in wavelength around the Bragg wavelength to yield a pass-band reflected amplitude spectrum and a stop-band transmitted amplitude spectrum, as depicted in Figure 4.2 for a 1 cm long FBG. In fact, at each refractive index discontinuity along the fiber axis, a weak Fresnel reflection is generated. These add in phase at the Bragg wavelength, yielding an important reflection band surrounded by side lobes.

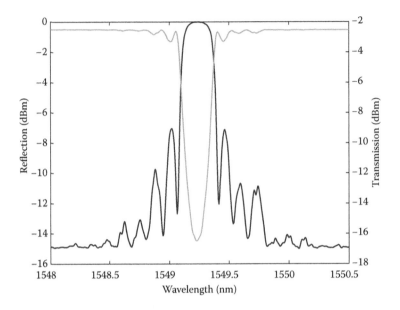

FIGURE 4.2   Reflected and transmitted amplitude spectra of a 1 cm long uniform FBG.

4.2.2 TEMPERATURE SENSITIVITY OF UNIFORM FBGS

In practice, the effective refractive index of the core and the spatial periodicity of the grating are both affected by changes in strain and temperature. In particular, the effective refractive index is modified through the thermo-optic and strain-optic effects, respectively. Hence, from Equation 4.1, the Bragg wavelength shift $\Delta\lambda_B$ due to strain $\Delta\varepsilon$ and temperature ΔT variations is given by

$$\Delta\lambda_B = 2\left(\Lambda\frac{dn_{\text{eff}}}{dT} + n_{\text{eff}}\frac{d\Lambda}{dT}\right)\Delta T + 2\left(\Lambda\frac{dn_{\text{eff}}}{d\varepsilon} + n_{\text{eff}}\frac{d\Lambda}{d\varepsilon}\right)\Delta\varepsilon \qquad (4.2)$$

The first term in Equation 4.2 represents the effect of temperature on the Bragg wavelength. The Bragg wavelength shift due to thermal expansion comes from the modification of the grating spacing and the refractive index. The relative wavelength shift due to a temperature change ΔT can be written as

$$\frac{\Delta\lambda_B}{\Delta T} = \lambda_B\left(\frac{1}{n_{\text{eff}}}\frac{dn_{\text{eff}}}{dT} + \frac{1}{\Lambda}\frac{d\Lambda}{dT}\right) \qquad (4.3)$$

where:
$1/n_{\text{eff}} \ dn_{\text{eff}}/dT$ is the thermo-optic coefficient
$1/\Lambda \ d\Lambda/dT$ is the thermal expansion coefficient of the optical fiber

The thermo-optic coefficient is approximately equal to 8.6×10^{-6}/K for germanium-doped silica core optical fiber, whereas the thermal expansion coefficient

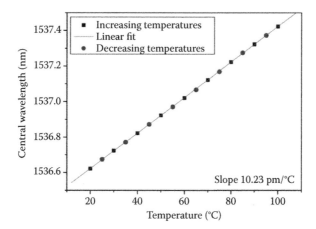

FIGURE 4.3 Evolution of the Bragg wavelength as a function of temperature changes.

is approximately equal to 0.55×10^{-6}/K for silica, so that the refractive index change is by far the dominant effect [2]. The order of magnitude of the temperature sensitivity of the Bragg wavelength is 10 pm/°C around 1550 nm.

Figure 4.3 displays the experimental evolutions of the Bragg wavelength as a function of temperature for a 1 cm long uniform FBG. The grating was placed inside an oven whose temperature was regulated with an accuracy of 0.1°C. The evolution is linear with a slope computed equal to 10.23 pm/°C. Experiments carried out for increasing and decreasing temperature values confirm the absence of hysteresis.

4.2.3 AXIAL STRAIN SENSITIVITY OF UNIFORM FBGS

The second term in Equation 4.2 represents the effect of longitudinal strain on an optical fiber. It corresponds to a change in the grating periodicity and the strain-optic-induced change in the refractive index. By defining the strain as $\varepsilon = \Delta\Lambda/\Lambda$, the change of the grating periodicity can be expressed by

$$\Lambda_S = \Lambda + \Delta\Lambda = \Lambda\left(1+\varepsilon\right) \tag{4.4}$$

where Λ_S represents the modified grating period after the application of the perturbation. The elasto-optic effect relates the refractive index change to the applied strain [8]:

$$\Delta\eta_{ij} = \left(\Delta\frac{1}{n_{\text{eff}}^2}\right)_{ij} = \sum_{i=1}^{6}\sum_{j=1}^{6}p_{ij}\varepsilon_{ij} \tag{4.5}$$

where:

$\Delta\eta_{ij}$ is the change in the electric impermeability tensor
ε_{ij} are the strain components
p_{ij} are the elements of the elasto-optic tensor

The shape of the elasto-optic tensor, but not the magnitude of the coefficients p_{ij}, can be derived from the symmetry of the material under consideration. In the most general case, there are 36 different coefficients p_{ij}. For the class of isotropic materials to which fused silica glass belongs, this number is reduced to two independent coefficients p_{11} and p_{12}, and the elasto-optic tensor becomes [8]

$$\begin{pmatrix} p_{11} & p_{12} & p_{12} & 0 & 0 & 0 \\ p_{12} & p_{11} & p_{12} & 0 & 0 & 0 \\ p_{12} & p_{12} & p_{11} & 0 & 0 & 0 \\ 0 & 0 & 0 & \frac{1}{2}(p_{11}-p_{12}) & 0 & 0 \\ 0 & 0 & 0 & 0 & \frac{1}{2}(p_{11}-p_{12}) & 0 \\ 0 & 0 & 0 & 0 & 0 & \frac{1}{2}(p_{11}-p_{12}) \end{pmatrix} \quad (4.6)$$

The magnitudes of the individual coefficients p_{ij} are dependent on the material considered. For pure bulk silica, it was found that typical values measured at 628 nm are $p_{11} = 0.121$ and $p_{12} = 0.270$ [8]. These values are often used when computing the influence of mechanical perturbations such as elongation and lateral compression of optical fibers. However, due to the presence of doping elements in the core, the effective values for fibers may be different from those for bulk silica. Measurements on single-mode optical fibers at 628 nm yielded values of $p_{11} = 0.113$ and $p_{12} = 0.252$ [9]. An error in these values of approximately 5% is possible because of the uncertainty in the value of the Poisson's ratio of the optical fiber.

Assuming that the grating is strained in the z-direction only (along the fiber axis) and that the fiber material follows Hooke's law, we obtain

$$\varepsilon_x = \varepsilon_y = -\nu\varepsilon_z = -\nu\varepsilon \quad (4.7)$$

where ν is Poisson's ratio.

Since all other strain components are null, Equation 4.5 combined with Equations 4.6 and 4.7 gives

$$\Delta n_x = -\frac{n_{eff}^3}{2}\left[-\nu p_{11} + (1-\nu)p_{12}\right]\varepsilon \quad (4.8)$$

$$\Delta n_y = -\frac{n_{eff}^3}{2}\left[-\nu p_{11} + (1-\nu)p_{12}\right]\varepsilon \quad (4.9)$$

so that the effect of axial strain is the same in the x- and y-directions (perpendicular to the optical fiber axis).

Taking into account Equation 4.1, the second term of Equation 4.2 can be rewritten as

$$\frac{\Delta\lambda_B}{\Delta\varepsilon} = \lambda_B \left(\frac{1}{n_{\text{eff}}} \frac{dn_{\text{eff}}}{d\varepsilon} + \frac{1}{\Lambda} \frac{d\Lambda}{d\varepsilon} \right) \tag{4.10}$$

This allows the so-called effective strain-optic constant p_e to be defined:

$$p_e = -\frac{1}{n_{\text{eff}}} \frac{dn_{\text{eff}}}{d\varepsilon} \tag{4.11}$$

such that under the assumption of small strain variations, the Bragg wavelength shift in response to axial changes is finally given by

$$\Delta\lambda_B = \lambda_B \left(1 - p_e \right) \Delta\varepsilon \tag{4.12}$$

$$p_e = \frac{n_{\text{eff}}^2}{2} \left[p_{12} - \upsilon \left(p_{11} + p_{12} \right) \right] \tag{4.13}$$

Substitution of parameters ($p_{11} = 0.113$, $p_{12} = 0.252$, $\upsilon = 0.16$, and $n_{\text{eff}} = 1.482$) in Equations 4.6 and 4.7 gives a strain-optic constant $p_e = 0.21$ and an axial strain sensitivity of the Bragg wavelength of 1.2 pm $\mu\varepsilon^{-1}$ around 1550 nm.

Figure 4.4 displays the evolution of the Bragg wavelength as a function of an axial strain change for a 1 cm long uniform FBG. To apply controlled axial strain values, the fiber containing the grating was fixed on two clamps, one of which was moved in the direction parallel to the fiber axis. As in the case of temperature, the evolution is linear and without hysteresis. The slope of the linear evolution was computed equal to 1.12 pm $\mu\varepsilon^{-1}$.

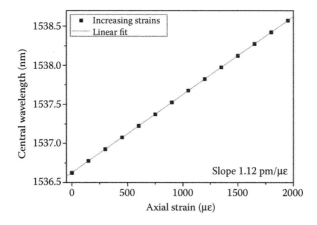

FIGURE 4.4 Evolution of the Bragg wavelength as a function of axial strain changes.

4.2.4 PRESSURE SENSITIVITY OF UNIFORM FBGs

Starting from Equation 4.1, it can be derived that a pressure p leads to a corresponding shift of the Bragg wavelength given by

$$\frac{\Delta\lambda_B}{\lambda_B} = \left(\frac{1}{n_{eff}} \frac{dn_{eff}}{dp} + \frac{1}{\Lambda} \frac{d\Lambda}{dp} \right) p \tag{4.14}$$

The strain components resulting from the pressure load are

$$\varepsilon_x = \varepsilon_y = \left(\frac{1-\nu}{E} \right) p \tag{4.15}$$

$$\varepsilon_z = \left(\frac{-2\nu}{E} \right) p \tag{4.16}$$

where E is the Young's modulus of the optical fiber. Substituting these values into Equation 4.5 gives

$$\Delta n_x = \Delta n_y = -\frac{n_{eff}^3}{2} \left[(1-\upsilon) p_{11} + (1-3\upsilon) p_{12} \right] \frac{p}{E} \tag{4.17}$$

Given that $\varepsilon_z = \dfrac{\Delta\Lambda}{\Lambda}$, the components of Equation 4.14 are given by

$$\frac{1}{\Lambda} \frac{d\Lambda}{dp} = \frac{-2\nu}{E} \tag{4.18}$$

$$\frac{1}{n_{eff}} \frac{dn_{eff}}{dp} = -\frac{n_{eff}^2}{2E} \left[(1-\upsilon) p_{11} + (1-3\upsilon) p_{12} \right] \tag{4.19}$$

The wavelength-pressure sensitivity can be formulated as

$$\frac{\Delta\lambda_B}{\lambda_B} = \left[\frac{-2\nu}{E} - \frac{n_{eff}^2}{2E} \left[(1-\upsilon) p_{11} + (1-3\upsilon) p_{12} \right] \right] p \tag{4.20}$$

After substitution of the numerical values given in Section 4.2.3, we obtain

$$\frac{\Delta\lambda_B}{\lambda_B} = -0.57 \frac{p}{E} \tag{4.21}$$

The typical hydrostatic pressure sensitivity of a uniform FBG around 1550 nm is about -4 pm MP^{-1}.

4.2.5 Transverse Strain Sensitivity of Uniform FBGs

Let us consider now that the applied stress σ is parallel to the y axis (perpendicular to the optical fiber axis), so that the strain components related to the stress field are given by

$$\varepsilon_y = \frac{\sigma}{E} \tag{4.22}$$

$$\varepsilon_x = \varepsilon_z = -\nu \varepsilon_y \tag{4.23}$$

The change of the Bragg wavelength due to a transverse strain is given by the following relationship:

$$\frac{\Delta \lambda_B}{\lambda_B} = \left(\frac{1}{n_{\text{eff}}} \frac{dn_{\text{eff}}}{d\sigma} + \frac{1}{\Lambda} \frac{d\Lambda}{d\sigma} \right) \sigma \tag{4.24}$$

Substituting the values of strain components into Eq. (4.24) gives

$$\Delta n_x = -\frac{n_{\text{eff}}^3}{2} \left[-\upsilon p_{11} + (1 - \upsilon) p_{12} \right] \frac{\sigma}{E} \tag{4.25}$$

$$\Delta n_y = -\frac{n_{\text{eff}}^3}{2} \left[p_{11} - 2\upsilon p_{12} \right] \frac{\sigma}{E} \tag{4.26}$$

Hence, due to the asymmetry of the load with respect to the (x,y) plane, dissimilar refractive index changes are obtained for the x- and y-directions. Consequently, the shift of the Bragg wavelength will be dependent on the polarization state of the light coupled into the optical fiber:

$$\left(\frac{\Delta \lambda_B}{\lambda_B} \right)_x = \left(-\nu - \frac{n_{\text{eff}}^3}{2} \left[-\upsilon p_{11} + (1 - \upsilon) p_{12} \right] \right) \frac{\sigma}{E} \tag{4.27}$$

$$\left(\frac{\Delta \lambda_B}{\lambda_B} \right)_y = \left(-\nu - \frac{n_{\text{eff}}^3}{2} \left[p_{11} - 2\upsilon p_{12} \right] \right) \frac{\sigma}{E} \tag{4.28}$$

In other words, transverse strain leads to birefringence, which is spectrally manifested by a broadening of the main reflection mode prior to its splitting into two different bands corresponding to both polarization modes. It is the reason why uniform FBGs written into polarization maintaining fibers (highly birefringent fibers) are most often preferred in practice to measure transverse force effects since the resonance bands corresponding to the x and y polarization modes are well resolved.

It now becomes apparent that any change in wavelength associated with the action of an external perturbation to the grating is the sum of strain and temperature terms.

Therefore, in sensing applications for which only one perturbation is of interest, or when the two perturbations have to be simultaneously and unambiguously measured, it is required to separate the effects of temperature and strain. This is not possible with a single grating written into a standard single-mode optical fiber as it exhibits a strong sensitivity to both perturbations. Several techniques have therefore been developed to achieve temperature compensation by measuring the strain and temperature effects simultaneously. They will be reviewed in Section 4.4.

As already mentioned in Section 4.1, the main advantage of FBG sensors is that the information about the perturbation is wavelength encoded. This property makes the sensor self-referencing and independent of fluctuating light levels. The system is therefore immune to source power and connector losses that affect many other types of optical fiber sensors. The very low insertion loss and narrowband wavelength reflection of FBGs provide convenient serial multiplexing along a single-mode optical fiber. There are further advantages of FBGs over conventional electrical strain gauges, such as linearity in response over many orders of magnitude. Furthermore, FBGs can be easily embedded into materials to provide damage detection or internal strain field mapping. FBG sensors are therefore very important components for the development of smart structure technology and for monitoring composite material curing and response.

4.2.6 OTHER KINDS OF FIBER GRATINGS OF INTEREST TO BE EMBEDDED INTO COMPOSITE MATERIALS

Tilted FBGs (TFBGs) belong to the family of short-period gratings, since the periodicity of the refractive index modulation is of the order of 500 nm [10]. They are characterized by grating planes blazed by a certain angle θ with respect to the fiber axis. Tilting the grating planes results in light (which is otherwise guided in the fiber core) being coupled into cladding or radiation modes. The tilt angle of the grating planes and the strength of the refractive index modulation determine the coupling efficiency and the bandwidth of the light that is outcoupled from the fiber core.

In the frame of sensing, the most interesting feature of TFBGs is their transmitted amplitude spectrum. It is composed of cladding mode resonances below the Bragg wavelength. Each cladding mode resonance corresponds to the coupling between the forward-going core mode and a backward-going cladding mode. TFBGs are governed by phase matching conditions that give the wavelength positions of the resonance bands corresponding to the couplings between two modes. The resonance wavelength λ_B and the wavelength at which the discrete coupling to the jth cladding mode occurs ($\lambda_{c,j}$) (characterized by an effective refractive index $n_{c,j}$) are given by the following set of equations:

$$\lambda_{\text{Bragg}} = 2n_{\text{eff}}\Lambda \tag{4.29}$$

$$\lambda_{c,j} = \left(n_{\text{eff}} + n_{c,j}\right)\Lambda \tag{4.30}$$

Figure 4.5 depicts the transmitted amplitude spectrum of a 3° TFBG.

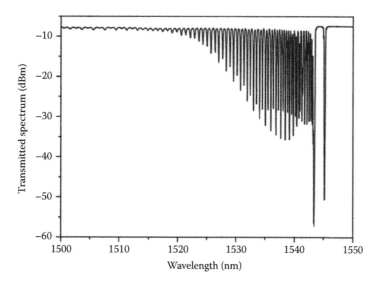

FIGURE 4.5 Transmitted amplitude spectrum of a 3° TFBG.

Obviously, TFBGs are sensitive to temperature and strain. It was shown in [11] that the cladding mode resonances present a similar temperature sensitivity to the Bragg resonance. However, the axial strain sensitivity tends to increase with the cladding mode order, so that a temperature-insensitive axial strain sensor can be obtained with such gratings.

More interestingly, TFBGs are sensitive to the surrounding refractive index, which is not the case for uniform FBGs, for which light is confined inside the core and is therefore insensitive to outside changes. It was demonstrated in 2001 that, when the surrounding refractive index increases in the range 1.30–1.45, a progressive smoothing of the transmitted spectrum starting from the shortest wavelengths is obtained [12]. This smoothing can be explained by the fact that one can assign to any resonance $\lambda_{c,j}$ a discrete cladding mode with an effective refractive index, denoted $n_{c,j}$, which decreases while $\lambda_{c,j}$ decreases. So when the external refractive index rises and reaches the value $n_{c,j}$, this mode becomes weakly guided due to the decrease of the overlapping integral between the fundamental guided mode and the jth cladding mode, thereby reducing the amplitude of the coupling coefficient and consequently the amplitude of the cladding mode resonance. When the surrounding refractive index matches the $n_{c,j}$ value, the cladding mode is no longer guided, and the coupling occurs with a continuum of radiation modes.

Long period fiber gratings (LPFGs) are also able to sense surrounding refractive index changes. They are characterized by a grating period up to 1000 times higher than that of uniform FBGs. They are diffractive gratings that couple the core light into forward-going cladding modes. Their amplitude spectrum is composed of several discrete broadband cladding mode resonances ranging over a scale of a few hundreds of nanometers [13], as depicted in Figure 4.6.

LPFGs manufactured inside standard single-mode optical fiber exhibit temperature sensitivities in the range 30–100 pm/°C, depending on the grating

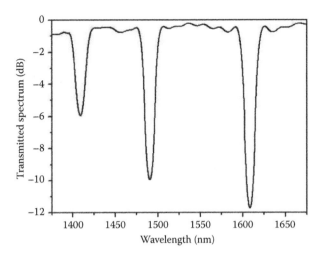

FIGURE 4.6   Transmitted amplitude spectrum of an LPFG.

periodicity and the central wavelength [14]. This is thus one order of magnitude larger than the sensitivity of fiber Bragg grating sensors. Temperature-compensated LPFGs were demonstrated for periodicities smaller than 100 μm. For example, an LPFG of period equal to 40 μm exhibited an attenuation band with sensitivity of 1.8 pm/°C, which is one order of magnitude smaller than that of a uniform fiber Bragg grating [15]. On the contrary, a very pronounced temperature sensitivity of 2.75 nm/°C was reported for an LPFG inscribed in a photosensitive boron germanium co-doped optical fiber [16]. A number of techniques have also been presented in order to enhance the temperature sensitivity of long period fiber gratings. Among them, we can quote the use of different fiber compositions, different fiber geometries, and the use of polymer coatings with a large thermo-optic coefficient. An LPFG with period equal to 340 μm exhibited an attenuation band with a strain sensitivity equal to 0.04 pm/με, which is more than one order of magnitude lower than that of uniform fiber Bragg gratings [15]. An LPFG with period equal to 40 μm exhibited a large negative strain sensitivity equal to −2.2 pm/με, since for $\Lambda < 100$ μm, the two contributions to strain sensitivity are negative [15]. Also, the different attenuation bands of a single transmitted spectrum present a different, linear response to the applied axial strain.

The dependence of the sensitivity to strain and temperature on the order of the cladding mode is an important feature of LPFGs. Indeed the simultaneous measurement of the wavelength shifts of the different attenuation bands can allow the decoupling of temperature and strain effects, which overcomes the main drawback of uniform FBGs. Moreover, for gratings with periodicities <100 μm, temperature-insensitive axial strain measurements are possible.

We will summarize in Section 4.4 how these different kinds of fiber gratings can be advantageously used inside composite materials. Prior to this, the following section presents the main demodulation techniques used to measure the Bragg wavelength shift in the amplitude spectrum of embedded FBGs.

4.3 DEMODULATION TECHNIQUES USED TO RETRIEVE THE BRAGG WAVELENGTH SHIFT

Different interrogation techniques are available to demodulate the spectral response of FBGs embedded into composite materials. Their choice arises from considerations mainly related to their resolution, sensitivity, accuracy, dynamic range, cost, and performances in (quasi-)static (<10 Hz) or dynamic (>10 Hz) measurements. In this section, we review the main techniques that are used to demodulate FBGs embedded into composite materials when they are subject to external perturbations.

4.3.1 TECHNIQUES BASED ON WAVELENGTH DETECTION

These techniques are used to measure the reflected/transmitted amplitude spectrum, allowing the Bragg wavelength shift induced by the external perturbation to be directly computed. For this class of instrumentation, a distinction can be made between active and passive systems, depending on the use of a wavelength sweeping mechanism. Passive demodulation techniques make use of an optical filter whose transfer function depends on the wavelength and converts the Bragg wavelength shift into an amplitude change [17]. Such a filter presents a bandwidth several times higher than that of the FBG, as sketched in Figure 4.7a. It can be a TFBG operating in radiation mode [18], a chirped FBG [19], or an LPFG [20]. The use of intensity referencing is necessary, as the light intensity may fluctuate with time. This could occur due to unwanted power fluctuations of the light source, disturbances in the light-guiding path, or dependency of the light source intensity on the wavelength. Hence, robust demodulation techniques based on an edge filter use a ratiometric scheme to make them independent of intensity fluctuations. In this case, the input light is split into two paths, one passing through the wavelength-dependent filter and one used as the reference arm. The wavelength is accurately determined from the ratio between the powers measured at both arms [17]. Such a system is robust and can be set up at a relatively low cost. However, it can only resolve a single FBG.

Active systems make use of a tunable filter (with a bandwidth narrower than that of the FBG, as illustrated in Figure 4.7b) to scan the FBG amplitude spectrum, which is compatible with wavelength-division multiplexing (WDM) [21]. With similar performances in terms of wavelength resolution and scanning speed, a tunable laser source and a photodetector can be combined to demodulate the FBG's amplitude

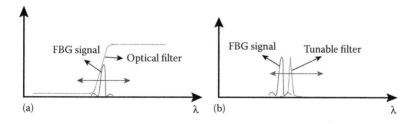

FIGURE 4.7   Operating principle of the edge filter (a) and tunable filter (b) techniques.

spectrum. Commercially available solutions based on such an implementation typically offer a 1 pm wavelength resolution and a frequency repetition that can reach 1 kHz for the most advanced systems. The use of an optical spectrum analyzer is another possibility for demodulation. It is essentially reserved for laboratory experiments, as such a system is slow (several seconds to scan a wavelength range of a few tens of nanometers), cumbersome, and expensive.

WDM techniques can be used to demodulate several tens of FBGs [22]. The operating ranges of the optical source and the detector limit the number of sensors that can be cascaded along a single optical fiber, considering that a sufficiently wide channel (2–3 nm) should be left for each FBG to avoid crosstalk.

4.3.2 REFLECTOMETRIC DEMODULATION TECHNIQUES

To overcome the limitations of WDM systems in terms of bandwidth, other techniques such as time-division multiplexing (TDM), spatial-division multiplexing (SDM), and their combination have been used [23–25]. It has been recently demonstrated that a wavelength-tunable optical time domain reflectometer (OTDR) can be used to interrogate a cascade of identical FBGs (i.e., same Bragg wavelengths under identical operating conditions and reflectance below 10%) [26–30]. An OTDR launches optical pulses in the fiber under test and detects the backscattered and reflected signals. As it shows the magnitude and location of any losses and reflections along the fiber length, detection and localization of tens of sensors along a unique optical fiber is achievable. However, a standard commercial OTDR is not directly suitable for demodulating FBG sensors: its internal optical source is too broad (several tens of nanometers) and not wavelength tunable, so that it cannot track the wavelength shift of perturbed FBGs. Because of these two limitations, it is required to modify a classical OTDR into a wavelength-tunable one with a much narrower spectral width (typically a few tens of picometers) [28]. In [29], an external tunable laser source and some active components (e.g., pulse generator, acousto-optic modulator) are used for this purpose. The method presented in [30] is a cost-effective and passive solution based on the use of an appropriately designed edge filter and a classical OTDR. However, techniques based on OTDRs can only be used for static measurements (traces have to be accumulated over a few seconds at least to filter out the measurement noise). Also, several meters of fiber have to be left between consecutive gratings due to the limited spatial resolution. OTDRs are thus marginally used for strain measurements in composite materials [31].

OFDRs (optical frequency domain reflectometers) are much more interesting in practice, offering both a high spatial resolution and a fast scanning rate [32–34]. A coherent OFDR exploits a frequency-modulated continuous-wave interference produced by a tunable laser source, whose frequency can be swept continuously in time without mode hopping, and an optical interferometer (Michelson) comprising a reference path and a measurement path. The fiber under test is connected to the measurement path, while the reference path, terminated by a mirror, is used as a local oscillator. Interferences between the two paths are electrically detected, and a Fourier transform allows the visualization of beat frequencies. If the optical frequency of the tunable source is modulated at a constant rate, beat frequencies are

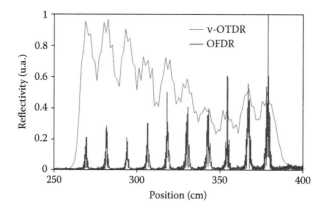

FIGURE 4.8   OFDR trace for 10 identical FBGs cascaded in an optical fiber embedded into a composite material fabric and corresponding photon-counting OTDR trace.

proportional to the optical path differences between the reflections in the fiber under test and the reference path. OFDRs can provide millimetric spatial resolution over a fiber length of a few hundreds of meters.

In [35], the OFDR developed in [34] was used to demonstrate its potential for the demodulation of closely spaced FBGs embedded into a composite material fabric. Ten FBGs (3 mm in length) were embedded with a spacing between them of approximately 10 cm. Figure 4.8 depicts the corresponding trace in the frequency domain converted into absolute position. A comparison with the same cascade measured by a photon-counting OTDR (high-resolution OTDR, as explained in [36]) is also displayed. One can readily see the superior spatial resolution offered by the coherent OFDR. To recover the reflection spectrum of each FBG in the array, the peak extracted by way of a band-pass filter is inversely Fourier transformed. As a result, Bragg wavelength shifts can be recorded for all the gratings.

4.4   DISCRIMINATION BETWEEN TEMPERATURE AND STRAIN EFFECTS

As they both lead to a shift of the Bragg wavelength, it is convenient to implement solutions able to discriminate between temperature and axial strain effects. According to Equation 4.2, a single FBG cannot achieve such discrimination. Two fiber gratings exhibiting different temperature and/or strain sensitivities are used in practice. Here is a detailed list of the more widespread solutions used for temperature–strain discrimination:

- Use of FBGs photo-inscribed in different types of fiber [37,38]. The chemical content of the optical fiber influences the FBG's sensitivity through a different response of the effective refractive index shift. This solution is limited for use in composite materials due to the splice, which may weaken the sensor integrity.

- Use of FBGs photo-inscribed at two different wavelengths in a single optical fiber [39]. A difference of sensitivity mainly arises from a difference in grating period in this case. Table 4.1 summarizes temperature and strain sensitivities for a standard optical fiber at different wavelengths, taken from [40].
- Exploitation of the first and second diffraction orders of a single FBG [41]. As in the previous solution, using different wavelengths provides different sensitivities. In both cases, a broadband optical source or different narrowband optical sources are needed, increasing the cost of the demodulation technique.
- Use of specialty optical fibers (polarization-maintaining fibers [42,43] or photonic crystal fibers [44,45]) that yield a very well-conditioned system. Here, attention has to be paid to the splicing between specialty and standard fibers as well as to the injection of light. The input state of polarization has to be tightly controlled.
- Processing of the gratings, such as encapsulation, decrease of the optical fiber diameter, or adjunction of a special coating [46–48]. These operations are time consuming and could weaken the optical fiber or, conversely, make it too thick for a proper encapsulation without a possible buckling effect.
- Coupling a Fabry–Perot interferometer to an FBG [49]. This solution requires using two different kinds of interrogators.
- Use of hybrid gratings such as a uniform FBG and an LPFG. They can be colocated to yield a localized discrimination between temperature and strain [50,51]. Here, the drastic difference between the inherent sensitivities of these two kinds of gratings is used to obtain a very well-conditioned system. However, the interrogation should be made in transmission over quite a wide broadband range (up several hundreds of nanometers).
- Use of FBGs of different types, such as type I and type IA (or type IIA) gratings [52,53].
- Use of TFBGs.

In Sections 4.4.1 through 4.4.3, different sensing configurations used with gratings embedded into composite materials will be detailed.

TABLE 4.1
FBG Strain and Temperature Sensitivities at Different Wavelengths

Wavelength (nm)	Strain Sensitivity (pm/$\mu\varepsilon$)	Temperature Sensitivity (pm/°C)
830	0.64	6.8
1300	1.00	10.0
1550	1.20	13.0

4.4.1 FBG Pairs with One of Them Embedded into a Glass Capillary

The simplest way to isolate temperature and strain effects consists in using two FBGs, one of which is embedded in a capillary and is therefore immune to strain changes. In the following, we present the results of an experiment conducted on a 2.8 mm thick composite material sample made by stacking together eight plain-weave layers of carbon reinforcing fibers. Manual stratification was operated with a matrix made of a mix between epoxy resin XB 3585 and hardener XB 3403 (Huntsman) whose curing cycle is 8 h at 80°C. Two FBGs were sandwiched between the fourth and fifth layers. One of them was embedded into a glass capillary (500 μm inner diameter and 850 μm outer diameter). The capillary was sealed at both ends to prevent resin intrusion. The final dimensions of the composite sample were 8 × 25 cm. The sample was subjected to temperature changes in the range between –20°C and 60°C (below the polymerization temperature) and to axial strains in the range between 0 and 2000 daN.

Reflected amplitude spectra were measured using a FiberSensing FS2200 interrogator with a 1 pm wavelength resolution. The Bragg wavelength shift was tracked in both spectra as a function of the external parameter. Figure 4.9 depicts the obtained evolutions as a function of strain and temperature. One can notice that the FBG embedded into the capillary is totally insensitive to axial strain. Also, the temperature sensitivity of the FBG in direct contact with the composite material is approximately 50% higher than that of the embedded FBG. This results from the fact that this grating undergoes the temperature change supplemented by a thermally induced mechanical deformation. These results can be expressed using the following expression [54]:

$$\begin{pmatrix} \Delta\lambda_{B,bare} \\ \Delta\lambda_{B,capillary} \end{pmatrix} = \begin{pmatrix} K_{\varepsilon,bare} & K_{T,bare} \\ K_{\varepsilon,capillary} & K_{T,capillary} \end{pmatrix} \begin{pmatrix} \varepsilon \\ \Delta T \end{pmatrix} \tag{4.31}$$

where the K-matrix contains the strain and temperature coefficients of both FBGs (bare and embedded into the capillary). Its determinant reflects the decoupling

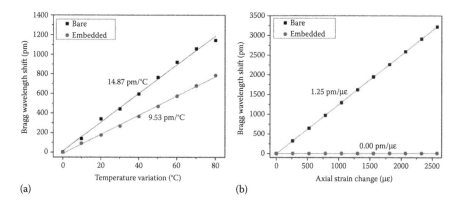

(a)   (b)

FIGURE 4.9 Bragg wavelength shifts as a function of temperature (a) and strain (b) for both FBGs integrated into the composite material sample.

efficiency. The latter increases with the determinant value. According to [55], uncertainties on strain and temperature can be computed from the uncertainty on the Bragg wavelength determination ($\delta\lambda_b$):

$$|\delta T| \leq \frac{|K_{\varepsilon,\text{bare}}||\delta\lambda_B| + |K_{\varepsilon,\text{capillary}}||\delta\lambda_B|}{|D|} \tag{4.32}$$

$$|\delta\varepsilon| \leq \frac{|K_{T,\text{bare}}||\delta\lambda_B| + |K_{T,\text{capillary}}||\delta\lambda_B|}{|D|} \tag{4.33}$$

where D is the determinant of the 2×2 matrix.

4.4.2 FBG Pair Comprising Type I and Type IA FBGs

As defined in [56–58], type I FBGs are standard uniform gratings, while type IA FBGs are regenerated. The regeneration process occurs in hydrogen-loaded photosensitive (core co-doped with boron and germanium) optical fiber under prolonged exposure to the UV laser beam. With the accumulation of the laser power, the type I grating is erased, and another grating is then reconstructed at a higher wavelength (typically between 5 and 15 nm). The reconstructed grating is called type IA. Its higher wavelength results from an increase of the core refractive index above its mean value. Type IA gratings can be conveniently produced by photo-bleaching the optical fiber with a uniform UV beam (a process known as the pre-exposition technique) prior to its exposition with the phase mask technique [56].

For the results reported hereafter, both types of FBGs were produced in hydrogen-loaded photosensitive single-mode optical fiber (PHOBBDCDC15 from POFC). The writing of the type IA FBG was done in two steps, as explained above. The pre-exposition process under a constant UV power of 80 mW lasted roughly 10 min. Both gratings were 4 mm in length and spaced by only 1 cm so as to behave as much as possible as a point sensor. Both gratings were annealed at 100°C for 24 h. The composite material sample was subject to similar perturbations as those reported in Section 4.3.1. Figure 4.10 depicts the measured wavelength shifts and the corresponding temperature and strain sensitivities. A relationship similar to Equation 4.8 can also be derived here.

4.4.3 Weakly Tilted FBG

In Section 4.4.3, we report results obtained on a 3 mm long 2° TFBG whose transmitted spectrum exhibits two main resonances, the Bragg mode and the ghost mode (resulting from the overlap of low-order cladding modes). This grating was embedded in a plain-weave composite material sample in the same way as the one reported in Section 4.3.1. These two resonances present different temperature sensitivities, as observed in the wavelength shifts shown in Figure 4.11. The axial strain sensitivities are roughly the same.

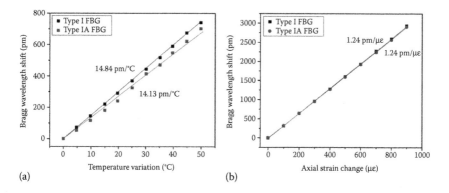

FIGURE 4.10 Bragg wavelength shifts of both grating types as a function of temperature (a) and strain (b).

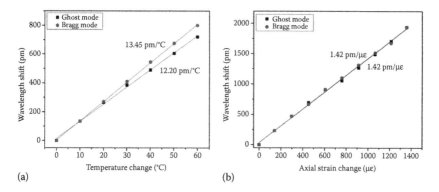

FIGURE 4.11 Wavelength shifts as a function of temperature (a) and strain (b) for the Bragg and Ghost Mode resonances.

4.4.4 COMPARISON OF PERFORMANCES

In this section, we evaluate the performances of the three solutions in Sections 4.4.1–4.4.3. The determinant of the K-matrix, which can be written in the general form $D = K_{\varepsilon,1}K_{T,2} - K_{\varepsilon,2}K_{T,1}$ (where subscripts 1 and 2 refer to the two resonances used for the purpose of discrimination), has to be different from 0 to ensure the decoupling. Table 4.2 compares the relative performances of the proposed solutions, where the uncertainties were computed according to Equations 4.9 and 4.10, using 1 pm for the wavelength resolution of the measurement device. These theoretical uncertainties depend, of course, on the measurement device.

In [51], it was proposed to estimate the efficiency E of the decoupling method based on the following relationship, which does not take into account the interrogation system:

$$E = \frac{D}{\sqrt{\left(K_{T,1}^2 + K_{T,2}^2\right)\left(K_{\varepsilon,1}^2 + K_{\varepsilon,2}^2\right)}} \qquad (4.34)$$

TABLE 4.2

Comparison of Performances between the Three Investigated Solutions for Temperature–Strain Discrimination

Method	Determinant ($pm^2/\mu\epsilon°C$)	Temperature Uncertainty (°C)	Strain Uncertainty ($\mu\epsilon$)
Bare + capillary	−11.912	0.1	2
Type I + type IA	−0.880	2.8	33
Ghost + Bragg	−1.775	1.6	14

This formula yields an efficiency of 54% for the first solution based on the capillary, and 2.5% and 6% for the other two configurations. Hence, based on this single parameter, the first solution remains the most competitive. However, when other parameters are taken into account, such as the ease of use, the integration into the composite material sample, or the multiplexing capability, using a capillary is definitely less attractive than other configurations. With the increase of the sensor diameter, the glass capillary unfortunately maximizes the distortion of the composite material around its location. Hence, there is no unique solution to the discrimination issue between temperature and axial strain effects. And the best solution is certainly not the one that maximizes the decoupling efficiency. One has to find a good trade-off between metrological and physical performances with respect to the target application and to the composite material composition and geometry.

Besides these techniques, other interesting configurations have been demonstrated, not necessarily in composite materials. Table 4.3 lists them with proper references to seminal works. When possible, it shows the K-matrix that allows the performance of the proposed technique to be evaluated.

4.5 RESIDUAL STRAIN SENSING DURING THE MANUFACTURING PROCESS

Residual strains appear during the fabrication process of composite materials and may drastically weaken the final structure, increasing the risk of sudden breakage. During the heating process followed to ensure the proper polymerization of the matrix, both the reinforced fibers and the matrix undergo thermally and chemically induced dimensional changes that are responsible for the appearance of residual strains [114–116]. They find their origin either inside the layer stacking (intrinsic residual strains) or at the interface between the piece and the mold (extrinsic residual strains). In 2006, Wisnom et al. identified three main causes for the appearance of residual strain [117]: thermal expansion of the resin [118], volume shrinkage of the resin [114,119], and differential thermal expansions between the mold and the sample [120].

There are numerous techniques to measure residual strains, which can be classified into destructive techniques (hole-drilling method [121], layer removal [122]) and nondestructive techniques (X-ray diffraction [123,124], bending measurement [125],

TABLE 4.3

Performances of the Temperature–Strain Discriminating Techniques

Technique		Sensitivity Matrix $\begin{bmatrix} K_{T1}\left(\dfrac{pm}{°C}\right) & K_{\varepsilon1}\left(\dfrac{pm}{\mu\varepsilon}\right) \\ K_{T2}\left(\dfrac{pm}{°C}\right) & K_{\varepsilon2}\left(\dfrac{pm}{\mu\varepsilon}\right) \end{bmatrix}$	Ref.
Single Grating			
Silica optical fiber Bare grating	TFBG (Bragg peak – ghost mode)	$\begin{bmatrix} 11.538 & 0.902 \\ 10.867 & 0.899 \end{bmatrix}$	[59,60]
	TFBG (Bragg peak – cladding mode)	$\begin{bmatrix} 11.096 & 0.923 \\ 10.677 & 0.846 \end{bmatrix}$	[61,62]
	Chirped FBG	N.A.	[63]
	First- and second-order FBG	$\begin{bmatrix} 9.70 & 1.092 \\ 4.89 & 0.472 \end{bmatrix}$	[41,64]
	Partially etched FBG	$\begin{bmatrix} 8 & 0.582 \\ 8 & 2.46 \end{bmatrix}$	[65]
	Two modes LPFG	$\begin{bmatrix} 93 & 1.942 \\ 37 & -0.0032 \end{bmatrix}$	[14]
	First- and second-order LPFG	$\begin{bmatrix} -509 & 1.461 \\ -251 & 0.647 \end{bmatrix}$	[66]
Silica optical fiber Special packaging	Fixed on two different metals	$K_{T1} = -0.4\,\dfrac{pm}{°C}$	[67]
	Partially coated	$\begin{bmatrix} 6.26 & 3.29 \\ 21.49 & 2.66 \end{bmatrix}$	[68]
	Embedded in a special design packaging	N.A.	[69]
Specialty optical fiber	HiBi single mode optical fiber (Panda, Bow-Tie…)	N.A.	[43,70,71]
	HiBi photonics crystal optical fiber	$\begin{bmatrix} 4.11 & 1.25 \\ 4.07 & 1.21 \end{bmatrix}$	[44]
	Few modes polymer optical fiber	$\begin{bmatrix} -9.8 & 1.193 \\ -11.1 & 1.163 \end{bmatrix}$	[72]

(*Continued*)

TABLE 4.3 (CONTINUED)

Performances of the Temperature–Strain Discriminating Techniques

Sensitivity Matrix

$$\begin{bmatrix} K_{T1}\left(\dfrac{pm}{°C}\right) & K_{\varepsilon1}\left(\dfrac{pm}{\mu\varepsilon}\right) \\[2mm] K_{T2}\left(\dfrac{pm}{°C}\right) & K_{\varepsilon2}\left(\dfrac{pm}{\mu\varepsilon}\right) \end{bmatrix}$$

Technique		Sensitivity Matrix	Ref.
Several Gratings			
Single optical fiber Bare fiber	Dual-wavelength FBG (superimposed or not)	$\begin{bmatrix} 8.72 & 0.96 \\ 6.30 & 0.59 \end{bmatrix}$	[73,74]
	Different types of FBGs (I–Ia)	$\begin{bmatrix} 9.56 & 1.076 \\ 8.84 & 1.073 \end{bmatrix}$	[53, 75]
	Different types of FBGs (Ia–IIa)	$\begin{bmatrix} 7.37 & 1.074 \\ 10.02 & 1.075 \end{bmatrix}$	[58]
	Different types of FBGs (I–IIa)	$\begin{bmatrix} 9.56 & 1.072 \\ 10.20 & 1.072 \end{bmatrix}$	[52,58]
	Dual LPFGs	$\begin{bmatrix} -310 & -0,3 \\ -500 & -0,02 \end{bmatrix}$	[76]
	Superstructure gratings	$\begin{bmatrix} 11.3\dfrac{pm}{°C} & 1.06\dfrac{pm}{\mu\varepsilon} \\[2mm] 0.026\dfrac{dB}{°C} & -0.0015\dfrac{dB}{\mu\varepsilon} \end{bmatrix}$	[77]
	FBG–LPFG (superimposed or not)	$\begin{bmatrix} 13 & 1.104 \\ 37.8 & -0.207 \end{bmatrix}$	[78,58,51]
	Two spacing identical FBG (FP)	N.A.	[79,80]
Single optical fiber Special packaging	FBGs inside a polymeric material	$\begin{bmatrix} 14.15 & 3.98 \\ 10.89 & 1.92 \end{bmatrix}$	[81]
	FBGs with two different coatings	$\begin{bmatrix} 11.433 & 1.228 \\ 11.333 & 1.170 \end{bmatrix}$	[82]
	Embedded into composite material with a special design (holed, triangular, chamfered plate)	$\begin{bmatrix} 12 & 1.555 \\ 12 & 0.745 \end{bmatrix}$	[83]

TABLE 4.3 (CONTINUED)
Performances of the Temperature–Strain Discriminating Techniques

Technique		Sensitivity Matrix $\begin{bmatrix} K_{T1}\left(\dfrac{pm}{°C}\right) & K_{\varepsilon1}\left(\dfrac{pm}{\mu\varepsilon}\right) \\ K_{T2}\left(\dfrac{pm}{°C}\right) & K_{\varepsilon2}\left(\dfrac{pm}{\mu\varepsilon}\right) \end{bmatrix}$	Ref.
	Embedded into hybrid composite laminates	$\begin{bmatrix} 14.1 & 1.08 \\ 10.9 & 0.26 \end{bmatrix}$	[84]
	One FBG fixed on an optical fiber	$\begin{bmatrix} 10.07 & 1.34 \\ 10.14 & 0.47 \end{bmatrix}$	[85]
	Encapsulated (glass or metal tube)	$\begin{bmatrix} 9.10 & 1.23 \\ 9.10 & 0 \end{bmatrix}$	[86,87]
Different optical fibers	FBGs in standard and specialty silicate optical fibers	$\begin{bmatrix} 8.43 & 0.969 \\ 7.37 & 0.949 \end{bmatrix}$	[38]
	FBGs in silicate and polymer optical fibers	$\begin{bmatrix} 10.47 & -149 \\ 1.17 & 1.5 \end{bmatrix}$	[88,89]
	FBGs in different diameter silicate optical fibers	$\begin{bmatrix} 6.99 & 0.81 \\ 5.73 & 0.42 \end{bmatrix}$	[90,91]
	FBGs in Er-doped and in Er/Yb co-doped fibers	$\begin{bmatrix} -678\,\dfrac{kHz}{°C} & 8.75\,\dfrac{kHz}{\mu\varepsilon} \\ -1142\,\dfrac{KHz}{°C} & 6.42\,\dfrac{kHz}{\mu\varepsilon} \end{bmatrix}$	[92]
	FBGs in single-mode fibers and microstructured optical fibers (MOF) silicate optical fibers	$\begin{bmatrix} 4.11 & 1.25 \\ 10.45 & 1.13 \end{bmatrix}$	[44]
	LPFGs in standard and specialty silicate optical fibers	$\begin{bmatrix} 100 & 460 \\ -590 & 460 \end{bmatrix}$	[93]

Hybrid Sensors and Techniques

Hybrid sensors	FBG/erbium-doped fiber amplifier (EDFA)	$K_{T,FBG}, K_{\varepsilon,FBG}, K_{P,EDFA} = -0.0425\,\dfrac{dBm}{°C}$	[94]

(Continued)

TABLE 4.3 (CONTINUED)
Performances of the Temperature–Strain Discriminating Techniques

Technique		Sensitivity Matrix $\begin{bmatrix} K_{T1}\left(\dfrac{pm}{°C}\right) & K_{\varepsilon 1}\left(\dfrac{pm}{\mu\varepsilon}\right) \\ K_{T2}\left(\dfrac{pm}{°C}\right) & K_{\varepsilon 2}\left(\dfrac{pm}{\mu\varepsilon}\right) \end{bmatrix}$	Ref.
	FBG encapsulated – extrinsic Fabry–Perot Interferometer	$\begin{bmatrix} 0\,\dfrac{pm}{°C} & 10.3\,\dfrac{pm}{\mu\varepsilon} \\ -13.2\,\dfrac{nm}{°C} & 27.2\,\dfrac{nm}{\mu\varepsilon} \end{bmatrix}$	[95,128,96,97]
	FBG – Mach–Zehnder interferometer	$\begin{bmatrix} 33 & 0.74 \\ 30.9 & -0.325 \end{bmatrix}$	[98]
	Chirped polarization maintaining fiber Bragg grating (PM FBG) – polarimetric	$\begin{bmatrix} -3.5 & 0.027 \\ 136 & -0.41 \end{bmatrix}$	[99]
	single-mode fiber Mach Zehnder interferometer (SMF MZI) – Polarization	N.A.	[100] [101, 102]
	maintaining giber Mach Zehnder interferometer (PCF MZI FBG) – modal interferometer (two sections of PCF)	$\begin{bmatrix} 13.5 & 1.05 \\ 56.5 & -0.325 \end{bmatrix}$	
	In-line Mach–Zehnder interferometer	$K_T = 51\,\dfrac{pm}{°C},\ K_\varepsilon = -0.023\,\dfrac{dBm}{\mu\varepsilon}$	[103]
	Clover microstructured fiber loop mirror	$\begin{bmatrix} 6.6 & -1.78 \\ 38.4 & -1.49 \end{bmatrix}$	[104]
	FBG in PM fiber – fiber loop mirror	$\begin{bmatrix} 11.2 & -1602.5 \\ 1.1 & 25.3 \end{bmatrix}$	[105]
	Optical fiber sensor and thermocouple	N.A.	[106]
	Optical fiber sensor and strain gauge	N.A.	[107]
Hybrid techniques	Raman/Brillouin optical time domain analysis (OTDA)	N.A.	[108]
	Brillouin optical time domain analysis (BOTDA)/OFDR	N.A.	[109]
	OFDR with two optical fibers (one free from stress)	N.A.	[110]

TABLE 4.3 (CONTINUED)

Performances of the Temperature–Strain Discriminating Techniques

Technique	Sensitivity Matrix $\begin{bmatrix} K_{T1}\left(\dfrac{pm}{°C}\right) & K_{\varepsilon 1}\left(\dfrac{pm}{\mu\varepsilon}\right) \\ K_{T2}\left(\dfrac{pm}{°C}\right) & K_{\varepsilon 2}\left(\dfrac{pm}{\mu\varepsilon}\right) \end{bmatrix}$	Ref.
Hybrid Brillouin–Rayleigh system	$\begin{bmatrix} 1.09e^{-3}\dfrac{GHz}{°C} & 5e^{-5}\dfrac{GHz}{\mu\varepsilon} \\ -1.498\dfrac{GHz}{°C} & -1.525e^{-1}\dfrac{GHz}{\mu\varepsilon} \end{bmatrix}$	[111]
Brillouin frequency shift and birefringence in a PM fiber	$\begin{bmatrix} 1.058\dfrac{MHz}{°C} & 0.0394\dfrac{MHz}{\mu\varepsilon} \\ -55.813\dfrac{MHz}{°C} & 0.8995\dfrac{MHz}{\mu\varepsilon} \end{bmatrix}$	[112,113]

ultrasound waves [126], etc.). FBG sensors, of course, belong to the nondestructive category. They have been used to follow the curing cycle [127,128], the resin polymerization process [129,130], and the formation of residual strain [131]. Studies have first focused on the measurement of strain along the fiber axis, neglecting the influence of transverse strains. Colpo et al. [132] used a 24 mm long FBG interrogated by an optical low coherence reflectometer (OLCR) to measure the axial strain distribution of the resin during the polymerization process. Eum et al. have reported the use of a 10 cm long FBG interrogated by an OFDR to follow the residual stress appearing during a vacuum-assisted resin transfer molding process [133]. In [134], BOTDA was used for a similar purpose. In all these works, the focus was only on the measurement of longitudinal strains. Transverse strains are mentioned in [135] without deep investigation of their value. Their precise measurement during the fabrication process is of great importance, because they are associated with the most fragile direction and can be at the origin of delamination.

As mentioned in Section 4.2.5, transverse strain leads to birefringence. This mechanically induced birefringence superimposes to the pristine fiber birefringence (due to the presence of asymmetries in the cross section) and to the photo-induced one during the writing process [136]. While this induced birefringence is scarcely perceived in the grating amplitude response, it leads to significant polarization-dependent loss (PDL) [137]. Birefringence (Δn) is defined as the difference in refractive index between two orthogonal polarization modes (or eigenmodes), labeled x and y modes. The refractive indices associated with these modes are defined by $n_{eff,x} = n_{eff} + \Delta n/2$ and $n_{eff,y} = n_{eff} - \Delta n/2$, where n_{eff} is the core effective refractive index without birefringence. Due to Δn, the x and y modes undergo different couplings through the grating. Therefore, the total FBG spectrum is the combination of the x and y mode signals. The wavelength spacing between both modes is defined by $2\Lambda\Delta n$ [138]. Hence, for increasing birefringence values, a broadening of the reflection band is first obtained before the polarization modes become well separated [139–143]. At this stage, the transverse strain value can be estimated from the measurement of the wavelength spacing between the peaks. This can be achieved for transverse force values exceeding a few tens of newtons [144,145]. To overcome this limitation and allow for residual transverse stress measurement, different configurations have been reported:

- FBGs photo-inscribed in polarization-maintaining (bow-tie) fibers [146] with an amplitude spectrum displaying two peaks.
- FBGs manufactured in highly birefringent photonic crystal fibers [147,148], further improving the resolution provided by standard polarization-maintaining fibers thanks to an increase of the wavelength spacing between the peaks and a temperature insensitivity. Polarization-maintaining fibers exhibit a strong orientation-dependent sensitivity. Accurate measurements can only be achieved after a correct orientation of the fiber.
- Standard FBGs with measurement of their spectral broadening for small transverse strain values [149].
- Standard FBGs with exploitation of their polarization-dependent loss spectrum. It was shown in [150,151] that the PDL spectrum contains much more information than the amplitude spectrum, which can be accurately used to detect transverse strain buildup due to matrix polymerization. A differential transverse strain of 60 $\mu\varepsilon$ was measured for cross-ply laminates.

Additional considerations on multi-axis strain sensing can be found in [152,153]. The state-of-the-art review again confirms that there is no universal solution to demodulate FBG sensors during and after the curing process. Table 4.4 summarizes

---

TABLE 4.4

Performances of Techniques Used for Strain Buildup Measurement during the Curing Process

Technique	Sensing Modalities	Requirements
Standard FBGs (Amplitude spectrum)	• Sensitive to axial strain • Sensitive to temperature	• Need for temperature compensation (cf. Section 4.3)
Standard FBGs (PDL spectrum)	• Sensitive to axial and transverse strains • Temperature compensation	• Use of polarized light
Hi-Bi FBGs (Amplitude spectrum)	• Sensitive to axial and transverse strains	• Use of polarized light • Need for proper fiber orientation • Need for temperature compensation
Hi-Bi PCF FBGs (Amplitude spectrum)	• Sensitive to axial and transverse strains • Immune to temperature changes	• Use of polarized light • Need for proper fiber orientation • Need for controlled splices

the metrological performances of the solutions available to follow the curing process. With the use of dedicated interrogators for FBG sensors, they are quite competitive in terms of wavelength resolution and cost, the main difference being related to whether or not polarized light is used.

4.6 CONCLUSION

Taking into account their miniaturized dimensions (which are further reinforced by using thinner optical fibers of external diameter of 80 μm or even 50 μm) and relative robustness, FBGs appear as ideal sensors for integration between the plies of composite materials. They can be used not only for structural health monitoring but also to reveal the residual stress buildup during the fabrication process. Despite these inherent specific advantages, the proper use of FBGs requires skill and practice. Numerous solutions have been implemented and disseminated in the scientific literature for many years now. This chapter has focused on three major issues arising from the use of FBGs embedded into composite material samples: the setup of a reliable demodulation technique, the discrimination between temperature and strain effects, and the residual stress monitoring. These challenges were investigated in depth, and while most of the issues were tackled, there is still some room for progress, especially regarding monitoring of the curing cycle. The latter is essential for an industrialization of the process. As yet, it has not revealed its full potential. Numerical simulations made with a finite element mode solver can surely provide insight to fully understand and interpret the amplitude spectrum modifications resulting from the curing process. Besides these possible routes for improvement, the issue of connectorization remains to be tackled. Currently, two antagonistic approaches appear very competitive: free-space connection, on the one hand, and designing a dedicated connector, on the other hand. No doubt, future research will enable the provision of robust, cost-effective, and reliable solutions to this problematic.

REFERENCES

1. Yu, F.T.S.; Yin, S.; Yu, Y.T.S. *Fiber Optic Sensors*, 3rd ed.; Marcel Dekker: New York, 2002.
2. Othonos, A.; Kalli, K. *Fiber Bragg Gratings: Fundamentals and Applications in Telecommunications and Sensing*; Artech House: New York, 1999.
3. Kersey, A.; Davis, M.A.; Patrick, H.J.; Leblanc, M. Fiber grating sensors. *Journal of Lightwave Technology* **1997**, 15, 1442–1463.
4. Luyckx, G.; Voet, E.; Lammens, N.; Degrieck, J. Strain measurements of composite laminates with embedded fibre Bragg gratings: Criticism and opportunities for research. *MDPI Sensors* **2011**, 11, 384–408.
5. Kinet, D.; Mégret, P.; Goossen, K.W.; Qiu, L.; Heider, D.; Caucheteur, C. Fiber Bragg grating sensors toward structural health monitoring in composite materials: challenges and solutions. *MDPI Sensors* **2014**, 14, 7394–7419.
6. Jensen, D.W.; Sirkis, J.S. Integrity of composite structures with embedded optical fibers. In *Fiber Optic Smart Structures*; Udd, E., Ed.; Wiley: New York, 1995; pp. 109–129.
7. Shivakumar, K.; Emmanwori, L. Mechanics of failure of composite laminates with an embedded fiber optic sensor. *Journal of Composite Materials* **2004**, 38, 669–680.

8. Kashyap, R. *Fiber Bragg Gratings*. Academic Press: San Diego, 1999.
9. Barlow, A.; Payne, D. The stress-optic effect in optical fibers. *Journal of Quantum Electronics* **1983**, 19, 834–839.
10. Albert, J.; Shao, L.Y.; Caucheteur, C. Tilted fiber Bragg grating sensors. *Laser & Photonics Reviews* **2013**, 7, 83–108.
11. Chen, C.; Albert, J. Strain-optic coefficients of individual cladding modes of single-mode fibre: Theory and experiment. *Electronics Letters* **2006**, 42, 1027–1028.
12. Laffont, G.; Ferdinand, P. Tilted short-period fibre-Bragg-grating induced coupling to cladding modes for accurate refractometry. *Measurement Science and Technology* **2001**, 12, 765–770.
13. Vengsarkar, A.; Lemaire, P.; Judkins, J.; Bhatia, V.; Erdogan, T.; Sipe, J. Long-period fiber gratings as band-rejection filters. *Journal of Lightwave Technology* **1996**, 14(1), 58–65.
14. Bhatia, V. Applications of long-period gratings to single and multi-parameter sensing. *Optics Express* **1999**, 4, 457–466.
15. Bhatia, V.; Campbell, D.; Sherr, D.; D'Alberto, T.; Zabaronick, N.; Ten Eyck, G.; Murphy, K.; Claus, R. Temperature-insensitive and strain-insensitive long-period grating sensors for smart structures. *Optical Engineering* **1997**, 36, 1872–1876.
16. Shu, X.; Allsop, T.; Gwandu, B.; Zhang, L.; Bennion, I. High-temperature sensitivity of long-period gratings in B-Ge codoped fibre. *Photonics Technology Letters* **2001**, 13, 818–820.
17. Melle, S.M.; Liu, K.; Measures, R.M. A passive wavelength demodulation system for guided-wave Bragg grating sensors. *Photonics Technology Letters* **1992**, 4, 516–518.
18. Guo, T.; Tam, H.-Y.; Albert, J. Chirped and tilted fiber Bragg grating edge filter for in-fiber sensor interrogation. *Proceedings of the IEEE Conference on Lasers and Electro-Optics Society*, **2011**, 1–6 May, Baltimore.
19. Liu, Y.; Zhang, L.; Bennion, I. Fabricating fibre edge filters with arbitrary spectral response based on tilted chirped grating structures. *Measurement Science and Technology* **1999**, 10, L1.
20. Fallon, R.W.; Zhang, L.; Everall, L.A.; Williams, J.A.R. All fiber optical sensing system: Bragg grating sensor interrogated by a long period grating. *Measurement Science and Technology* **1998**, 9, 1969–1973.
21. Xu, M.G.; Geiger, H.; Archambault, J.L.; Reekie, L.; Dakin, J.P. Novel interrogating system for fibre Bragg grating sensors using an acousto-optic tunable filter. *Electronics Letters* **1993**, 29, 1510–1511.
22. Sano, Y.; Yoshino, T. Fast optical wavelength interrogator employing arrayed waveguide grating for distributed fiber Bragg grating sensors. *Journal of Lightwave Technology* **2003**, 21, 132–136.
23. Xiao, G.Z.; Zhao, P.; Sun, F.G.; Lu, Z.G.; Zhang, Z.; Grover, C.P. Interrogating fiber Bragg grating sensors by thermally scanning a de-multiplexer based on arrayed waveguide gratings. *Optics Letters* **2004**, 29, 2222–2224.
24. Rao, Y. J.; Ribeiro, A.B.L.; Jackson, D.A.; Zhang, L.; Bennion, I. Simultaneous spatial, time and wavelength division multiplexed in-fibre grating sensing network. *Optics Communications* **1996**, 125, 53–58.
25. Hongwei, G.; Hongmin, L.; Bo, L.; Hao, Z.; Jianhua, L.; Ye, C.; Shuzhong, Y.; Weigang, Z.; Giuyun, K.; Xiaoyi, D. A novel Fiber Bragg Grating sensors multiplexing technique. *Optics Communications* **2005**, 251, 361–366.
26. Valente, L.C.G.; Braga, A.M.B.; Ribeiro, A.S.; Regazzi, R.D.; Ecke, W.; Chojetzki, C.; Willsch, R. Combined time and wavelength multiplexing technique of optical fiber grating sensor arrays using commercial OTDR equipment. *IEEE Sensors Journal* **2003**, 3, 31–35.

27. Zhang, P.; Cerecedo-Nunez, H.H.; Qi, B.; Pickrell, G.; Wang, A. Optical time domain reflectometry interrogation of multiplexing low reflectance Bragg-grating-based sensor system. *Optical Engineering* **2003**, 42, 1597–1603.

28. Crunelle, C.; Caucheteur, C.; Wuilpart, M.; Mégret, P. Quasi-distributed temperature sensor combining FBGs and temporal reflectometry technique interrogation. *Optics and Lasers in Engineering* **2009**, 47, 412–418.

29. Crunelle, C.; Caucheteur, C.; Wuilpart, M.; Mégret, P. Original interrogation system for quasi-distributed FBG-based temperature sensor with fast demodulation technique. *Sensors and Actuators A: Physical* **2009**, 150, 192–198.

30. Voisin, V.; Caucheteur, C.; Kinet, D.; Mégret, P.; Wuilpart, M. Self-referenced photon counting OTDR technique for quasi-distributed fiber Bragg gratings sensors. *IEEE Sensors Journal* **2012**, 12, 118–123.

31. Claus, R.O.; Jackson, B.S.; Bennett, K.D. Nondestructive testing of composite materials by OTDR in imbedded optical fibers. *Proceedings of the SPIE Conference on Fiber Optic and Laser Sensors III*, **1985**, San Diego.

32. Eickhoff, W.; Ulrich, R. Optical frequency domain reflectometry in single-mode fiber. *Applied Physics Letters* **1981**, 39, 693–695.

33. Soller, B.; Gifford, D.; Wolfe, M.; Froggatt, M. High resolution optical frequency domain reflectometry for characterization of components and assemblies. *Optics Express* **2005**, 13, 666–674.

34. Yüksel, K.; Moeyaert, V.; Mégret, P.; Wuilpart, M. Complete analysis of multi-reflection and spectral shadowing crosstalks in a quasi-distributed fiber sensor interrogated by OFDR. *IEEE Sensors Journal* **2012**, 12, 988–995.

35. Kinet, D.; Yüksel, K.; Caucheteur, C.; Garray, D.; Wuilpart, M.; Narbonneau, F.; Mégret, P. Structural health monitoring of composite materials with fibre Bragg gratings interrogated by optical frequency domain reflectometer. *Proceedings of the European Conference on Composite Materials*, **2012**, Venice, Italy.

36. Healey, P.; Hensel, P. Optical time domain reflectometry by photon counting. *Electronics Letters* **1980**, 16, 631–633.

37. Guan, B.-O.; Tam, H.Y.; Ho, S.-L.; Chung, W.-H.; Dong X.-Y. Simultaneous strain and temperature measurement using a single fibre Bragg grating. *Electronics Letters* **2000**, 36, 1018–1019.

38. Cavaleiro, P.M.; Araujo, F.M.; Ferreira, L.A.; Santos, J.L.; Farahi, F. Simultaneous measurement of strain and temperature using Bragg gratings written in germanosilicate and boron-codoped germanosilicate fibers. *Photonics Technology Letters* **1999**, 11, 1635–1637.

39. Xu, M.G.; Archambault, J.-L.; Reekie, L.; Dakin J.P. Discrimination between strain and temperature effects using dual-wavelength fibre grating sensors. *Electronics Letters* **1994**, 30, 1085–1087.

40. Rao, Y.-J. In-fibre Bragg grating sensors. *Measurement Science and Technology* **1997**, 8, 355–375.

41. Echevarria, J.; Quintela, A.; Jauregui, C.; Lopez-Higuera, J.M. Uniform fiber Bragg grating first- and second-order diffraction wavelength experimental characterization for strain-temperature discrimination. *Photonics Technology Letters* **2001**, 13, 696–698.

42. Chen, G.; Liu, L.; Jia, H.; Yu, J.; Xu, L.; Wang W. Simultaneous pressure and temperature measurement using Hi-Bi fiber Bragg gratings. *Optics Communications* **2003**, 228, 99–105.

43. Sudo, M.; Nakai, M.; Himeno, K.; Suzaki, S.; Wada, A.; Yamauchi R. Simultaneous measurement of temperature and strain using panda fiber grating. In *12th International Conference on Optical Fiber Sensors*, **1997**, Optical Society of America, Williamsburg, VA, p. OWC7.

44. Frazão, O.; Carvalho, J.P.; Ferreira, L.A.; Araújo, F.M.; Santos J.L. Discrimination of strain and temperature using Bragg gratings in microstructured and standard optical fibres. *Measurement Science and Technology* **2005**, 16, 2109.
45. Sonnenfeld, C.; Sulejmani, S.; Geernaert, T.; Eve, S.; Lammens, N.; Luyckx, G.; Voet, E.; Degrieck, J.; Urbanczyk, W.; Mergo, P.; Becker, M.; Bartelt, H.; Berghmans, F.; Thienpont, H. Microstructured optical fiber sensors embedded in a laminate composite for smart material applications. *Sensors* **2011**, 11, 2566–2579.
46. Xu, H.; Dong, X.; Yang, Z.; Ni, K.; Shum, P.; Lu, C.; Tam, H.Y. Simple FBG sensor head design for strain-temperature discrimination. In *14th OptoElectronics and Communications Conference*, 2009, Hong Kong, China, pp. 1–2.
47. Yoon, H.-J.; Costantini, D.M.; Michaud, V.; Limberger, H.G.; Manson, J.-A.; Salathe, R.P.; Kim, C.-G.; Hong, C.S. *In situ* simultaneous strain and temperature measurement of adaptive composite materials using a fiber Bragg grating based sensor. In *Smart Structures and Materials 2005: Smart Sensor Technology and Measurement Systems* **2005**, 5758, 62–69.
48. Mondal, S.K.; Tiwari, U.; Poddar, G.C.; Mishra, V.; Singh, N.; Jain, S.C.; Sarkar, S.N.; Chattoypadhya, K.D.; Kapur, P. Single fiber Bragg grating sensor with two sections of different diameters for longitudinal strain and temperature discrimination with enhanced strain sensitivity. *Review of Scientific Instruments* **2009**, 80, 103106.
49. de Oliveira, R.; Ramos, C.A.; Marques, A.T. Health monitoring of composite structures by embedded FBG and interferometric Fabry-Perot sensors. *Computers & Structures* **2008**, 86, 340–346.
50. Patrick, H.J.; Williams, G.L.M.; Kersey, A.D.; Pedrazzani, J.R.; Vengsarkar, A.M. Hybrid fiber Bragg grating/long period fiber grating sensor for strain-temperature discrimination. *Photonics Technology Letters* **1996**, 8, 1223–1225.
51. Triollet, S.; Robert, L.; Marin, E.; Ouerdane, Y. Discriminated measures of strain and temperature in metallic specimen with embedded superimposed long and short fibre Bragg gratings. *Measurement Science and Technology* **2011**, 22, 015202.
52. Frazão, O.; Lima, M.J.N.; Santos, J.L. Simultaneous measurement of strain and temperature using type I and type IIA fibre Bragg gratings. *Journal of Optics A: Pure and Applied Optics* **2003**, 5, 183.
53. Filograno, M.L.; Chluda, C.; Kinet, D.; Caucheteur, C.; Garray, D.; Corredera, P.; Mégret, P. Temperature and strain effects discrimination into composite materials with embedded dual type I-IA fibre Bragg gratings. In *15th European Conference on Composite Materials, Proceedings*, **2012**. Venice, Italy,
54. Butter, C.D.; Hocker, G.B. Fiber optics strain gauge. *Applied Optics* **1978**, 17, 2867–2869.
55. Jin, W.; Michie, W.; Thursby, C.; Konstantaki, G.M.; Culshaw, B. Simultaneous measurement of strain and temperature: Error analysis. *Optical Engineering* **1997**, 36, 598–609.
56. Simpson, G.; Kalli, K.; Zhou, K.; Zhang, L.; Bennion, I. Blank beam fabrication of regenerated type IA gratings. *Measurement Science and Technology* **2004**, 15, 1665.
57. Canning, J. Fibre gratings and devices for sensors and lasers. *Laser & Photonics Reviews* **2008**, 2, 275–289.
58. Shu, X.; Zhao, D.; Zhang, L.; Bennion, I. Use of dual-grating sensors formed by different types of fiber Bragg gratings for simultaneous temperature and strain measurements. *Applied Optics* **2004**, 43, 2006–2012.
59. Chehura, E.; James, S.W.; Tatam, R.P. Temperature and strain discrimination using a single tilted fibre Bragg grating. *Optics Communications* **2007**, 275, 344–347.
60. Kinet, D.; Garray, D.; Mégret, P.; Caucheteur, C. Temperature and strain effects discrimination inside composite materials with embedded weakly tilted fibre Bragg grating. *Proceedings of SPIE 8794, Fifth European Workshop on Optical Fibre Sensors*, **2013**, Krakow, Poland, p. 87942R.

61. Jovanovic, N.; Williams, R.J.; Thomas, J.; Steel, M.J.; Marshall, G.D.; Fuerbach, A.S.; Nolte, A.T.; Withford, M.J. Temperature and strain discriminating sensor based on the monitoring of cladding modes of a single femtosecond inscribed grating. *Proceedings SPIE 7503, 20th International Conference on Optical Fibre Sensors* **2009**, Edinburgh, United Kingdom, p. 750336.

62. Miao, Y.; Liu, B.; Zhao, Q. Simultaneous measurement of strain and temperature using tilted fibre Bragg grating. *Electronics Letters* **2008**, 44(21), 1242–1243.

63. Xu, M.G.; Dong, L.; Reekie, L.; Tucknott, J.A.; Cruz, J.L. Temperature-independent strain sensor using a chirped Bragg grating in a tapered optical fibre. *Electronics Letters* **1995**, 31(10), 823–825.

64. Brady, G.P.; Kalli, K.; Webb, D.J.; Jackson, D.A.; Reekie, L.; Archambault, J.L. Simultaneous measurement of strain and temperature using the first- and second-order diffraction wavelengths of Bragg gratings. *IEE Proceedings, Optoelectronics* **1997**, 144, 156–161.

65. Mondal, S.K.; Mal, S.; Tiwari, U.; Sarkar, S.; Mishra, V.; Poddar, G.C.; Mehla, N.S.; Jain, S.C.; Kapur, P. Simultaneous measurement of strain and temperature using partially etched single fiber Bragg grating sensor. *ICOP 2009-International Conference on Optics and Photonics* **2009**, Chandigarh, India, pp. 1–3.

66. Allsop, T.; Zhang, L.; Webb, D.J.; Bennion, I. Discrimination between strain and temperature effects using first and second-order diffraction from a long-period grating. *Optics Communications* **2002**, 211, 103–108.

67. Tian, K.; Liu, Y.; Wang, Q. Temperature-independent fiber Bragg grating strain sensor using bimetal cantilever. *Optical Fiber Technology* **2005**, 11(4), 370–377.

68. Sengupta, D.; Sai Shankar, M.; Umesh, T.; Kishore, P.; Saidi Reddy, P.; Sai Prasad, R.L.N.; Vengal Rao, P.; Narayana, K.S. Strain-temperature discrimination using a single FBG at cryogenic region. *Sensors & Transducers Journal* **2011**, 131(8), 36–42.

69. Geernaert, T.; Sulejmani, S.; Sonnenfeld, C.; Luyckx, G.; Chah, K.; Areias, L.; Mergo, P.; Urbanczyk, W.; Van Marcke, P.; Coppens, E.; Thienpont, H.; Berghmans, F. Microstructured optical fiber Bragg grating-based strain and temperature sensing in the concrete buffer of the Belgian supercontainer concept. *Proc. of SPIE 9157, 23rd International Conference on Optical Fibre Sensors*, **2014**, Santander, Spain, p. 915777-1-4.

70. Ferreira, L.A.; Araújo, F.M.; Santos, J.L.; Farahi, F. Simultaneous measurement of strain and temperature using interferometrically interrogated fiber Bragg grating sensors. *Optical Engineering* **2000**, 39(8), 2226–2234.

71. Oh, S.T.; Han, W.T.; Paek, U.C.; Chung, Y. Discrimination of temperature and strain with a single FBG based on the birefringence effect. *Optics Express* **2004**, 12(4), 724–729.

72. Qiu, W.; Cheng, X.; Luo, Y.; Zhang, Q.; Zhu, B. Simultaneous measurement of temperature and strain using a single Bragg grating in a few-mode polymer optical fiber. *Photonics Technology Letters* **2013**, 31(14), 2419–2425.

73. Xu, M.G.; Archambault, J.L.; Reekieet, L.; Dakin, J.P. Discrimination between strain and temperature effects using dual-wavelength fiber grating sensors. *Electronics Letters* **1994**, 30(13), 1085–1087.

74. Wu, M.-C.; Prosser, W.H. Simultaneous temperature and strain sensing for cryogenic applications using dual-wavelength fiber Bragg gratings. *Proceedings of SPIE 5191, Optical Diagnostics for Fluids, Solids, and Combustion II*, **2003**, San Diego, CA, pp. 208–213.

75. Simpson, A.G.; Kalli, K.; Zhang, L; Zhou, L; Bennion, I. Type 1A fibre Bragg grating photosensitivity and the development of optimum temperature invariant type I—type IA strain sensors. *Proceedings of SPIE Vol. 5459, Optical Sensing*, **2004**, San Diego, CA, pp. 118–127.

76. Bey, A.K.; Sun, T.; Grattan, K. Simultaneous measurement of temperature and strain with long period grating pairs using low resolution detection. *Sensors and Actuators A* **2008**, 144, 83–89.

77. Guan, B.-O.; Tam, H.-Y.; Tao, X.-M.; Dong, X.-Y. Simultaneous strain and temperature measurement using a superstructure fiber Bragg grating. *Photonics Technology Letters* **2000**, 12(6), 675–677.

78. Wang, M.; Li, C.T.; Zhao, Y.; Wei, H.; Tong, Z.; Jian, S. Simultaneous measurement of strain and temperature using dual-period fiber grating. *Proc. SPIE 4579, Optical Fiber and Planar Waveguide Technology* **2001**, Beijing, China, pp. 265–268.

79. Wang, Z. Intrinsic Fabry-Perot interferometric fiber sensor based on ultra-short Bragg gratings for quasi-distributed strain and temperature measurements. PhD Thesis, Blacksburg, Virginia, **2006**.

80. Islam, Md. R.; Ali, M.M.; Lai, M.-H.; Lim, K.-S.; Ahmad, H. Chronology of Fabry-Perot interferometer fiber-optic sensors and their applications: A review. *Sensors (Basel)* **2014**, 14, 7451–7488.

81. Dreyer, U.; de Morais Sousa, K.; Somenzi, J.; de Lourenço Junior, I., Cardozo da Silva, J.C.; de Oliveira, V.; Kalinowski, H.J. A technique to package Fiber Bragg Grating Sensors for strain and temperature measurements. *Journal of Microwaves, Optoelectronics and Electromagnetic Applications* **2013**, 12(2), 638–646.

82. Lu, P.; Men, L.; Chen, Q. Resolving cross sensitivity of fiber Bragg gratings with different polymeric coatings. *Applied Physics Letters* **2008**, 92, 171112-1-3.

83. Rodriguez-Cobo, L.; Marques, A.T.; Lopez-Higuera, J.M.; Santos, J.L., Frazão, O. Simplified sensor design for temperature-strain discrimination using Fiber Bragg Gratings embedded in laminated composites. *Proceedings of SPIE Vol. 8794, Fifth European Workshop on Optical Fibre Sensors*, **2013**, Krakow, Poland, 87942W-1-4.

84. Ferreira, M.S.; Vieira, J.C.; Frias, C.; Frazão, O. Strain and temperature discrimination using FBG sensors embedded in hybrid composite laminates. *16th International Conference on Composite Structures ICCS 16* **2011**, Porto, Portugal, pp. 1–2.

85. Frazao, O.; Marques, L.; Marques, J.M.; Baptista, J.M.; Santos, J.L. Simple sensing head geometry using fibre Bragg gratings for strain-temperature discrimination. *Optics Communications* **2007**, 279, 68–71.

86. Xu, M.G.; Geiger, H.; Dakin, J.P. Fibre grating pressure sensor with enhanced sensitivity using a glass-bubble housing. *Electronics Letters* **1996**, 32(2), 128–129.

87. Song, M.; Lee, S.B.; Choi, S.S.; Lee, B. Simultaneous measurement of temperature and strain using two fiber Bragg gratings embedded in a glass tube. *Optical Fiber Technology* **1997**, 3(2), 194–196.

88. Liu, H.B.; Liu, H.Y.; Peng, G.D.; Chu, P.L. Strain and temperature sensor using a combination of polymer and silica fibre Bragg gratings. *Optics Communications* **2003**, 219(1–6), 139–142.

89. Rajan, G.; Ramakrishnan, M.; Semenova, Y.; Ambikairajah, E.; Farrell, G.; Peng, G.-D. Experimental study and analysis of a polymer fiber Bragg grating embedded in a composite material. *Journal of Lightwave Technology* **2014**, 32(9), 1726–1732.

90. James, S.W.; Dockney, M.L.; Tatam, R.P. Simultaneous independent temperature and strain measurement using in-fibre Bragg grating sensors. *Electronics Letters* **1996**, 32, 1133–1134.

91. Song, M.; Lee, B.; Lee, S.B.; Choi, S.S. Interferometric temperature insensitive strain measurement with different diameter fiber Bragg gratings. *Optics Letters* **1997**, 22(11), 790–792.

92. Tan, Y.-N.; Zhang, Y.; Jin, L.; Guan, B.-O. Simultaneous strain and temperature fiber grating laser sensor based on radio-frequency measurement. *Optics Express* **2011**, 19(21), 20650–20656.

93. Han, Y.G.; Lee, S.B.; Kim, C.-S.; Kang, J.U.; Paek, U.-C.; Chung, Y. Simultaneous measurement of temperature and strain using long period fiber gratings with controlled temperature and strain sensitivities. *Optics Express* **2003**, 11(5), 476–481.

94. Jung, J.; Nam, H.; Lee, J.H.; Park, N.; Lee, B. Simultaneous measurement of strain and temperature by use of a single-fiber Bragg grating and an erbium-doped fiber amplifier. *Applied Optics* **1999**, 38(13), 2749–2751.

95. Ferreira, L.A.; Santos, A.B.; Farahi, J.L. Simultaneous measurement of displacement and temperature using a low finesse cavity and a fiber Bragg grating. *Photonics Technology Letters* **1996**, 8(11), 1519–1521.

96. Xiong, L.; Zhang, D.; Li, L.; Guo, Y. EFPI-FBG hybrid sensor for simultaneous measurement of high temperature and large strain. *Chinese Optics Letters* **2014**, 12(12), 120605-1-5.

97. Kang, H.-K.; Ryu, C.-Y.; Hong, C.-S.; Kim, C.-G. Simultaneous measurement of strain and temperature of structures using fiber optic sensor. *Journal of Intelligent Material Systems and Structures* **2001**, 12, 277–281.

98. Kang, Z.; Wen, X.; Li, C.; Sun, J.; Wang, J.; Jian, S. Up-taper-based Mach-Zehnder interferometer for temperature and strain simultaneous measurement. *Applied Optics* **2014**, 53(12), 2691–2695.

99. Bieda, M.S.; Lesiak, P.; Szeląg, M.; Kuczkowski, M.; Domański, A.W.; Woliński, T.R.; Farrell, G. Polarimetric and fiber Bragg grating reflective hybrid sensor for simultaneous measurement of strain and temperature in composite material. *Proceedings of SPIE Vol. 9506, Optical Sensors* **2015**, Prague, Czech Republic, p. 95060D-1-6.

100. Najari, M.; Javan, A.M.; Amiri, N. Hybrid all-fiber sensor for simultaneous strain and temperature measurements based on Mach–Zehnder interferometer. *Optik* **2015**, 126, 2022–2025.

101. Yu, S.; Pei, L.; Liu, C.; Wang, Y., Weng, S. Simultaneous strain and temperature measurement using a no-core fiber-based modal interferometer with embedded fiber Bragg grating. *Optical Engineering* **2014**, 53(8), 087105-5.

102. Li, C.; Ning, T.; Wen, X.; Li, J.; Zheng, J.; You, H.; Chen, H.; Zhang, C.; Jian, W. Strain and temperature discrimination using a fiber Bragg grating and multimode interference effects. *Optics Communications* **2015**, 343, 6–9.

103. Zhou, J.; Liao, C.; Wang, Y.; Yin, G.; Zhong, X.; Yang, K.; Sun, B.; Wang, G.; Li, Z. Simultaneous measurement of strain and temperature by employing fiber Mach-Zehnder interferometer. *Optics Express* **2014**, 22(2), 1680–1686.

104. Pérez-Herrera, R.A.; André, R.M.; Silva, S.F.; Becker, M.; Schuster, K.; Kobelke, J.; Lopez-Amo, M.; Santos, J.L.; Frazão, O. Simultaneous measurement of strain and temperature based on clover microstructured fiber loop mirror. *Measurement* **2015**, 65, 50–53.

105. Zhou, D.-P.; Wei, L.; Liu, W.-K.; Lit, J.W.Y. Simultaneous measurement of strain and temperature based on a fiber Bragg grating combined with a high-birefringence fiber loop mirror. *Optics Communications* **2008**, 281(18), 4640–4643.

106. Shen, X.; Lin, Y. Measurements of temperature and residual strain during fatigue of a CFRP composite using FBG sensors. *International Conference on Measuring Technology and Mechatronics Automation*, **2009**, Zhangjiajie, China, pp. 35–38.

107. Horoschenkoff, A.; Klein, S.; Haase, K.-H. Structural integration of strain gages, **2006**, HBM. https://www.hbm.com/fileadmin/mediapool/hbmdoc/technical/s2182.pdf.

108. Taki, M.; Muanenda, Y.; Nannipieri, T.; Signorini, A.; Oton, C.J.; Di Pasquale, F. Advanced coding techniques for long-range Raman/BOTDA distributed strain and temperature measurements. *Optical Fiber Communications Conference and Exhibition*, **2015**, Anaheim, CA, pp. 1–3.

109. Zhou, D.-P.; Li, W.; Chen, L.; Bao, X. Distributed temperature and strain discrimination with stimulated Brillouin scattering and rayleigh backscatter in an optical fiber. *Sensors (Basel)* **2013**, 13(2), 1836–1845.

110. Maurin, L.; Rougeault, S.; Dewynter-Marty, V.; Périsse, J.; Villani, D.; Macé, J.-R.; Ferdin, P. OFDR distributed temperature and strain measurements with optical fibre sensing cables: Application to drain pipeline monitoring in a nuclear power plant. *7th European Workshop on Structural Health Monitoring*, **2014**, Nantes, France, pp. 363–370.

111. Ito, Y.; Fujimoto, K.; Minakuchi, S.; Mizutani, T.; Takeda, N.; Koinuma, H.; Shimizu, T. Cure monitoring of carbon/epoxy composite by optical-fiber-based distributed strain/temperature sensing. *International Conference on Composite Materials*, Jeju Island, South Korea, **2011**, pp. IF1491.

112. Zou, W.; He, Z.; Hotate, K. Complete discrimination of strain and temperature using Brillouin frequency shift and birefringence in a polarization-maintaining fiber. *Optics Express* **2009**, 17(3), 1248–1255.

113. Hotate, K.; Zou, W.; Yamashita, R.K.; He, Z. Distributed discrimination of strain and temperature based on Brillouin dynamic grating in an optical fiber. *Photonic Sensors* **2013**, 3(4), 332–344.

114. Li, C.; Potter, K.; Wisnom, M.R.; Stringer, G. *In-situ* measurement of chemical shrinkage of MY750 epoxy resin by a novel gravimetric method. *Composites Science and Technology* **2004**, 64, 55–64.

115. Hill, R.R.; Muzumdar, S.V.; Lee, L.J. Analysis of volumetric changes of unsaturated polyester resins during curing. *Polymer Engineering & Science* **1995**, 35, 852–859.

116. Barnes, J.A.; Byerly, G.E. The formation of residual stresses in laminated thermoplastic composites. *Composites Science and Technology* **1994**, 51, 479–494.

117. Wisnom, M.R.; Gigliotti, M.; Ersoy, N.; Campbell, M.; Potter, K.D. Mechanisms generating residual stresses and distortion during manufacture of polymer-matrix composite structures. *Composites Part A: Applied Science and Manufacturing* **2006**, 37, 522–529.

118. Thomas, H.H. Residual stresses in polymer matrix composite laminates. *Journal of Composite Materials* **1976**, 10, 266–278.

119. Nawab, Y.; Tardif, X.; Boyard, N.; Sobotka, V.; Casari, P.; Jacquemin, F. Determination and modelling of the cure shrinkage of epoxy vinylester resin and associated composites by considering thermal gradients. *Composites Science and Technology* **2012**, 73, 81–87.

120. de Oliveira, R.; Lavanchy, S.; Chatton, R.; Costantini, D.; Michaud, V.; Salathé, R.; Månson, J.-A. E. Experimental investigation of the effect of the mould thermal expansion on the development of internal stresses during carbon fibre composite processing. *Composites Part A: Applied Science and Manufacturing* **2008**, 39, 1083–1090.

121. Sicot, O.; Gong, X.L.; Cherouat, A.; Lu, J. Determination of residual stress in composite laminates using the incremental hole-drilling method. *Journal of Composite Materials* **2003**, 37, 831–844.

122. Paterson, M.W.A.; White, J.R. Layer removal analysis of residual stress. *Journal of Materials Science* **1989**, 24, 3521–3528.

123. Benedikt, B.; Rupnowski, A.; Kumosa, L.; Sutter, J.K.; Predecki, P.K.; Kumosa, M. Determination of interlaminar residual thermal stresses in a woven 8HS graphite/PMR-15 composite using X-ray diffraction measurements. *Mechanics of Advanced Materials and Structures* **2002**, 9, 375–394.

124. Meske, R.; Schnack, E. Particular adaptation of X-ray diffraction to fiber reinforced composites. *Mechanics of Materials* **2003**, 35, 19–34.

125. Cowley, K.D.; Beaumont, P.W.R. The measurement and prediction of residual stresses in carbon-fibre/polymer composites. *Composites Science and Technology* **1997**, 57, 1445–1455.

126. Ben, B.S.; Ben, B.A.; Vikram, K.A.; Yang, S.H. Damage identification in composite materials using ultrasonic based Lamb wave method. *Measurement* **2013**, 46, 904–912.

127. Murukeshan,V.M.; Chan, P.Y.; Ong, L.S.; Seah, L.K. Cure monitoring of smart composites using fiber Bragg grating based embedded sensors. *Sensors and Actuators A: Physical* **2000**, 79, 153–161.

128. Kang, H.-K.; Kang, D.-H.; Bang, H.-J.; Hong, C.-S.; Kim, C.-G. Cure monitoring of composite laminates using fiber optic sensors. *Smart Materials and Structures* **2002**, 11, 279.

129. Vacher, S.; Molimard, J.; Gagnaire, H.; Vautrin, A. A Fresnel's reflection optical fiber sensor for thermoset polymer cure monitoring. *Polymers and Polymer Composites* **2004**, 12, 269–276.

130. Buggy, S.J.; Chehura, E.; James, S.W.; Tatam, R.P. Optical fibre grating refractometers for resin cure monitoring. *Journal of Optics A: Pure and Applied Optics* **2007**, 9, S60.

131. Kalamkarov, A.L.; Fitzgerald, S.B.; MacDonald, D.O. The use of Fabry Perot fiber optic sensors to monitor residual strains during pultrusion of FRP composites. *Composites Part B: Engineering* **1999**, 30, 167–175.

132. Colpo, F.; Humbert, L.; Botsis, J. Characterisation of residual stresses in a single fibre composite with FBG sensor. *Composites Science and Technology* **2007**, 67, 1830–1841.

133. Eum, S.H.; Kageyama, K.; Murayama, H.; Uzawa, K.; Ohsawa, I.; Kanai, M.; Kobayashi, S.; Igawa, H.; Shirai, T. Structural health monitoring using fiber optic distributed sensors for vacuum-assisted resin transfer molding. *Smart Materials and Structures* **2007**, 16, 2627.

134. Minakuchi, S.; Takeda, N.; Takeda, S.-I.; Nagao, Y.; Franceschetti, A.; Liu, X. Life cycle monitoring of large-scale CFRP VARTM structure by fiber-optic-based distributed sensing. *Composites Part A: Applied Science and Manufacturing* **2011**, 42, 669–676.

135. Nielsen, M.W.; Schmidt, J.W.; Høgh, J.H.; Waldbjørn, J.P.; Hattel, J.H.; Andersen, T.L.; Markussen, C.M. Life cycle strain monitoring in glass fibre reinforced polymer laminates using embedded fibre Bragg grating sensors from manufacturing to failure. *Journal of Composite Materials* **2013**, 48, 1–17.

136. Erdogan, T.; Mizrahi, V. Characterization of UV-induced birefringence in photosensitive Ge-doped silica optical fibers. *Journal of the Optical Society of America B* **1994**, 11, 2100–2105.

137. Zhu, Y.; Simova, E.; Berini, P.; Grover, C.P. A comparison of wavelength dependent polarization dependent loss measurements in fiber gratings. *IEEE Transactions on Instrumentation and Measurement* **2000**, 49, 1231–1239.

138. Bette, S.; Caucheteur, C.; Wuilpart, M.; Mégret, P.; Garcia-Olcina, R.; Sales, S.; Capmany, J. Spectral characterization of differential group delay in uniform fiber Bragg gratings. *Optics Express* **2005**, 13, 9954–9960.

139. Okabe, Y.; Yashiro, S.; Tsuji, R.; Mizutani, T.; Takeda, N. Effect of thermal residual stress on the reflection spectrum from fiber Bragg grating sensors embedded in CFRP laminates. *Composites Part A: Applied Science and Manufacturing* **2002**, 33, 991–999.

140. Guemes, J.A.; Menéndez, J.M. Response of Bragg grating fiber-optic sensors when embedded in composite laminates. *Composites Science and Technology* **2002**, 62, 959–966.

141. Emmons, M.C.; Carman, G.P.; Mohanchandra, K.P.; Richards, W.L. Characterization and birefringence effect on embedded optical fiber Bragg gratings. *Proceedings of the SPIE Conference on Health Monitoring of Structural and Biological Systems*, **2009**, 72950C, San Diego.

142. Luyckx, G.; Voet, E.; Lammens, N.; DeWaele, W.; Degrieck, J. Residual strain-induced birefringent for multi-axial strain monitoring of composite laminates. *NDT & E International* **2013**, 54, 142–150.

143. Caucheteur, C.; Bette, S.; Garcia-Olcina, R.; Wuilpart, M.; Sales, S.; Capmany, J.; Mégret, P. Influence of the grating parameters on the polarization properties of fiber Bragg gratings. *Journal of Lightwave Technology* **2009**, 27, 1000–1010.

144. Gafsi, R.; El-Sherif, M.A. Analysis of induced-birefringence effects on fiber Bragg gratings. *Optical Fiber Technology* **2000**, 6, 299–323.

145. Caucheteur, C.; Bette, S.; Garcia-Olcina, R.; Wuilpart, M.; Sales, S.; Capmany, J.; Mégret, P. Transverse strain measurements using the birefringence effect in fiber Bragg gratings. *Photonics Technology Letters* **2007**, 19, 966–968.

146. Chehura, E.; Skordos, A.A.; Ye, C.-C.; James, S.W.; Partridge, I.K.; Tatam, R.P. Strain development in curing epoxy resin and glass fibre/epoxy composites monitored by fibre Bragg grating sensors in birefringent optical fibre. *Smart Materials and Structures* **2005**, 14, 354.

147. Geernaert, T.; Luyckx, G.; Voet, E.; Nasilowski, T.; Chah, K.; Becker, M.; Bartelt, H.; Urbanczyk, W.; Wojcik, J.; De Waele, W.; Degrieck, J.; Terryn, H.; Berghmans, F.; Thienpont, H. Transversal load sensing with fiber Bragg gratings in microstructured optical fibers. *IEEE Photonics Technology Letters* **2009**, 21, 6–8.

148. Sonnenfeld, C.; Luyckx, G.; Collombet, F.; Grunevald, Y.-H.; Douchin, B.; Crouzeix, L.; Torres, M.; Geernaert, T.; Sulejmani, S.; Eve, S.; Gomina, M.; Chah, K.; Mergo, P.; Thienpont, H.; Berghmans, F. Embedded fiber Bragg gratings in photonic crystal fiber for cure cycle monitoring of carbon fiber-reinforced polymer materials. *Proceedings of the SPIE Conference on Micro-structured and Specialty Optical Fibres II*, **2013**, 877500, Prague, Czech Republic.

149. Minakuchi, S.; Umehara, T.; Takagaki, K.; Ito, Y.; Takeda, N. Life cycle monitoring and advanced quality assurance of L-shaped composite corner part using embedded fiber-optic sensor. *Composites Part A* **2013**, 48, 153–161.

150. Luyckx, G.; Kinet, D.; Lammens, N.; Chah, K.; Caucheteur, C.; Mégret, P.; Degrieck, J. Temperature-insensitive cure cycle monitoring of cross-ply composite laminates using the polarization dependent loss property of FBG. *Proceedings of the 15th European Conference on Composite Materials*, **2012**, Venice, Italy.

151. Lammens, N.; Kinet, D.; Chah, K.; Luyckx, G.; Caucheteur, C.; Degrieck, J.; Mégret, P. Residual strain monitoring of out-of-autoclave cured parts by use of polarization dependent loss measurements in embedded optical fiber Bragg gratings. *Composites Part A: Applied Science and Manufacturing* **2013**, 52, 38–44.

152. Luyckx, G.; Voet, E.; De Waele, W.; Degrieck, J. Multi-axial strain transfer from laminated CFRP composites to embedded Bragg sensor: I. Parametric study. *Smart Materials and Structures* **2010**, 19, 105017.

153. Voet, E.; Luyckx, G.; De Waele, W.; Degrieck, J. Multi-axial strain transfer from laminated CFRP composites to embedded Bragg sensor: II. Experimental validation. *Smart Materials and Structures* **2010**, 19, 105018.

5 Structural Health Monitoring of Composite Materials Using Distributed Fiber-Optic Sensors

Hideaki Murayama

CONTENTS

5.1 Introduction 106
    5.1.1 Distributed Sensing and Performance Parameters 106
    5.1.2 Measurands and Sensing Technologies 109
    5.1.3 Issues in Practical Implementation 115
    5.1.4 Application to SHM of Composite Materials 118
5.2 Distributed Fiber-Optic Sensing 120
    5.2.1 Raman-Based Distributed Temperature Sensing 120
    5.2.2 Brillouin-Based Distributed Temperature/Strain Sensing 122
    5.2.3 Rayleigh-Based and FBG-Based Distributed Temperature/Strain
        Sensing 125
    5.2.4 Development of Distributed Temperature and Strain Sensing 127
    5.2.5 Summary 128
5.3 SHM of Composite Materials Using Distributed Fiber-Optic Sensors 129
    5.3.1 Applications of Distributed Temperature Sensing 131
        5.3.1.1 Temperature Monitoring of Composite Structures 131
        5.3.1.2 Temperature Monitoring of Manufacturing Processes 133
        5.3.1.3 Application to NDT Methods 133
    5.3.2 Applications of Distributed Strain Sensing 134
        5.3.2.1 Strain Monitoring of Composite Materials and Structures ... 136
        5.3.2.2 Strain Monitoring in Steep Strain Distributions 140
        5.3.2.3 Deformation/Load Identification Based on
        Measurement Data 144
5.4 Concluding Remarks and Future Prospects.. 148
References.. 150

5.1   INTRODUCTION

Among the advantages of intrinsic fiber-optic sensors, distributed sensing is often cited as one of the principal potential benefits [1]. Distributed sensing is the ability to return the value of a measurand as a function of the linear position along an optical fiber. Fiber-optic sensors with such ability are called *distributed fiber-optic sensors* or *fiber-optic distributed sensors*. They inherit useful characteristics for structural health monitoring (SHM) from fiber-optic sensors, such as immunity to electromagnetic interference, durability, and the capability to be embedded into composite materials. They are still being actively developed into a unique form of sensor for which, in general, there may be no counterpart based on conventional sensor technologies [2]. This also means there is something unique and unfamiliar that characterizes the sensing technique. This introduction provides a brief overview of distributed fiber-optic sensors to outline their fascinating characteristics. Their sensing techniques and their application to the SHM of composite materials, as well as some issues to be considered for practical implementation, are also described briefly in this section and will be described in depth in Sections 5.2 and 5.3. This chapter focuses on the sensing process in SHM rather than the diagnosis and prognosis processes.

5.1.1   Distributed Sensing and Performance Parameters

Figure 5.1 shows basic single-ended and loop system configurations for distributed sensing with an example of measurement output. In general, a sensing instrument consists of laser sources, detectors, modulators, and couplers or circulators. Off-the-shelf systems, additionally, have data acquisition and processing modules to acquire measurement data from the detector signal and an interface to output or display

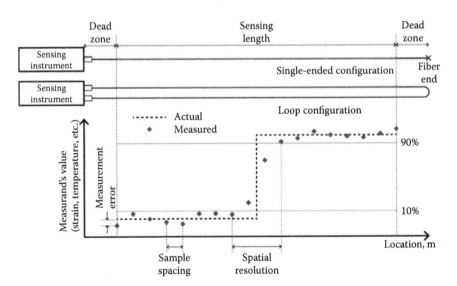

FIGURE 5.1   Schematic of distributed fiber-optic sensors.

the data. Some techniques for distributed sensing require the loop configuration to generate counter-propagating beams in the sensing fiber, while in the single-ended configuration, a light, which is scattered or reflected at a location and propagates backward in the sensing fiber, is captured at the detector. For distributed strain or temperature sensing based on nonlinear scattering, that is, Raman and Brillouin, the loop configuration is able to decrease measurement error due to higher signal-to-noise ratio (SNR). This, however, might have several drawbacks. First, it increases cost and complexity in the sensing instrument by requiring two high-performance laser sources or added components as well as those in the implementation for SHM by making the loop line [3]. Furthermore, when there is a point at which the optical fiber is broken or bent with a large curvature to prevent light propagation, the loop configuration does not allow us to continue the measurements. In such a situation, however, the data between the sensing instrument with the single-ended configuration and the breaking point are still available.

Measurement results are usually represented as a graph, termed a *trace*, as shown in Figure 5.1, with the measurand's value on one axis and the location along the sensing fiber on the other. This enables us to clearly understand the specific definition of distributed sensing mentioned in Section 5.1 and the difference between the concepts of distributed sensing and multipoint sensing or quasi-distributed sensing, where the measurement is taken at specific locations with point sensors, such as fiber Bragg gratings (FBGs) [4].

In distributed sensing, one can obtain a value of the measurand at a sample point that is associated with a single point along the sensing fiber. The distance between two consecutive sample points along the location axis is called *sample spacing*. Sample spacing is often referred to as *readout resolution* or *sampling resolution*. There are dead zones in the sensing fiber, where strong signals from, for instance, Fresnel reflections or fiber connectors cause a saturation of the detector and form locations that cannot be measured. When some optical devices, coatings, or transducers along the sensing fiber are associated with a function of distributed sensing, also the locations where those are missing can be the dead zones. SHM for civil structures or the oil and gas industry sometimes requires distributed sensors with a longer sensing coverage length of more than several kilometers for application to monitoring of a large-scale structure or remote monitoring. In the case of SHM for aerospace and other vehicle industries, a sensing length of less than 1 km would cover most requirements, but a higher spatial resolution would be needed. The spatial resolution is the smallest length of fiber over which any detectable change in the spatial variation of the measurand can be detected within the specified measurement error [3]. The measurement error is the difference between a measurand's actual value and a measured one. Since the actual value is generally unknown, in performance tests it can be substituted by a reference value measured by other sensors or derived from theoretical analysis. For sensing instruments, in which the input light is a pulse signal, the spatial resolution is usually defined as the pulse width. When distributed sensing is applied to the measurement of strain or temperature distributions in a structure, however, it might be preferred that the spatial resolution is defined as a *response delay in length*, as shown in Figure 5.1. In this case, it is calculated as 10%–90% rise distance of a transition of the measurand, which actually distributes in a stepwise pattern [5,6]. Determining the spatial resolution experimentally might give conservative results, because it is difficult to realize a perfect step distribution of measurands, such as strain and temperature, in the sensing fiber. The spatial resolution defined by this method obviously gives the capability to discretize a gradient and identify localized changes in the distribution of the measurand [7]. Therefore, when one verifies whether one can detect damage or anomaly in a local area based on measurement data, the spatial resolution is the essential performance parameter of the sensing instrument to be examined. It should also be noted that using the readout resolution instead of the spatial resolution can be misleading.

Those who apply distributed sensing to SHM have a strong interest in the temporal resolution as well as the spatial resolution and the resolution of measurement of the measurand. When a trace, as shown in Figure 5.1, which is a set of all of a measurand's values separated by the sample spacing within a sensing length, is required for the data of a measurement, measurement time can be defined as a *constant time interval between two consecutive traces obtained in successive measurements* (Figure 5.2). In the case that all values of a trace are detected at the same time in a shorter measurement time, in other words, that they are synchronized in high-speed measurements, dynamic changes in the external environment or the structural response can be analyzed based on the time-varying trace. For instance, strain distributions dynamically measured along a beam-like structure can enable us to identify the vibration modes and to use vibration-based damage identification methods [8,9]. When looking at the specifications of various sensing instruments, one would notice the trade-off between the performance parameters of the distributed sensing. In the development of distributed sensing systems, shortening the measurement time is probably the most difficult and costly performance task compared with others such as maximum sensing length, measurand measurement range, and spatial resolution.

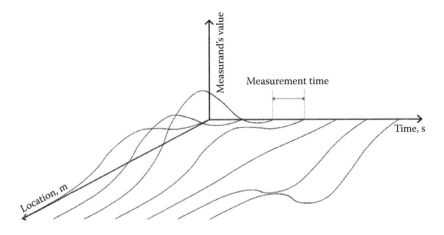

FIGURE 5.2   Time-varying traces in successive measurements.

5.1.2 Measurands and Sensing Technologies

In the context of assessing structural integrity or estimating structural reliability based on stress-strength analysis, measurands to be monitored and processed in SHM should be related to the load/environment applied to the structure or the strength/stiffness belonging to it [6,10]. In Figure 5.3, the load/environment and the strength/stiffness are represented as stress and strength, respectively. We can say that when the probability that the strength is larger than the applied stress is high, the structure has a higher reliability. However, damage and structural deterioration can make the inherent strength smaller and then make the probability of failure larger over time, as shown in Figure 5.3. The initial strength can be smaller than expected due to a design flaw or manufacturing defects. Furthermore, unexpected stress caused by an additional weight or an eccentric operation can exceed the present strength. The goal of SHM is to monitor the current safety margin and confirm whether it can be controlled at the proper level.

This doctrine has given us an incentive to explore methods of measuring physical parameters relating to stress and strength, such as pressure, temperature, strain, and vibration, along an optical fiber. Sensing techniques measuring pressure, temperature, strain, and vibration in the manner of distributed sensing with optical fibers are often referred to as *distributed pressure sensing* (DPS), *distributed temperature sensing* (DTS), *distributed strain sensing* (DSS) and *distributed acoustic sensing* (DAS), respectively. Technological developments over the last several decades have explored methods to determine these physical parameters continuously along the sensing fiber [7,11–13]. Now, however, it is known through a number of mechanisms that a desired physical parameter causes the changes in amplitude, frequency, phase, or polarization of the propagating light in the fiber, as well as the methods that allow us to determine a location, where the changes occur along the fiber, with an appropriate sensitivity and spatial resolution.

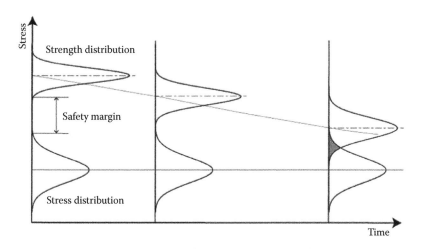

FIGURE 5.3   Stress-strength model vs. age.

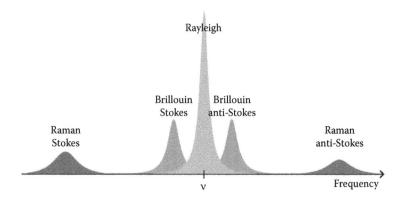

FIGURE 5.4 Schematic of elastic and inelastic scattering spectra in an optical fiber.

There are three types of scattering in optical fibers that can be exploited in distributed sensing [14]. These are Rayleigh, Raman, and Brillouin scattering. As shown in Figure 5.4, Rayleigh scattering is an example of an elastic scattering mechanism whose wavelength is the same as that of the incident light, while Raman and Brillouin scatterings are two prime examples of inelastic scattering mechanisms with both low-frequency (red-shifted) components (the Stokes band) and high-frequency (blue-shifted) components (the anti-Stokes band) [15]. The frequency shift of Raman scattering in silica is in the range of about 13 THz, whereas that of Brillouin scattering is about 10 GHz in the near-infrared wavelength band. DTS using the temperature dependence of the amplitude of Raman scattering is well known and has been developed into a commercial instrument for the past 30 years [5,16]. In addition, the fact that the peak offset frequency of Brillouin scattering is shifted by temperature or strain has attracted many researchers to the development of DTS or DSS using Brillouin scattering since the end of the 1980s [17,18].

In an optical fiber, some of the scattered light is recaptured by the waveguide and sent in a backward direction as the forward-propagating light travels, termed *backscattering*. Typical off-the-shelf DTS systems with a single-ended configuration detect Raman backscattering continuously along the sensing fiber and measure the change of its amplitude due to the temperature at each sample point. To measure the amplitude distribution of the backscattering along the sensing fiber, optical time domain reflectometry (OTDR) technology is used, in which radar-type signal processing is adopted to separately identify the amplitudes at different sample points along the sensing fiber and then to allow the DTS system to provide temperature traces at a constant measurement time.

OTDR was originally developed to measure attenuation characteristics in an optical fiber [19]. Figure 5.5 shows the typical arrangement of a sensing instrument with Rayleigh-based OTDR. In OTDR, an accurately timed light pulse is injected from the laser source with the laser driver into the fiber, and backscattering is observed by the detector. The signal from the detector is acquired by an analog-to-digital converter (ADC) and analyzed by a digital signal processor (DSP) to monitor the level of the backscattering mapped in the sensing fiber. The spatial information is acquired by determining the location derived from the time of flight of the light

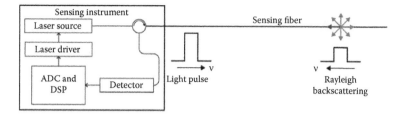

FIGURE 5.5 Schematic diagram of a sensing instrument with Rayleigh-based OTDR.

pulse traveling to and from the sample point. When the time of flight, which is the delay between the launch of the light pulse and the time at which the backscattering from a sample point is received, is t, the location of the sample point, z, is given by

$$z = \frac{ct}{2n_{\text{eff}}} \tag{5.1}$$

where:

  c is the speed of light

  n_{eff} is the effective refractive index of the fiber

If the fiber is homogeneous and is subjected to a uniform environment, the backscattering intensity decays exponentially with the location due to the intrinsic loss in the fiber [2]. Figure 5.6 shows an example of OTDR measurement in which microbending induces a drop of the intensity in the sensing fiber. If there is a connector or a breaking of the fiber at a location, an intensity peak, due to the Fresnel reflection and a relatively larger drop behind the peak due to a loss, would be observed around the location. The accuracy of finding the location of such an imperfection in the sensing fiber significantly depends on the spatial resolution. The spatial resolution, δz, of the sensing instrument of conventional OTDR is determined by

$$\delta z = \frac{c\tau}{2n_{\text{eff}}} \tag{5.2}$$

where τ is the pulse duration.

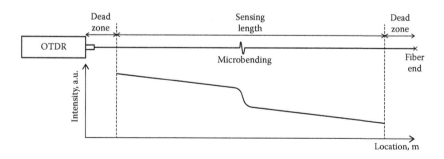

FIGURE 5.6 Distributed loss sensing by OTDR.

The effective refractive index is approximately 1.46 for a standard silica optical fiber. Therefore, in order to achieve a spatial resolution of 1 m, a pulse duration of 10 ns is needed. This resolution is usually referred to as the *highest resolution of DTS, or DSS based on the conventional OTDR*, and can be high enough for application to some large civil structures and oil/gas field development. In order to detect damage or degradation in composite materials applied to vehicles, however, higher spatial resolution would be required [20]. In addition, considerable averaging of the backscattering signal detected by OTDR is required to achieve a good SNR ratio, and, consequently, it results in an inconvenient longer measurement time, because Brillouin scattering is typically two orders of magnitude weaker than Rayleigh scattering, which is three to four orders of magnitude weaker compared with the incident radiation. Raman scattering is even weaker in comparison with Brillouin scattering [13].

The spatial resolution of distributed fiber-optic sensors using Brillouin scattering has improved notably over a decade in laboratory conditions, as well as in commercial systems [4,21,22]. Most of the Brillouin-based sensing instruments with high spatial resolution exploit the loop configuration and are based on Brillouin optical time domain analysis (BOTDA). Resolutions of centimeter order in the maximum sensing length of more than 1 km have been reported. When it comes to DTS based on Raman scattering, photon-counting techniques enable resolutions of several tens of centimeter of centimeter-ordering in the research [23,24]. However, the maximum sensing length is restricted to less than several meters [25]. Although Brillouin- or Raman-based sensors have achieved such higher spatial resolutions, using high-specification components makes the system expensive and complicated [7].

One alternative solution is optical frequency domain reflectometry (OFDR), which can be used for DTS and DSS [26,27]. OFDR was originally proposed as a technique to analyze the distribution of losses along an optical fiber, as well as OTDR [28]. A basic configuration of OFDR with an auxiliary interferometer is shown in Figure 5.7. The heart of this instrument is two Michelson interferometers [29,30]. During the measurement, the frequency, ν, of the narrow line width laser is continuously scanned by a tunable laser source (TLS) and the reflections in the sensing fiber interfere at the detector with a fixed reflection, namely, the local oscillator (LO). OFDR relies on the principle that the frequency of the interference fringe,

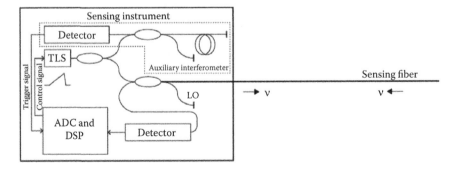

FIGURE 5.7   Schematic diagram of a sensing system of OFDR with auxiliary interferometer.

generated at the detector as the TLS swept, depends on the distance traveled by the light in the sensing fiber compared with the length of the reference path [31]. Let us consider a single reflection coming from a location, z, in the sensing fiber. The location, z, is a distance from the LO. The alternating current (ac) at the detector, $I(v)$, is given by the interference between the reflection and the LO as follows:

$$I(v) \propto I_0 \cos\left(\frac{4\pi n_{eff} v}{c} z - \varphi\right) \tag{5.3}$$

where:
  I_0 is the amplitude of the laser
  φ is a phase

This equation shows that, for a reflection at a given distance, z, the intensity of the detector is a sinusoidal function of v that is tuned by the TLS. The frequency $2 n_{eff} z/c$ of this function is given by the distance between the reflection and the LO. Therefore, each reflection in the sensing fiber corresponding to different lengths can be distinguished by means of a Fourier transform. The function of the auxiliary interferometer is to generate a trigger signal for the ADC to acquire the signals at equally spaced optical frequencies [29]. In this way, all data points acquired from the detector of the measurement interferometer with the sensing fiber are spaced by precisely the same optical frequency variation, which is given by

$$\Delta v_{acq} = \frac{c}{2 l_{aux} n} \tag{5.4}$$

where l_{aux} is the distance between the reflectors in the auxiliary interferometer.

This function of the auxiliary interferometer is important to compensate the nonlinearity of the frequency sweep by the TLS and then to use a DSP, such as a fast Fourier transform (FFT), of the beat signal acquired from the detector in the measurement interferometer. Figure 5.8 shows an example of OFDR measurements that determine the locations of the reflectors generating Fresnel reflections in the sensing fiber. Equation 5.3 means that light reflected from farther locations generates higher-frequency fringes than that from closer locations. In this case, the signal obtained from the detector has the frequency components corresponding to the locations of two reflectors. Therefore, FFT of the signal at the DSP provides the amplitude distribution of the reflected light along the sensing fiber and two peaks can be seen at the locations of the reflectors. The sample spacing of the amplitude distributions, Δz, is determined by the spectral bandwidth of the optical frequency sweep range of the TLS, Δv_{TLS}, as follows [29,32]:

$$\Delta z = \frac{c}{2 n_{eff} \Delta v_{TLS}} \tag{5.5}$$

where $\Delta v_{TLS} \cong \Delta \lambda_{TLS} / 2\lambda^2$.

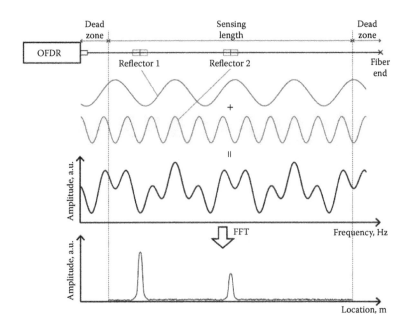

FIGURE 5.8   Distributed reflection sensing by OFDR.

In addition, $\Delta\lambda_{TLS}$ and λ are the wavelength sweep range of the TLS and the center wavelength of the sweep range, respectively. The corresponding sample spacing for a 40 nm wavelength scan centered at 1550 nm is about 20 μm. In a conventional way, to use OFDR of this value is also called the *spatial resolution* [7]. The maximum sensing length is determined by l_{aux} or the differential delay, τ_g, in the auxiliary interferometer using the Nyquist sampling criteria by [32,33]:

$$L_{max} = \frac{l_{aux}}{2} = \frac{c\tau_g}{4n_{eff}} \tag{5.6}$$

When using OFDR for DTS or DSS, the Rayleigh backscattering, or the reflection light from fiber gratings, is observed, and the change in the frequency/wavelength spectrum is measured at each sample point. The high spatial resolution of millimeter and submillimeter order for DTS and DSS could be achieved in the sensing length of several tens of meters and less than 10 meters, respectively, in the literature [31,33].

According to the performance chart for distributed sensors shown in [7], we can see that the spatial resolution has been much improved, as previously mentioned in this section. The measurement time of the Brillouin-based system, however, seems to be almost the same as that of about 15 years ago (i.e., the high-resolution measurements by Brillouin-OTDR [BOTDR] required about 5 min [34]). This can be due to the difference in the sensing length or the measurement accuracy. Also, for Raman OTDR (ROTDR) using photon counting, it generally takes several minutes to obtain a temperature trace with high resolution. DAS based on OTDR is able to detect strain changes along a 1 km optical fiber at the sampling rate of 100 kHz with the spatial resolution from 1 to 10 m, which is determined by the length of the light pulse as well as that of the conventional OTDR technique [35]. In this case, the measurement time is restricted by the period for which the light pulse travels back and forth in the sensing fiber. For DAS, the sampling rate is defined as the *inverse of the measurement time*. However, the time to constantly obtain from an arbitrary sample point in the sensing fiber is sometimes referred to also as the *sampling rate* [36]. In the case of this definition, the measurement time can be more than 1 s, when the sampling rate, the sensing length, and the sample spacing are 1 kHz, 100 m, and 10 cm, respectively. In order to shorten the measurement time of OFDR, an ADC with a high sampling rate and a laser source with a high scan speed are required. A measurement time of about 1.25 ms in strain measurements has been achieved with a spatial resolution of less than 1 mm and a sensing length of 1 m [8]. By using such state-of-the-art distributed sensing techniques with an ability for high-speed measurements, the applicability of vibration-based damage identification methods to SHM of composite materials would be enhanced significantly.

5.1.3 ISSUES IN PRACTICAL IMPLEMENTATION

The diagnosis process, which is one of the primary goals of SHM and provides information on the strength/stiffness (i.e., damage) or the load/environment (i.e., operating conditions) that affect the safety/performance of a structure, strongly depends on the measurement data from the sensing process. Sensing systems should be implemented to collect and pass data according to the requirements of the diagnosis process, as without diminishing some sources of uncertainty, a convenient condition for measurements is broken, and one cannot rely on the measurement data. In addition, the prognosis process that predicts possible future states generally requires long-term and stable measurement data to correctly predict development of damage or changes in operating conditions.

When choosing and implementing distributed fiber-optic sensors for a SHM system, the key issues to consider in practical implementation are (1) the physical parameter to be measured, (2) the variation of measure and (3) the locations to be monitored in a structure, (4) the environment around the sensing fiber and the cross-sensitivity effect, (5) the operating conditions of the sensing instrument, (6) the data transfer and storage, and (7) the cost. The approaches to consider these issues will now be discussed.

When the measurands and their variable natures in a structure to be monitored are given, the applicable sensing techniques are investigated. Temperature is measured by intrinsic fiber-optic sensors based on Raman, Brillouin, and Rayleigh scattering processes and FBG sensors. The Raman process cannot be applied to strain measurements but the others can. From the early stages of research, combinations of fiber-optic sensors and transducers have been explored to create more variation in measurands. Intensity modulation induced by microbending loss in multimode fibers (MMF) can be used as a transduction mechanism to measure pressure, acceleration, and magnetic/electric fields [37]. In the case of DPS, mechanical transducers, such as diaphragms, might be employed to make strain sensors more sensitive to pressure. Evanescent field radioactive loss due to external refractive index changes has been applied to leakage monitoring [38]. These kinds of technologies and concepts could expand the multifunctional ability to measure various measurands in one sensing fiber line and the applicability to monitoring of manufacturing processes [39]. However, we should keep in mind that the ability to provide passive sensing without any external transducers is useful for making the sensing system more robust. When looking at sensing instruments, we are able to obtain some necessary information from the specifications published by companies. The performance charts in the literature for distributed fiber-optic sensors would additionally be useful to investigate the comprehensive knowledge on distributed fiber-optic sensors [4,7,25]. Unfortunately, the terminology in literature and the performance tests for products are not consistent. Recently, standardizations and guidelines have been developed to help users to choose favorable sensing techniques and instruments [40,41].

The measurand conditions, in other words, how the measurand varies spatially and temporally in areas to be monitored, are essential issues for users to consider when choosing a suitable sensing instrument. Conditions include looking at the measurement accuracy, the spatial resolution, the measurement time, and other performance parameters according to the requirements of the diagnosis process. Unfortunately, at present, users do not have the opportunity to choose from many options. In addition, there is a trade-off between the performance parameters. Therefore, it is important to find out which performance parameter is dominant for minimizing measurement errors. Simulation techniques that take a holistic approach to considering the measurement conditions can be powerful tools when evaluating the performance of distributed fiber-optic sensors [6,20]. When a complicated and steep strain field occurs around an embedded fiber-optic sensor, so-called measurement simulation helps us to confirm whether the sensor is capable of detecting it and to interpret measurement results accurately [42,43].

The locations to be monitored restrict the placement and the installation of the sensing fibers. The placement has a strong relation to information density and quantity, whereas the installation has an influence on the measurement errors and the reliability/durability of the sensing system. The information density and quantity of the measurement data obtained from a well-designed fiber-optic sensor would far exceed those obtained from the sum of point sensors installed with a supreme effort. In the case of distributed fiber-optic sensors, an optical fiber has a roll of the transmission pass of the light for a considerable distance. It is extremely important and difficult to keep sufficient transmission ability in a whole length of the sensing fiber, especially when the fiber is embedded into composite materials. For instance, improper handling in the installation, or forces acting on the fiber in the curing process, may introduce significant deformation or curvature and, subsequently, a large loss at a local point of the sensing fiber, so that the light signal will be sharply reduced, causing considerable deterioration in the sensing performance [44]. Ingress–egress points of the fiber then become critical locations [45]. Additionally, an error in the mapping process, for which every sample point within the sensing fiber is associated with a point in the structure, may cause a mismatch between the actual and estimated conditions of the monitored area. If the measurand is not a scalar, it is also necessary to correctly identify the component measured at the point in the mapping process. These issues would make the installation a critical and time-consuming one in SHM.

Although the installation of distributed fiber-optic sensors must be a formidable challenge, it is good news that they usually do not need any of the shields that are often used for conventional electrical sensors according to their environment. Therefore, there is no need to bother with shields, which significantly increase cost, size, and weight as well as decreasing sensitivity. In the case of OFDR, however, vibration of the fibers in the interferometers causes significant fluctuation in measurement data. In terms of cost, it is also important that standard single-mode fibers (SMF), MMF, polarization-maintaining fibers (PMF), or FBG are used as the sensing fiber. These silica-optical fibers and in-fiber devices have coatings made of polymers, such as acrylate, silicon, and polyimide. When it comes to SHM of composite materials and the condition of the surface-mounted or embedded sensing fiber, the thin and hard polyimide coating with good properties of heat resistance (i.e., 300°C) and adhesion (e.g., silica and epoxy friendly) enables sensitive temperature and strain measurements via direct contact with some types of glue or polymer matrix.

Among the sources of uncertainty that cause measurement errors, cross-sensitivity has practical limitations, because this effect leads to ambiguity in the measurements. Fortunately, the Raman-based DTS technique has very little cross-sensitivity to other physical parameters, such as strain and pressure. Indeed, this useful characteristic must be a key to success, as it has been used in a wide range of practical and commercial applications. On the other hand, Brillouin-based, Rayleigh-based, and FBG-based DSS techniques suffer from temperature changes that affect measurement accuracy. Since they also have an ability to provide absolute measurements when discriminating temperature and strain properly, a lot of researchers have been addressing this fundamental issue [21,46]. When absolute strain measurements are implemented for long periods, a huge amount of spatial–temporal information associated with structural conditions can be used to determine wear-out failure, that is, fatigue damage and degradation due to corrosion or creep. For a distributed strain sensor bonded to a structure, adjacent secondary temperature sensors, which may be encased and bonded to the structure in a small tube to isolate them from mechanical strain and which have the same temperature as that of the structure, could be used as a typical solution to this problem. This approach, however, would make the installation process more complicated and formidable. Recently, a number of attractive approaches to enable simultaneous temperature and strain measurements by a single fiber have been proposed. But it remains a major challenge to discriminate temperature and strain in various field conditions while maintaining the performance parameter at a high level [47].

There is always the question as to whether a sensing instrument is able to show the same performance in field conditions as it does in a laboratory. Some sensing instruments seem to be too complicated to survive in harsh conditions. Temperature, humidity, and vibration are major concerns in the operating conditions of the sensing instrument. A loop configuration, which can operate by the single-ended configuration to improve the redundancy or the accuracy of the sensing system, is sometimes adopted for a sensing instrument.

Generally, measurement data provided from a sensing instrument to a host system are a set of traces. As described in the next section, distributed fiber-optic sensing needs data processing between analog signals acquired from detectors and trace data in a measurement. Sometimes, the by-product (e.g., spectrum data) can provide critical information used for SHM [20]. The transfer rate is determined by the requirements of the diagnosis process, but it should be high enough to carry out an express evaluation on structural conditions based on the measurement data. Even for the traces, however, the data size of distributed sensors can be immediately too big to deal with. Data formatting for distributed fiber-optic sensors is an issue that needs to be standardized.

When developing high-density or wide-area sensor networks, one can reduce costs for the measurement system or the interrogator, which is usually one of the dominant influences on the total cost, by using distributed sensing techniques. A significant limitation of such systems, however, is in the deployment, since over 75% of the installation time is attributable to the laying of wires in a structure, with installation costs representing up to 25% of the system total cost for larger-scale structures, such as long-span bridges [48]. Furthermore, wires installed in a combat vessel are vulnerable to detriments, such as heat, moisture, and toxic chemicals, common in harsh military operational environments [49].

5.1.4 APPLICATION TO SHM OF COMPOSITE MATERIALS

In the early applications of distributed fiber-optic sensing to SHM of composite materials in a laboratory environment, OTDR was used to identify the location with a steep strain distribution or damage in a sensing fiber embedded into fiber-reinforced polymers (FRP) [50,51]. The location could be identified by associating strange losses or reflections in the sensing fiber with such an anomaly. These applications are based on the idea that such a loss or reflection comes from microbending or fiber breakage in the sensing fiber, and such conditions of the fiber are caused by failure/imperfection of a host structure. As shown in Figure 5.9, Grace et al. embedded the sensing fibers (i.e., M1 and M2) between the layers of prepreg materials to make meshes in which the location of impact damage could be determined by measuring the length to the Fresnel reflection in M1 and M2 [51]. In the test, however, M1 did not, unfortunately, work for damage detection, so damage detection was carried out by using M2, whose sensing length was 15.9 m.

Murayama et al. first applied BOTDR and ROTDR to strain and temperature measurements in full-scale composite structures [34]. They measured the strain or temperature distributions in the hulls of a 24 m racing yacht (International America's Cup Class [IACC] yacht) and in the fuselage of a 13 m full-scale model of a reentry vehicle (H-II Orbiting Plane-Experimental [HOPE-X]). Figure 5.10a shows the racing yacht on the stand, and Figure 5.10b shows the reentry vehicle on the test frame. The SMFs were bonded on the inside surface of the hull of the IACC yacht to give a total length of 59.2 m of sensing fibers for strain measurements. In the case of HOPE-X, the SMF and MMF were bonded parallel to each other on the outer surface of the fuselage over the length of 54.8 m to monitor temperature and strain distributions during the manufacturing process and the structural test. Some of the success obtained by the temperature and strain distributions along such long sensing fibers was due to the fact that the structures were composite monocoque structures with smooth and wide surfaces that provided good conditions to install the optical fibers without critical imperfection. Strain monitoring of the IACC yacht at the end of the

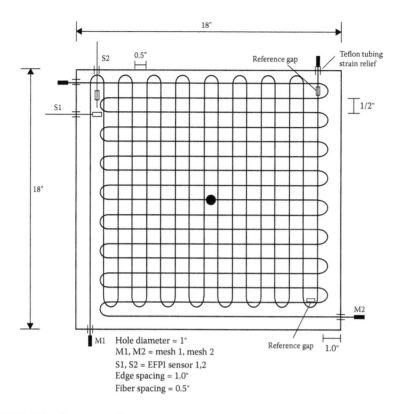

FIGURE 5.9 Geometry of a smart panel test specimen. (From Gifford, D.K., et al., *Proceedings of SPIE*, 6770, 67700F, 2007.)

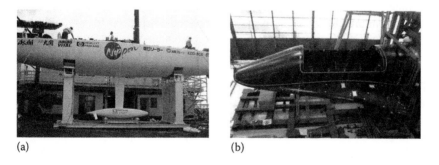

(a)   (b)

FIGURE 5.10 Full-scale composite structures ((a), IACC yacht; (b), HOPE-X) monitored using DSS and DTS.

1990s was the first challenge in the application of distributed fiber-optic sensors using the BOTDR technology to a full-scale composite structure in practical use. The sensors, covering a widespread area, provided measurement data that could be mapped onto the hull in a three-dimensional coordinate and represented the overall deformation of the yacht structure. It can be noted that the sensing system had been employed for the regular inspection for about eight months without any serious trouble. The details of these research projects will be discussed again in Section 5.3.2.

These early works using DTS and DSS with the OTDR technology and taking simple approaches were important to show the intuitively understandable and fascinating concepts of SHM using distributed fiber-optic sensors. At the same time, it was a challenge to embed optical fibers into composite structures, as it was not possible to eliminate imperfections in the sensing system completely. Additionally, it has been found that improving the spatial resolution and diminishing the cross-sensitivity problem are critical development issues, because the measurement errors due to these insufficient performances considerably affect the applicability of the diagnosis process in SHM.

During the past decade, distributed fiber-optic sensors with high spatial resolution have provided a new opportunity to explore their application to SHM of composite materials. Also, a number of fascinating works addressing the cross-sensitivity problem in distributed fiber-optic sensors have been undertaken by researchers. Some of these works will be described in the following sections.

5.2   DISTRIBUTED FIBER-OPTIC SENSING

Distributed fiber-optic sensing is still being actively developed for new and interesting applications to various fields. There are many techniques to measure a physical parameter continuously along an optical fiber using fiber-optic sensors. The OTDR technique is more popular for long–sensing length applications. On the other hand, when high-spatial-resolution measurement is the essential requirement, the OFDR technique may be the only solution because of its ability to implement distributed sensing with the centimeter- or millimeter-spatial resolution, even if there is a compromise in the sensing length.

Raman, Brillouin, and Rayleigh scattering mechanisms and fiber gratings enable us to perform pressure, temperature, strain, or acoustic distributed sensing by using the OTDR or OFDR technique. Among all combinations, some of them have not yet been demonstrated, just proposed. To the author's knowledge, there have been off-the-shelf sensing instruments of DTS based on Raman scattering and OTDR, DTS/DSS based on Brillouin scattering and OTDR, DTS/DSS based on Rayleigh scattering/FBG and OFDR, and DAS based on Rayleigh scattering and OTDR. In addition, some DTS/DSS instruments with loop configuration based on Brillouin scattering, such as BOTDA and Brillouin optical frequency domain analysis (BOFDA), have been available in the commercial marketplace. This section will review the working principles of typical distributed sensing techniques among these combinations and the development of distributed temperature and strain sensing techniques. Applications to SHM of composite materials will follow in Section 5.3.

5.2.1   Raman-Based Distributed Temperature Sensing

Raman scattering is an inelastic scattering mechanism in which a photon interacts with an optical phonon. When a monochromatic light beam propagates in an optical fiber, spontaneous Raman scattering occurs [52]. The Raman scattering process produces two components with different frequency shifts. The component with the lower frequency is called Raman Stokes (red-shifted component), whereas the other one with the higher frequency is called Raman anti-Stokes (blue-shifted component) (Figure 5.4). When a short light pulse is directed down to an optical fiber, there is a small amount of backscattered light in the optical fiber. This backscattering includes the components of Raman scattering, that is, Raman backscattering, and it is known that the intensity ratio of the two components of Raman backscattering (i.e., Raman Stokes and Raman anti-Stokes) is a function of temperature. The OTDR technique is able to resolve the two components from each sample point and then calculate their intensity ratio along the sensing fiber. This ratio is given by [53,54]

$$\frac{I_a}{I_s} = \left(\frac{v_a}{v_s}\right)^4 \exp\left(-\frac{h\Delta v}{kT}\right)$$

$$h v_s = h\left(v - \Delta v\right)$$

$$h v_a = h\left(v + \Delta v\right) \tag{5.7}$$

where:

I_s and I_a	are the intensities of Raman Stokes and Raman anti-Stokes, respectively
v, v_s, and v_a	are the frequencies of the incident light, Raman Stokes, and Raman anti-Stokes, respectively
Δv	is the frequency of the optical phonon
h	is Planck's constant
k	is Boltzmann's constant
T	is absolute temperature

Figure 5.11 shows the schematic diagram of a sensing instrument of ROTDR [55]. When a light pulse is entered from the laser source with a laser driver into the optical fiber, the detectors start to detect the intensity of Raman backscattering at the ADC through the optical filter, which is able to separate the Raman Stokes and Raman anti-Stokes components. As shown in Figure 5.12, the sensing instrument determines the temperature distribution along the sensing fiber by measuring the

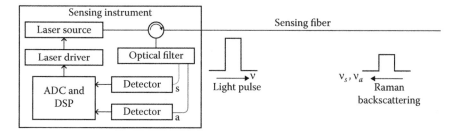

FIGURE 5.11 Schematic diagram of a sensing instrument of ROTDR.

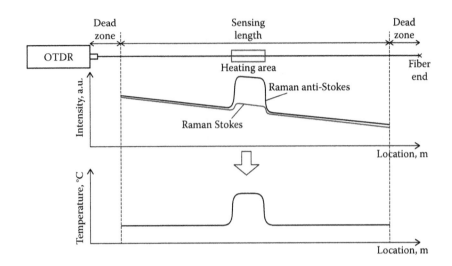

FIGURE 5.12   Distributed temperature sensing by ROTDR.

intensities of Raman Stokes and Raman anti-Stokes and calculating their ratio at each sample point by the DSP. The intensities of both Raman Stokes and Raman anti-Stokes increase with temperature. Raman anti-Stokes has higher temperature dependency than Raman Stokes, while the anti-Stokes scattering is weaker than the Stokes scattering process (Figure 5.4). The temperature dependency of Raman anti-Stokes intensity is reported to be 0.8%/°C at room temperature in an optical fiber [5]. The intensity of Stokes scattering rises to a constant value and that of anti-Stokes decays to zero at low temperature [52].

Raman-based DTS has the great advantage that it is sensitive only to temperature, and conventional MMF and SMF can be used. To maximize the light that is spontaneously scattered requires the use of MMF [56]. Typical off-the-shelf sensing instruments have a temperature resolution of 0.02°C at the sensing length of 10 km for the spatial resolution of 1 m. This measuring condition requires a measurement time of about 10 min, but compromising on the accuracy or the sensing length enables a second-order measurement time. The spatial resolution of ROTDR, as well as that of OTDR, is given by Equation 5.2. The m-order spatial resolution usually restricts the application of ROTDR to SHM of large-scale structures.

5.2.2   BRILLOUIN-BASED DISTRIBUTED TEMPERATURE/STRAIN SENSING

As well as Raman scattering, Brillouin scattering is also an inelastic scattering mechanism. It occurs due to Bragg-type scattering from propagating acoustic waves (acoustic phonons) [52]. The forward and backward acoustic waves produce two backscattering components with different frequency shifts, namely, Brillouin Stokes (red-shifted component) and Brillouin anti-Stokes (blue-shifted component) (Figure 5.4). So, Brillouin frequency shift (BFS) originates from the photon–acoustic phonon interaction, while Raman frequency shift is due to photon–optical phonon interaction. Brillouin scattering occurs only in a backward direction, whereas Raman scattering can occur both backward and forward. The BFS, v_B, is approximately given by

$$v_B = \frac{2n_{\text{eff}}V_a}{\lambda} \tag{5.8}$$

where:
  V_a is the acoustic velocity
  λ is the free-space wavelength

The shift is about 11 GHz at a wavelength of 1550 nm. It was reported that the BFS in silica fibers varies greatly with strain and temperature [17,18]. The temperature and strain coefficients of the v_B change are given by [57]

$$\frac{dv_B}{dT} = 1.00 \left[\text{MHz/}^{\circ}\text{C}\right], \quad \frac{dv_B}{d\varepsilon} = 4.93\times10^{-2}\left[\text{MHz/}\mu\varepsilon\right] \tag{5.9}$$

and we can use these dependences of the BFS on DTS and DSS. BOTDR measures local changes in the BFS along a sensing fiber by OTDR and spectral analysis.

Figure 5.13 shows the schematic diagram of a sensing instrument of BOTDR. A continuous coherent light wave with the frequency, v, is divided into probe and reference light waves. The probe light wave is modulated into a pulse with a frequency shift, v_m, that is approximately equal to the BFS, v_B. The spontaneous Brillouin backscattering in the sensing fiber is directed into a coherent heterodyne receiver, in which the reference light wave is used as an optical local oscillation. By changing the probe frequency shift, v_m, the electrical beat signal with the frequency $v_m - v_B$ is obtained as the output of the heterodyne receiver to reconstruct the Brillouin spectrum at each sample point along the sensing fiber [20]. The Brillouin spectrum can be described using the Lorentzian function with full width at half maximum (FWHM), Δv_B (Brillouin linewidth), center frequency, v_B (BFS), peak gain, and g_B (Brillouin scattering coefficient), as follows:

$$g(v_m) = \frac{g_B}{1+4(v_m - v_B)^2 / \Delta v_B^2} \tag{5.10}$$

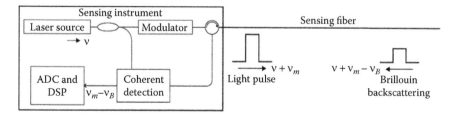

FIGURE 5.13   Schematic diagram of a sensing instrument of BOTDR.

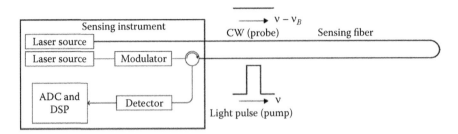

FIGURE 5.14 Schematic diagram of a sensing instrument of BOTDA.

Figure 5.14 shows the schematic diagram of a sensing instrument of BOTDA. In a conventional BOTDA, a continuous wave that is detected as a probe by a receiver and a pulse that acts as a pump to create stimulated Brillouin scattering (SBS) in an optical fiber are counterpropagated through the optical fiber with a frequency difference near the BFS. The Brillouin spectrum is measured at each sample point along the sensing fiber by scanning the frequency difference between the pump and probe waves as the probe wave is detected. BOTDA is able to generate strong scattering signals to reduce averaging times, that is, the measurement time. A loop configuration with one single laser source is available for BOTDA to use to its advantage.

As shown in Figure 5.15, the sensing instrument determines the longitudinal strain distribution along the sensing fiber by obtaining the Brillouin spectrum and then finding the BFS at each sample point. The spatial resolution of BOTDR, as well as that of OTDR, is given by Equation 5.2. Although the equation shows that pulse narrowing improves the spatial resolution, the shorter pulse is detrimental to the frequency resolution of the measurements, namely, for the temperature and strain

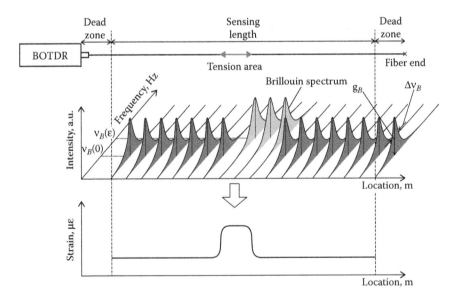

FIGURE 5.15 Distributed strain sensing by BOTDR.

measurement accuracies [58,59]. In this figure, attenuation effects along the sensing fiber are neglected.

The typical instrument of BOTDR has temperature and strain accuracies of about $\pm 1°C$ and ± 15 µε, respectively, for the spatial resolution of 1 m and the sensing length of 1 km. A sensing instrument based on the BOTDA technique provides higher spatial resolution and measurement accuracies. Some instruments have measurement accuracies of 0.5°C and 10 µε, approximately, for the spatial resolution of 0.2 m and the sensing length of 1 km. When a higher temperature/strain measurement range is required in addition to these measurement conditions, the measurement times of BOTDR and BOTDA can be more than several minutes and several tens of seconds, respectively. It is an advantage to normally use standard telecommunication-grade SMF and components in both BOTDR and BOTDA. Furthermore, the cross-sensitivity effect should be considered seriously for applications in which temperature varies with strain measurements, or vice versa.

5.2.3   RAYLEIGH-BASED AND FBG-BASED DISTRIBUTED TEMPERATURE/STRAIN SENSING

Rayleigh scattering is an elastic scattering mechanism. It is caused by refractive index variations, local defects, or distortions of waveguide geometry. It has the same frequency as that of incident light (Figure 5.4), and the scattered power is proportional to the power of the incident light.

Froggatt and Moore proposed Rayleigh-based DSS using the Rayleigh backscattering that occurs as a result of random fluctuations of the refractive index in the core of a standard SMF [60]. These random fluctuations along the core could be modeled as a Bragg grating with a random variation of amplitude and phase along the grating length, while FBG normally has a specific variation, that is, a periodic variation in the refractive index of the fiber core. The strains at six cascaded 30 cm sections in the 180 cm sensing fiber were determined by measuring the shift in the Rayleigh backscattering spectrum of each section. In order to measure the Rayleigh-backscattering spectrum, coherent OFDR was used [61]. The shift in frequency/wavelength was calculated by observing the maximum value of the cross correlation between the Rayleigh backscattering spectra of the different strain levels at the same section. Finally, strain at each section (the difference between the two strain levels) could be determined by using the shift measured and the strain sensitivity, which is almost the same as that of FBG [31]. Basically, this method detects relative changes from each of the reference fiber conditions.

Figure 5.16 shows the schematic diagram of a sensing instrument of Rayleigh-based DTS/DSS [32,43]. The sensing system consists of a TLS, two Mach–Zehnder interferometers, and a polarization-sensitive receiver. One interferometer (the measurement interferometer) includes a sensing fiber. Another interferometer (the auxiliary interferometer) generates trigger signals. The receiver has a polarization beam splitter (PBS), three detectors, and a high-speed ADC. The PBS separates the s and p polarization components of the light returning from the sensing fiber by reflecting the p-component, while allowing the s-component to pass. This propagation diversity technique is used to mitigate signal fading due to misalignment between

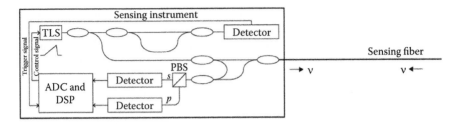

FIGURE 5.16   Schematic diagram of a sensing instrument of Rayleigh-based DTS/DSS.

the polarization states of the reference light and the probe light. When the laser frequency of the TLS is tuned, interference fringes that can be related to the reflectivity along the sensing fiber are observed at the detectors through the PBS, as the auxiliary interferometer generates the trigger signals that are used as an external clock of ADC to obtain each component signal from the detectors at a constant frequency interval. The laser tuning can be controlled by a signal from a data acquisition card (DAQ) in the ADC and DSP.

DTS is carried out by OFDR, as shown in Figure 5.17. The FFT amplitude distributions of the s- and p-components in the location domain are shown, as well as that of the vector sum of the s and p data set. A polarization oscillation of the light in the sensing fiber could cause the perturbation of the amplitudes of the

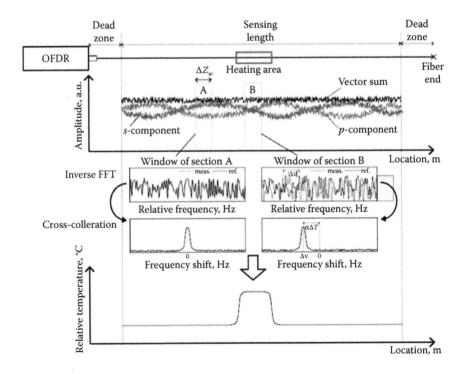

FIGURE 5.17   Distributed temperature sensing by Rayleigh-based DTS.

s- and *p*-components with the location, as shown in Figure 5.17. Then, the Rayleigh-backscattering spectrum of a section is calculated by windowing a section, Δz_w, in the location-domain data and performing an inverse FFT. Figure 5.17 includes the spectra of Section A, in which there is no difference between the temperature of the reference data and the measurement data, and Section B, in which temperature changes by ΔT, respectively. A cross correlation of the reference and measurement spectra yields a peak that is shifted from the center by the frequency shift induced by the temperature change. This frequency shift (i.e., Rayleigh backscattering spectral shift) is converted to temperature using calibration coefficients that are dependent on the laser source and the optical fiber used. This process is repeated for all sections while sliding the window at an interval along the sensing fiber. This interval becomes the sample spacing of the temperature trace obtained from the sensing instrument. The section length selected, Δz_w, will determine the effective spatial resolution of the distributed sensing. This length also has an effect on the resolution of the frequency shift calculation and, in turn, the temperature resolution. The longer Δz_w is, the higher the temperature resolution, but the lower the spatial resolution.

This technology fills gaps in sensitivity and spatial resolution between conventional FBG-based distributed sensing, which is well suited to distributed sensing over short lengths or at discrete points, and Raman/Brillouin scattering technology, which is ideally suited to sensing over very long lengths at a reduced spatial resolution [62]. The typical off-the-shelf instrument based on OFDR has temperature and strain accuracies of $\pm 0.2°C$ and ± 2 µε, respectively, for a spatial resolution of less than 1 cm, a measurement time of about 0.4 s, and a sensing length of 50 m. An advantage of this technique is that it does not require specialty fiber. Therefore, measurements can be made in preinstalled SMF or MMF.

Igawa et al. proposed FBG-based DSS that measures strain at an arbitrary position along low-reflectivity FBGs (weak FBGs) serially written in cascade along an optical fiber [63]. They demonstrated distributed strain sensing along a sensing fiber that had fifteen 100 mm long FBGs in cascade to make the sensing length of 1500 mm [64]. Recently, an optical fiber that contains densely spaced low reflective (<0.1%) gratings has become available in the marketplace and can be used as distributed fiber-optic sensors with OFDR [65,66]. Although the cross-sensitivity effect should be considered, as well as BOTDR, an FBG-based distributed sensing system is able to employ various methods developed for FBG sensors in order to discriminate temperature and strain [46,47].

5.2.4   Development of Distributed Temperature and Strain Sensing

Many efforts have been made to remove the significant limitations of DSS due to its cross-sensitivity. The most fascinating approach is to extend the ability of DSS to simultaneously measure temperature and strain while keeping the fundamental performance.

One approach to implementing simultaneous measurements is to use reference sensors along the sensing fiber, that is, sensors that are in thermal contact with the structure but do not respond to local strain changes. The simplest way to make a reference sensor is by putting a sensing fiber into a small tube (e.g., a polyimide tube) to make stress-free conditions for the sensing fiber and to use the same sensing technique as for DSS. This technique can provide a good compensation for point sensors (i.e., the temperature and strain uncertainties of 0.1°C and 2 με, respectively, for FBG sensors) [67] or short–sensing length applications of DSS. This approach, however, would make the installation process more complicated and formidable when the simultaneous measurement is required along a long sensing fiber.

Another approach for simultaneous measurements is to have two sensing mechanisms with different responses to strain ($K_{\varepsilon 1}$, $K_{\varepsilon 2}$) and temperature (K_{T1}, K_{T2}) at a sample point along one sensing fiber. Then, a matrix equation

$$\begin{pmatrix} \Delta X_1 \\ \Delta X_2 \end{pmatrix} = \begin{pmatrix} K_{\varepsilon 1} & K_{T1} \\ K_{\varepsilon 2} & K_{T2} \end{pmatrix} \begin{pmatrix} \Delta \varepsilon \\ \Delta T \end{pmatrix} \tag{5.11}$$

can be written and inverted to yield strain and temperature changes ($\Delta \varepsilon$, ΔT) from measurements of the changes in optical parameters (ΔX_1, ΔX_2).

When the change of the BFS (ΔX_1) and the Rayleigh backscattering spectral shift (ΔX_2) are measured by the prepumping BOTDA and OFDR techniques, respectively, temperature and strain accuracies of ±1.2°C and ±15 με are achieved with 92 m maximum sensing length and 0.5 m spatial resolution [68]. Kumagai et al. developed a prototype of a sensing instrument that measures the BFS and the dynamic grating spectrum (DGS) from the Brillouin dynamic grating (BDG) along a PMF by using the Brillouin optical correlation domain analysis (BOCDA) technique and calculates the temperature and strain based on the changes of BFS and DGS [69]. Temperature and strain accuracies of 1°C and 60 με for 500 m maximum sensing length and 0.3 m spatial resolution were shown in the specification of the prototype. It should be noted that the performance shown in the specification could not have been achieved in a field test.

In the case of using FBG-based OFDR technique, a number of strain and temperature discrimination methods developed for FBG sensors are basically exploited. Among them, a method using FBGs written in a PMF (PMF-FBGs) enables simultaneous measurements without major modifications and enhancements in the sensing instrument. A PMF-FBG reflects two Bragg wavelengths corresponding to two polarization modes, that is, fast and slow modes, due to the birefringence effect in the PMF. These two Bragg wavelengths differ from each other in their sensitivities toward temperature and strain. By this method, with the OFDR technique, accuracies of about ±5.9°C and ±85 με are obtained for the maximum sensing length of 25 m and the spatial resolution of 1.6 mm [47]. Froggatt et al. reported that the error in temperature and strain data was estimated to be 3.5°C and 35 με, respectively, for the spatial resolution of 2 cm over a potential sensing length of 70 m, by using Rayleigh backscattering spectral shift in the sensing fiber of PMF [70].

5.2.5 SUMMARY

Record highs in performance parameters of distributed fiber-optic sensing have been updated in research year after year. Also, various techniques to implement distributed fiber-optic sensing have been proposed in the last two decades. For instance, the spatial resolution of Brillouin-based DTS/DSS has been improved from 1 m to 1.6 mm [71], and several types of Brillouin-based DTS/DSS are now available in the commercial marketplace. Furthermore, it is interesting that Rayleigh-based techniques have recently experienced a resurgence, although the conventional Rayleigh-based OTDR had been applied to initial basic research on SHM in a limited way. This kind of technique has been intensively applied to SHM of composite materials. In addition, the coherent-OTDR technique with Rayleigh scattering is creating sensing platforms capable of continuous DAS along a fiber path of tens of kilometers. Applications of composite materials to large-scale structures in the fields of civil engineering and the oil and gas industries may further drive the development of SHM of composite materials using DAS.

This section is concluded by Table 5.1, which shows a performance chart of off-the-shelf sensing instruments for distributed sensing fiber-optic sensors. Here, no attempt at completeness is made, and the references to this table do not constitute endorsement, either expressed or implied, by the author. It should also be noted that it may not be fair to compare performance only by this table, since the performance tests are not consistent. For instance, the measurement time and repeatability would depend on the temperature or strain measurement range and standard deviation used (i.e., σ, 2σ, or 3σ), respectively. However, it can be seen that the variety of technologies complements the applicability of distributed fiber-optic sensing for SHM.

5.3 SHM OF COMPOSITE MATERIALS USING DISTRIBUTED FIBER-OPTIC SENSORS

At the beginning of the development of distributed fiber-optic sensors that could measure temperature or strain, they were applied mostly to large-scale structures for monitoring of the overall condition in a structure [78,79]. The recent performance increase, especially in terms of the spatial resolution, has extended the application range of DTS and DSS techniques to not only monitoring of the overall condition but also detection of a local change in structural components.

Distributed fiber-optic sensors with high spatial resolution are actually capable of detecting damage or manufacturing defects in various kinds of composite materials and structures, for example, FRP, sandwich structures, 3D-woven composites, and jointed composite structures. They are also useful for revealing complicated stress states or unpredictable behaviors due to inhomogeneity and anisotropy of composite materials. Especially when a sensing fiber is embedded into the material, we can really feel that the optical fiber is working as a nerve and letting us know the state inside the material that is not visible externally.

In this section, the early challenges in applying DTS and DSS to composite structures will be described. These are still good examples to show the applicability of distributed fiber-optic sensors to large-scale structures. Additionally, recent advances in temperature and strain measurements using the high-spatial-resolution sensors, and their application to composites, as well as some practical issues, will be discussed.

TABLE 5.1

Performance Chart of the Off-the-Shelf Sensing Instruments for Distributed Fiber-Optic Sensing

Configuration	Single-End					Loop	
Measurand	DTS	DTS/DSS		DAS		DTS/DSS	
Mechanism	Raman	Brillouin	Rayleigh (FBG)	Rayleigh		Brillouin	
Method	OTDR	OTDR	OFDR	OTDR	OTDA	OTDA w/Prepumping	OFDA
Spatial resolution	1 m	1 m	<0.01 m	1–10 m	1 m	0.2 m	0.2 m
Measurement time	10 s	3 min	0.4 s	10 μs	10 min	25 s	20 s
Sensing length	1 km	1 km	50 m	1 km	2 km	1 km	2 km
Temperature repeatability	0.5°C	±0.8°C	±0.2°C	N.A.	0.2°C	0.5°C	0.1°C
Strain repeatability	N.A.	±16 με	±2 με	N.A.	4 με	10 με	2 με
References	[72]	[73]	[74]	[35]	[75]	[76]	[77]

5.3.1   APPLICATIONS OF DISTRIBUTED TEMPERATURE SENSING

Temperature distributions in composite materials and the structural components fabricated from them may give us critical information relating to their reliability and safety. First, this is because temperature is the most important of the environmental factors affecting the behavior of composite materials, especially polymeric composites, which are more sensitive to temperature than metals [80]. For instance, thermal cycling, which is common for aircraft and spacecraft, can induce microcracking in some composites [81]. The interfacial shear strength between fiber and matrix also decreases with increasing temperature, which affects the mechanical performance of composite materials [82]. Second, the fabrication of composite structures is usually accompanied by more or less intensive heating (e.g., curing), and further cooling induces thermal stresses and strains. This means that monitoring of temperature distributions in not only operational conditions but also manufacturing processes would be beneficial to SHM concepts. And third, existing fiber-optic sensor networks mounted on a composite structure may be applied to nondestructive testing (NDT) methods, such as thermography [83]. The basic research reported on these three topics is discussed in Sections 5.3.1.1–5.3.1.3.

5.3.1.1   Temperature Monitoring of Composite Structures

Figure 5.18 shows a rectangular composite panel made of carbon fiber-reinforced plastic (CFRP), which is equipped with an integrated fiber-optic sensor network. The sensor network is comprised of three sensing fibers of SMF (SMF1, SMF2, and SMF3) and one sensing fiber of MMF (MMF1). MMF1 was bonded on the panel, arranged like rib bone, and connected to a sensing instrument of Raman-based DTS. After all sensing fibers were bonded, a glass fiber–reinforced plastic (GFRP) sheet was overlaid on the CFRP panel to improve the strength and durability of

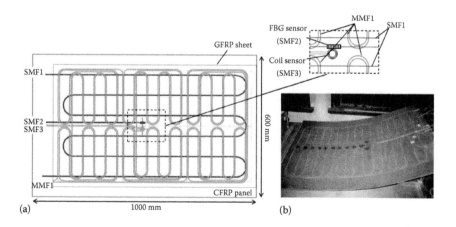

FIGURE 5.18   CFRP panel with fiber-optic sensor network: (a) temperature distribution along MMF1 and (b) 2D-data mapping.

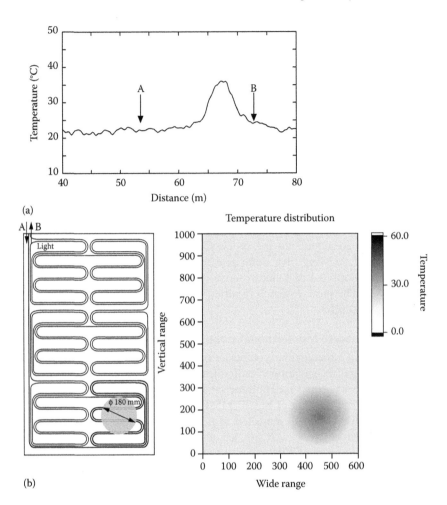

FIGURE 5.19   (a) Temperature distribution monitoring of sensing fiber MMF1. (b) Mapping of measurement data on panel model [84].

the sensing fibers. The instrument based on ROTDR could measure temperature along MMF1 at a sample spacing of 0.2 m for a spatial resolution of 2 m. Because of the low spatial resolution, the panel was divided into six zones, and the layout of MMF1 was determined to put a portion of the sensing fiber with length of more than 2 m into each zone. A circular area whose diameter was 180 mm was heated by an incandescent lamp. Figure 5.19a shows the temperature distribution of MMF1, and it can be seen where the section of MMF1 approximately corresponding to the heated zone was warmed. The measurement data were mapped on the panel model, as shown in Figure 5.19b, and the hot spot can be found at a glance. This kind of visualization tool is necessary unless the monitored structure has a linear geometry.

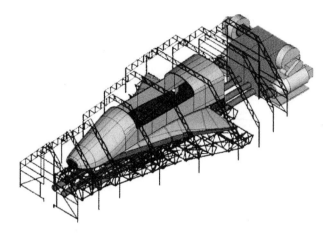

FIGURE 5.20 The fuselage in an oven during the postcure process. (From Murayama, H., et al., *Journal of Intelligent Material Systems and Structures*, 14(1), 3–13, 2003.)

5.3.1.2 Temperature Monitoring of Manufacturing Processes

Most components of the full-scale model of HOPE-X, shown in Figure 5.10, were made of sandwich CFRP structures and cured at 100°C [34]. These parts were then joined by adhesive bonding to the fuselage and cured again at 180°C in a freestanding state. Figure 5.20 shows the freestanding state of the assembled fuselage in the oven without walls. An MMF coated with Teflon® was surface mounted on the outer skin of the fuselage to monitor temperature distributions and confirm whether the temperature in the oven was being controlled adequately during the cure process. At the same time and location, a conventional single mode with ultraviolet-curable acrylate coating was also bonded to measure strain distribution and find unexpected deformation and damage. The locations of the sensing fibers and the thermocouples on the fuselage are shown in Figure 5.21. As shown in Figure 5.22a, the temperature values obtained from the sensing instrument with ROTDR agree with those of the thermocouples, and the distributions are almost uniform. In Figure 5.22b, the strain distributions measured using BOTDR are also shown, and we can see the residual strain is about −200 με at 42°C after the cure.

5.3.1.3 Application to NDT Methods

Wu et al. developed a technique for thermographic detection of delamination in composites by performing temperature measurements using the OFDR technique [84]. The investigated structure, whose dimensions were 31.75 × 31.75 × 0.19 cm, was a 10-ply quasi-isotropic composite panel with fabricated square delaminations of various sizes and depths. Temperatures were monitored at 32 FBGs with a separation distance of 5.08 cm, and each FBG, except several used for reference, was placed at the center of each square delamination. By application of a cyclic thermal heat flux to the surface of the investigated structure, the temporal temperature variations were measured by each FBG, and the depth of the delamination under the

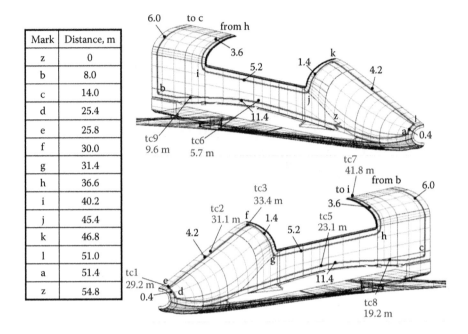

Mark	Distance, m
z	0
b	8.0
c	14.0
d	25.4
e	25.8
f	30.0
g	31.4
h	36.6
i	40.2
j	45.4
k	46.8
l	51.0
a	51.4
z	54.8

FIGURE 5.21   Locations of the sensing fiber (red lines) and the thermocouples (tc1–tc9) on the fuselage.

FBG was identified by comparing the model predictions. These could provide better estimations for the shallow delaminations.

The spatial resolution of DTS used for these first two applications was 2 m. This was why the sensing fiber ran several times along a line to make a complicated layout on the panel, as shown in Figure 5.15. Therefore, it might be difficult to come up with the idea that the conventional DTS technique generates much denser data than a conventional one using thermocouples for the cure process monitoring of each part, although it is able to provide an effective way to monitor the overall temperature state in a large-scale structure. The recent high-spatial-resolution sensing techniques, mentioned in Sections 5.1.2 and 5.3.2.1, could increase the information density as well as making the sensor layout simpler and providing more flexible applicability of temperature monitoring for SHM. Additionally, although the quasi-distributed sensing technique was used in the last application, it is no wonder that the fully distributed sensing technique can provide more informative data for such active sensing and NDT approaches, if not only the spatial resolution but also the performance parameters are adequate for them.

5.3.2   Applications of Distributed Strain Sensing

Strain components are related to stress components through constitutive relationships, such as Hooke's law. They can also be variables in failure criteria and can be used in failure analysis, as well as stresses. A strain distribution directly relates

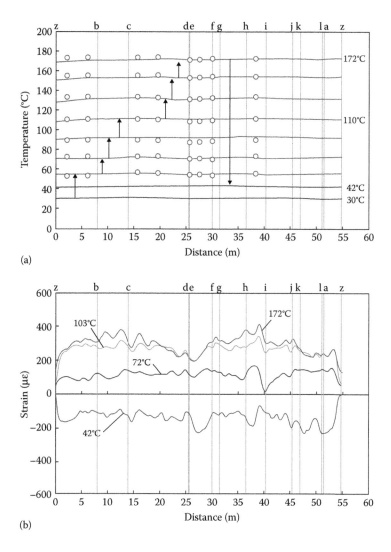

FIGURE 5.22   (a) Temperature distributions of the fuselage during the postcure process. (b) Strain distributions of the fuselage during the postcure process.

to deformation in a structure. Then, the magnitude, mode, and time variation of deformation relate to the stiffness of the structure. If changes in deformation are observed in the same load and boundary conditions, they are first suspected to be due to stiffness degradation caused by defects. Furthermore, a local defect, such as a crack, delamination, or bolt loosening, may cause a steep or noticeable strain change in measured strain distributions, whereas it can sometimes have little measurable influence on deformation modes, that is, on the global stiffness. In a relatively simple way, loads applied to the structure can be derived from the strain level when the stiffness and the load distribution are known. Otherwise, they can be estimated based on strain distributions by using a computational method. When the

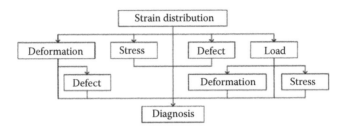

FIGURE 5.23   Diagnosis procedures based on strain distribution.

loads are determined, the overall stress and deformation may be reconstructed by using structural models. In these various ways, strain distributions can be employed for diagnosis in SHM either directly or through some processing. Figure 5.23 shows a simple flowchart that starts from strain distribution measurements and then proceeds to diagnosis through some processes. Fortunately, strain is one of the primary measurands for distributed fiber-optic sensing. Distributed strain sensing techniques applied to mainly composite materials and structures will now be discussed.

5.3.2.1   Strain Monitoring of Composite Materials and Structures

The Japanese IACC yacht in practical use for America's Cup 2000 was the first full-scale composite structure equipped with distributed fiber-optic sensors for SHM. The simple diagnosis approach to detect deterioration of the stiffness and unusual deformation is shown in Figure 5.24 [34]. The yacht was subjected to large static forces by the stays and shrouds, and the longitudinal bending moment became maximal at the position of the mast. In periodic inspections, those forces were simulated in the yacht on the stand, and the latest strain distributions were compared with the previous ones measured under the same load and boundary conditions to detect changes over time in the strain distributions. The arrangement of sensors was determined such that the changes of the longitudinal stiffness and the stress state under the mast could be accessed. As shown in Figure 5.25, the sensing fiber (a–h) was fixed onto the inside of the hull with CFRP sheets, and it traveled back and forth in the longitudinal direction. Two lines (a–b and g–h) were located above the neutral axis on which the stresses were zero in longitudinal bending, while another two lines (c–d and e–f) were located below the axis. Figure 5.26a shows the strain distributions measured by using the BOTDR technique and calculated by finite element analysis (FEA) along the longitudinal sensing fiber. At that time, the spatial resolution and the accuracy for the strain measurement by BOTDR were 1.0 m and $\pm 30\ \mu\varepsilon$, respectively. It seems that the actual strain distributions of the applied structure were not overly complicated, so that good agreement between the measured results and the calculated ones could be obtained. The visualization application mapping the estimated strain data onto the hull was developed so that the longitudinal stiffness of the yacht could be assessed as compared with the FEA result (Figure 5.26b). Continuous monitoring for eight months and damage simulations resulted in confirming the necessity for a temperature compensation method that could reduce the error in a whole length of the sensing fiber, and the higher spatial resolution that would enable detection of possible and critical damage, respectively, for more precise and reliable diagnosis.

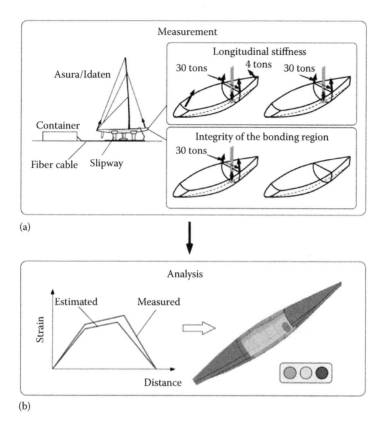

(a)

(b)

FIGURE 5.24   Outline of SHM for the IACC yacht showing (a) the measurement condition and (b) the simple diagnosis approach. (From Murayama, H., et al., *Journal of Intelligent Material Systems and Structures*, 14(1), 3–13, 2003.)

Several interesting reports about strain monitoring of full-scale composite structures are also seen in the literature [85,86].

Generally, the principal strains/stresses and directions are required to apply failure criteria for laminated composite materials to diagnose in SHM. Basically, fiber-optic sensors measure longitudinal strain along the fiber. This means that a surface-mounted or embedded fiber-optic sensor measures a normal strain component in a specific direction. So, at least three strains with different directions should be measured to find the principal strains/stresses and directions, just like using a typical rosette strain gauge. When the high spatial resolution allows us to approximate the curved parts in a sensing fiber by discrete straight lines, design flexibility in the geometry of rosette gauges using distributed fiber-optic sensors is given, as shown in Figure 5.27 [87]. Figure 5.28a shows a steel cruciform specimen with a surface-mounted fiber-optic sensor that was comprised of 100 mm long FBG and formed a circle-shaped rosette, shown in Figure 5.27, at the center of the specimen. When strain distributions are measured along the circle-shaped rosette by using a high-spatial-resolution sensing technique and a specific strain state is assumed in the

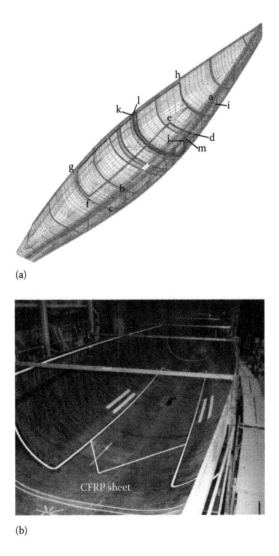

(a)

(b)

FIGURE 5.25 The sensing fibers installed to the yacht structure: (a) location of the sensing fibers and (b) longitudinal sensing fiber covered with CFRP sheets. (From Murayama, H., et al., *Photonic Sensors*, 3(4), 355–376, 2013; Murayama, H., et al., *Journal of Intelligent Material Systems and Structures*, 14(1), 3–13, 2003.)

area including the rosette, the strain follows a sinusoidal pattern along the sensing fiber, that is, the angle. Figure 5.28b shows the strain distribution of the sensing fiber in the rosette measured by the OFDR technique when the specimen was subjected to a biaxial tensile load. The principal strains and directions can be determined by curve fitting. In this method, exact mapping is essential for accurate measurements. Mapping procedures can be much more complicated and difficult, especially when sensing fibers are embedded into composite materials in a winding, three-dimensional way [88].

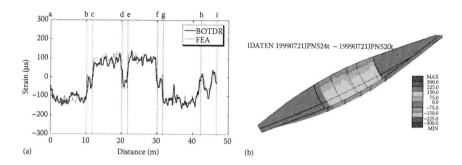

FIGURE 5.26 Strain monitoring in the IACC yacht: (a) strain distributions along the longitudinal sensing fiber and (b) 3D-data mapping. (From Murayama, H., et al., *Photonic Sensors*, 3(4), 355–376, 2013; Murayama, H., et al., *Journal of Intelligent Material Systems and Structures*, 14(1), 3–13, 2003.)

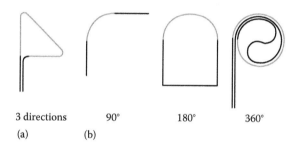

FIGURE 5.27 Rosette strain gauges using distributed fiber-optic sensors (red lines): (a) specimen with FBG-rosette gauge and (b) strain distributions along FBG-rosette gauge. (From Kobayashi, S., Application of optical fiber strain sensors with high spatial resolution, Graduate thesis, The University of Tokyo, Tokyo, 2006.)

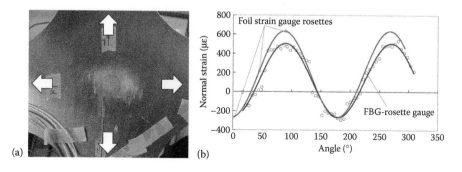

FIGURE 5.28 Cruciform specimen (a) with FBG-rosette gauge and (b) measurement data under a biaxial tensile load.

5.3.2.2 Strain Monitoring in Steep Strain Distributions

In general, the location where a detected defect occurs is not previously known, and the noticeable influence caused by the defect on strain distributions is limited to a relatively small area because of Saint-Venent's principle. So, sensors are required to enable high-density and wide-area monitoring in structural components. The layout of distributed fiber-optic sensors can be designed to provide a wider observation area (e.g., Figures 5.9 and 5.18), while the spatial resolution has the most significant effect on the information density in measurement data.

The effects of spatial resolution on detection of damage and stress concentration zones are described in reference to the research on SHM. Figure 5.29a shows a sandwich beam that had 300 mm debonding between the CFRP skin and the aluminum honeycomb core and sensing fibers bonded on both skin surfaces in the longitudinal direction [20]. When the sandwich beam was subjected to the load in four-point bending tests with the support span of 2000 mm, the debonding area of the upper skin buckled, as shown in Figure 5.29b. Figure 5.30a shows the strain distributions along the sensing fibers calculated by FEA and the strains measured by strain gauges at the center. The calculated result shows steep strain gradients in the debonding area, whereas in the strain distributions measured by using BOTDR with 1 m spatial resolution, in Figure 5.30b, there are no such gradients, and also obvious differences do not exist between the distributions of the two surfaces. Consequently, damage in the sandwich beam could not be detected based on the measured strain

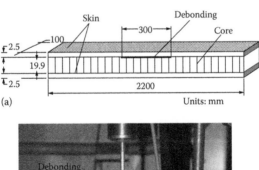

(a)

(b)

FIGURE 5.29 (a) Schematic of a CFRP sandwich beam with debonding. (b) Photo of the buckling of the upper skin in the four-point bending test. (From Murayama, H., et al., *Journal of Intelligent Material Systems and Structures*, 15(1), 17–25, 2004.)

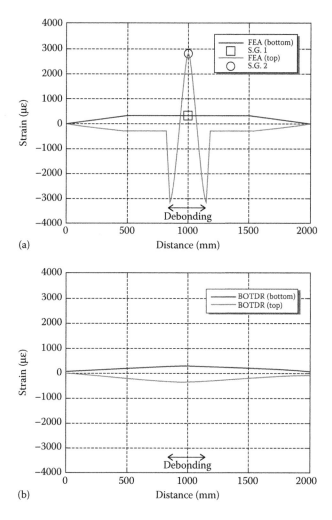

FIGURE 5.30 Strain distributions of the lower and upper skins (a) calculated by FEA and measured by strain gauges and (b) measured by BOTDR.

distributions because of the insufficient spatial resolution. Effectively, in the distributed sensing, the measurement information was presented as a series of values, each value representing the magnitude of measurand averaged over a section of fiber of length of the spatial resolution [3]. It was actually confirmed by the measurement simulation that the measured strain distributions would be almost equal to the moving average of the FEA data, which has the calculation period corresponding to the spatial resolution, and the strain distributions of the intact sandwich beam would be almost equal to those of the damaged beam in all locations except for the debonding area [89].

The effects of fiber coating on the measurements of steep strain distributions will now be discussed. Microcracking can propagate to the interface between the fiber and its coating, or between the host composite materials, if the coating stiffness is much less than that of the fiber and the host or if the adhesion between the fiber and the various constituents is poor. Actually, ultraviolet-curable coatings are particularly susceptible to interfacial microcracking [90]. In addition, a local deformation of coating materials, due to high temperature or shrinkage effects occurring in the cooling-down phase, causes interfacial delamination [91]. Such interfacial defects can cause strain transfer problems in distributed strain measurements, which apparently provide similar results to the spatial resolution problem, shown in Figure 5.30. Figure 5.31a shows an aluminum tensile specimen with holes. It has a 50 mm long FBG with acrylic coating bonded to a surface and 10 mm long FBG with polyimide coating bonded to the opposite surface [92]. These FBGs were aligned parallel to the tensile loads, which were located adjacent to the holes, and the strain distributions along the FBGs were measured by the OFDR technique, that is, the FBG-based DSS. As shown in Figure 5.31b, steep strain variations can be seen in the strain distribution measured by the polyimide-coated FBG around the hole with 4 mm diameter, and these fit the variations in the distributions obtained from the measurement simulation and the FEA. The measurement simulation did not take into account any imperfection in the coating, and it showed a spatial resolution of less than 1 mm

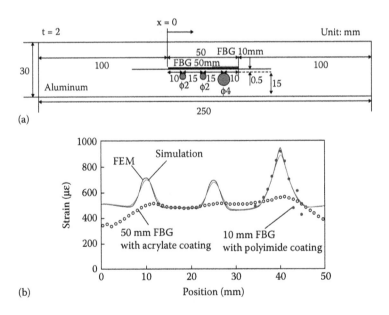

FIGURE 5.31 Strain transfer problem caused by an inappropriate fiber coating. (a) Tensile specimen with holes and FBG sensors. (b) Strain distributions adjacent to the holes. (From Murayama, H., et al., *Proceedings of Asia-Pacific Workshop on Structural Health Monitoring*, 304–310, 2006.)

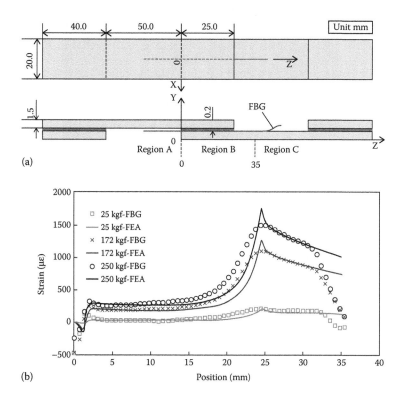

FIGURE 5.32 Strain distribution monitoring of a single-lap joint with the embedded FBG sensor. (a) Schematic of single-lap joint specimen. (b) Strain distributions of the sensing fiber. (From Ning, X., et al., *Smart Materials and Structures*, 23(10), 105011, 2014.)

[6]. On the other hand, the acrylate-coated FBG could not detect such variations, but only the direct-current (DC) component in the tensile strain distribution. So, it is believed that this deviation was caused by an interfacial imperfection between the fiber and the acrylate coating.

Indeed, when sensors have the appropriate coating and sufficient spatial resolution, the applicability of distributed fiber-optic sensors to damage detection based on measurement data can be improved considerably. Figure 5.32a shows a schematic of a single-lap joint specimen [93]. Two CFRP substrates were bonded by epoxy resin to form the joint, and one of the substrates had a long-length FBG embedded onto its surface. Figure 5.32b shows the strain distributions along the FBG measured by the OFDR technique and calculated by FEA at three loads under sinusoidal cyclic loading with a frequency of 0.5 Hz and a load ratio of 0.1. In this study, the coating material was polyimide resin, and the measurements were carried out continuously at a spatial resolution of millimeter-order and a measurement time of 0.2 s. The measured data seem to reasonably follow the actual steep strain gradients along the sensing fiber. During the test, cracks initiated from both the overlap ends and propagated to the center according to the cycle number, as shown in Figure 5.33a. The accurate strain distributions continuously measured by the high-spatial-resolution

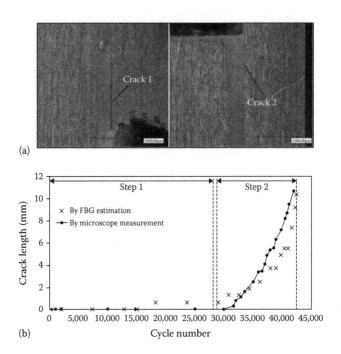

(a)

(b)

FIGURE 5.33 Crack propagation monitoring based on measured strain distributions: (a) cracks observed in the cyclic test and (b) estimated and observed crack lengths. (From Ning, X., et al., *Smart Materials and Structures*, 23(10), 105011, 2014.)

sensor enabled a good estimation of the crack-tip location in real time, as shown in Figure 5.33b.

5.3.2.3 Deformation/Load Identification Based on Measurement Data

Deformation shape prediction plays an important role in the SHM and control of aerospace structures [45]. Various studies have been reported regarding the estimation of displacement starting from strain measurement, and some approaches have been validated by using fiber-optic sensors. Nishio et al. applied BOTDA with a prepumping technique to the strain measurements of a laminated composite plate in a cantilever plate-bending test [94]. The authors embedded an optical fiber network into a panel that was 0.90 m long and 0.45 m wide to form 39 lines along which strain distributions would be measured, as shown in Figure 5.34a. The shape under a static bending load was identified by a shape reconstruction algorithm using a finite element (FE) model and measurement data, as shown in Figure 5.34b.

The NASA Dryden Flight Research Center has developed a shape-sensing methodology based on fiber-optic strain sensing [95]. By using quasi-distributed fiber-optic sensing techniques, they performed validation tests in which a swept plate made of an aluminum alloy was bolted into a cantilever fixture and subjected to static bending loads from 12 points located along both side edges [96]. The plate was approximately 1.30 m in span and 0.31 m in width. Three surface-mounted sensing fibers (with 100 FBG sensors each) were located at the top and bottom surfaces, and strain

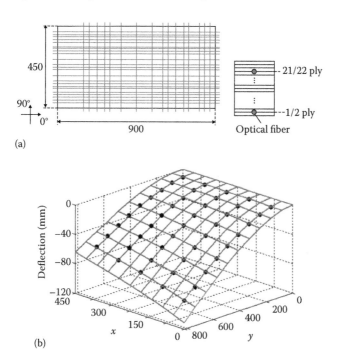

(a)

(b)

FIGURE 5.34   Composite laminate panel with (a) a fiber-optic sensor network and (b) reconstructed deflection. (From Nihio, M., et al., *Smart Materials and Structures*, 19(3), 035011, 2010.)

measurements were carried out to identify the deformations at various load cases. Comparing data at the maximum wing tip deflection, the root-mean-square (RMS) error was 2.28% and 0.106 inches. Rapp et al. reconstructed the dynamic displacements of an acrylic cantilever plate that was 0.5 m in span and 0.3 m in width and had eight surface-mounted FBGs on the topside and eight FBGs on the bottom side [97]. Strain measurements were carried out by an interrogating unit with 200 Hz sampling frequency and four channels as applying the sinusoidal excitation to the plate by a modal shaker. Comparing the displacement values of the reference measurement system with the estimated displacements at the corresponding locations, the estimated RMS error was 0.44% at 7.14 Hz excitation and 2.20% at 31.08 Hz excitation.

Although there are a few reports about deformation shape reconstruction in dynamic responses by using distributed fiber-optic sensors [98,99], Wada et al. demonstrated dynamic deformation monitoring of a GFRP cantilever beam that was 0.4 m in span and 0.02 m in width [8]. The sensing fiber, which was co-cured, was comprised of five cascaded 100 mm long FBGs, and strain measurements were carried out at a measurement time of about 1.25 ms to identify the bending deformation at resonant vibrations. The measured strain distributions and calculated deformations of the second vibrational mode for 0.1 s are shown in Figure 5.35, and the vibration frequencies obtained from the fiber-optic sensor and the laser displacement sensor were 62.9 and 62.6 Hz, respectively.

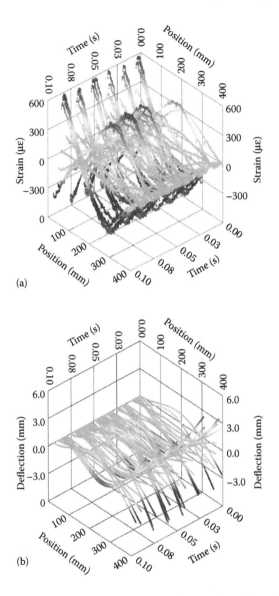

FIGURE 5.35 Dynamic deformation monitoring of a GFRP cantilever beam in the second vibrational mode: (a) time history of strain distribution and (b) time history of deflection. (From Wada, D., et al., *Proceedings of 20th International Conference on Composite Materials*, IEEE, Denmark, pp. 4120–4123, 2015.)

Knowledge of the operational loading of a structure is also valuable for a variety of aerospace applications [96]. Transfer functions constructed in advance by using analytical models and measurement data enable real-time identifications of loads distributed over a structure [100]. Figure 5.36 shows a composite wing box that was 6 m long and made of CFRP and the test equipment for limit load (LL) tests [86].

(a)

(b)

FIGURE 5.36 (a) A 6 m composite wing structure. (b) Test equipment for structural tests. (From Murayama, H., et al., Strain monitoring of 6 m composite wing structure by fiber-optic distributed sensing system with FBGs, in *Proceedings of 15th European Conference on Composite Materials*, Venice, Italy, 2012.)

In total, 246 10 mm long FBGs, eight FBG sensors with 300 mm sensing length, and six FBG sensors with 500 mm sensing length were bonded on either the inner or the outer skins of the wing box. The 10 mm FBG sensor arrays and the FBG sensors with the long sensing lengths were used to monitor the overall deformation of the wing box and measure strain distributions around stress concentration zones by using OFDR techniques. The composite wing box was fixed at the root, and the load was applied to 16 points through the flames by four actuators. Figure 5.37a shows the strain distributions measured by a sensor array with 42 FBGs (solid circles) and calculated by FEA (solid lines) at the six load cases. Both results show the stiffness changes due to discrete thickness at around 2 and 4 m. Measurement data obtained from the seven FBG arrays (each array includes about 30–40 FBGs) were used to identify the 16-point loads. In Figure 5.37b, we can see that the estimated loads at the points near the root depart from the actual loads. In this study, it is believed that insufficient sensors, especially the absence of sensors near the root, resulted in the errors. Loads identified based on strain information could also be employed for structural analysis to determine deformation or stress.

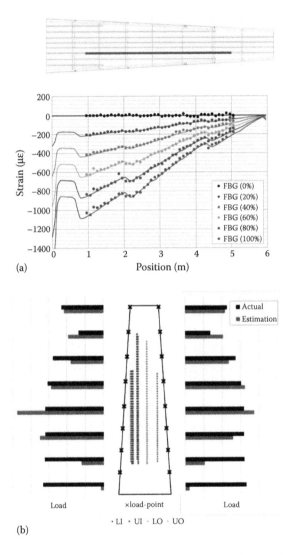

FIGURE 5.37 Load identification based on measurement data: (a) strain distributions along a sensing fiber and (b) results of load estimation (80% of LL). (From Murayama, H., et al., Strain monitoring of 6-m composite wing structure by fiber-optic distributed sensing system with FBGs, in *Proceedings of 15th European Conference on Composite Materials*, Venice, Italy 2012.)

5.4   CONCLUDING REMARKS AND FUTURE PROSPECTS

It should come as no surprise that sensing technologies to detect one measurand or more at an arbitrary position along a hair-thin fiber continue to fascinate not only those who develop sensors but also those who use sensors. State-of-the-art distributed fiber-optic sensors, indeed, are based on such sensing technologies. The research on distributed fiber-optic sensing has been one of the most active topics in international conferences on fiber-optic sensors for many years. SHM of large-scale structures in civil engineering and well/reservoir monitoring in the oil and gas industry have created one of the biggest demands for the development of distributed fiber-optic sensors. The improved spatial resolution, which can be comparable to the typical gauge length of conventional point sensors, is additionally driving a demand for the distributed fiber-optic sensors used in SHM of composite materials. Finally, in summarizing the current situation, some future prospects are given:

- At present, temperature and strain are fundamental measurands for distributed fiber-optic sensors. For the monitoring of manufacturing process or load/environment, pressure can be another primary measurand. Passive sensing of acoustic emissions and active sensing using ultrasonic waves (i.e., sensing slight and high-frequency strain variations) are useful techniques, but applying the distributed fiber-optic sensors to them would be a big challenge, especially in terms of sensitivity, spatial resolution, and measurement time.
- The improvement of spatial resolution has been critically important in extending the applicability of distributed fiber-optic sensors so the measurement data can be comparable to analytical results based on multiscale models of composite materials (i.e., macro, meso, and, hopefully, microscales). Thinner fibers may be preferable to get into the smaller world.
- If the measurement time is improved in addition to the spatial resolution, a vast number of existing vibration-based damage identification methods with multiple-point sensors can be used directly or could be further extended to enable more precise diagnosis in a wider area.
- Optimization of the sensor network topology can mitigate the limitation derived from the shorter sensing length. Effective and reliable installation methods to deploy surface-mounted or embedded sensors are essential to exploit the longer sensing length (i.e., wider sensing area).
- When a distributed fiber-optic sensor for strain measurements is used in a field environment, the measurement accuracy can be greatly affected by the cross-sensitivity between strain and temperature. A realistic and typical way is still to use two sensors, that is, one fixed on the structure and another in a stress-free condition. It must be one of the most important challenges to develop a reasonable method for a single fiber to achieve the same accuracy as the typical method using two fibers.
- It can be a great advantage to use a bare (standard) fiber in terms of cost. However, the returns of the optimized fibers for distributed fiber-optic sensing may consequently be large in terms of not only performance but also cost.
- The completeness of the sensing fiber to avoid significant optical losses and harm to the host structure is a critical issue in practical use, as well as reliability and durability of the sensing instrument for steady measurements. Therefore, both the consideration of real structure in system design and more field-testing of the system need to be addressed.

REFERENCES

1. Culshaw B., Optical fiber sensor technologies: Opportunities and-perhaps-pitfalls, *Journal of Lightwave Technology*, 22(1), 39–50, 2004.
2. Kersey, A.D., Distributed and multiplexed fiber optic sensors, in E. Udd, W.B. Spillman (eds.), *Fiber Optic Sensors: An Introduction for Engineers and Scientists*, 2nd edn, New York: Wiley, pp. 277–314, 2011.
3. Rogers, A., Distributed optical-fibre sensing, *Measurement Science and Technology*, 10(8), R75–R99, 1999.
4. Perez-Herrera, R.A., Lopez-Amo, M., Fiber optic sensor networks, *Optical Fiber Technology*, 19(6), 689–699, 2013.
5. Hartog, A.H., Leach, A.P., Gold, M.P., Distributed temperature sensing in solid-core fiber, *Electronics Letters*, 21(23), 1061–1062, 1985.
6. Murayama, H., Wada, D., Igawa, H., Structural health monitoring by using fiber-optic distributed strain sensors with high spatial resolution, *Photonic Sensors*, 3(4), 355–376, 2013.
7. Bao, X., Chen, L., Recent progress in distributed fiber optic sensors, *Sensors*, 12, 8601–8639, 2012.
8. Wada, D., Murayama, H., Tamaoki, T., Ogawa, D., Kageyama, K., Dynamic deformation monitoring of GFRP beam using optical fiber distributed sensing system based on optical frequency domain reflectometry, in *Proceedings of 20th International Conference on Composite Materials*, Copenhagen, Denmark, pp. 4120–4123, 2015.
9. Fan, W., Qiao, P., Vibration-based damage identification methods: A review and comparative study, *Structural Health Monitoring*, 10(1), 83–111, 2011.
10. Murayama, H., Structural health monitoring based on strain distributions measured by fiber-optic distributed sensors, in *Proceedings of the 20th Opto-Electronics and Communications Conference*, Shanghai, China, pp. 1–2, 2015.
11. Wang, J., Distributed pressure and temperature sensing based on stimulated Brillouin scattering, Master's thesis, Virginia Polytechnic Institute and State University, Blacksburg, VA, 2013.
12. Kersey, A.D., Fiber optic sensors shine bright: Industrial applications where FOS bring differentiated performance/value, *Proceedings of SPIE*, 8421, 84210F, 2012.
13. Srinivasan, B., Venkitesh, D., Distributed fiber-optic sensors and their applications, in G. Rajan (ed.), *Optical Fiber Sensors: Advanced Techniques and Applications*, Tokyo: CRC Press, pp. 309–358, 2015.
14. Horiguchi, T., Rogers, A., Michie, W.C., Stewart, G., Culshaw, B., Distributed sensors: Recent developments, in J. Dakin, B. Culshaw (eds.), *Optical Fiber Sensors: Applications, Analysis, and Future Trends, Volume Four*, Norwood, MA: Artech House, pp. 309–368, 1997.
15. Fang, Z., Chin, K.K, Qu, R., Cai, H., *Fundamentals of Optical Fiber Sensors*, New York: Wiley, pp. 278–350, 2012.
16. Dakin, J.P, Pratt, D.J., Bibby, G.W., Ross, J.N., Distributed optical fibre Raman temperature sensor using a semiconductor light source and detector, *Electronics Letters*, 21(13), 569–570, 1985.
17. Culvershouse, D. Farahi, F., Pannell, C.N., Jackson, D.A., Potential of stimulated Brillouin scattering as sensing mechanism for distributed temperature sensor, *Electronics Letters*, 25(14), 913–915, 1989.
18. Horiguchi, T., Kurashima, T., Tateda, M., Tensile strain dependence of Brillouin frequency shift in silica optical fibers, *Photonics Technology Letters*, 1(5), 107–108, 1989.
19. Barnoski, M.K., Jensen, S.M., Fiber waveguides: A novel technique for investigating attenuation characteristics, *Applied Optics*, 15(9), 2112–2115, 1976.

20. Murayama, H., Kageyama, K., Naruse, H., Shimada, A., Distributed strain sensing from damaged composite materials based on shape variation of the Brillouin spectrum, *Journal of Intelligent Material Systems and Structures*, 15(1), 17–25, 2004.

21. Bao, X., Chen, L., Recent progress in Brillouin scattering based fiber sensors, *Sensors*, 11, 4152–4187, 2011.

22. Nishiguchi, K., Che-Hien, L., Kishida, K., High-performance Brillouin distributed sensing using pulse compression technique, *IEICE Technical Report*, OFT2008-91, 108(450), 19–24, 2009 (in Japanese).

23. Höbel, M., Ricka, J., Wüthrich, J.M., Binkert, T., High-resolution distributed temperature sensing with the multi photon-timing technique, *Applied Optics*, 34(16), 2955–2967, 1995.

24. Tanner, M.G., Dyer, S.D., Baek. B., Hadfield, R.H., Woo Nam, S., High-resolution single-mode fiber-optic distributed Raman sensor for absolute temperature measurement using superconducting nanowire single-photon detectors, *Applied Physics Letters*, 99, 201110, 2011.

25. Chen, Y., Hartog, A.H., March, R.J., Hilton, I.M., Hadley, M.R., Ross, P.A., A fast high-spatial-resolution Raman distributed temperature sensor, *Proceedings of SPIE*, 9157, 91575M, 2014.

26. Gifford, D.K., Soller, B.J., Wolfe, M.S., Froggatt, M.E., Distributed fiber-optic temperature sensing using Rayleigh backscatter, in *ECOC 2005 Proceedings*, 3, Glasgow, UK, 511–512, 2005.

27. Igawa, H., Murayama, H., Kasai, T., Yamaguchi, I., Kageyama, K., Ohta, K., Measurements of strain distributions with a long gauge FBG sensors using optical frequency domain reflectometry, *Proceedings of SPIE*, 5855, 547–550, 2005.

28. Eickhoff, W., Ulrich, R., Optical frequency domain reflectometry in single-mode fiber, *Applied Physics Letters*, 39(9), 693–695, 1981.

29. Passy, R., Gisin, N., von der Weid, P., Gilgen, H.H., Experimental and theoretical investigations of coherent OFDR with semiconductor laser sources, *Journal of Lightwave Technology*, 12(9), 1622–1630, 1994.

30. Oberson, P. Huttner, B., Guinnard, O., Guinnard, L., Ribordy, G., Gisin, N., Optical frequency domain reflectometry with a narrow line width fiber laser, *Photonics Technology Letters*, 12(7), 867–869, 2000.

31. Gifford, D.K., Kreger, S.T., Sang, A.K., Froggatt, M.E., Duncan, R.G., Wolfe, M.S., Soller, B.J., Swept-wavelength interferometric interrogation of fiber Rayleigh scatter for distributed sensing applications, *Proceedings of SPIE*, 6770, 67700F, 2007.

32. Soller, B.J., Gifford, D.K., Wolfe, M.S., Froggatt, M.E., High resolution optical frequency domain reflectometry for characterization of components and assemblies, *Optics Express*, 13(2), 666–674, 2005.

33. Murayama, H., Kageyama, K., Uzawa, K., Ohara, K., Igawa, H., Strain monitoring of a single-lap joint with embedded fiber-optic distributed sensors, *Structural Health Monitoring*, 11(3), 325–344, 2011.

34. Murayama, H., Kageyama, K., Naruse, H., Shimada, A., Uzawa, K., Application of fiber-optic distributed sensors to health monitoring for full-scale composite structures, *Journal of Intelligent Material Systems and Structures*, 14(1), 3–13, 2003.

35. Parker, T., Shatalin, S., Farhadiroushan, M., Distributed acoustic sensing: A new tool for seismic applications, *First Break*, 32(2), 61–69, 2014.

36. Song, K.Y., Hotate, K., Distributed fiber strain sensor with 1-kHz sampling rate based on Brillouin optical correlation domain analysis, *Photonics Technology Letters*, 19(23), 1928–1930, 2007.

37. Lagakos, N., Cole, L.H., Bucaro, J.A., Microbend fiber-optic sensor, *Applied Optics*, 26(11), 2171–2180, 1987.

38. Sumida, S., Okagzaki, S., Asakura, S., Nakagawa, H., Murayama, H., Hasegawa, T., Distributed hydrogen determination with fiber-optic sensor, *Sensors and Actuators B*, 108, 508–514, 2005.

39. Woerdeman, D.L., Spoerre, J.K., Flynn, K.M., Parnas, R.S., Cure monitoring of the liquid composite molding process using fiber optic sensors, *Polymer Composites*, 18(1), 133–150, 1997.

40. Habel, W.R., Daum, K.K., Standardization in fibre-optic sensing for structural safety: Activities in ISHMII and IEC, *Proceedings of SPIE*, 9436, 94360S-1-7, 2015.

41. SAE international, Fiber optic sensors for aerospace applications, AIR6358, SAE international, 2015.

42. Wada, D., Murayama, H., Analytical investigation of response of birefringent fiber Bragg grating sensors in distributed monitoring system based on optical frequency domain reflectometry, *Optics and Lasers in Engineering*, 52, 99–105, 2014.

43. Ning, X., Structural health monitoring of adhesive bonded single-lap joints in composite materials by fiber-optic distributed sensing system, PhD thesis, The University of Tokyo, Tokyo, 2013.

44. Xu, P., Dong, Y., Zhang, J., Zhou, D., Jiang, T., Xu, J., Zhang, H., Zhu, T., Lu, Z., Chen, L., Bao, X., Bend-insensitive distributed sensing in singlemode–multimode–single mode optical fiber structure by using Brillouin optical time-domain analysis, *Optics Express*, 23(17), 22714–22722, 2015.

45. Di Sante, R., Fiber optic sensors for structural health monitoring of aircraft composite structures: Recent advances and applications, *Sensors*, 15(8), 18666–18713, 2015.

46. Zhao, Y., Liao, Y., Discrimination methods and demodulation techniques for fiber Bragg grating sensors, *Optics and Lasers in Engineering*, 41(1), 1–18, 2004.

47. Wada, D., Murayama, H., Igawa, H., Kageyama, K., Uzawa, K., Omichi, K., Simultaneous distributed measurement of strain and temperature by polarization marinating fiber Bragg grating based on optical frequency domain reflectometry, *Smart Materials and Structures*, 20(8), 085028, 2011.

48. Lynch, J.P., Law, K.H., Kiremidjian, A.S., Carryer, E., Farrar, C.R., Allen, D.W., Nadler, B., Wait, J.R., Design and performance validation of a wireless sensing unit for structural monitoring applications, *Structural Engineering and Mechanics*, 17(3), 394–423, 2004.

49. Swarts, R.A., Zimmerman, A.T., Lynch, J.P., Rosario, J., Brady, T., Salvino, L., Law, K.H., Hybrid wireless hull monitoring system for naval combat vessels, *Structure and Infrastructure Engineering*, 8(7), 621–638, 2012.

50. Claus, R.O., Jackson, B.S., Bennett, K.D., Nondestructive testing of composite materials by OTDR in imbedded optical fibers, *Proceedings of SPIE*, 566, 243–248, 1985.

51. Grace, J.L., Poland, S.H., Murphy, K.A., Claus, R.O., Abraham, P., Sridharan, K., Embedded fiber optic sensors for structural damage detection, *Proceedings of SPIE*, 2718, 196–201, 1996.

52. Singh, S.P., Gangwar, R., Singh, N., Nonlinear scattering effects in optical fibers, *Progress in Electromagnetics Research*, 74, 379–405, 2007.

53. Rogers, A.J., Nonlinear optics and optical fibers, in L.S. Grattan, B.T. Meggitt (eds.), *Optical Fiber Sensor Technology, Advanced Applications: Bragg Gratings and Distributed Sensors*, Dordrecht, the Netherlands: Kluwer Academic, pp. 1889–240, 2000.

54. Zhang, Z., Wang, J., Li, Y., Gong, H., Yu, X., Liu, H., Jin, Y., et al., Recent progress in distributed optical fiber Raman photon sensors at China Jiliang University, *Photonic Sensors*, 2(2), 127–147, 2012.

55. Fukazawa, T., DTS200 distributed temperature sensor for oil and gas production, *Yokogawa Technical Report English Edition*, 55(2), 77–80, 2012.

56. Thévenaz, L., Review and progress on distributed fiber sensing, *Optical Fiber Sensors, OSA Technical Digest (CD) (Optical Society of America, 2006)*, paper ThC1, 2006.

57. Kurashima, T., Hogari T., Matsuhashi, S., Horiguchi, T., Koyamada, Y., Wakui, Y., Hirano, H., Measurement of distributed strain in frozen cable and its potential for use in predicting cable failure, in *Proceedings of the International Wire and Cable Symposium*, Atlanta, GA, pp. 593–600, 1994.

58. Fellay, A., Thévenaz, L., Facchini, M., Niklès, M. Robert, P., Distributed sensing using stimulated Brillouin scattering: Towards ultimate resolution, in *Proceedings of the OFS'97, 12th International Conference on Optical Fiber Sensors*, IEEE, Washington, DC, pp. 324–327, 1997.

59. Naruse, H., Tateda, M., Trade-off between the spatial and frequency resolutions in measuring the power spectrum of the Brillouin backscattered light in an optical fiber, *Applied Optics*, 38(31), 6516–6521, 1999.

60. Froggatt M., Moore, J., High-spatial-resolution distributed strain measurement in optical fiber with Rayleigh scatter, *Applied Optics*, 37(10), 1735–1740, 1998.

61. Glombitza, U., Brinkmeyer, E., Coherent frequency-domain reflectometry for characterization of single-mode integrated-optical waveguides, *Journal of Lightwave Technology*, 11(8), 1377–1384, 1993.

62. Santos, J.L., Farahi, F., Optical sensors: Final thoughts, in J.L. Santos, F. Farahi (eds.), *Handbook of Optical Sensors*, Boca Raton, FL: CRC Press, pp. 631–683, 2015.

63. Igawa, H., Ohta, K., Kasai, T., Yamaguchi, I., Murayama, H., Kageyama, K., Distributed measurements with long gauge FBG sensor using optical time domain reflectometry (1st report, system in investigation using optical simulation model), *Journal of Solid Mechanics and Materials Engineering*, 2(9), 1242–1252, 2008.

64. Igawa, H., Murayama, H., Nakamura, T., Yamaguchi, I., Kageyama, K., Uzawa, K., Measurement of distributed strain and load identification using 1500 mm gauge length FBG and optical frequency domain reflectometry, *Proceedings of SPIE*, 7503, 750351, 2009.

65. Lindner, E., Canning, J., Chojetzki, C., Brückner, S., Becker, M., Rothhardt, M., Bartelt, H., Regenerated draw tower grating (DTG) temperature sensors, *Proceedings of SPIE*, 7753, 77536M, 2011.

66. Matsumoto, R., Murayama, H., Wada, D., Igawa, H., Kageyama, K., Bending deformation measurement using high resolution optical fiber sensor, in *Proceedings of JCOSSAR 2015*, 267–271, 2015 (in Japanese).

67. Kinet, D., Mégret, P., Goosen, K.W., Qiu, L., Heider, D., Caucheteur, C., Fiber Bragg grating sensors toward structural health monitoring, in composite materials: Challenges and solutions, *Sensors*, 14(4), 7394–7419, 2014.

68. Zhou, D-P, Li, W., Chen, L., Bao, X., Distributed temperature and strain discrimination with stimulated Brillouin scattering and Rayleigh backscattering in an optical fiber, *Sensors*, 13(2), 1836–1845, 2013.

69. Kumagai, Y., Matsuura, S., Yari, T., Saito, N., Hotate, K., Kishi, M., Yoshida, M., Fiber-optic distributed strain and temperature sensor using BOCDA technology at high speed and with high spatial resolution, *Yokogawa Technical Report English Edition*, 56(2), 83–86, 2013.

70. Froggatt, M., Gifford, D., Kreger, S., Wolfe, M., Soller, B., Distributed strain and temperature discrimination unaltered polarization maintaining fiber, *Optical Fiber Sensors, OSA Technical Digest (CD) (Optical Society of America, 2006)*, paper ThC5, 2006.

71. Song, KY., He, Z., Hotate, K., Distributed strain measurement with millimeter-order spatial resolution based on Brillouin optical correlation domain analysis, *Optics Letters*, 31(17), 2526–2528, 2006.

72. Yokogawa, General Specifications—Yokogawa: http://web-material3.yokogawa.com/ TI39J06B45-01E.pdf (Accessed 29 November 2015).

73. Fiber optic distributed strain and temperature sensor (DSTS): BOTDA+BOTDR Combo Module, Ozotopics, http://www.ozoptics.com/ALLNEW_PDF/DTS0139.pdf (Accessed 29 November 2015).

74. Optical distributed sensor interrogator (Models ODiSI A10 and A50), Luna, http://lunainc.com/wp-content/uploads/2012/11/LT_DS_ODiSI-A_Data-Sheet_Rev-07.pdf (Accessed 29 November 2015).

75. Omnisens VISION Dual data sheet: http://www.omnisens.com/ditest/download.php?content=DT-208%20%28VISION%20Dual%20%20TM%29%20en-01.pdf (Accessed 29 November 2015).

76. Specifications NBX-6065, Neubrex, http://www.neubrex.jp/pdf/NBX_6065.pdf (Accessed 29 November 2015).

77. fTB 2505: Fiber-optic sensor system for distributed strain and temperature monitoring, fibrisTerre, http://www.fibristerre.de/files/fibrisTerre_flyer.pdf (Accessed 29 November 2015).

78. Thévenaz, L., Facchini, M., Fellay, A., Robert, P., Monitoring of large structure using distributed Brillouin fibre sensing, *Proceedings of SPIE*, 3764, 345–348, 1999.

79. Ohno, H., Naruse, H., Kihara, M., Shimada, A., Industrial applications of the BOTDR optical fiber strain sensor, *Optical Fiber Technology*, 7(1), 45–64, 2001.

80. Vasiliev, V.V., Morozov, E., Environmental, special loading, and manufacturing effects, in V. Vasiliev, E. Morozov (eds.), *Mechanics and Analysis of Composite Materials*, Amsterdam, the Netherlands: Elsevier Science, pp. 301–364, 2001.

81. Whitaker, A.F., Finckenor, M.M., Dursch, H.W., Tennyson, R.C., Young, P.R., Environmental effects on composites, in S.T. Peters (ed.), *Handbook of Composites*, London: Chapman & Hall, pp. 810–821, 1998.

82. Zhuang, H., Wightman, J.P., The influence of surface properties on carbon fiber/epoxy matrix interfacial adhesion, *The Journal of Adhesion*, 62(1–4), 213–245, 1997.

83. Ley, O., Godinez, V., Non-destructive evaluation (NDE) of aerospace composites: Application of infrared (IR) thermography, in V. Karbhari (ed.), *Non-Destructive Evaluation (NDE) of Polymer Matrix Composites*, Cambridge, MA: Woodhead, pp. 309–334, 2013.

84. Wu, M-C., Winfree, W.P., Fiber optic thermal detection of composite delaminations, *Proceedings of SPIE*, 8013, 801314, 2011.

85. Pedrazzani, J.R., Klute, S.M., Gifford, D.K., Sang, A.K., Froggatt, M.E., Embedded and surface mounted fiber optic sensors detect manufacturing defects and accumulated damage as a wind turbine blade is cycled to failure, in *SAMPE Technical Conference, Proceedings of Emerging Opportunities: Materials and Process Solutions*, Baltimore, MD, 2012.

86. Murayama, H., Tachibana, K., Igawa, H., Hirano, Y., Aoki, Y., Kageyama, K., Uzawa, K., Strain monitoring of 6-m composite wing structure by fiber-optic distributed sensing system with FBGs, in *Proceedings of 15th European Conference on Composite Materials*, Venice, Italy, 2012.

87. Kobayashi, S., Application of optical fiber strain sensors with high spatial resolution, Graduate thesis, The University of Tokyo, Tokyo, 2006 (in Japanese).

88. Castellucci, M., Klute, S., Lally, E.M., Froggatt, M.E., Lowry, D., Three-axis distributed fiber optic strain measurement in 3D woven composite structures, *Proceedings of SPIE*, 8690, 869006, 2013.

89. Murayama, H., PhD. thesis, The University of Tokyo, Tokyo, 2001 (in Japanese).

90. Sirkis, J.S., Dasgupta, A., Optical fiber/composite interaction mechanics, in E. Udd (ed.), *Fiber Optic Smart Structures*, New York: Wiley, pp. 61–107, 1995.

91. Habel, W.R., Schukar, V.G., Kusche, N., Fibre-optic strain sensors are making the leap from lab to industrial use: Reliability and validation as a precondition for standards, *Measurement Science and Technology*, 24(9), 094006, 2013.

92. Murayama, H., Igawa, H., Kageyama, K., Ohsawa, I., Uzawa, K., Kanai, M., Kobayashi, S., Shirai, T., Distributed strain sensing with long gauge FBG sensors based on optical frequency domain reflectometry, in *Proceedings of Asia-Pacific Workshop on Structural Health Monitoring*, Yokohama, Japan, pp. 304–310, 2006.

93. Ning, X., Murayama, H., Kageyama, K., Wada, D., Kanai, M., Ohsawa, I., Igawa, H., Dynamic strain distribution measurement and crack detection of an adhesive-bonded single-lap joint under cyclic loading using FBG, *Smart Materials and Structures*, 23(10), 105011, 2014.

94. Nihio, M., Mizutani, T., Takeda, N., Structural shape reconstruction with consideration of the reliability of distributed strain data from a Brillouin-scattering-based optical fiber sensor, *Smart Materials and Structures*, 19(3), 035011, 2010.

95. Ko. W., Richards, W., Tran, V., Displacement theories for in-flight deformed shape predictions of aerospace structures, NASA/TP-2007-214612, 2007.

96. Bakalyar, J., Jutte, C., Validation tests of fiber optic strain-based operational shape and load measurements, in *Proceedings of 53rd AIAA/ASME/ASCE/AHS/ASC Structures, Structural Dynamics and Materials Conference*, Honolulu, HI, AIAA 2012-1904, 2012.

97. Rapp., S., Kang, L-H., Han, J-H., Mueller U.C., Bajer, H., Displacement field estimation for a two-dimensional structure using fiber Bragg grating sensors, *Smart Materials and Structures*, 18(2), 025006, 2009.

98. Kreger, S.T., Sang, A.K., Garg, N., Michel, J., High resolution, high sensitivity, dynamic distributed structural monitoring using optical frequency domain reflectometry, *Proceedings of SPIE*, 8722, 87220D, 2013.

99. Peled, Y., Motil, A., Kressel, I., Tur, M., Monitoring the propagation of mechanical waves using an optical fiber distributed and dynamic strain sensor based on BOTDA, *Optics express*, 21(9), 10697–10705, 2013.

100. Murayama, H., Akiyama, G., Igawa, H., Nakamura, T., Kageyama, K., Uzawa, K., Hirano, Y., Aoki, Y. Application of inverse analysis of distributed load with strain sensors to wing structures, in *Proceedings of the 7th International Workshop on Structural Health Monitoring*, Stanford, CA, pp. 1, 75–82, 2009.

6 Importance of Strain Transfer Effects for Embedded Multiaxial Strain Sensing and Optical Fiber Coating Optimization

Nicolas Lammens, Geert Luyckx, Thomas Geernaert, Damien Kinet, Francis Berghmans, Christophe Caucheteur, and Joris Degrieck

CONTENTS

6.1 Introduction 158
6.2 Strain Transfer Research 159
    6.2.1 Surface-Mounted Bragg Gratings.. 159
    6.2.2 Embedded Bragg Gratings 160
6.3 Coordinate Systems for Composite Materials and Optical Fiber Sensors ... 161
6.4 Transfer Coefficient Matrix 162
    6.4.1 Analytical Approach... 163
        6.4.1.1 Application to an Embedded Sensor in an M18/M55J
                 Carbon Fiber Composite Host 166
    6.4.2 Numerical Approach... 168
        6.4.2.1 Finite Element Model and Loading Conditions................ 168
        6.4.2.2 Parametric Study.. 169
    6.4.3 Experimental Approach.. 176
        6.4.3.1 Experimental Setup.. 176
        6.4.3.2 Experimental Results... 179
6.5 Coating Optimization 179
    6.5.1 State of the Art 181
    6.5.2 Transfer Matrix Tool.. 184
        6.5.2.1 Model Description 184
        6.5.2.2 TC Matrix Determination.. 185
        6.5.2.3 FE Implementation 186

     6.5.3   Limits of Applicability 188
               6.5.3.1   Part Size 189
               6.5.3.2   Composite Layup 190
               6.5.3.3   Coating Eccentricity 191
     6.5.4   Examples... 192
               6.5.4.1   Pure Axial or Transverse Load.................................. 192
               6.5.4.2   Combined Axial and Transverse Load 193
               6.5.4.3   Global Optimum 194
6.6   Conclusions... 195
References... 196

6.1   INTRODUCTION

Composite materials have attracted growing interest over the past decades as high-performance construction materials for lightweight applications requiring stiff and strong application under harsh operating conditions. However, their anisotropic nature results in significantly different behavior of the material compared with conventional isotropic construction materials. Health monitoring and strain sensing have received much attention for the safe design and operation of composite structures. Among other techniques, optical fiber sensing has attracted a lot of interest from the research community [1–3]. The fibrous nature and multiaxial sensitivity of optical fiber Bragg gratings make them perfect candidates for embedding purposes in (fibrous) composite materials during the production step [4–9]. Embedded inside the host material during manufacturing, they enable the users to monitor the strain state and development of residual strains during the entire life cycle of the composite structure [10–13]. This significant benefit has resulted in an extensive amount of research being performed on the use of optical fiber sensors in composite materials.

Assuming perfect bonding between the optical fiber sensors and the surrounding host material, axial strains are almost perfectly transferred from the host to the sensor, allowing a direct readout of axial strain from the sensor response. However, when multiaxial sensing is envisaged, one must account for the differences in stiffness between the host material and the optical fiber sensor (and optionally its coating). Mismatches in material properties such as Young's modulus, Poisson ratio, or coefficient of thermal expansion (CTE) will inevitably lead to localized stress/strain concentrations surrounding the sensor and a redistribution of the stress/strain field in the vicinity. This distortion of the strain field is localized around the optical fiber sensor. At a distance of a couple of optical fiber diameters [14–16], this distortion has almost completely vanished, and a uniform strain field is obtained. These uniform strains will be referred to as *far-field strains*. A result of the mismatch in material properties is that the strain experienced at the core of the optical fiber sensor (the *near-field strain*) rarely corresponds to the far-field strains in the composite host. To perform any meaningful multiaxial strain sensing in composite materials, it is therefore necessary to account for these strain transfer effects between the composite host and the optical fiber sensors.

This chapter will illustrate the approach needed to account for the necessary strain transfer effects. Theoretical calculations, numerical simulations, and experimental simulations will be shown, illustrating different possible approaches to modeling these strain transfer effects. In addition, the strain transfer principle can be employed to study the interfacial stresses and strains occurring at the boundary between the optical fiber sensor and the composite host. As these stresses and strains are dependent on the precise mechanical properties (Young's modulus, Poisson ratio, CTE, etc.) of both the host and the sensor, an optimization study can be performed for coated optical fiber sensors in which the coating properties can be tuned to achieve the optimal strain field around the sensor, avoiding the onset of early damage in the composite structure.

6.2 STRAIN TRANSFER RESEARCH

6.2.1 SURFACE-MOUNTED BRAGG GRATINGS

In its most direct and simplest application, an optical fiber sensor can be surface mounted onto a surface to register the strains occurring at the surface of the structure. In this application, a thin layer of adhesive is applied as an interface between the optical fiber sensor and the host material. The adhesive layer thickness and material properties will influence the amount of strain transferred from the host to the sensor [17]. Li et al. [18] further indicated that the host's Young's modulus and thickness have a pronounced influence on the strain transfer. While thick and stiff hosts result in an almost perfect strain transfer to the sensor, a thin and flexible host will be locally affected by the stiffening due to the optical fiber sensor and result in a lower near-field strain than the far-field strains in the host. Besides requiring the user to account for these strain transfer effects, this also indicates that the presence of the sensor affects the behavior of the host, which is usually undesirable. Wan et al. [19] found that not only bond thickness has an effect on strain transfer, but also bond length plays an important part in the amount of strain transferred between host and sensor (Figure 6.1). The investigation contained both experimental and numerical

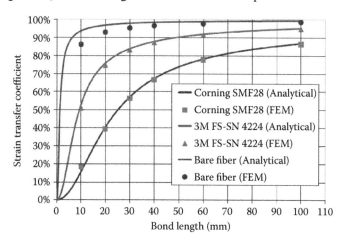

FIGURE 6.1 Variation of strain transfer coefficient of different fiber types as a function of bond length (bottom thickness = 0.1 mm, side width 2 mm, top thickness = 0.15 mm). (From Wan, K.T. et al., *Smart Mater Struct*, 17(3), 2008.)

analysis, showing good correspondence. In his study, Wan considered three different types of fiber, with the bare fiber giving the best strain transfer in both numerical simulations and experimental tests.

6.2.2 EMBEDDED BRAGG GRATINGS

The concepts of bond length and thickness become meaningless when optical fibers are embedded inside the composite host material. Chang et al. [20] studied the existence of strain gradients from the host to the sensor, and found that only the first few millimeters at the entry point of the embedded fiber are exposed to strain gradients. The precise length was found to be dependent on the material properties of the host and sensor, with more flexible material (lower Young's modulus) leading to longer areas of strain gradients near the ingress and egress of the optical fiber sensor.

Fan et al. [21] performed an experimental investigation to determine the influence of induced radial strain in embedded optical fiber sensors. In an unrestricted optical fiber sensor, measured wavelength shifts and applied axial strains are commonly related by the so-called strain gauge factor $(1 - P)$ [22]:

$$\frac{\Delta\lambda}{\lambda_0} = (1 - P)\varepsilon \tag{6.1}$$

where:

$\Delta\lambda$ is the measured wavelength shift relative to the initial (unstrained) wavelength λ_0

ε represents the applied axial strain to the optical fiber sensor

The strain-optic coefficient P in Equation 6.1 can be determined as

$$P = \frac{1}{2}\bar{n}^2\left[p_{12} - v_f(p_{11} + p_{12})\right] \tag{6.2}$$

where:

$\bar{n}$ is the effective refractive index of the grating

p_{11} and p_{12} are the strain-optic coefficients of the fiber

$\#_f$ is the Poisson coefficient of the optical fiber

Fan et al. found that, once embedded, the traditional approach of using this theoretical gauge factor $(1-P)$ to translate wavelength shifts to axial strains was no longer accurate. This can be explained by the fact that the fiber is constrained by the surrounding host and can no longer contract freely, as is assumed in the derivation of the strain-optic coefficient P. The authors tried to quantify the mismatch by producing two $[0_8, 90_4]$ composite samples in which two fiber Bragg gratings (FBGs) were embedded. Errors of up to 8% were found when using the traditional strain gauge factor. To correct for these errors, the authors proposed using a coefficient that reflects the type of loading applied.

These results indicate that the optical FBG response is in fact sensitive to the full strain field applied to the sensor, and not solely dependent on the axial strain applied. Research activities have pointed out the need to determine the total strain field response of an embedded FBG [23,24] rather than focusing on the axial strain response. To accurately measure strains using optical FBGs, it therefore becomes important to consider the multiaxial strain sensitivity of the fiber, given by [25]

$$\frac{\Delta\lambda_{B,1}}{\lambda_{B,0}} = \varepsilon_{3'} - \frac{1}{2}\bar{n}^2\left[p_{11}\varepsilon_{1'} + p_{12}\left(\varepsilon_{2'} + \varepsilon_{3'}\right)\right]$$

$$\frac{\Delta\lambda_{B,2}}{\lambda_{B,0}} = \varepsilon_{3'} - \frac{1}{2}\bar{n}^2\left[p_{11}\varepsilon_{2'} + p_{12}\left(\varepsilon_{1'} + \varepsilon_{3'}\right)\right] \tag{6.3}$$

in which the 3-axis is aligned to the axis of the optical fiber, while the 1- and 2-axes are transverse directions to the optical fiber sensor. The prime symbol is used to indicate strains measured at the core of the sensor according to the convention defined in the next section, which are different from those in the surrounding host due to mismatches in material properties between the two.

The coefficients p_{11} and p_{12} are the strain-optic coefficients. Bertholds and Dandliker measured these values to be [26]

$$p_{11} = 0.113$$

$$p_{12} = 0.252 \tag{6.4}$$

Bosia et al. [27] studied the response of an embedded FBG sensor in an epoxy host exposed to a biaxial strain field. Finite element analysis revealed differences between the strain field as sensed by the FBG and those found in the far-field region. As a result, strain transfer mechanisms between host and optical fiber sensor need to be taken into account to provide accurate strain sensing of a composite host.

6.3   COORDINATE SYSTEMS FOR COMPOSITE MATERIALS AND OPTICAL FIBER SENSORS

Optical fiber sensors and composite materials have both evolved over time. When combining both technologies, certain conflicts are to be expected in terminology and notations. A common issue in this field is the use of different conventions for coordinate systems.

Composites usually adhere to a convention whereby the 1-direction corresponds to the fiber reinforcement direction, the 2-direction is perpendicular to the reinforcements and aligned to the plane of the ply, and the 3-direction is the out-of-plane transverse direction (Figure 6.2a).

Optical fiber sensors, on the other hand, use a convention whereby the 3-axis corresponds to the longitudinal axis of the fiber, and the 1–2 plane is perpendicular to the fiber (Figure 6.2b).

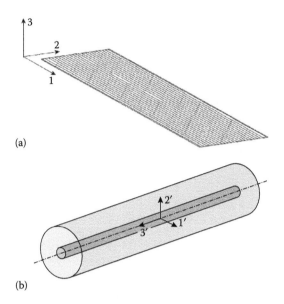

FIGURE 6.2   Coordinate system convention for (a) composites and (b) optical fiber sensors.

To avoid confusion within this chapter, the composite convention is given by the standard 1, 2, 3-coordinate system, while the optical fiber convention is indicated by the 1′, 2′, 3′ coordinate system (as in Equation 6.3).

6.4   TRANSFER COEFFICIENT MATRIX

Assuming a linear material response and small deformations of the host structure, the strain transfer can be described in its most general form through the use of a transfer coefficient (TC) matrix, defined as

$$
\begin{bmatrix}
\varepsilon_1^c \\
\varepsilon_2^c \\
\vdots \\
\varepsilon_6^c \\
\Delta T^c
\end{bmatrix}
=
\begin{bmatrix}
TC_{11} & TC_{12} & \cdots & TC_{16} & TC_{17} \\
TC_{21} & TC_{22} & \cdots & TC_{26} & TC_{27} \\
\vdots & \vdots & \ddots & \vdots & \vdots \\
TC_{61} & TC_{62} & \cdots & TC_{66} & TC_{67} \\
TC_{71} & TC_{72} & \cdots & TC_{76} & TC_{77}
\end{bmatrix}
\begin{bmatrix}
\varepsilon_1^s \\
\varepsilon_2^s \\
\vdots \\
\varepsilon_6^s \\
\Delta T^s
\end{bmatrix}
\tag{6.5}
$$

where:
   ε^c refers to far-field strains measured in the composite host (in contracted notation according to [28])
   ε^s refers to the strain field in the FBG sensor core (in contracted notation according to [28])

ΔT is the change in temperature (which will affect the strain transfer as a result of differences in CTE)

TC_{ij} represent the transfer coefficients between the sensor and the host strains

In its most general form (i.e., Equation 6.4), the TC matrix contains a total of 49 independent coefficients, which will represent the strain transfer behavior between sensor and host. Considering the small dimensions of the optical fiber sensors, it is usually reasonable to assume that ΔT is identical in the sensor and the surrounding host, reducing the TC matrix bottom row to all zeros except for $TC_{77}=1$. It must be emphasized that, in its most general form, the temperature parameter must remain in the expression, as differences in CTE between sensor and host may result in the creation of additional strains. As a result, the last column of the TC matrix does not reduce to zeros, and therefore, the temperature parameters must remain. If effects of moisture expansion (or other effects) are deemed important for the application, these parameters need to be included as well.

The process of determining the necessary TC coefficients remains identical, regardless of whether or not temperature (or moisture) effects are accounted for. In the remainder of this chapter, only effects of mechanical strain transfer (i.e., those resulting from mechanically applied strain) will be considered. Using this assumption, the temperature parameter can be eliminated from the TC matrix, reducing the number of independent TC coefficients to a total of 36:

$$\begin{bmatrix} \varepsilon_1^c \\ \varepsilon_2^c \\ \vdots \\ \varepsilon_6^c \end{bmatrix} = \begin{bmatrix} TC_{11} & TC_{12} & \cdots & TC_{16} \\ TC_{21} & TC_{22} & \cdots & TC_{26} \\ \vdots & \vdots & \ddots & \vdots \\ TC_{61} & TC_{62} & \cdots & TC_{66} \end{bmatrix} \begin{bmatrix} \varepsilon_1^s \\ \varepsilon_2^s \\ \vdots \\ \varepsilon_6^s \end{bmatrix} \tag{6.6}$$

In the following section, an analytical approach to determining the strain transfer (and the TC matrix) is presented based on the work by Kollar and Van Steenkiste [16] and Van Steenkiste et al. [29,30]. In the subsequent sections, the analytical TC matrix (with its limitations) will then be compared with numerical results for different scenarios (different laminates, optical fiber diameters, etc.) and experimentally obtained results.

6.4.1 ANALYTICAL APPROACH

To obtain analytical expressions for the strain field in a composite material with an embedded sensor, certain simplifications and assumptions must be made. The optical fiber sensor is assumed to be perfectly aligned with the reinforcement fibers in the surrounding composite host, as this avoids the creation of lenticular resin pockets, which would make the analytical treatment of the problem impossible. The optical fiber sensor is considered to be an infinitely long circular inclusion inside a homogeneous and orthotropic host (Figure 6.3). Perfect bonding between all constituents is assumed, as well as small deformations and a linear elastic material response. Temperature changes are assumed to be uniform throughout the host and sensor.

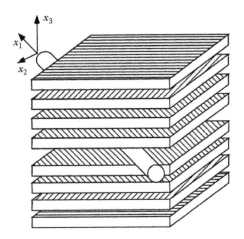

FIGURE 6.3 Sensor in a laminated composite material as analyzed by Kollar et al. (From Kollar, L.P. and Van Steenkiste, R.J., *J Compos Mater*, 32(18), 1647–79, 1998.)

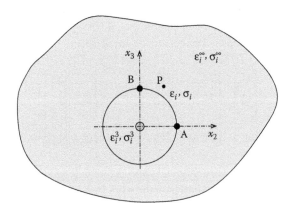

FIGURE 6.4 A hole or inclusion in a laminate alters the local stress and strain fields in the material. The far-field strains differ from those near the inclusion, such as at point P. (From Kim, K.S. et al., *J Compos Mater*, 27(17), 1618–62, 1993.)

Kollar assumes that a hole or inclusion in the laminate will alter the stress and strain fields in its immediate vicinity. The stresses, strains, and displacements at a point P (Figure 6.4) in the near field can be written as

$$\sigma_i = \sigma_i^\infty + \sigma_i^*$$

$$\varepsilon_i = \varepsilon_i^\infty + \varepsilon_i^*$$

$$u_i = u_i^\infty + u_i^* \tag{6.7}$$

where

∞   superscript refers to the far-field quantities

*   refers to the local perturbation in the vicinity of the inclusion or hole

For circular and elliptical holes, expressions for σ^*, ε^*, and u^* have been presented by Lekhnitskii [15].

The strains and displacements in Equation 6.7 are related by [31]

$$\varepsilon_1 = \varepsilon_{11} = \frac{\partial u_1}{\partial x_1} \qquad \varepsilon_4 = \gamma_{23} = \frac{\partial u_3}{\partial x_2} + \frac{\partial u_2}{\partial x_3}$$

$$\varepsilon_2 = \varepsilon_{22} = \frac{\partial u_2}{\partial x_2} \qquad \varepsilon_5 = \gamma_{13} = \frac{\partial u_3}{\partial x_1} + \frac{\partial u_1}{\partial x_3}$$

$$\varepsilon_3 = \varepsilon_{33} = \frac{\partial u_3}{\partial x_3} \qquad \varepsilon_6 = \gamma_{12} = \frac{\partial u_2}{\partial x_1} + \frac{\partial u_1}{\partial x_2} \qquad (6.8)$$

The stresses and strains are related by the orthotropic stiffness matrix C:

$$\sigma_i = C_{ij}\varepsilon_j \qquad (6.9)$$

The stiffness matrix C is the inverse of the compliance matrix S, which for an orthotropic material is given as

$$S = \begin{bmatrix} \dfrac{1}{E_1} & \dfrac{-v_{12}}{E_1} & \dfrac{-v_{13}}{E_1} & 0 & 0 & 0 \\[2mm] \dfrac{-v_{21}}{E_2} & \dfrac{1}{E_2} & \dfrac{-v_{23}}{E_2} & 0 & 0 & 0 \\[2mm] \dfrac{-v_{31}}{E_3} & \dfrac{-v_{32}}{E_3} & \dfrac{1}{E_3} & 0 & 0 & 0 \\[2mm] 0 & 0 & 0 & \dfrac{1}{G_{23}} & 0 & 0 \\[2mm] 0 & 0 & 0 & 0 & G_{31} & 0 \\[2mm] 0 & 0 & 0 & 0 & 0 & G_{12} \end{bmatrix} \qquad (6.10)$$

Using Equation 6.7, it is possible to determine the stress, strain, and displacement field at the inclusion or hole in the laminate. In addition, Kim et al. [32] and Van Steenkiste et al. [29] have shown that for an uncoated sensor embedded in an anisotropic material, the stresses and strains are constant in the sensor. To obtain the stresses and strain at the optical fiber core, it is then necessary to express continuity conditions at the interface between the sensor and the host.

It is sufficient to invoke continuity conditions at two distinct points on the interface to automatically satisfy the continuity requirements on the entire sensor surface. In this work, points A and B are chosen (Figure 6.4).

The first continuity condition states that the displacements of the sensor must be equal to those of the surrounding material:

$$u_i^s = u_i \quad \text{with} \quad i = 1,2,3 \tag{6.11}$$

where

u^s is the displacement field of the sensor

u is the displacement field of the host, given by Equation 6.7

The second continuity requirement states that tractions at the interface between sensor and host must be identical. Expressed at points A and B, this leads to

$$
\begin{aligned}
\sigma_2^s &= \sigma_2 & \sigma_3^s &= \sigma_3 \\
\sigma_4^s &= \sigma_4 \text{ at point A and } & \sigma_4^s &= \sigma_4 \text{ at point B} \\
\sigma_6^s &= \sigma_6 & \sigma_5^s &= \sigma_5
\end{aligned}
\tag{6.12}
$$

The combination of the continuity conditions given by Equations 6.11 and 6.12, combined with the assumption of a uniform stress and strain field inside the sensor, then, fully define the stress and strain fields as a function of far-field stresses and strains.

6.4.1.1 Application to an Embedded Sensor in an M18/ M55J Carbon Fiber Composite Host

The different transfer matrix approaches described in this chapter will all be applied to a concrete combination of composite host and optical fiber sensor to allow direct comparison of the obtained strain transfer description.

In this chapter, an M18/M55J carbon fiber–reinforced epoxy will be used as the host material. The relevant material properties for this carbon fiber–reinforced polymer (CFRP) are given in Table 6.1.

The optical fiber is considered to be an isotropic medium with properties as given in Table 6.2.

The theoretical analysis described in Section 6.4.1 reveals that far-field normal stresses (strains) only result in normal stresses at the optical fiber core, and far-field shear stresses (strains) only result in shear stresses at the optical fiber core. Hence, as optical fibers are only sensitive to normal strain (see Equation 6.3), the TC matrix given by Equation 6.6 can be further reduced to

TABLE 6.1

Elastic Material Properties for M18/M55J Carbon/Epoxy Composite

E_{11} (GPa)	$E_{22}=E_{33}$ (GPa)	$G_{12}=G_{13}$ (GPa)	G_{23} (GPa)	$\nu_{12}=\nu_{13}$ (–)	$\nu_{23}=\nu_{32}$ (–)	$\nu_{21}=\nu_{31}$ (–)
300	6.30	4.30	2.28	0.32	0.38	0.0017

TABLE 6.2

Elastic Material Properties for a Silica Optical Fiber

E (GPa)	$\nu(-)$
70	0.17

Source: De Waele, W., *Structural Monitoring of Composite Elements Using Optical Fibres with Bragg Sensors*, Faculty of Engineering, Ghent University, Ghent, Belgium, 2002 [33].

$$\begin{bmatrix} \Delta\varepsilon_1^c \\ \Delta\varepsilon_2^c \\ \Delta\varepsilon_3^c \end{bmatrix} = \begin{bmatrix} TC_{11} & TC_{12} & TC_{13} \\ TC_{21} & TC_{22} & TC_{23} \\ TC_{31} & TC_{32} & TC_{33} \end{bmatrix} \begin{bmatrix} \Delta\varepsilon_1^s \\ \Delta\varepsilon_2^s \\ \Delta\varepsilon_3^s \end{bmatrix} \tag{6.13}$$

Using the method described in Section 6.4.1, and applying it to the combination of a silica fiber in an M18/M55J CFRP composite, the TC matrix can be found to be

$$TC = \begin{bmatrix} 1 & 0 & 0 \\ 1.035 & 7.538 & -0.829 \\ 0.878 & -0.964 & 6.809 \end{bmatrix}$$

$$TC^{-1} = \begin{bmatrix} 1 & 0 & 0 \\ -0.154 & 0.135 & 0.016 \\ -0.151 & 0.019 & 0.149 \end{bmatrix} \tag{6.14}$$

The value of $TC_{11} = 1$ indicates a perfect axial strain transfer between the host and the sensor in an infinitely long optical fiber, which corresponds to results published in the literature by Chang et al. [20]. The other diagonal elements, TC_{22} and TC_{33}, are significantly larger than unity, indicating that optical fibers only sense a fraction of the transverse strains applied to the host material (about 14% according to the inverse of the TC matrix), hence reducing their sensitivity to these strain components. In addition, the inverse of the TC matrix given in Equation 6.14 indicates that a transverse strain in the host will lead to a three-dimensional strain field inside the optical fiber sensor. This is a result of the differences in Young's modulus and Poisson ratio between the host and the sensor material. Finally, it can be seen that the presence of transverse strains in the host will induce axial strains in the optical fiber sensor. Hence, it is necessary to always account for the multiaxial response of the Bragg gratings, as given by Equation 6.3, as well as the strain transfer response, as described by the TC matrix given in Equation 6.14. Failing to account for these effects may lead to significant errors in determining strain fields within the composite host, as reported by Fan et al. [21].

6.4.2 Numerical Approach

The analytical method presented in Section 6.4.1 has the significant benefits of being fully analytic and computationally fast. Hence, it allows the user to quickly investigate the influence of several parameters on the strain transfer behavior between host and sensor. As a result, it is a very powerful tool in a first study of the strain transfer behavior. However, the approach does pose some stringent assumptions, such as an infinitely long sample in an infinitely wide homogenized orthotropic host. As a result, it does not account for effects resulting from finite dimensions of the composite host or boundary effects occurring at changes in laminate layup.

An alternative technique to the analytical approach is the use of finite element (FE) analysis. The FE method allows the effects of finite part sizes and boundary effects in composite layups containing differently oriented plies to be taken into account.

6.4.2.1 Finite Element Model and Loading Conditions

The FE analysis results shown in this chapter were all obtained using the FE software package ABAQUS by Dassault. Only rectangular plates with embedded optical fibers are considered in this work, although more complex shapes can just as easily be analyzed. As a result of symmetry conditions, only one eighth of the total part is simulated. Figure 6.5 shows the model used for the analysis of strain transfer. The model represents a 24-layer laminate with total dimensions of $20 \times 2.4 \times 20$ mm. It was meshed using a total of 17,600 linear hexahedral elements of type C3D8R. The mesh is locally refined around the optical fiber sensor. The total number of mesh elements is dependent on the layup and the total thickness of the laminate analyzed.

The material properties used in the model correspond to the values used in the analytical approach and are given in Tables 6.1 and 6.2.

Based on Equation 6.13, a total of nine unknown TC coefficients need to be determined to model the strain transfer between composite and sensor. For each independent load case simulated in ABAQUS, a total of three independent equations can be derived relating the far-field composite strains to those at the core of the optical fiber. As a result, to determine all nine TC matrix coefficients, a total of three independent load conditions need to be simulated.

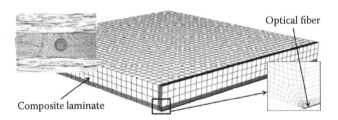

FIGURE 6.5 FE model used for simulation of the strain transfer between composite host and embedded optical fiber. (From Luyckx, G. et al., *Smart Mater Struct*, 19(10), 2010.)

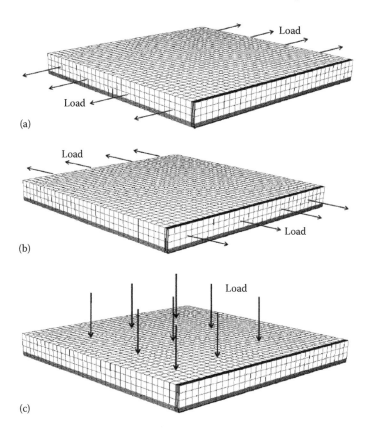

FIGURE 6.6 FE load cases for strain transfer analysis. (From Luyckx, G. et al., *Smart Mater Struct,* 19(10), 2010.)

While any combination of three independent loads can be analyzed, the most straightforward load cases will be considered in this work: (i) longitudinal load (Figure 6.6a), (ii) transverse in-plane load (Figure 6.6b), and (iii) transverse out-of-plane load (Figure 6.6c). The loads are obtained by defining a displacement of the end-faces of the part.

6.4.2.2 Parametric Study

The FE method described in Section 6.4.2.1 allows the influence of several parameters on the strain transfer mechanism (characterized by the TC matrix) and on the correspondence to the analytical transfer coefficient matrix to be studied.

In this work, the following parameters will be investigated briefly to illustrate their effect on the strain transfer response:

- Layup of the composite laminate (layer thickness and stacking sequence)
- Position of the sensor inside a certain layer of the composite laminate (at the mid-plane or near the boundary)
- Type of optical fiber sensor (traditional, bow-tie, and microstructured fibers)

6.4.2.2.1 Layup of the Composite Laminate

6.4.2.2.1.1 Unidirectional Laminate The analytical tool discussed in Section 6.4.1 assumes the composite host to be infinitely wide. In reality, however, thin laminates are common, and the influence of finite dimensions on the strain transfer response should be investigated.

To analyze the effect of laminate dimensions, a unidirectional (UD) laminate $[0_x]$ is simulated, changing the number of layers as denoted by x. The sensor is assumed to be located at the mid-plane of the laminate. Using the simulation approach discussed in the previous section, the TC matrices can be extracted for different laminate thicknesses.

A couple of observations can be made from the results in Table 6.3. In thin laminates such as the $[0_2]$ laminate, a significant difference exists between the response to transverse strains in the in-plane and out-of-plane directions. This is identified by the differences in coefficients in the second and third rows of the TC matrix. As the laminate increases in thickness, the response to both transverse directions becomes almost identical, and the TC matrix becomes symmetric. This is to be expected due to the transversely isotropic nature of the UD laminate. At low thicknesses, edge effects affect the TC matrix response in the through-thickness direction. The effect becomes less pronounced as the laminate thickness increases, and the response to both transverse load directions becomes almost identical.

TABLE 6.3

TC Matrices for a Glass Optical Fiber Sensor Embedded in a Unidirectional $[0_x]$ M55J/M18 CFRP Laminate

$[0_2]$

$$\begin{bmatrix} 1.000 & 0.000 & 0.001 \\ 1.112 & 8.509 & -0.404 \\ 0.394 & -2.118 & 6.210 \end{bmatrix}$$

$[0_4]$

$$\begin{bmatrix} 1.000 & 0.000 & 0.001 \\ 0.896 & 7.915 & -1.045 \\ 0.705 & -1.509 & 7.315 \end{bmatrix}$$

$[0_8]$

$$\begin{bmatrix} 1.000 & 0.000 & 0.000 \\ 0.838 & 7.732 & -1.193 \\ 0.791 & -1.309 & 7.587 \end{bmatrix}$$

$[0_{16}]$

$$\begin{bmatrix} 1.000 & 0.000 & 0.000 \\ 0.825 & 7.699 & -1.231 \\ 0.811 & -1.262 & 7.647 \end{bmatrix}$$

$[0_{24}]$

$$\begin{bmatrix} 1.000 & 0.000 & 0.000 \\ 0.823 & 7.708 & -1.248 \\ 0.814 & -1.265 & 7.668 \end{bmatrix}$$

$[0_{32}]$

$$\begin{bmatrix} 1.000 & 0.000 & 0.000 \\ 0.822 & 7.702 & -1.25 \\ 0.817 & -1.256 & 7.675 \end{bmatrix}$$

$[0_{48}]$

$$\begin{bmatrix} 1.000 & 0.000 & 0.000 \\ 0.816 & 7.648 & -1.233 \\ 0.821 & -1.23 & 7.679 \end{bmatrix}$$

Figure 6.7 shows the transverse strain fields for an out-of-plane transverse load in a $[0_{24}]$ laminate, while Figure 6.8 shows the strain field for the same load case in a $[0_4]$ laminate.

The results in Figures 6.7 and 6.8 show that the disturbance caused by the embedded optical fiber in a thin laminate extends to the boundaries of the laminate. For a thick laminate, on the other hand, the disturbance is completely enclosed within a (small) portion of the laminate. Obviously, adding extra layers to a thick laminate will have no further influence on the strain disturbance and will thus have only a

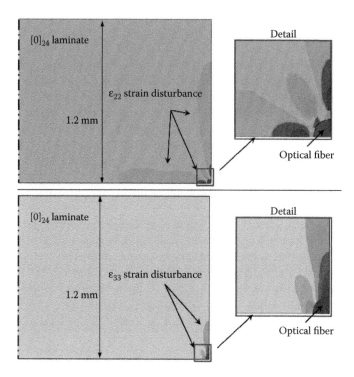

FIGURE 6.7 Transverse strain fields for a $[0_{24}]$ laminate under an out-of-plane transverse load.

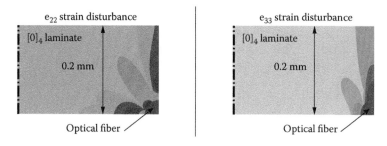

FIGURE 6.8 Transverse strain fields for a $[0_4]$ laminate under an out-of-plane transverse load. (From Luyckx, G. et al., *Smart Mater Struct,* 19(10), 2010.)

minimal influence on the TC matrix coefficients, as can be seen in Table 6.3 for laminates of [0₁₆] and up.

A second observation that can be made is that laminate thickness has no measurable effect on the axial strain transfer. For all laminate sizes investigated, the axial strain transfer remains at unity.

Finally, comparing the results of Table 6.3 with the analytical TC matrix obtained in Equation 6.14, it can be seen that as the laminate becomes thicker, the TC matrix converges to the TC matrix found analytically. For thinner laminates, however, there is a pronounced difference between the FE results and the analytical result. As thin laminates are common in practical applications, the FE analysis therefore presents a significant added value over the analytical approach.

6.4.2.2.1.2 Cross-Ply Laminate   Another benefit of the FE approach is that it allows the study of layups other than a purely UD laminate. In this section, the TC matrix for cross-ply laminates of the type $[90_x, 0_x]_{ns}$ will be investigated for different laminate sizes. The optical fiber is embedded in the mid-plane of the laminate. Due to the cross-ply layup, additional stresses and strain can be expected at the interfaces where the ply orientation is altered in the laminate. It is to be expected that for thin laminates (or more precisely, the layers of the same orientation), these disturbances will affect the strain transfer response compared with a purely UD laminate or the analytical TC matrix. The resulting TC matrices for different cross-ply laminates are given in Table 6.4.

Again, the results show that the TC matrix loses its symmetry for thin laminates, where a difference in response will occur depending on whether an in-plane or out-of-plane transverse load is applied. The imbalance diminishes as the individual layers grow thicker and the TC matrix becomes more symmetric. For the thickest cross-ply laminate analyzed ($[90_4 0_4]_{2s}$), the TC matrix approaches the TC matrix found for a thick UD laminate and the analytical TC matrix.

TABLE 6.4

TC Matrices for a Glass Optical Fiber Sensor Embedded in a Cross-Ply $[90_x/0_x]_{2s}$ M55J/M18 CFRP Laminate

$[90/0]_{2s}$

$$\begin{bmatrix} 1.000 & 0.000 & 0.000 \\ 0.812 & 7.431 & -1.048 \\ 0.780 & -1.115 & 7.311 \end{bmatrix}$$

$[90_2/0_2]_{2s}$

$$\begin{bmatrix} 1.000 & 0.000 & 0.000 \\ 0.818 & 7.631 & -1.206 \\ 0.809 & -1.221 & 7.592 \end{bmatrix}$$

$[90_3/0_3]_{2s}$

$$\begin{bmatrix} 1.000 & 0.000 & 0.000 \\ 0.817 & 7.657 & -1.235 \\ 0.816 & -1.239 & 7.652 \end{bmatrix}$$

$[90_4/0_4]_{2s}$

$$\begin{bmatrix} 1.000 & 0.000 & 0.000 \\ 0.818 & 7.672 & -1.248 \\ 0.818 & -1.248 & 7.671 \end{bmatrix}$$

Figures 6.9 and 6.10 show the strain fields for a $[90_4/0_4]_{2s}$ and a $[90/0]_{2s}$ laminate, respectively, exposed to an out-of-plane transverse load.

Similarly to the observation for UD laminates, the disturbance caused by the presence of the optical fiber sensor in the $[90/0]_{2s}$ laminate (Figure 6.10) stretches out to the boundary with the adjacent 90-layers. As a result, the strain field and the

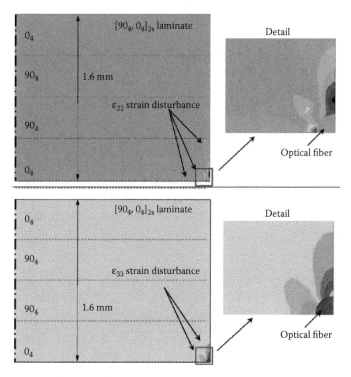

FIGURE 6.9   Transverse strain fields for a $[90_4/0_4]_{2s}$ laminate under an out-of-plane transverse load. (From Luyckx, G. et al., *Smart Mater Struct,* 19(10), 2010.)

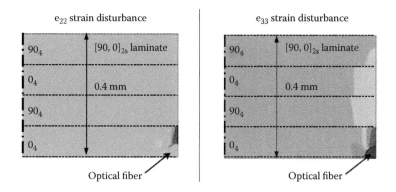

FIGURE 6.10   Transverse strain fields for a $[90/0]_{2s}$ laminate under an out-of-plane transverse load. (From Luyckx, G. et al., *Smart Mater Struct,* 19(10), 2010.)

resulting TC matrix are affected by the cross-ply layup in this case. For the $[90_4/0_4]_{2s}$ laminate, on the other hand, the disturbance is fully enclosed within the 0-layers (Figure 6.9), and hence, no significant influence of the adjacent 90-layers can be found in the strain field and the TC matrices. Under these circumstances, the TC matrix is nearly identical to the one obtained for a thick UD laminate (either through FE analysis or by using the analytical approach).

6.4.2.2   Effect of Sensor Position

Assuming the optical fiber to be embedded at the mid-plane of the laminate is rarely representative of actual applications, where the embedding location will be dictated by the desired quantity to be measured (for example, if bending strains in a turbine blade are envisaged, embedding near the neutral line of the laminate is pointless). The flexibility of the FE approach enables the influence of embedding position on the resulting TC matrix to be investigated.

A $[90_4/0_4]_{2s}$ laminate is used in this analysis with the optical fiber embedded in the central 0 layer, but with an offset distance from the centerline indicated by x. This is illustrated schematically in Figure 6.11.

The results of this study for different values of offset x are given in Table 6.5. The offset is increased in increments of 0.1 mm corresponding to the ply thickness of the M55J/M18 laminate.

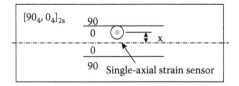

FIGURE 6.11   Schematic of an optical fiber sensor embedded away from the center line of a $[90_4/0_4]_{2s}$ M18/M55J cross-ply CFRP laminate. (From Luyckx, G. et al., *Smart Mater Struct,* 19(10), 2010.)

TABLE 6.5
TC Matrices for Different Offset Positions of the Optical Fiber from the Center Line in a $[90_4/0_4]_{2s}$ Cross-Ply CFRP Laminate

x=0.0mm

$$\begin{bmatrix} 1.000 & 0.000 & 0.000 \\ 0.818 & 7.672 & -1.248 \\ 0.818 & -1.248 & 7.671 \end{bmatrix}$$

x=0.1mm

$$\begin{bmatrix} 1.000 & 0.000 & 0.000 \\ 0.814 & 7.642 & -1.242 \\ 0.821 & -1.235 & 7.679 \end{bmatrix}$$

x=0.2mm

$$\begin{bmatrix} 1.000 & 0.000 & 0.000 \\ 0.814 & 7.635 & -1.235 \\ 0.819 & -1.232 & 7.661 \end{bmatrix}$$

x=0.3mm

$$\begin{bmatrix} 1.000 & 0.000 & 0.000 \\ 0.808 & 7.543 & -1.186 \\ 0.812 & -1.187 & 7.574 \end{bmatrix}$$

As the offset is increased, the off-diagonal coefficients TC_{23} and TC_{32} decrease simultaneously. In addition, the diagonal terms TC_{22} and TC_{33} decrease slightly, and the difference between both increases. This is again a consequence of the strain disturbance around the optical fiber sensor reaching the boundary of the 0 layers, and as a result affecting the strain distribution and the TC matrix.

6.4.2.2.3 Optical Fiber Type

Equation 6.3 indicates that a standard optical fiber sensor is inherently sensitive to transverse strains, which will result in birefringence of the sensor. However, when the differences in transverse strain components ε_1' and ε_2' are small, the difference in reflected Bragg wavelength is minimal and difficult to measure using traditional readout equipment. One solution to overcoming this issue is to exploit the different polarization state of both Bragg peaks through the polarization-dependent loss technique [10]. However, this requires specialized equipment and is currently only available for low sampling rates. For applications where traditional readout equipment is used or where high sampling frequencies are required, an alternative exists in the use of specialty optical fibers known as polarization-maintaining (PM) fibers and microstructured optical fibers (MOF). Figure 6.12 shows both a PM fiber and an MOF fiber (designed by VUB-TONA).

In Figure 6.12a, stress-applying parts (indicated as SAP) are used with different thermal and mechanical properties as the surrounding optical fiber material. During the cool-down phase of the manufacturing process, the SAPs result in the creation of a transverse strain field in the optical fiber core, which will lead to a large birefringence of the FBG, even in the absence of external transverse strains.

In the MOF design, on the other hand (Figure 6.12b), birefringence is a result of the air-hole pattern manufactured in the optical fiber cladding. These will create a strain field inside the optical fiber core, again leading to birefringence of the fiber. Although more costly than traditional optical fiber sensors, a well-chosen MOF design can result in an increase in transverse sensitivity of the optical fiber sensor [34].

Using the same FE technique as presented in Section 6.4.2, the influence of optical fiber structure on the transfer coefficient matrix can be investigated. Table 6.6 shows the resulting TC matrices for a standard optical fiber sensor, a bow-tie fiber, and an MOF fiber embedded in a $[90_2/02]_{2s}$ M18/M55J cross-ply laminate. The optical fiber

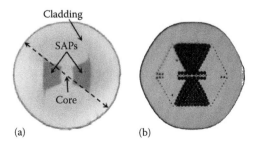

FIGURE 6.12 Photos of (a) a bow-tie fiber and (b) a microstructured optical fiber © VUB-TONA.

TABLE 6.6

TC Matrices for Different Types of Optical Fiber Sensors

Traditional			Bow-Tie			MOF		
1.000	0.000	0.000	1.000	0.000	0.000	1.004	0.027	0.009
0.818	7.631	−1.206	0.822	8.082	−0.969	−8.758	−47.948	−15.527
0.809	−1.221	7.592	0.805	−1.160	6.867	10.651	58.108	20.829

is assumed to be embedded at the mid-plane of the laminate. Table 6.6 shows the resulting TC matrices for a traditional, a bow-tie, and an MOF fiber.

As can be seen from the results, a significantly different TC matrix is obtained depending on the type of optical fiber sensor used. The difference is most pronounced for the MOF fiber, where the air-hole pattern is responsible for a significant modification of the stresses and strains inside the optical fiber core.

6.4.3 EXPERIMENTAL APPROACH

The technique applied in the FE approach to determine the TC matrix coefficients can equally be used in an experimental setup, using the embedded FBG as the sensor for sensed strains and using theoretical assumptions (or FE analysis) for the applied far-field strains. In this way, the numerical results can be validated and compared with experimental test data.

6.4.3.1 Experimental Setup

Several different composite samples with embedded optical fiber sensors were manufactured for the purpose of experimentally determining the transfer coefficient matrix.

As the FBG response will be used to determine the sensed strains, Equation 6.3 should be inverted to obtain the sensed strains from the measured wavelength shifts. However, three unknown strain values cannot be determined from the two wavelength shifts obtained from a single (birefringent) optical fiber sensor. Hence, two different types of composite samples were created.

In the first type of sample, a pure optical fiber is embedded inside the material and exposed to the full strain field of the composite host. In the second type of sample, the FBG is first encapsulated in a glass capillary, effectively isolating it from transverse strains in the host material. In this case, no birefringence can occur in the sensor, and the transverse strains are linked to the axial strain by the Poisson's coefficient of the fiber (as it is free to contract within the capillary). For this configuration, Equation 6.3 simplifies to

$$\frac{\Delta\lambda_{B,\text{cap}}}{\lambda_{0,\text{cap}}} = \varepsilon_{3'} - \frac{1}{2}\bar{n}^2\left[p_{11}\nu\varepsilon_{3'} + p_{12}\left(\nu\varepsilon_{3'} + \varepsilon_{3'}\right)\right] \tag{6.15}$$

This second sensor can then be used to determine the axial strain in the sensor. Combined with the wavelength shifts recorded by the first type of sample, the entire sensed strain field ε_i' can then be derived from the measured wavelength shifts. Combining Equation 6.3 for the second sensor and Equation 6.15 for the encapsulated sensor, the relationship between strains and wavelength shifts can be written as

$$
\begin{bmatrix} \Delta\varepsilon_1^s \\ \Delta\varepsilon_2^s \\ \Delta\varepsilon_3^s \end{bmatrix} = K^{-1} \begin{bmatrix} \dfrac{\Delta\lambda_{B,1}}{\lambda_0} \\ \dfrac{\Delta\lambda_{B,2}}{\lambda_0} \\ \dfrac{\Delta\lambda_{B,\mathrm{cap}}}{\lambda_{0,\mathrm{cap}}} \end{bmatrix}
\tag{6.16}
$$

with K given by

$$
K = \begin{bmatrix} 1 - \dfrac{\bar{n}^2}{2} p_{12} & -\dfrac{\bar{n}^2}{2} p_{11} & -\dfrac{\bar{n}^2}{2} p_{12} \\ 1 - \dfrac{\bar{n}^2}{2} p_{12} & -\dfrac{\bar{n}^2}{2} p_{12} & -\dfrac{\bar{n}^2}{2} p_{11} \\ \left(1 - \dfrac{\bar{n}^2}{2} p_{12}\right) + v\dfrac{\bar{n}^2}{2}(p_{11}+p_{12}) & 0 & 0 \end{bmatrix}
\tag{6.17}
$$

Using $v = 0.17$, $p_{11} = 0.113$, $p_{12} = 0.252$, and $\bar{n} = 1.456$, the K-matrix reduces to

$$
K = \begin{bmatrix} 0.733 & -0.120 & -0.267 \\ 0.733 & -0.268 & -0.120 \\ 0.799 & 0 & 0 \end{bmatrix}
\tag{6.18}
$$

The relationship between far-field composite strains and measured wavelength shift is then finally given by

$$
\left[\varepsilon^c\right] = \left[TC\right]\left[K\right]^{-1}\left[\dfrac{\Delta\lambda}{\lambda}\right]
\tag{6.19}
$$

Figure 6.13 shows the different sample configurations used in the determination of the TC matrix. All samples have a $[90_2/0_2]_{2s}$ layup. This layup has the benefit of creating birefringence in the optical fiber sensor during manufacturing as a result of residual strains. Hence, traditional optical fiber sensors could be used to determine the entire sensed strain field.

All samples were tested along their length direction, which corresponds to in-plane transverse loading for the type 1 samples and axial loading for the type 2 samples. The loading was achieved by placing the samples in an INSTRON 8801 hydraulic tensile machine with a 100 kN load cell (Figure 6.14a). The samples were loaded to roughly 120 MPa, and the test was repeated three times for each sample.

The third load case is achieved by loading both sample types in the through-thickness direction in compression. A ball-hinge was used to ensure a proper and

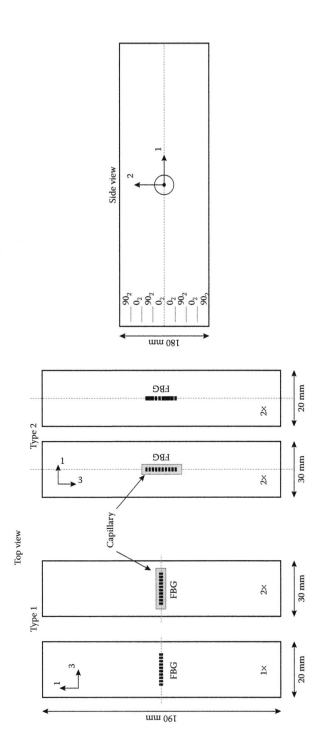

FIGURE 6.13 Different sample geometries used for the determination of the experimental TC matrix. (From Lammens, N., in Boller, C., Janocha, H., editors, *New Trends in Smart Technologies*, Fraunhofer, Germany, 2013 [35].)

(a) (b)

FIGURE 6.14 Loading of composite samples with embedded optical fiber sensors to experimentally determine the TC matrix: (a) axial and in-plane transverse loading, (b) out-of-plane transverse compressive loading.

uniform stress distribution over the FBG (Figure 6.14b). The samples were loaded to approximately 20 MPa compressive stress.

Using the wavelength shifts of the embedded optical fibers under these loading conditions, the sensed strains for all three load cases can be extracted. The far-field host strains are extracted from FE simulations corresponding to the applied load levels. Combining these strains then enables the TC matrix to be calculated for the tested material and layup.

6.4.3.2 Experimental Results

Figure 6.15 shows the measured wavelength shifts of the embedded optical fiber sensors for the different loading conditions applied to the samples.

These results indicate a reproducible, linear response of the sensor to the applied loads. Using Equation 6.16, it is possible to convert these wavelength shifts into sensed strain values. By combining these sensed strains with the simulated far-field strains corresponding to the applied loads, the TC matrix is found to be

$$TC_{exp} = \begin{bmatrix} 0.960 & -0.014 & -0.006 \\ 0.707 & 7.651 & -2.060 \\ 0.788 & -2.515 & 8.035 \end{bmatrix} \tag{6.20}$$

Compared with the results found through FE simulations, given in Table 6.4, a good correspondence can be observed between both approaches. These results therefore suggest that FE analysis can indeed be used to accurately predict the strain transfer behavior from a composite structure to its embedded sensor.

6.5 COATING OPTIMIZATION

So far, this chapter has illustrated that the process of embedding an optical fiber in a composite host will lead to a local distortion of the stress and strain field surrounding the optical fiber sensor, and as a result, will affect how far-field strains occurring in

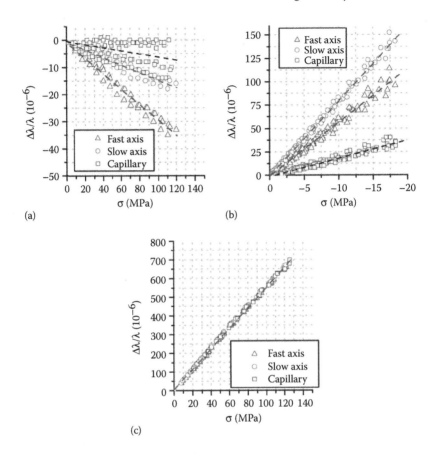

FIGURE 6.15 Wavelength shifts for different loading conditions on an embedded optical fiber sensor in a $[90_2/0_2]_{2s}$ cross-ply CFRP laminate: (a) in-plane transverse load, (b) out-of-plane transverse load, (c) axial load. (From Lammens, N., in Boller, C., Janocha, H., editors, *New Trends in Smart Technologies*, Fraunhofer, Germany, 2013 [35].)

the composite are transferred to the optical fiber core, where the strain field is sensed by the inscribed FBG. The previous sections have presented analytical, numerical, and experimental techniques for modeling this strain transfer response to obtain high-quality, reliable strain measurements from embedded optical fiber sensors.

As the mismatch in material properties between the embedded optical fiber sensor and the surrounding composite host becomes smaller (i.e., the properties are more comparable), the distortion resulting from the embedding will become less pronounced. Ultimately, when the optical fiber and composite material properties are perfectly matched, no distortion will occur. As these localized stress concentrations surrounding the optical fiber sensor will have a detrimental effect on the performance (e.g., onset of cracking and failure strength), any effort to reduce the amount of mismatch in material properties will be beneficial to the performance of the final "smart" structure. The most straightforward approach to matching the material properties is clearly to change the type of composite material used, as this allows a wide range of stiffness values to be obtained. However, the choice of composite material is usually dictated by the specific applications and its requirements, and in general, it cannot be changed freely.

An alternative way of matching the sensor and composite properties is through modifications of the optical fiber coating material. This coating material acts as a buffer between the host and the sensor. An adequate choice of coating properties (Young's modulus, Poisson's ratio, thickness, etc.) can have a significant effect on the resulting stress concentrations occurring in the host material.

This section will present a method, closely linked to the TC matrix approach, allowing the investigation of the effect of coating properties on interfacial stresses between the sensor and the composite host. As such, it allows an optimization study to be performed to determine the best-matched properties for the coating material to minimize the impact on the composite host.

6.5.1 STATE OF THE ART

Many publications are available documenting the influence of an embedded optical fiber on the composite host performance [36–41]. Shivakumar et al. [36] studied the influence of the optical fiber embedding orientation on the structural behavior of an instrumented composite. While tensile properties were only minimally affected, severe reductions in compressive properties were observed. In addition, the severity of impact on the properties was found to be related to the orientation of the optical fiber relative to the reinforcements. The largest impact was observed when the optical fiber was embedded perpendicular to the reinforcements, in which case typical lenticular resin pockets occur, which cause deterioration in the response of the structure [36,39,40,42–47]. Carman et al. [37] studied the impact of an optical fiber on the transverse strength of a composite. Similarly to the findings of Shivakumar et al., significant reductions in transverse strength of up to 50% were reported for traditional optical fibers (125 µm diameter), while less pronounced effects were reported for smaller-diameter fibers. Roberts et al. [38] studied both acrylate and polyimide-coated fibers and measured no significant influence on the static properties of the composite host, except for the compressive strength, for which reductions up to 26% were found. Lee et al. [39] and Surgeon et al. [40] investigated the behavior of a composite with an embedded optical fiber sensor under both static and fatigue experiments. It was reported that only small influences could be observed in static tensile properties, while some data suggested a substantial degradation of the composite performance under fatigue loading.

These publications all serve as evidence of the distortion of the composite host as a consequence of embedding an optical fiber sensor. Zolfaghar et al. [41] proposed either minimizing the size of the mismatch by using smaller sensors (as demonstrated by Kister et al. [48,49]) or tuning the coating properties of the optical fiber to achieve better performance of the host. Several researchers have investigated this second possibility and studied the influence of coating material properties on the mechanical behavior of the composite host and sensor [18,29,41,50–55].

Zolfaghar et al. [41] investigated the fiber–matrix interfacial shear strength for different types of coating material in composite hosts. Since strains are transferred from the host to the sensor through shear strains, proper shear strength was found beneficial to the longevity of the sensor and for applications where high strain levels are to be expected. In their research, it was found that a silane-coated fiber provided the best shear strength in a pure epoxy matrix. The shear strength of traditional coatings (e.g., acrylate or polyimide) was found to be slightly lower and resulting in thicker coating diameters, and thus being more disruptive. Barton et al. [54] and Wang et al. [55] studied the effect of coating properties on a cross-ply composite laminate. Two different coatings with significantly different material properties were explored, giving similar results under static loading conditions. It was concluded that edge effects were more important than the obtrusive effects of the sensor under these loading conditions. Under fatigue loads, however, a decrease in crack growth was observed, attributed to the local stiffening of the laminate at the optical fiber locations. Van Steenkiste et al. [29] investigated the strain transfer from a composite host to an embedded optical fiber, finding a clear distinction in transmitted strains between a coated and an uncoated optical fiber sensor. These publications show that the stresses and strains surrounding an optical fiber sensor are indeed affected by the precise coating material properties, and an optimal coating should exist.

Based on the premise that embedding an optical fiber inside a composite host will inevitably affect the mechanical properties of the host material, Dasgupta and Sirkis [56], Hadjiprocopiou et al. [57], and Barton et al. [58] set out to determine an optimum set of coating material properties to minimize its effect on the composite host.

Dasgupta and Sirkis [56] presented an analytical approach to determine the stress field in the composite host caused by the inclusion of a coated optical fiber sensor, aligned parallel to the reinforcements of a unidirectional composite under axial loading. Knowing the analytical expressions for the stress field, it was shown to be possible to optimize the coating material properties (Young's modulus, Poisson's ratio, and/or coating thickness) to achieve different objectives (minimal stresses in the fiber, the coating, or the composite host). It was shown that under certain circumstances, it was possible to completely eliminate transverse stresses in the host, effectively removing the influence of the embedded optical fiber sensor. Hadjiprocopiou et al. performed a similar investigation using finite element modeling of an embedded sensor in a unidirectional composite. In contrast to Dasgupta and Sirkis's research, Hadjiprocopiou et al. studied the optimal coating properties for a UD laminate exposed to a transverse strain rather than the axial load studied by Dasgupta and Sirkis. Under these transverse load circumstances, it is generally no longer possible to completely eliminate the influence of the optical fiber sensor, and an optimization criterion is required. It was found that in thick coatings, the highest stress at the coating–composite interface is located at 90° relative to the loading direction, while for thin coatings, the highest stress occurs in-line with the loading direction. Based on this data, Hadjiprocopiou et al. used an optimization criterion $\sigma_{rr}^{host}(0°) = \sigma_{\theta\theta}^{host}(90°)$, as already proposed by other authors [59,60]. In this criterion, $\theta = 0°$ represents the direction of the applied load perpendicular to the optical fiber axis, while $\theta = 90°$ is perpendicular to both the load direction and the optical fiber axis, as shown in Figure 6.16. This criterion is essentially identical to minimizing the maximal principal stress at the coating–composite interface.

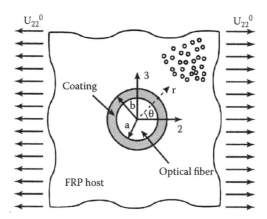

FIGURE 6.16 Schematic of a coated, embedded optical fiber under transverse loading showing the definition of coordinate system used. (From Lammens, N. et al., *Smart Mater Struct*, 24(11), 2015.)

Barton et al. [58] performed finite element simulations of a coated optical fiber embedded between the 0 plies of a cross-ply E-glass/epoxy laminate and investigated the influence of fiber position through the thickness of the laminate on optimal coating material properties. It was found that the FE results corresponded closely to analytical equations for an embedded fiber in a UD composite, which suggests that if the ply into which the optical fiber is embedded is sufficiently thick, the interfacial stresses and strains become independent of the part layup. A similar observation was made in Section 6.4.2.2.1, where the TC matrix of a cross-ply laminate was found to correspond to that of a UD laminate for sufficiently thick laminates.

While the work performed by Dasgupta, Hadjiprocopiou, and Barton presents valuable information on optimal coating properties, the load cases and composite layups considered have very limited applicability in real-world applications. In this work, a novel technique based on the TC matrix approach presented in Section 6.4 is described. It allows the coating–composite interfacial strains to be predicted for any load case (within the linear elastic regime where the superposition principle is valid) and for any composite layup provided the layer thickness into which the fiber is embedded is sufficiently thick. Combining this approach with a design space exploration (i.e., variations on coating's Young's modulus, Poisson's ratio, or thickness) and an optimization criterion, the methodology can be used to determine optimal coating properties for any given combination of host material, coating material, and load case(s) with a minimum amount of user inputs.

The work discussed in the following sections was performed within the EU FP7 "SmartFiber" project [61]. Within this project, one of the objectives was to optimize the coating thickness of the optical fiber sensor to make it better suited for embedding and sensing purposes. As a result, the examples presented in this chapter will be focused on optimizing the coating thickness. However, the method can just as easily be used to optimize any other property available to the user (e.g., Young's modulus, Poisson's ratio, coating thickness, or even any combination of these parameters).

6.5.2 TRANSFER MATRIX TOOL

6.5.2.1 Model Description

The coating optimization tool relies on the same basic principle of the TC matrix as presented in Section 6.4. However, in contrast to the previous TC matrix approach, the TC matrices used here do not map the far-field strains to the optical fiber core strains, but rather, to different points on the coating–composite interface, as shown in Figure 6.17.

For each point on the coating–composite interface, the relationship between local strains measured at the interface ($\varepsilon^s(\theta)$) and far-field strains (ε^∞) in the composite host can then be expressed as

$$\left[\varepsilon^s(\theta)\right] = \left[TC(\theta)\right]\left[\varepsilon^\infty\right] \tag{6.21}$$

Note that this method requires the strain field in the vicinity of the optical fiber to be uniform. If nonuniform strains exist at the embedding location, Equation 6.21 can no longer be used accurately, as a single point measurement of $\varepsilon\infty$ is no longer sufficient to describe the strain state of the pure sample. Additionally, it is assumed that the optical fiber is embedded parallel to the surrounding reinforcing fibers, as this avoids the creation of lenticular resin pockets, which would severely affect the composite strength and lead to varying strain fields at the coating–composite interface [36,39,40,42–47].

The TC matrices calculated by Equation 6.21 will be dependent on the precise coating material properties. Therefore, the TC matrices need to be calculated for the entire range of coating properties to be considered. Once all TC matrices are known (one for each angle θ and over the entire range of coating properties considered), Equation 6.21 can be used to calculate the interfacial strains $\varepsilon^s(\theta)$ for any given combination of far-field strains $\varepsilon\infty$ by application of the superposition principle. Using this knowledge of the coating–composite interfacial strain field, combined with any optimization criterion of choice (max. principal stress, Tsai-Wu, etc.), allows the most optimal set of

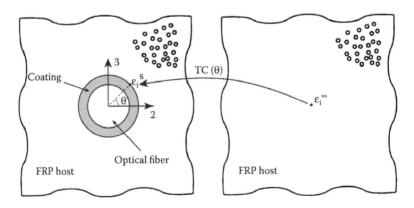

FIGURE 6.17 Transfer coefficient (TC) matrix principle for coating optimization. (From Lammens, N. et al., *Smart Mater Struct*, 24(11), 2015.)

coating properties to be determined for the given load cases (defined by its far-field strain field ε^∞). The limits of applicability of this method (related to laminate thickness, laminate layup, and coating eccentricity) will be considered later in this chapter.

6.5.2.2 TC Matrix Determination

In their most general form (excluding temperature effects), the TC matrices take on the form given by Equation 6.6 and represent 6×6 matrices. To determine all unknown TC coefficients for a given angle θ would therefore require the simulation of six independent load cases. However, using the assumption of transverse isotropy, this number of simulations can be reduced.

First of all, it should be obvious that in a transversely isotropic material, tensile loading along the 3-direction (with a coordinate system as defined in Section 6.3) produces identical interfacial stresses and strains as loading in the 2-direction, only rotated over an angle of 90° in the 2–3 plane. The same observation can be made for shear loading in the 1–2 and the 1–3 plane, thereby effectively reducing the number of required independent simulations from six to four.

Additionally, shear in the 2–3 plane only causes a rotation of the stresses and strains around the optical fiber axis (the 1-axis) without affecting their magnitude. Conceptually, this can be understood by considering that the stresses and strains should be independent of choice of coordinate system. As a consequence, the stresses and strains at a random point (defined by a radial distance r from the center of the optical fiber and an angle θ measured from the 2-axis toward the 3-axis) under pure τ_{23} shear load (Figure 6.18a) should be equal to those found at that same point using a coordinate system rotated over 45° along the 1-axis. In this rotated coordinate system, the shear load is translated into pure tensile loading along the new 2- and 3-axes (Figure 6.18b).

This assumption was validated using FE simulations, and as a consequence, it is always possible to work in a coordinate system for which $\tau_{23}^\infty = 0$ (the principal coordinate system in the 2–3 plane). Note that as τ_{23}^s is not necessarily zero, the TC matrix is reduced to a 6×5 matrix. The number of independent simulations required is then reduced to only three simulations.

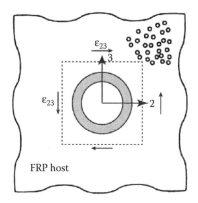 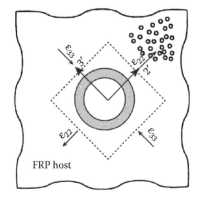

FIGURE 6.18 Influence of shear loading in the 2–3 plane on interfacial stresses and strains. (From Lammens, N. et al., *Smart Mater Struct*, 24(11), 2015.)

6.5.2.3 FE Implementation

The FE simulations are performed using ABAQUS finite element software. The model geometry is shown in Figure 6.19. The model contains a pure host material as well as a host with an embedded optical fiber sensor.

Although not necessary for simple load conditions, simulating the pure material as well allows the extraction of the far-field strains for any desired loading condition. Considering the simple load conditions used in this work, the simulated far-field strains are used only as confirmation that the proper boundary conditions were successfully applied to the model.

Exploiting symmetry, one quarter of the total model is required, thereby reducing computational efforts. The model has been partitioned in such a way that mesh refinement could be included at the location where the strain fields are extracted. The width and height of the sample are chosen to be five times the coating radius (indicated as b in Figure 6.16). The length is chosen to be 50 times the coating radius. These dimensions approximate an infinitely wide host, avoiding influences of edge effects on the determined TC matrices.

Three distinct load cases are simulated: (i) pure axial loading, (ii) pure transverse loading, and (iii) shear loading in the 1–2 plane. These loads are achieved by applying the correct displacements to the applicable faces, as shown in Figure 6.20.

As linear elasticity is assumed, the precise values of the displacement are not relevant. However, to avoid any possible geometrical nonlinearity, the displacements are limited to obtain a strain level of approximately 0.1%.

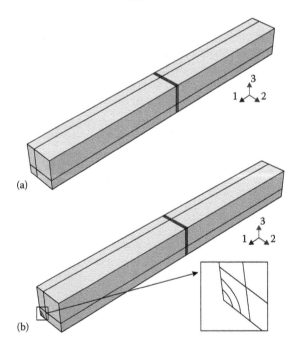

FIGURE 6.19 Model geometry for the determination of TC matrix coefficients: (a) pure host material, (b) embedded optical fiber. (From Lammens, N. et al., *Smart Mater Struct,* 24(11), 2015.)

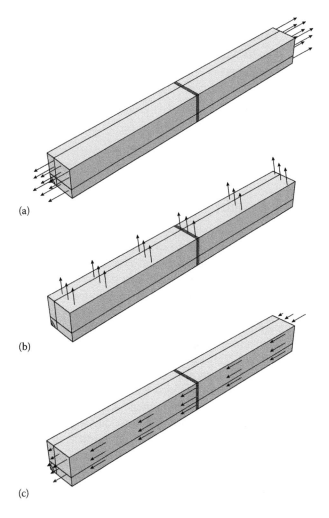

FIGURE 6.20   (a) Pure axial loading, (b) pure transverse loading, and (c) shear loading in the 1–2 plane. (From Lammens, N. et al., *Smart Mater Struct,* 24(11), 2015.)

The material properties of the host and the optical fiber are according to the data given in Tables 6.1 and 6.2. The coating is assumed to be Ormocer, for which the material properties are given in Table 6.7.

The simulations were performed using ABAQUS/Standard. Both composite parts were meshed using approximately 42.000 C3D8R elements, which was found to be the most suitable trade-off between calculation times and accuracy. The far-field strains and interfacial strains were exported and processed using MATLAB to determine the TC matrices over the entire coating interface. The coating thickness was varied between b/a = 1.1 (with b the coating outer radius and a the inner radius according to Figure 6.16) and b/a = 3.0 in steps of 0.01 to investigate the optimal coating thickness. The level of mesh refinement results in 23 nodes at which strain data is available at

TABLE 6.7
Elastic Material Properties for an Ormocer

E (GPa)	$\nu(-)$
1.5	0.32

the coating–composite interface. This results in 23 TC matrices per b/a ratio, leading to a total of 4393 TC matrices for the entire analysis of optimal coating thickness.

6.5.3 Limits of Applicability

As was stated previously, the dimensions of the FE part have been chosen such that edge effects will not have an effect on the strain distribution surrounding the optical fiber sensor. This results in part dimensions that approximate an infinitely wide host. For practical applications, however, an infinite host is a very large simplification and is not representative of experimental samples having limited dimensions.

The FE analysis allows the part dimensions to be changed to reflect true sample dimensions, to compare the actual (simulated) strain field in a part having finite dimensions with those obtained using the TC matrix approach (representing an infinite host). This analysis then allows us to determine the error made by assuming an infinitely wide host and thus to determine the limits of applicability of the method.

The TC matrices described in this section contain a total of 30 coefficients, and a separate TC matrix is calculated for each individual angle θ. An approach is necessary to quantify the error made by assuming an infinitely wide host as a single value.

Figure 6.21 shows the approach chosen. For each far-field strain component, the strains in a finite host (Step 1) and an infinite host (Step 2) are calculated. The largest

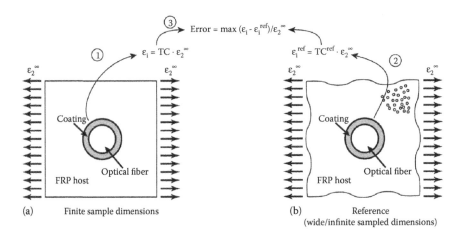

FIGURE 6.21 Approach to the calculation of the TC matrix error, applied to a far-field load ε_2^∞. (From Lammens, N. et al., *Smart Mater Struct*, 24(11), 2015.)

difference in strain component, normalized to the applied far-field strain, is stored as the error. This results in an error indicator for each individual far-field strain component. A reported error of 1% for ε_1^∞ then indicates that all calculated strains using the TC matrix are accurate to within 1% of the applied far-field strain.

6.5.3.1 Part Size

The error calculation allows the effect of part dimensions on the accuracy of the TC matrix approach to be investigated. Figure 6.22 shows the terminology used in the analysis of part dimensions in a UD laminate.

The part dimensions (c) are changed gradually in comparison to the optical fiber coating (b). For each individual c/b ratio, the TC matrix error is calculated as described in Section 6.5.3. The results of this analysis applied to an Ormocer-coated optical fiber (b/a ratio of 1.1) embedded in a M55J/M18 CFRP composite are shown in Figure 6.23. Similar results are obtained for b/a = 3.0.

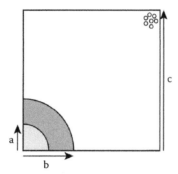

FIGURE 6.22 Dimensions for error analysis in a UD laminate. (From Lammens, N. et al., *Smart Mater Struct,* 24(11), 2015.)

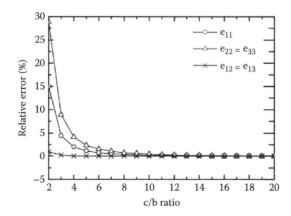

FIGURE 6.23 TC matrix error for a UD laminate with finite dimensions. (From Lammens, N. et al., *Smart Mater Struct,* 24(11), 2015.)

As expected, the TC matrix error gradually decreases to 0 as the part becomes larger and larger. At a c/b ratio of 4, the error on all far-field strain components is already lower than 5%. Assuming an 80 μm optical fiber sensor, this would equal a laminate thickness of 350 μm. As a single M55J/M18 prepreg layer has an approximate thickness of 100 μm, this means that a $[0_4]$ laminate can already be analyzed using the TC matrix approach with error below 5%.

6.5.3.2 Composite Layup

Similarly to the analysis of part size, FE simulations allow the effect of composite layup to be investigated. Figure 6.24 shows the dimensions used to study the effect of a $[90_x/0_x]_s$ cross-ply laminate.

The layer thickness (c) is gradually increased, and the TC matrix error is calculated for each c/b ratio. The results of this analysis for an Ormocer-coated optical fiber (b/a = 1.1) embedded in a M55J/M18 CFRP composite are given in Figure 6.25.

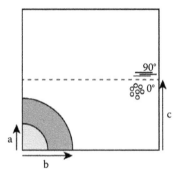

FIGURE 6.24 Dimensions for error analysis in a $[90_x/0_x]_s$ cross-ply laminate. (From Lammens, N. et al., *Smart Mater Struct,* 24(11), 2015.)

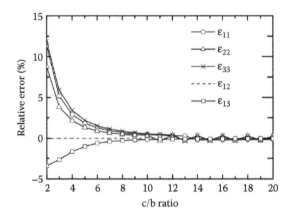

FIGURE 6.25 TC matrix error for a cross-ply laminate with finite dimensions. (From Lammens, N. et al., *Smart Mater Struct.* 24(11), 2015.)

While the decrease to 0% is slower than in the case of a UD laminate, increasing the layer thickness results in a decrease in TC matrix error. This means that, for sufficiently thick laminates, the TC matrix approach may be used regardless of the actual layup. This is an important benefit of the method, as it thus becomes independent of the specific layup of the part and does not require dedicated calculations for each type of layup considered.

For a c/b ratio of 3, the TC matrix error is approximately 5%. Assuming an 80 μm optical fiber, this is equivalent to a $[90_2/0_2]_s$ laminate with the optical fiber embedded in the mid-plane of the composite.

For an optical fiber embedded eccentric from the mid-plane of the layer into which it is embedded, this analysis can still be used to estimate the error. In this case, the lowest value of c (i.e., the distance to the nearest change in ply orientation) should be used, giving an overestimation of the error made using the TC matrix.

Note that these values represent a worst-case scenario and ensure that all errors in predicted strains are below 5% of the actual value. It should be noted that these, sometimes high, values are in part an artifact of the TC matrix error calculation. The maximal relative error focuses on the strain component that has the largest deviation from the reference strain. However, this does not necessarily coincide with the most dominant strain components. It should be emphasized that the error indicator can potentially overestimate the true error between different TC matrices. However, given the 30 coefficients in each TC matrix, any single value representing the "error" will have its downsides.

6.5.3.3 Coating Eccentricity

Besides working under the assumption of an infinitely wide host, the TC matrix approach also assumes the coating and optical fiber to be perfectly concentric. In reality, this is difficult to achieve, and the effect of coating eccentricity should be taken into account.

Using FE analysis, it is again possible to study the effect of coating eccentricity on the TC matrix error. Eccentricity is defined as shown in Figure 6.26.

The resulting TC matrix error for this analysis in a UD laminate with b/a = 1.1 is given in Figure 6.27.

These results show that the total error is limited for reasonable values of coating eccentricity (<10%). However, at large coating eccentricities, the errors become excessively large, and the TC matrix approach will no longer yield reliable results.

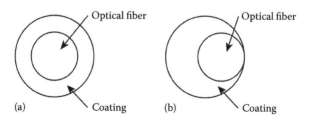

FIGURE 6.26   Definition of coating eccentricity: (a) 0% eccentricity, (b) 100% eccentricity. (From Lammens, N. et al., *Smart Mater Struct,* 24(11), 2015.)

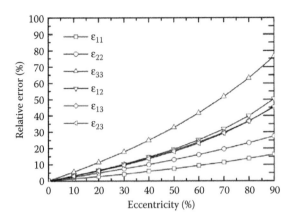

FIGURE 6.27 TC matrix error for different values of coating eccentricity. (From Lammens, N. et al., *Smart Mater Struct.* 24(11), 2015.)

The TC matrix error curves presented here can be used as a first indication of whether the TC matrix approach can be used safely or a more detailed investigation is required.

6.5.4 EXAMPLES

To illustrate the power and flexibility of the presented method, and to validate the results obtained against the limited amount of data available from the literature, a couple of examples are given for the specific case of an Ormocer-coated optical fiber sensor embedded in an M55J/M18 composite host.

6.5.4.1 Pure Axial or Transverse Load

Dasgupta and Sirkis [56] presented a technique to optimize the optical fiber coating under purely axial load conditions in a UD laminate. Hadjiprocopiou et al. [57] presented an FE technique for the optimization of coating properties in a UD laminate under transverse loading conditions. The methods described by these authors serve as a benchmark to test the validity and accuracy of the TC matrix approach. The analysis is applied to an Ormocer-coated optical fiber (properties as given in Tables 6.2 and 6.7) embedded in an M18/M55J UD composite (properties as given in Table 6.1). The results according to the literature and the TC matrix approach are given in Table 6.8.

Clearly, the results in Table 6.8 show a perfect correspondence between the methods described in the literature and the results achieved using the TC matrix approach presented. In addition, these results indicate the strong dependence of optimal coating properties (b/a ratio) on the type of load applied to the composite host. As such, a tool such as the TC matrix is very important, as it allows the problem of optimal coating properties to be studied under a variety of representative load conditions, rather than being limited to academic load conditions such as a perfect axial or transverse load.

TABLE 6.8

Optimal b/a Ratios according to the Literature and the TC Matrix Approach

	Optimal b/a Ratios	
	Literature	TC Matrix
Axial load	2.61	2.61
Transverse load	1.27	1.27

Source: Data for axial load from Dasgupta, A. and Sirkis, J.S., *AIAA J*, 30(5), 1337–43, 1992, and data for transverse load from Hadjiprocopiou, M. et al., *Smart Mater Struct*, 5(4), 441–8, 1996.

6.5.4.2 Combined Axial and Transverse Load

As the TC matrix approach relies on the principle of superposition of individual load components, the calculations required to determine the TC matrix coefficients only need to be performed once for a given combination of host material and coating material. Once known, the determined TC matrices can be used to determine the interfacial strain field for any combination of far-field strains. As a result, the TC matrix allows the evolution of optimal b/a ratio to be studied for a wide variety of loads with a minimum amount of computational effort involved. As an example of the capabilities and value of the TC matrix approach, the evolution of optimal b/a ratio with a load evolving from purely transverse (u_t) to purely axial (u_a) is studied. The applied load is expressed as $u = \alpha u_a + (1-\alpha)u_t$, with α ranging from 0 to 1. The calculation of the TC matrices is according to the description given in Section 6.5.2.2. Minimization of the maximal principal strain over the coating–composite interface is chosen as an optimization criterion, as this should converge to the optimization criteria used in the literature by Dasgupta and Sirkis [56] and Hadjiprocopiou et al. [57]. The results of this optimization study are shown in Figure 6.28.

As expected, the results for a purely transverse load ($\alpha = 0\%$) converge to the result of Hadjiprocopiou et al. [57], while a purely axial load ($\alpha = 100\%$) corresponds to the results obtained by Dasgupta and Sirkis [56]. For results between a purely axial and a purely transverse load, however, the axial and transverse contributions in the strain field interfere and lead to intuitively unpredictable results, which could not be anticipated solely based on the results of an axial and transverse optimization.

Note that the curve shown in Figure 6.28 represents the optimal b/a ratio for a range of load conditions, resulting in the lowest principal strains over the coating–composite interface. However, this does not mean that large deviations in b/a ratio will necessarily result in significant fluctuations of the strains and stresses surrounding the optical fiber. For example, changing the b/a ratio from b/a = 2.61 for a purely axial load ($\alpha = 100\%$) to a value of b/a = 1.27 will only lead to radial stresses equal to $\varsigma_r = 6.77 \times 10^{-4}\sigma_a$, where σ_a is the applied axial stress, which represents a negligibly small value. Under a purely

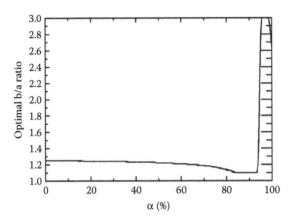

FIGURE 6.28 Optimal b/a ratio for evolving loads from purely transverse ($\alpha=0\%$) to purely axial ($\alpha=100\%$) considering b/a ratio in the range 1.1–3.0. (From Lammens, N. et al., *Smart Mater Struct,* 24(11), 2015.)

transverse load ($\alpha=0\%$), on the other hand, switching from an optimal b/a$=1.27$ to b/a$=2.61$ will increase the stress concentration factor from 1.25 to 1.80. The sensitivity of stress and strain levels to changes in b/a ratio is thus strongly dependent on the specific load case as well as the material combination studied.

6.5.4.3 Global Optimum

In many practical applications, load scenarios can be expected to vary as a function of time or location on the composite structure. In addition, changing the optical fiber b/a ratio for each specific application may not always be feasible or economically reasonable. It is therefore useful to study the optimal coating properties, assuming that all far-field strains (or any linear combination of these strains) are equally likely to occur (as a function of either time or space):

$$\varepsilon = \sum_{i=1}^{6} \alpha_i \varepsilon_i^{\infty} \tag{6.22}$$

with $\alpha_i=-1.1$. The optimal b/a ratio in this case would then be the one for which the maximal principal strain over all possible loads and over the entire interface is minimal. The results of this analysis considering only principal strains ($i=1\ldots3$), only shear strains ($i=4\ldots6$), and all strain components ($i=1\ldots6$) are given in Table 6.9.

The results show that a value-minimal coating thickness is needed when only shear strains are considered. In the case where principal loads are included, the optimal coating b/a ratio increases to b/a$=1.17$ (for the studied combination of coating and host properties). The computations involved in the determination of the results given in Table 6.9 equated to a total of 11 million simulations to be considered, which required less than 60 min to compute using the TC matrix approach. It should be clear that no other technique is currently available in the literature capable of

TABLE 6.9
Global Optimum b/a Ratios

	b/a Ratio
Principal strains	1.17
Shear strains	1.10
All strains	1.17

considering this number of load cases in a reasonable amount of time, which emphasizes the benefits and value of the TC matrix approach.

A similar approach to the one described in Section 6.5 can be used to determine the optimal b/a ratio for a subset of far-field strains (corresponding to the load cases expected for a given application) or for a subset of achievable b/a ratios. Similarly, the methodology can be used to determine the optimal value for any other property that can be manipulated by the user.

6.6   CONCLUSIONS

This chapter has focused on the concept of strain transfer between a composite host and an embedded optical fiber sensor. It was shown to be critically important to account for strain transfer effects to obtain relevant and reliable measurements from an embedded sensor system, and several methodologies were presented that enabled accounting for these strain transfer effects. The different approaches presented (analytical, numerical, and experimental) were compared with each other, showing good correspondence between the different approaches. The benefits and downsides of the different methodologies were discussed, allowing the reader to make an informed decision concerning which approach is best suited for a specific application.

Continuing on the strain transfer principle and the fact that an optical fiber will locally distort the surrounding stress and strain field in the composite host, an extension of the TC matrix approach was presented to study the concept of optimal coating properties. It has been stated in the literature that tuning the material properties of the optical fiber coating could have beneficial effects on the interference of the sensor with the composite host [29,41,48,49,55–59]. The number of methodologies allowing the optimal coating properties concept to be studied is limited to a couple of papers, applicable to very academic loading conditions (purely axial or purely transverse). By extending the TC matrix approach, a methodology was derived that allows the study of optimal coating properties for any arbitrary load case with a minimal amount of computational effort. The concept and FE implementation of the methodology were discussed, and the limits of applicability were derived. These results show that the TC matrix approach could be used for a wide range of laminate sizes, layups, and coating eccentricities without significant influence on the accuracy of the method. As the diameter of optical fibers will continue to decrease, the TC matrix will be usable for almost any conceivable type of composite structure. Finally, to illustrate and validate the performance of the novel TC matrix approach, a couple of examples were discussed. The TC matrix approach was shown to be capable of determining a globally optimal b/a ratio, giving optimal results when the precise load conditions are unknown (or equally, when all load cases can be expected). This is a unique capability of the TC matrix approach, which would be inconceivable with any other modeling approach currently available in the literature.

The methods presented in this chapter represent a critically important tool-set for the development of accurate and reliable strain-sensing solutions using optical FBGs in composite host structures while minimizing the influence of the sensing solution on the structural performance of the host. Techniques leading to a more accurate interpretation of the sensor response, as well as techniques leading to a reduction of the impact on the structural performance of the host, have been presented. To obtain the optimum solution, both methodologies should be combined to determine the global optimum sensor design, readout technique, and embedding path for optimal performance of the resulting smart structure.

REFERENCES

1. Degrieck J, De Waele W, Verleysen P. Monitoring of fibre reinforced composites with embedded optical fibre Bragg sensors, with application to filament wound pressure vessels. *NDT & E Int* 2001;34(4):289–96.
2. de Oliveira R, Ramos CA, Marques AT. Health monitoring of composite structures by embedded FBG and interferometric Fabry–Pérot sensors. *Comput Struct* 2008;86(3–5):340–6.
3. De Baere I, Voet E, Van Paepegem W, Vlekken J, Cnudde V, Masschaele B, et al. Strain monitoring in thermoplastic composites with optical fiber sensors: Embedding process, visualization with micro-tomography, and fatigue results. *J Thermoplast Compos* 2007;20(5):453–72.
4. Kalamkarov AL, Fitzgerald SB, MacDonald DO, Georgiades AV. The mechanical performance of pultruded composite rods with embedded fiber-optic sensors. *Compos Sci Technol* 2000;60(8):1161–9.
5. Li XH, Zhao CS, Lin J, Yuan SF. The internal strain of three-dimensional braided composites with co-braided FBG sensors. *Opt Laser Eng* 2007;45(7):819–26.
6. Yuan SF, Huang R, Rao YJ. Internal strain measurement in 3D braided composites using co-braided optical fiber sensors. *J Mater Sci Technol* 2004;20(2):199–202.
7. Jung K, Kang TJ. Cure monitoring and internal strain measurement of 3-D hybrid braided composites using fiber Bragg grating sensor. *J Compos Mater* 2007;41(12):1499–519.
8. Jung K, Kang TJ, Lee B, Kim Y. Estimation of residual strain and stress in interply hybrid composite using fiber Bragg grating sensor. *Fiber Polym* 2007;8(4):438–42.
9. Mulle M, Collombet F, Olivier P, Grunevald YH. Assessment of cure residual strains through the thickness of carbon-epoxy laminates using FBGs, Part I: Elementary specimen. *Compos Part A-Appl S* 2009;40(1):94–104.
10. Lammens N, Kinet D, Chah K, Luyckx G, Caucheteur C, Degrieck J, et al. Residual strain monitoring of out-of-autoclave cured parts by use of polarization dependent loss measurements in embedded optical fiber Bragg gratings. *Compos Part A-Appl S* 2013;52:38–44.
11. Minakuchi S, Takeda N, Takeda S-i, Nagao Y, Franceschetti A, Liu X. Life cycle monitoring of large-scale CFRP VARTM structure by fiber-optic-based distributed sensing. *Compos Part A-Appl S* 2011;42(6):669–76.

12. Hernández-Moreno H, Collombet F, Douchin B, Choqueuse D, Davies P, González Velázquez JL. Entire life time monitoring of filament wound composite cylinders using Bragg grating sensors: II. Process monitoring. *Appl Compos Mater* 2009;16(4):197–209.

13. Hernandez-Moreno H, Collombet F, Douchin B, Choqueuse D, Davies P. Entire life time monitoring of filament wound composite cylinders using Bragg grating sensors: III. In-service external pressure loading. *Appl Compos Mater* 2009;16(3):135–47.

14. Timoshenko S. *Theory of Elasticity*. New York: McGraw-Hill; 1951.

15. Lekhnitskii SG. *Theory of Elasticity of an Anisotropic Body*. Moscow: Mir; 1981.

16. Kollar LP, Van Steenkiste RJ. Calculation of the stresses and strains in embedded fiber optic sensors. *J Compos Mater* 1998;32(18):1647–79.

17. Cheng CC, Lo YL, Pun BS, Chang YM, Li WY. An investigation of bonding-layer characteristics of substrate-bonded fiber Bragg grating. *J Lightwave Technol* 2005;23(11):3907–15.

18. Li WY, Cheng CC, Lo YL. Investigation of strain transmission of surface-bonded FBGs used as strain sensors. *Sensor Actuat A-Phys* 2009;149(2):201–7.

19. Wan KT, Leung CKY, Olson NG. Investigation of the strain transfer for surface-attached optical fiber strain sensors. *Smart Mater Struct* 2008;17(3).

20. Chang CC, LeBlanc M, Vohra S. Investigation of transverse stress measurements by using embedded fiber Bragg grating sensors subjected to host Poisson's effect. *P Soc Photo-Opt Ins* 2000;3986:472–9.

21. Fan Y, Kahrizi M. Characterization of a FBG strain gage array embedded in composite structure. *Sensor Actuat A-Phys* 2005;121(2):297–305.

22. Butter CD, Hocker GB. Fiber Optics Strain-Gauge. *Appl Opt* 1978;17(18):2867–9.

23. Voet E, Luyckx G, Degrieck J. *Response of Embedded Fibre Bragg Gratings: Strain Transfer Effects*. 2009. SPIE, Bellingham, WA, p. 75035N-N-4.

24. Luyckx G, Voet E, De Waele W, Van Paepegem W, Degrieck J, Vlekken J. Strain monitoring of FRP elements using an embedded fibre optic sensor. *Adv Sci Tech* 2009;56:435–40.

25. Voet E, Luyckx G, De Waele W, Degrieck J. Multi-axial strain transfer from laminated CFRP composites to embedded Bragg sensor: II. Experimental validation. *Smart Mater Struct* 2010;19(10).

26. Bertholds A, and R. Dandliker. Determination of the individual strain-optic coefficients in single-mode optical fibers. *J Lightwave Technol* 1988;6(1):17–20.

27. Bosia F, Giaccari P, Botsis J, Facchini M, Limberger HG, Salathe RP. Characterization of the response of fibre Bragg grating sensors subjected to a two-dimensional strain field. *Smart Mater Struct* 2003;12(6):925–34.

28. Dubbel H, Beitz W, Küttner KH, Davies BJ. *Handbook of Mechanical Engineering*. Berlin: Springer; 1994.

29. Van Steenkiste RJ, Kollár LP. Effect of the coating on the stresses and strains in an embedded fiber optic sensor. *J Compos Mater* 1998;32(18):1680–711.

30. Van Steenkiste RJ. *Strain and Temperature Measurement with Fiber Optic Sensors*. Routledge: Taylor & Francis; 1996.

31. Herakovich CT. *Mechanics of Fibrous Composites*. New York: Wiley; 1997.

32. Kim KS, Kollar L, Springer GS. A model of embedded fiber optic Fabry-Perot temperature and strain sensors. *J Compos Mater* 1993;27(17):1618–62.

33. De Waele W. *Structural Monitoring of Composite Elements Using Optical Fibres with Bragg-Sensors*. Ghent, Belgium: Ghent University, Faculty of Engineering; 2002.

34. Sonnenfeld C, Sulejmani S, Geernaert T, Eve S, Lammens N, Luyckx G, et al. Microstructured optical fiber sensors embedded in a laminate composite for smart material applications. *Sensors* 2011;11(3):2566–79.

35. Lammens N. Experimental determination of the multi-axial strain transfer from CFRP-laminates to embedded Bragg sensor. In: Boller C, Janocha H, editors. *New Trends in Smart Technologies*, Saarbrucken, Germany: Fraunhofer; 2013. pp. 151–5.

36. Shivakumar K, Emmanwori L. Mechanics of failure of composite laminates with an embedded fiber optic sensor. *J Compos Mater* 2004;38(8):669–80.

37. Carman GP, Paul C, Sendeckyj GP. *Transverse Strength of Composites Containing Optical Fibers*. Bellingham: SPIE - Int Soc Optical Engineering; 1993.

38. Roberts SSJ, Davidson R. *Mechanical-Properties of Composite-Materials Containing Embedded Fiber Optic Sensors*. Bellingham: SPIE - Int Soc Optical Engineering; 1991.

39. Lee DC, Lee JJ, Yun SJ. The mechanical characteristics of smart composite structures with embedded optical-fiber sensors. *Compos Struct* 1995;32(1–4):39–50.

40. Surgeon M, Wevers M. Static and dynamic testing of a quasi-isotropic composite with embedded optical fibres. *Compos Part A-Appl S* 1999;30(3):317–24.

41. Zolfaghar K, Folkes MJ. The effect of surface coatings of optical fibers on the interfacial shear strength in epoxy resins. *J Mater Sci Lett* 1999;18(24):2017–20.

42. Seo DC, Lee JJ. Effect of embedded optical-fiber sensors on transverse crack spacing of smart composite structures. *Compos Struct* 1995;32(1–4):51–8.

43. Melin LG, Levin K, Nilsson S, Palmer SJP, Rae P. A study of the displacement field around embedded fibre optic sensors. *Compos Part A-Appl S* 1999;30(11):1267–75.

44. Ling HY, Lau KT, Lam CK. Effects of embedded optical fibre on mode II fracture behaviours of woven composite laminates. *Compos Part B-Eng* 2005;36(6–7):534–43.

45. Hadzic R, John S, Herszberg I. Structural integrity analysis of embedded optical fibres in composite structures. *Compos Struct* 1999;47(1–4):759–65.

46. Choi C, Seo DC, Lee JJ, Kim JK. Effect of embedded optical fibers on the interlaminar fracture toughness of composite laminates. *Compos Interface* 1998;5(3):225–40.

47. Benchekchou B, Ferguson NS. The effect of embedded optical fibres on the fatigue behaviour of composite plates. *Compos Struct* 1998;41(2):113–20.

48. Kister G, Wang L, Ralph B, Fernando GF. Self-sensing E-glass fibres. *Opt Mater* 2003;21(4):713–27.

49. Kister G, Ralph B, Fernando GF. Damage detection in glass fibre-reinforced plastic composites using self-sensing E-glass fibres. *Smart Mater Struct* 2004;13(5):1166–75.

50. Her SC, Huang CY. Effect of coating on the strain transfer of optical fiber sensors. *Sensors (Basel)* 2011;11(7):6926–41.

51. Yuan LB, Zhou LM, Wu JS. Investigation of a coated optical fiber strain sensor embedded in a linear strain matrix material. *Opt Laser Eng* 2001;35(4):251–60.

52. Sirkis JS, Grande R. Non-linear analysis of ductile metal coated optical fiber sensors embedded in a unidirectional laminate. *J Compos Mater* 1997;31(10):1026–45.

53. Okabe Y, Tanaka N, Takeda N. Effect of fiber coating on crack detection in carbon fiber reinforced plastic composites using fiber Bragg grating sensors. *Smart Mater Struct* 2002;11(6):892–8.

54. Barton EN, Ogin SL, Thorne AM, Reed GT, Le Page BH. Interaction between optical fibre sensors and matrix cracks in cross-ply GRP laminates—part 1: Passive optical fibres. *Compos Sci Technol* 2001;61(13):1863–9.

55. Wang H, Ogin SL, Thorne AM, Reed GT. Interaction between optical fibre sensors and matrix cracks in cross-ply GFRP laminates. Part 2: Crack detection. *Compos Sci Technol* 2006;66(13):2367–78.

56. Dasgupta A, Sirkis JS. Importance of coatings to optical fiber sensors embedded in smart structures. *AIAA J* 1992;30(5):1337–43.

57. Hadjiprocopiou M, Reed GT, Hollaway L, Thorne AM. Optimization of fibre coating properties for fiber optic smart structures. *Smart Mater Struct* 1996;5(4):441–8.

58. Barton EN, Ogin SL, Thorne AM, Reed GT. Optimisation of the coating of a fibre optical sensor embedded in a cross-ply GFRP laminate. *Compos Part A-Appl S* 2002;33(1):27–34.

59. Case SW, Carman GP. Optimization of fiber coatings for transverse performance: An experimental study. *J Compos Mater* 1994;28(15):1452–66.

60. Carman GP, Averill RC, Reifsnider KL, Reddy JN. Optimization of fiber coatings to minimize stress concentrations in composite materials. *J Compos Mater* 1993;27(6):589–612.

61. SmartFiber. SmartFiber Final Technology white paper (http://www.smartfiber-fp7.eu).

7 Monitoring Process Parameters Using Optical Fiber Sensors in Composite Production

Edmon Chehura

CONTENTS

7.1 Introduction 202
7.2 Manufacturing Processes and Process Parameters 203
    7.2.1 Automated Fabric Placement..204
    7.2.2 Through-Thickness Reinforcement 204
    7.2.3 Resin Infusion..206
    7.2.4 Process Temperature..207
    7.2.5 Curing Process...208
    7.2.6 Commissioning: Postproduction Monitoring 209
    7.2.7 Summary of Current Monitoring Techniques 210
7.3 Principles of Optical Fiber Sensors 211
    7.3.1 Optical Fiber Fresnel Sensor 211
    7.3.2 Tapered Optical Fiber Sensor 215
    7.3.3 Optical Fiber Bragg Grating (FBG) 217
    7.3.4 Tilted Fiber Bragg Grating Sensor (TFBG) 222
7.4 Measurement Applications: Laboratory Manufacturing 225
    7.4.1 Resin Infusion and Flow Monitoring..225
    7.4.2 Process Temperature Monitoring 233
    7.4.3 Cure Monitoring of Fiber-Reinforced Preform: Refractive Index
        of Resin 238
    7.4.4 Cure Monitoring of Fiber-Reinforced Preform: Strain
        Development 243
7.5 Measurement Applications: Industrial Manufacturing 244
    7.5.1 Carbon Fiber–Composite Aircraft Tailcone Assembly 244
        7.5.1.1 Carbon Fiber Preform and Sensor Layup 245
        7.5.1.2 Infusion Monitoring..248
        7.5.1.3 Cure Monitoring: Development of Transverse Strain........250
        7.5.1.4 Cure Monitoring: Development of Longitudinal Strain 253
        7.5.1.5 Post-Fabrication Mechanical Testing: Strain Monitoring...256

     7.5.2   Superconducting Magnet Coil: Postfabrication Test Monitoring 260
7.6   Conclusions, Challenges, and Future Requirements 264
References...265

7.1 INTRODUCTION

There is a significant increase in the uptake of composite material technology across a number of industries, particularly aerospace, thanks to the development of composite materials with innovative characteristics, aided by advancements in manufacturing processes. Composite materials are lightweight and they have a large strength-to-weight ratio, making them important for energy conservation in a range of applications, including aerospace. The aerospace industry alone has experienced major strides in the use of composite materials for aircraft structures, where Airbus (A380 and A350 planes) and Boeing (787 Dreamliner plane) have pioneered commercial jet aircraft with significant composite material content. Industrial-scale production of composite material components does not, at present, incorporate sensors within the components to monitor the manufacturing process, and quality control is largely carried out empirically.

Unreliability or defects in composite materials arise from a number of issues, all of which are linked to the process by which they are made. The process of filling the preform with resin (i.e., infusion process) and the resin curing process directly influence the quality of the final part. Defects that can arise from the infusion process include nonuniformly filled preform, porosity of the preform, and dry spots. The curing process can cause degradation in the composite part during manufacture, which will lead to poor structural health for the component. Degradation can result from nonuniform cure across the composite part, which will lead to, for example, development of internal stresses, residual stresses, thickness variation, and warpage. Online monitoring of process parameters such as temperature or pressure across the entire composite part could be exploited by implementing active feedback control or adaptive manufacturing, which will enhance quality upgrade during the curing process.

Small-scale laboratory monitoring of some of the manufacturing processes is often performed using conventional nonoptical methods [1]. Dielectric sensors are used to monitor resin flow, void content, viscosity, and the degree of cure by measuring the dielectric permittivity and loss factor (i.e., dielectric sensitivity) of the composite material between two or more electrodes that are excited by an alternating current [2,3]. Direct current sensors are used to monitor resin flow and the degree of cure by sensing the resin/composite electrical resistance between two electrodes [1,4]. Ultrasonic transducers can also be used for monitoring resin flow, resin cure, and void content [5]. The dependency of the ultrasonic wave velocity on the density and elastic modulus means that the time of flight through the composite preform will change with the progress of the cure reaction. Thermocouples and infrared thermometers can be used to monitor resin flow when there is a difference in temperature between the mold and the resin during the infusion process [6]. Pressure transducers, often based on the piezoresistive effect, are used for monitoring resin flow by transducing resin pressure into electrical resistance [7]. The use of these sensors is restricted to laboratory applications, as the sensors are either not suitable for embedding into the composite or too large to be embedded into critical components without compromising the structural integrity.

Quality control in composite manufacturing processes can be implemented by the use of embedded sensors that are capable of providing real-time monitoring. Optical fiber sensors are well suited to this application, because they have many appropriate characteristics, including their small diameter, typically 40–125 µm. There are four categories of optical fiber sensors that can be considered for monitoring the processing of composite materials: evanescent field sensors, Fresnel sensors, interferometric sensors, and grating sensors. Resin infusion and cure can be monitored using sensors that are sensitive to changes in the refractive index of the surrounding medium. The arrival of resin at a location can thus be detected as the surrounding medium changes from air to resin with a concomitant significant increase in refractive index. The refractive index of the resin is known to be well correlated with the degree of cure of the resin, as will be shown in this chapter. Suitable sensors are based on detection of the perturbation of the evanescent field of the electromagnetic wave propagating in the core of the optical fiber, exploiting etched fiber, tapered fiber, and side polished fiber [8–10]. Alternatively, the refractive index dependence of the Fresnel reflection from a cleaved optical fiber end [11–14] can be exploited. Optical fiber interferometers, such as the extrinsic Fabry–Perot interferometer, have been used to monitor the development of internal strains during the curing of a composite [15]. The infusion, flow, and cure of resin can also be monitored using optical fiber grating sensors, such as the fiber Bragg grating (FBG), tilted fiber Bragg grating (TFBG), and long-period grating (LPG) sensors [12,16,17]. Grating sensors can also be easily configured to perform multi-parameter sensing, which is useful for simultaneous strain and temperature measurements [18–20].

This chapter presents the principles and applications of tapered optical fiber sensors, optical fiber Fresnel sensors, TFBG sensors, and FBG sensors fabricated in single mode (SM) and linearly high birefringent (HiBi) optical fiber for monitoring various process parameters during composite manufacture and out-of-production mechanical testing. Attention will be given to the monitoring of the tufting (through-thickness reinforcement) of a preform, progress of the infusion of resin, process temperature, progress of the degree of cure of resin, and the development of 3D components of strain during manufacture. Examples of both laboratory and industrial manufacturing applications will be provided, highlighting the challenges that exist.

7.2 MANUFACTURING PROCESSES AND PROCESS PARAMETERS

This section introduces basic manufacturing processes or procedures for composite materials, highlighting the need for optical fiber sensors to provide process monitoring. Detailed analysis of the manufacturing procedures of composites is beyond the scope of this chapter but can be found elsewhere [21,22].

(a)   (b)

FIGURE 7.1 Delivery head of the robotic system, i.e., part of an automated fabric placement system being used to apply a 90° ply of carbon fiber onto a tailcone assembly mold/tooling as used in the ADVITAC project, a European-funded research program (see Section 7.5.1). (a) Enlarged view; (b) whole view of the tailcone assembly mold. (From Coriolis Composites, Fiber placement. http://www.coriolis-composites.com/products/fiber-placement-process. html; ADVITAC, The Advitac Project. http://www.advitac.eu/ [24].)).

7.2.1 AUTOMATED FABRIC PLACEMENT

The process of manufacture starts with the placement of the carbon- or glass-reinforcement fibers (i.e., fabric) onto the required mold/tooling, and the fabric can either be preimpregnated (i.e., prepreg) with resin by a machine or can be dry fabric. The fabric, which can be woven or nonwoven, is either hand-laid or laid automatically using robotic machines onto a mold. Automated fiber-reinforcement (i.e., tape, tor, or filament) placement systems using robotic machines (Figure 7.1) are increasingly becoming the most preferred procedures when manufacturing large composite parts consisting of flat or complex geometries, particularly in the aerospace industry [22,23]. Automated fabric placement (AFP) ensures high production levels, precision, and repeatability. The robot delivery head is not only responsible for laying the fabric but also includes a heating system in the form of an infrared lamp, a diode laser, or a hot air system to preheat the fabric to ensure that the plies of the fabric adhere to each other during placement (Figure 7.1) [23]. An infrared camera fitted to the delivery head enables temperature sensing, which facilitates the control of temperature during fabric placement. The delivery head has a roller/shoe making contact with the fabric that has been placed onto the mold/tool, thus providing the required compaction pressure (Figure 7.1). The AFP system can be preprogrammed off-line such that it follows the computed trajectory automatically during the placement of fabric over the mold/tool.

Clearly, the preheating temperature and the compaction pressure are very important parameters, which should be precisely monitored and controlled if the uniformity of the laid preform is to be maintained. Nonuniform compaction pressure will lead to the formation of varying internal stresses within the laid preform. These process parameters (temperature and compaction pressure) will have to be monitored online, *in situ* and in real time, ideally by embedded sensors, making optical fiber sensors invaluable.

7.2.2 THROUGH-THICKNESS REINFORCEMENT

The fabric placement procedures discussed in Section 7.2.1 produce composites with filaments (or plies) lying in-plane only, and there are no fibers oriented in the out-of-plane (i.e., Z) direction to bridge the adjacent plies. Such composites are susceptible to interlaminar delamination on out-of-plane or impact loading. It is normal practice to introduce out-of-plane reinforcement to increase the resistance to delamination by a composite, for example, at a joint between two components. Tufting is an automated robotic system that is commonly used to implement the Z-reinforcement of the fabric after it has been laid up onto the mold/tooling but before the process of resin impregnation is carried out. Figure 7.2 shows the needle tufting a carbon-reinforced fiber preform and the principle of tufting as described by the sequences numbered from 1 to 4. During tufting, the needle first penetrates the fiber preform, inserting the thread, and then exits along the same path, leaving behind a thread loop (or *tuft*) that is held in place by friction, as shown in Figure 7.2b. A suitable supporting material (typically sacrificial foam) holds the thread on the underside of the preform, facilitating the formation of the tuft loops.

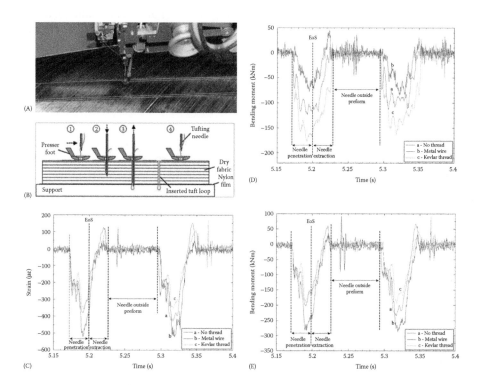

FIGURE 7.2 (A) Robot head driving the needle during the tufting of a carbon fiber reinforced preform and (B) schematic diagram showing the principle of tufting. The steps in the process are numbered sequentially. (C) A single tuft period showing (a) axial strain in the needle, (D) bending moment in the left–right plane, and (E) bending moment in the back–front plane of the needle, evaluated from the signals measured by four FBG sensors bonded to the needle. (From Dell'Anno, G. et al., *Structural Integrity*, **3**, 22–40, 2012; Chehura, E. et al., *Smart Materials and Structures*, **23**, 075001, 2014.) A four-layer quasi-isotropic preform was tufted at a speed of 500 rpm by a needle without thread, with metal wire, and with Kevlar® thread. EoS: end of stroke.

It is not always possible to tuft a dry preform successfully. Tight weave styles, use of sizing or binding agents, and excessive fiber compaction imposed during layup often make it difficult for the needle to penetrate the preform, resulting in either the tufting machine stopping altogether or, in the worst case, needle breakage. This is a manufacturing procedure that will clearly benefit from implementing an online monitoring sensor system for process parameters such as the stresses and strains experienced by the needle and the thread in real time during tufting. The process monitoring system will allow control of quality, control of repeatability, and control of the uniformity of the compaction pressure of the preform imposed by the robot head. The small dimensions of optical fiber sensors make them ideal for this task, as they can be, for example, bonded to the needle shaft [25]. In addition, optical fiber sensors embedded into the preform (during layup) as described in Section 7.2.1, for monitoring compaction pressure, could also be used together with the sensors embedded for resin infusion for the purposes of understanding and optimizing the flow of the resin through the preform. Figure 7.2 also shows strains and bending moments experienced by the needle during tufting, monitored online by four FBG sensors bonded to the needle. A detailed report on the monitoring of multiple strain components during tufting of carbon fiber–reinforced preform can be found elsewhere [14,25].

7.2.3 RESIN INFUSION

Resin infusion is the manufacturing process that follows the laying up of dry preform and/or the tufting process (Sections 7.2.1 and 7.2.2). Resin infusion involves the filling with resin of the dry preform laid onto a mold, in preparation for the curing process that follows. In common use is a vacuum bag (VB) technique [10], in which a pump and a vacuum bag installed over the preform are used to maintain a vacuum before the resin is drawn into the preform (Figure 7.3). Resin infusion may be directed in-plane or through the thickness of the preform. Resin transfer molding (RTM) is another common technique, which employs two molds clamped together to enclose the laid-up preform (Figure 7.4). A catalyzed, low-viscosity resin is subsequently pumped under pressure into the mold, displacing the air and releasing it

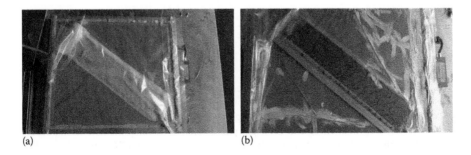

(a)   (b)

FIGURE 7.3 (a) Carbon fiber–reinforced preform laid onto a flat mold sealed with vacuum bags ready for resin infusion using the connected tubing and (b) cured composite showing pen markings recorded to show the progress of resin flow visually during the infusion process.

(a)   (b)

FIGURE 7.4   (a) Resin transfer molding (RTM) tool showing the surface over which a preform would be laid and three holes from which the resin and sensors can be fed through from the bottom and (b) resin flowing from right to left through the closed RTM with a glass top cover for visual monitoring of resin flow during the infusion process. Glass fiber–reinforcement preform is enclosed in the RTM.

through strategically located vents [26]. Resin infusion has a direct impact on the quality of the final composite part [27]. It is critical to ensure uniform distribution of the resin over the impregnated preform. Excess resin is undesirable, as it causes the manufactured part to become brittle and thus weaker. Insufficient resin impregnation can lead to void formation, which can significantly compromise the quality and performance of the final composite part. Nonuniformity in the distribution of resin within the preform can lead to thickness variations across the manufactured composite part [28]. The fundamental principles that govern the success of resin infusion are directly linked to the compaction pressure of the preform, the permeability of the preform, and the kinetics and viscosity of the resin system [1,29]. Compaction pressure is applied to the preform by the vacuum in the vacuum bagging system (Figure 7.3a), whereas the glass cover and the metal clamps above the sample provide the required compaction pressure in the RTM system (Figure 7.4b). Resin infusion is clearly a critical composite manufacturing process requiring online monitoring and control. At present, resin infusion relies mostly on visual inspection, which requires the top surface of the preform to be transparent (Figure 7.3). Optical fiber sensors can be embedded into the preform at locations deemed critical and can be exploited for detecting resin during the infusion process online and in real time. The sensor signals can be implemented within an active feedback control system that will regulate, in real time, resin infusion and flow, thus leading to an optimized infusion process.

7.2.4   Process Temperature

Temperature is one process parameter that directly influences the quality and properties of the final composite part. Composite manufacturing processes such as resin infusion and resin cure are defined by the cure cycle (i.e., the thermal or temperature cycle), which should be precisely adhered to during production [30]. Resin infusion is performed at a sufficiently low isothermal temperature to minimize any cure or cross-linking of the matrix while maintaining low viscosity for the resin. Curing, on the other hand, is often performed at temperatures higher than the infusion temperature, and the greater the temperature, the greater the degree of cure (or the greater the cross-linking density of the matrix) and the shorter the cure time [30,31]. It is crucial to maintain uniform temperature distribution across the entire preform during resin infusion and during the curing process. Nonuniform temperature distribution over the preform can lead to cross-link density variations in the composite, that is, different parts of the composite attaining a different final degree of cure. Such differences in the final degree of cure can lead to the formation of residual stresses and warpage in the final composite part. Thermocouple sensors are used to monitor temperature in the mold and in the heating element, but not in the composite part itself, and this is not ideal for producing high-quality composites. Optical fiber sensors offer the capability to be embedded within the preform and monitor temperature *in situ*, which also provides the opportunity to implement an active feedback control system in real time, leading to adaptive manufacturing of the composite.

7.2.5 CURING PROCESS

The curing of a composite resin involves the conversion of the matrix resin from a low–molecular weight viscous liquid or semisolid to a highly cross-linked rigid structure (solid). The properties (e.g., mechanical) of a curing composite change significantly. The composite exhibits regimes of expansion and shrinkage during the cure process, caused by a combination of temperature and chemical/material state changes [1,28,32]. The quality of the final composite part is directly affected by cure time, temperature, and pressure. Precise measurement and control of these cure process parameters is crucial for obtaining fully cured and high-quality composites at reduced manufacturing costs. The cure cycle is usually defined by temperature as a function of time, but some cure manufacturing processes, for example, the autoclave, require well-defined pressure specifications, as pressure affects the curing reaction thermodynamically. The curing of composites is generally performed by either out-of-autoclave or autoclave manufacturing. Examples of out-of-autoclave manufacturing include vacuum bagging, resin transfer molding, quickstep, pultrusion, and hot press [21]. Autoclave manufacturing is preferred in the aerospace industry because it has the capability to implement precisely prescribed temperature and pressure cycles during cure, which ensures the production of high-quality composites. Figure 7.5 shows typical industrial [33] and laboratory [34] autoclave systems for curing composites.

It is evident that a sensor system, deployed *in situ*, having the capability to monitor any of the cure process parameters, such as temperature and pressure, or changes in the composite properties, such as optical (e.g., refractive index of resin), chemical (e.g., degree of cure), and mechanical (e.g., strain), will be invaluable in composite manufacturing. Optical fiber sensors have the most desirable attributes for this application, including their multifunctional capability and multiplexability. They are suitable for embedding into the composites to perform real-time monitoring of the cure process, and their output signals can be exploited in active feedback control systems, leading to adaptive manufacturing procedures. Differential scanning

FIGURE 7.5 (a) Aerospace industrial autoclave, Boeing system. (From Boeing Australia, Boeing Australia component repairs. http://www.boeing.com.au/boeing-in-australia/subsidiaries/boeing-defence-australia/boeing-australia-component-repair-gallery.page?.) (b) Laboratory autoclave. (From Cranfield University, Composites. https://www.cranfield.ac.uk/About/Cranfield/Themes/Register-Online-Now/Composites.)

calorimetry (DSC) [35] is a cure monitoring technique used widely in the laboratory to monitor the degree of cure of small quantities of resin (a few milligrams) through the measurement of heat flow as a function of temperature and time or a measurement of the cure kinetics of the resin [36,37]. The DSC technique is not compatible with the online monitoring of the cure process during composite production, but can instead be used to calibrate optical fiber sensors before they are deployed for online monitoring.

7.2.6 Commissioning: Postproduction Monitoring

The manufacturing processes of composite parts are clearly complex, and this makes postproduction testing a necessity to provide a means of evaluating the fabricated part. Parameters often considered in postproduction testing include strength, stiffness, flexure, torsion, deformation, failure, and fatigue, which are normally obtained by monitoring the strains and stresses exhibited by a composite during, for example, applied mechanical loadings. Nondestructive testing or evaluation (NDT or NDE) of the manufactured composite parts that are intended for onward usage is an important requirement, but it is not unusual to test a composite to destruction in order to determine the operational limits for the part. Laboratory testing on small composite samples is often carried out using dynamic mechanical analysis (DMA), dynamic mechanical thermal analysis (DMTA), and tension and torsion test machines [38]. Conventional minimally intrusive or nonintrusive NDT techniques, such as digital image correlation (DIC), electronic speckle pattern interferometry (ESPI), shearography, and resistance foil strain gauges (RFSG), provide the surface evaluation of a composite [39]. Optical fiber sensors have the capability to provide localized as well as distributed measurements in the interior of a composite when embedded, making them suitable for the NDT of composites in postproduction. Additionally, optical fiber sensors can be configured to be multifunctional, such that they can perform more than one task. For example, a sensor embedded into the composite, originally to monitor the manufacturing process, can be used further in monitoring postproduction tests.

7.2.7 SUMMARY OF CURRENT MONITORING TECHNIQUES

Some of the measurement techniques used for monitoring various parameters during the processing of composite materials are summarized in Table 7.1.

TABLE 7.1
Summary of Experimental Monitoring Techniques for Composite Material Processing

Monitoring Technique	Layup Monitoring	Resin Infusion/Flow	Cure Monitoring	Postfabrication
Extrinsic Fabry–Perot interferometer (EFPI)		Temperature	Temperature, strain	
FBG	Strain, pressure, temperature	Temperature	Temperature, strain	Strain, temperature
TFBG, LPG	Strain, pressure, temperature	Attenuation, temperature	Temperature, strain, degree of cure	Strain, temperature
Optical fiber Fresnel sensor, optical fiber taper		Attenuation, temperature	Refractive index, degree of cure	
Dielectric analysis (DEA)		Dielectric permittivity, viscosity	Dielectric permittivity, viscosity, degree of cure	Dielectric permittivity
Direct current (DC) analysis		Resistance, viscosity	Resistance, viscosity, degree of cure	
Electrical time domain reflectometry (ETDR)		Velocity	Amplitude fluctuations, degree of cure	
Fourier transform infrared spectroscopy (FTIR)		Wavelength	Wavelength fluctuation, degree of cure	
Fluorescence spectroscopy		Wavelength	Wavelength fluctuation, degree of cure	
Raman spectroscopy		Wavelength	Wavelength fluctuation, degree of cure	
Thermometer		Temperature	Temperature, viscosity	Temperature
Piezoelectric transducer		Pressure	Pressure	
Ultrasonic transducer		Velocity, attenuation	Velocity	Velocity

7.3   PRINCIPLES OF OPTICAL FIBER SENSORS

This section introduces measurement principles of optical fiber sensors for the various process parameters of interest: resin infusion/presence, resin flow, refractive index of resin, strain, and degree of cure. The measurement principles of optical fiber tapered sensors, optical fiber Fresnel sensors, TFBG sensors, and optical FBG sensors are simple and appropriate for composite material processing. The extrinsic optical fiber Fabry–Perot interferometer (EFPI) and the optical fiber LPG sensor are not considered in this section. The EFPI sensor is bulky due to its design and packaging and is thus unsuitable for embedding in composites that are intended for application [15,40]. The handling of an LPG sensor is difficult because of its high sensitivity to bending [41], which makes the interpretation of the signal difficult [42].

7.3.1   OPTICAL FIBER FRESNEL SENSOR

The cure process of a resin is accompanied by volume shrinkage due to the chemical cure reaction [43]. The number of cross-links between functional chemical groups increases as the cure reaction progresses, and the resin undergoes phase transition from viscous liquid to solid, leading to an increase in density [32]. The changing density of a resin during cure results in a changing refractive index of the resin. The refractive index of a material is related to its density by the Lorentz–Lorenz formula [44], given by Equation 7.1.

$$\frac{n^2 - 1}{n^2 + 2} = \frac{A}{M}\rho \qquad (7.1)$$

where:

  n is the refractive index of the material
  ρ is the density
  A is the molar refractivity
  M is the molar mass

An optical fiber Fresnel sensor is sensitive to changes in the refractive index of the surrounding medium and is thus suitable for monitoring the infusion and cure of a resin. The sensor monitors the fluctuations in intensity of the light reflected off the end-face of an optical fiber in response to the variations in the surrounding refractive index [11,12]. A schematic diagram showing the principle of operation and a photograph of the sensing element are shown in Figure 7.6. Light propagating in the core of the optical fiber is both partially transmitted to the external medium and partially reflected back into the fiber at the interface between the optical fiber and the external medium (Figure 7.6b). Figure 7.6c is a plot of the Fresnel sensor reflectivity as a function of refractive index, calculated using Equation 7.3, assuming that the effective refractive index of the mode propagating through the optical fiber is 1.45. When the light reflected back into the fiber from the external medium is referenced to the light from the same source but reflected from a cleaved fiber end that is held in air, the refractive index of the external medium can be determined. The reference is required

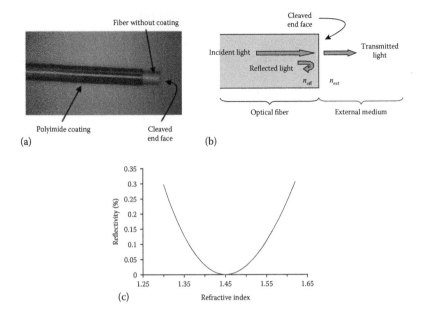

FIGURE 7.6 Optical fiber Fresnel sensor element: (a) image of sensor, (b) the sensing principle, and (c) the plot of reflectivity as a function of refractive index, calculated using Equation 7.3, assuming that the effective index of the mode propagating through the optical fiber is 1.45. n_{eff}, effective refractive index of the optical fiber; n_{ext}, external refractive index. The diameter of the optical fiber, with and without the polyimide coating, is 135 and 125 μm, respectively.

to account for changes in light-source intensity and down lead losses. The reflectivity of the cleaved end of the fiber is also influenced by the thermo-optical property of the fiber when the temperature of the optical fiber changes.

The use of another cleaved fiber end, placed in the same temperature environment as the sensing fiber but exposed to air, can be used to compensate for the thermo-optical effect. The external refractive index n_{ext}, measured by an optical fiber Fresnel sensor, is governed by the Fresnel equation [44]:

$$r_N = \frac{n_{eff} \cos \theta_i - n_{ext} \cos \theta_t}{n_{eff} \cos \theta_i + n_{ext} \cos \theta_t} \qquad r_T = \frac{n_{ext} \cos \theta_i - n_{eff} \cos \theta_t}{n_{ext} \cos \theta_i + n_{eff} \cos \theta_t} \tag{7.2}$$

where:

n_{eff} is the effective refractive index of the fiber core

r_N and r_T are the normal and transversal reflectance coefficients, respectively

θ_i and θ_t are the angles of incident and transmitted light respectively

It can be assumed that $\theta_i \approx \theta_t \approx 0°$ for the optical fiber Fresnel sensor; thus, the reflectivity of the sensor interface is expressed by

$$|r_N|^2 \approx |r_T|^2 \approx \left(\frac{n_{\text{eff}} - n_{\text{ext}}}{n_{\text{eff}} + n_{\text{ext}}} \right)^2 \tag{7.3}$$

The voltages produced by photodetectors that monitor the reflections from the reference optical fiber/air and the optical fiber/resin interfaces are V_{air} and V_{resin} respectively, given by

$$V_{\text{air}} \propto \left(\frac{n_{\text{eff}} - n_{\text{air}}}{n_{\text{eff}} + n_{\text{air}}} \right)^2 \tag{7.4}$$

$$V_{\text{resin}} \propto \left(\frac{n_{\text{eff}} - n_{\text{resin}}}{n_{\text{eff}} + n_{\text{resin}}} \right)^2 \tag{7.5}$$

When Equation 7.4 is divided by Equation 7.5 and then simplified, Equation 7.6 is obtained, where the new variables are defined in Equations 7.7 and 7.8. Equation 7.6 gives a measurement of the refractive index of the resin. Equation 7.8 takes account of the optical fiber coupling efficiencies, photodetector responsivities, fluctuations in intensity of the light source, and down lead losses. In Equation 7.8, the term in parentheses refers to the average of the reference and signal photodetectors when the reference and signal Fresnel sensors are both in air before the signal Fresnel sensor is exposed to the resin:

$$n_{\text{resin}} = n_{\text{eff}} \cdot \left(\frac{1 - \dfrac{\Delta}{\sqrt{R}}}{1 + \dfrac{\Delta}{\sqrt{R}}} \right) \text{ for } n_{\text{eff}} \rangle n_{\text{resin}}, \text{ and } n_{\text{resin}} = n_{\text{eff}} \cdot \left(\frac{1 + \dfrac{\Delta}{\sqrt{R}}}{1 - \dfrac{\Delta}{\sqrt{R}}} \right) \text{ for } n_{\text{eff}} \langle n_{\text{resin}} \tag{7.6}$$

where

$$\Delta = \frac{n_{\text{eff}} - n_{\text{air}}}{n_{\text{eff}} + n_{\text{air}}} \tag{7.7}$$

$$R = \frac{V_{\text{air}}}{\left(\dfrac{V_{\text{air}}}{V_{\text{resin}}} \right)_{\text{air}} \cdot V_{\text{resin}}} \tag{7.8}$$

Compensation for the thermo-optical effect of the optical fiber Fresnel sensor on the refractive index measurement can be performed by, for example, making use of process temperature data T_P obtained using an optical fiber sensor or a thermocouple embedded into the composite. Equation 7.9 gives a refractive index measurement that includes temperature compensation, where K is the thermo-optical coefficient of the optical fiber of the Fresnel sensor.

$$n_{resin} = n_{eff} \cdot \left(\frac{1 - \dfrac{\Delta}{\sqrt{R}}}{1 + \dfrac{\Delta}{\sqrt{R}}} \right) - K.T_P \quad \text{for } n_{eff} > n_{resin}$$

$$n_{resin} = n_{eff} \cdot \left(\frac{1 + \dfrac{\Delta}{\sqrt{R}}}{1 - \dfrac{\Delta}{\sqrt{R}}} \right) - K.T_P \quad \text{for } n_{eff} < n_{resin} \qquad (7.9)$$

Optical fiber Fresnel sensors, which are used in this chapter for monitoring the refractive index of resin during cure in carbon fiber–reinforced composite production, make use of Equation 7.9. A refractometer instrument was developed consisting of eight multiple optical fiber Fresnel sensors to monitor the infusion and the refractive index of the resin at multiple locations within a composite [16,20,45,46] as described by Figure 7.7. The instrument is made up of a network of directional couplers configured such that each photodetector (New Focus, 2011-FC) monitors a reflection from only one sensor. A broadband light source (BBS), having a central wavelength of 1550 nm (Covega 1005), is modulated at a specific frequency (e.g.,

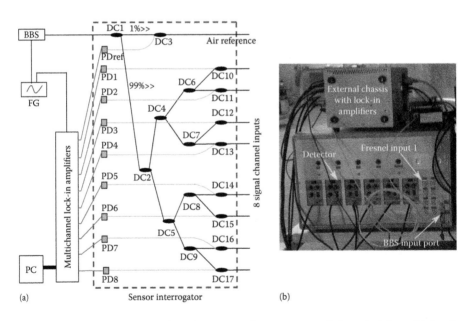

FIGURE 7.7   (a) Schematic of a refractometer instrument consisting of eight multiplexed Fresnel sensors and a reference Fresnel sensor and (b) photograph of the instrument. BBS: broadband light source centered at the wavelength of 1550 nm (Covega 1005); DC1–DC17: directional couplers; PD$_{ref}$ and PD1–PD8: photodetectors (New Focus, 2011-FC); FG: function generator (Stanford Research Systems, DS345); PC: personal computer. DC1 has a splitting ratio of 99/1, and the rest of the directional couplers have a 50/50 splitting ratio.

1 kHz) using a function generator (Stanford Research Systems, DS345) before being coupled into the sensor network. The signals reflected from the individual sensors are demodulated simultaneously using individual digital lock-in amplifiers (NI PXI-4462). The modulation of the light source and the subsequent demodulation using lock-in amplifiers enhances the signal-to-noise ratio of the detection.

7.3.2 TAPERED OPTICAL FIBER SENSOR

Optical fiber tapers enhance the interaction between the light propagating in the fiber and the surrounding environment. Tapering an optical fiber allows the evanescent field of the propagating core mode to penetrate into the surrounding environment, with the proportion of the power in the evanescent field increasing with decreasing diameter of the tapered region. This effect is accompanied by a local increase in attenuation of the fiber, which is sensitive to the refractive index of the surrounding medium. Optical fiber tapers have been applied to evanescent wave spectroscopy [47], to the measurement of refractive index of a surrounding medium [48], and to refractive index change–dependent chemical sensing [49]. One way of interrogating an optical fiber taper is to use coherent optical frequency domain reflectometry (OFDR) [9], which is a distributed measurement technique capable of identifying changes in attenuation of an optical fiber with high spatial resolution. OFDR is particularly suitable for interrogating serially multiplexed fiber tapers. Optical fiber tapers have been multiplexed in a single length of optical fiber before, but for ease of interrogation and to facilitate the identification of individual tapers during sensing, an FBG has been fabricated in the same optical fiber immediately after each taper, with all the FBGs having different center wavelengths [10].

This chapter discusses an OFDR implemented using optical back scatter reflectometry (OBR 4400, Luna Technologies) to monitor the infusion of resin using multiplexed optical fiber tapers that were embedded in carbon fiber–reinforced preform. Figure 7.8 is a schematic diagram of the principle of an optical fiber taper and a typical

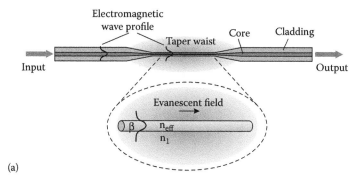

(a)

FIGURE 7.8   (a) Schematic diagram of a tapered optical fiber showing the propagation of the core mode.

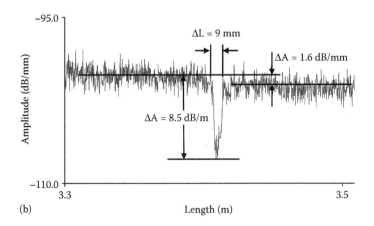

FIGURE 7.8 (Continued)　(b) A Rayleigh backscatter trace of an optical fiber (SM-1322, Optical Fibres) containing a tapered region (51.2 µm diameter) interrogated using optical backscatter reflectometry (OBR, Luna Technologies). ΔL: total length of the tapered region; ΔA: change in amplitude; n_{eff}: effective index of the core mode; n_t: refractive index of surrounding fiber.

Rayleigh backscatter trace from the taper when interrogated using OFDR (OBR 4400). The amplitude of the trace is characterized by significant attenuation localized at the tapered region, and the mean amplitude is generally smaller after the taper than it is before the taper. The multiplexing principle is demonstrated in Figure 7.9, where three tapered regions are serially multiplexed in a single length of an optical fiber. The amplitude of the backscatter trace decreases after each taper when the tapers are surrounded by oil of refractive index (e.g., RI = 1.48) greater than that of the optical fiber cladding.

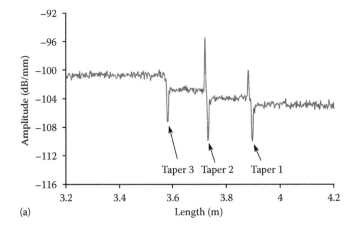

FIGURE 7.9　A Rayleigh backscatter trace of an optical fiber (SM-1322, Optical Fibres) containing a serial array of three tapered regions interrogated using optical backscatter reflectometry (OBR, Luna Technologies) when surrounded by (a) air.

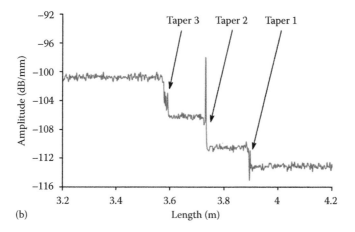

FIGURE 7.9 (Continued) (b) Oil of refractive index 1.48, showing increased attenuation after each tapered region.

7.3.3 Optical Fiber Bragg Grating (FBG)

An optical FBG is constituted by a periodic modulation in the refractive index of the core of an optical fiber. The refractive index modulation is produced by exposure to an output, usually from a UV laser, using either a phase mask or an interferometer [50,51]. Typical FBG lengths are of the order of 5 mm. When light from a BBS is coupled into the fiber, the FBG reflects a narrow wavelength band of full-width half-maximum of the order of 0.1 nm, centered at the Bragg wavelength (Figure 7.10). The FBG principle is governed by Equation 7.10, the phase matching condition, for an FBG fabricated in SM optical fiber.

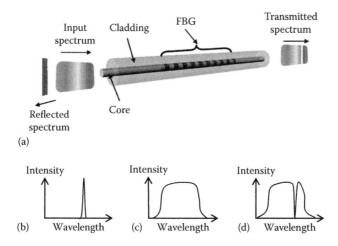

FIGURE 7.10 (a) The principle of an FBG, (b) the spectrum reflected by an FBG, (c) the input broadband spectrum, and (d) the spectrum transmitted through the FBG. The reflected and transmitted spectra are typical of an FBG fabricated in SM optical fiber.

$$\lambda_B = 2n_{eff}\Lambda \tag{7.10}$$

where:

λ_B is the reflected Bragg wavelength
n_{eff} is the effective refractive index of the fiber core
Λ is the period of the FBG

Measurands that interact with the optical fiber and alter the effective refractive index and/or the grating period will cause a shift in the reflected Bragg wavelength. A wide range of FBG sensor systems have been demonstrated for the measurement of parameters such as strain, temperature, and pressure [52]. When an FBG is exposed to longitudinal strain and temperature simultaneously, the Bragg wavelength response of the FBG is obtained by differentiating Equation 7.10 with respect to both strain and temperature, which produces

$$\Delta\lambda_B = \lambda_B \left[\left(\frac{1}{\Lambda}\frac{\partial\Lambda}{\partial\varepsilon} + \frac{1}{n_{eff}}\frac{\partial n_{eff}}{\partial\varepsilon} \right)\Delta\varepsilon + \left(\frac{1}{n_{eff}}\frac{\partial n_{eff}}{\partial T} + \frac{1}{\Lambda}\frac{\partial\Lambda}{\partial T} \right)\Delta T \right] \tag{7.11}$$

$$\Delta\lambda_B = \frac{\partial\lambda_B}{\partial\varepsilon}\Delta\varepsilon + \frac{\partial\lambda_B}{\partial T}\Delta T \tag{7.12}$$

where:

ε is the strain
T is the temperature
$\Delta\lambda_B$ is the Bragg wavelength shift
$\Delta\varepsilon$ is the change in strain
ΔT is the temperature change

Equation 7.11 is simplified to Equation 7.12 by rewriting the terms in parentheses if the optical fiber responsivity to strain and temperature is known. The strain coefficient, $\partial\lambda_B/\partial\varepsilon$, in Equation 7.12 may be obtained experimentally by loading known weights onto a freely suspended optical fiber containing an FBG held at constant temperature. Similarly, the temperature coefficient, $\partial\lambda_B/\partial T$, may be obtained experimentally by heating the FBG to known temperatures in the absence of applied strain.

When an FBG is fabricated in HiBi fiber, two Bragg peaks will exist simultaneously (Figure 7.11). The center wavelengths of the Bragg peaks are separated by typically 0.3 nm [53], one polarized along each of the orthogonal axes of the fiber. The center wavelengths of the two Bragg peaks are given by Equations 7.13 and 7.14 for the fast and slow axis, respectively. HiBi FBG sensors thus provide the capability to measure normal components of strain at the same spatial location [53–55]. Figure 7.11c shows the wavelengths reflected by two multiplexed HiBi FBG sensors. Each polarization Eigen-mode was monitored independently and simultaneously using a custom-designed interrogation system based on a scanning fiber Fabry–Perot (FFP) interferometer coupled to HiBi optical fiber, shown in Figure 7.11b [56]. The

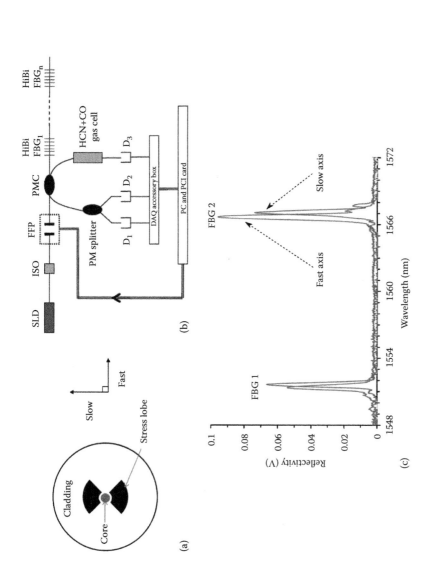

FIGURE 7.11 (a) The cross section of a HiBi optical fiber with bow-tie stress rods, (b) schematic of the interrogation system, and (c) the reflection spectrum from two multiplexed FBGs, FBG1 and FBG2, fabricated in this type of fiber (HB1500, Fibercore). SLD: superluminescent diode; FFP: scanning HiBi fiber–coupled Fabry–Perot tunable filter; PM Splitter, fiber optic polarization-maintaining splitter; DAQ: data acquisition; D_n: detectors; ISO: fiber optic–coupled Faraday isolator; PMC: polarization-maintaining fiber optic coupler.

system is capable of the simultaneous interrogation of wavelength-division multiplexed (WDM) sensors, separating the orthogonally polarized components of the spectra (Figure 7.11b). The interrogation system has a maximum scan rate of approximately 250 Hz and a wavelength resolution of 40 pm.

$$\lambda_{B,f} = 2n_{\mathrm{eff},f}\Lambda \tag{7.13}$$

$$\lambda_{B,S} = 2n_{\mathrm{eff},S}\Lambda \tag{7.14}$$

where:

$\lambda_{B,f}$ and $\lambda_{B,S}$ are the Bragg wavelengths reflected along the fast and slow axis, respectively

$n_{\mathrm{eff},f}$ and $n_{\mathrm{eff},S}$ are the effective refractive indices of the fiber core along the fast and slow axis, respectively

Λ is the period of the FBG

Equations 7.13 and 7.14 can be rewritten as Equations 7.15 and 7.16, respectively, when the FBG sensor is subjected to normal components of strain and temperature simultaneously in the absence of cross-sensitivity and shear strain.

$$\lambda_S = \lambda_{B,S} \cdot \left\{ 1 + \varepsilon_a - \frac{1}{2}n_0^2 \left[p_{11}\varepsilon_S + p_{12}\left(\varepsilon_a + \varepsilon_f \right) \right] + (\alpha_{0,S} + \xi_{0,S})T \right\} \tag{7.15}$$

$$\lambda_f = \lambda_{B,f} \cdot \left\{ 1 + \varepsilon_a - \frac{1}{2}n_0^2 \left[p_{11}\varepsilon_f + p_{12}\left(\varepsilon_a + \varepsilon_S \right) \right] + (\alpha_{0,f} + \xi_{0,f})T \right\} \tag{7.16}$$

where:

λ_i and $\lambda_{B,i}$ $(i=s,f)$ are the new wavelengths and the original wavelengths (at quiescent), respectively, for the slow (s) and fast (f) axes of the fiber

p_{11} and p_{12} are the Pockels strain-optic coefficients

$\alpha_{0,i}$ $(i=s,f)$ are the thermal expansion coefficients

$\xi_{0,i}$ $(i=s,f)$ are the thermo-optical coefficients in the transverse directions

n_0 is the average refractive index along the two orthogonal Eigen-axes of the fiber

T is the temperature of the fiber

$\varepsilon_i(i=a,s,f)$ are the normal strain components in the axial, slow, and fast axis directions, respectively

Equations 7.15 and 7.16 can be simplified to represent wavelength shifts, $\Delta\lambda_s$ and $\Delta\lambda_f$, along the slow and fast axes, respectively. The resulting equations have a common axial strain term. The contribution to the wavelength shifts due to the temperature can be compensated for in both equations, given the temperature history of the sensor measured by other means, for example, a thermocouple. Thus, the two equations can be considered to have three unknowns: ε_f (transverse strain along the fast axis), ε_s (transverse strain along the slow axis), and ε_a (axial strain). The axial strain term is eliminated by performing a subtraction between the two equations, resulting in an effective transverse strain, given by [54,55].

$$\varepsilon_f - \varepsilon_s = \frac{2}{n_o^2 \left(p_{12} - p_{11} \right)} \left[\frac{\Delta\lambda_f}{\lambda_{B,f}} - \frac{\Delta\lambda_s}{\lambda_{B,s}} \right] \tag{7.17}$$

FBG sensors can be embedded in composite materials for the purposes of monitoring strain development during the cure of resin [54] and for structural health monitoring via strain measurement [53]. FBG sensors fabricated in SM fiber are generally used for longitudinal strain measurement [52], while HiBi FBGs can be used to measure transverse strain [55]. In this chapter, HiBi FBG sensors are used to monitor the development of transverse strain during the cure of resin in carbon fiber–reinforced preform.

An ideal HiBi FBG sensor is one that will be able to measure orthogonal components of, for example, strain by exploiting the two Bragg wavelengths in Equations 7.13 and 7.14 reflected along the polarization Eigen modes of the HiBi optical fiber. For the single HiBi FBG to measure two such parameters, that is, two orthogonal strains (or strain and temperature), the two Bragg peaks should exhibit significantly different sensitivities to each of the two parameters for accurate measurements to be possible. If the single HiBi FBG is required to measure longitudinal strain and temperature, this is equivalent to solving Equations 7.15 and 7.16 simultaneously. Thus, in principle, to measure four physical parameters, such as axial strain, two orthogonal transverse strains, and temperature, two HiBi FBG sensors will be required, which can be fabricated such that they are superimposed or colocated. This measurement scenario generates four equations, which can be solved simultaneously to obtain the four unknown variables (i.e., physical parameters). These four equations are represented in simple form by (a matrix) Equation 7.18 for a linear sensing system without cross-sensitivity.

$$\begin{bmatrix} \Delta\lambda_{s,1} \\ \Delta\lambda_{f,1} \\ \Delta\lambda_{s,2} \\ \Delta\lambda_{f,2} \end{bmatrix} = \begin{bmatrix} K_{11} & K_{12} & K_{13} & K_{14} \\ K_{21} & K_{22} & K_{23} & K_{24} \\ K_{31} & K_{32} & K_{33} & K_{34} \\ K_{41} & K_{42} & K_{43} & K_{44} \end{bmatrix} \begin{bmatrix} \varepsilon_a \\ \varepsilon_s \\ \varepsilon_f \\ \Delta T \end{bmatrix} \tag{7.18}$$

The matrix on the left-hand side represents the measured wavelength shifts from the two HiBi FBG sensors (subscripts 1 and 2), and the slow and fast axes are represented by subscripts s and f, respectively. The second matrix (K-matrix) consists of the sensitivity coefficients of each of the four Bragg peaks to each of the four physical parameters. Each entry to the K-matrix is a result of an experimental calibration of one Bragg peak (out of the four peaks) to one physical parameter only, that is, in the absence of the other three physical parameters. The third matrix on the right-hand side of the equation is a set of the four unknown physical parameters, which are axial strain, transverse strains along the slow and fast axes, and temperature. The inverse of Equation 7.18 gives the required solution for the four measurands. This measurement approach is currently just a proposition [57]. This chapter uses measurements from HiBi FBG sensors to determine effective transverse strain using Equation 7.17.

7.3.4 TILTED FIBER BRAGG GRATING SENSOR (TFBG)

The refractive index modulation of an FBG can be purposely tilted with respect to the fiber axis to produce a TFBG [18,58]. A TFBG couples light from a forward-propagating mode of the fiber core into the numerous backward-propagating cladding modes of the fiber (Figure 7.12a). The proximity of the cladding mode resonances to the fiber's outside environment means that they can be influenced by the refractive index of the medium surrounding the optical fiber. The effect can lead to the center wavelengths of the cladding mode resonances shifting and/or the cladding mode resonances diminishing in their amplitude or visibility. The spectral response of a TFBG is governed by the phase matching condition,

$$\lambda_i = \left(n_{\text{eff}}^{co} + n_i^{\text{clad}} \right) \cdot \frac{\Lambda_B}{\cos \theta} \tag{7.19}$$

where:

λ_i is the ith cladding mode resonance wavelength
n_{eff}^{co} is the effective index of the core
n_i^{clad} is the effective index of the ith cladding mode
θ is the tilt angle
Λ_B is the grating period

    The exact form of the transmission spectrum of a TFBG is highly dependent on the tilt angle employed [59]. Figures 7.12b and c show the typical transmission spectra from TFBGs of relatively weak and strong tilt angles, characterized by a Bragg peak, a strongly coupled cladding mode resonance (or ghost mode), and the numerous cladding mode resonances. TFBGs have been demonstrated to be good candidates for refractive index measurements for refractive indices that are smaller than that of the optical fiber [12,58]. It is shown in this chapter that TFBGs also have the capability to detect the presence of resin during the infusion process [16] and to monitor the process temperature during the cure of a resin in a carbon fiber–reinforced preform. The significant difference between the temperature coefficients of the dominant cladding mode resonance (ghost mode) and the Bragg peak [18] is exploited in this chapter for temperature measurements. It is shown in this chapter that the longitudinal strain coefficients of the ghost mode and the Bragg peak have similar magnitudes. Equation 7.20 is used in this chapter to discriminate temperature from longitudinal strain, thus providing a measurement of the process temperature of the resin curing in a composite.

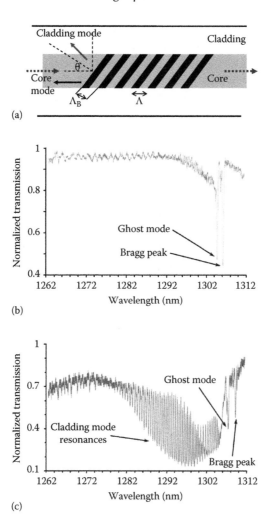

FIGURE 7.12 (a) Schematic diagram of a TFBG and the TFBG transmission spectra for (b) a 1.5° and (c) a 3° tilt angle fabricated in SM fiber (PS1250, Fibercore). Λ_B: grating period; Λ: grating period along the fiber axis; θ: tilt angle. Dashed arrows represent the forward-propagating core mode, and solid arrows show coupling to the contrapropagating core mode and cladding modes.

$$\begin{pmatrix} \Delta\lambda_{core} \\ \Delta\lambda_{clad} \end{pmatrix} = \begin{pmatrix} K_{core,\varepsilon} & K_{core,T} \\ K_{clad,\varepsilon} & K_{clad,T} \end{pmatrix} \begin{pmatrix} \varepsilon \\ T \end{pmatrix} \qquad (7.20)$$

where:

$\Delta\lambda_{core}$ and $\Delta\lambda_{clad}$ are the wavelength shifts of the core-core mode resonance and of the core-cladding mode resonance respectively

ε is longitudinal strain

T is temperature

the entries for the *K*-matrix are the respective sensitivity coefficients for the core-core and the core-cladding mode resonances for strain and temperature

The sensitivity coefficients are determined experimentally prior to using Equation 7.20 for monitoring the process temperature during the cure of a resin in a composite preform. A well-conditioned *K*-matrix is required for accurate temperature and strain discrimination, as the solution to Equation 7.20 involves the use of the inverse of the matrix. The influence of the conditioning of the *K*-matrix on strain and temperature discrimination was analyzed previously, showing that the smaller the conditioning number, the better the performance of the measurement system [18].

A TFBG sensor embedded in dry fiber–reinforced preform is surrounded by air, which has a lower refractive index (RI ~ 1) than that of the TFBG cladding (RI ~ 1.45 at 1330 nm). During the infusion, the arrival of resin at the TFBG can be detected as the surrounding medium changes from air to resin, with a concomitant significant increase in refractive index (RI = 1.59 for RTM6 resin). The effect of the change from a surrounding medium of air to resin on the TFBG transmission spectrum is shown in Figure 7.13. Significant attenuation is experienced by the cladding mode

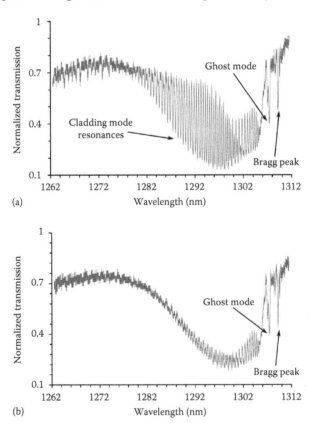

FIGURE 7.13 The transmission spectrum of a TFBG fabricated at 3° tilt angle, when the TFBG is surrounded by (a) air (refractive index = 1) and (b) RTM6 resin (refractive index = 1.59). The TFBG is fabricated in SM fiber (PS1250, Fibercore).

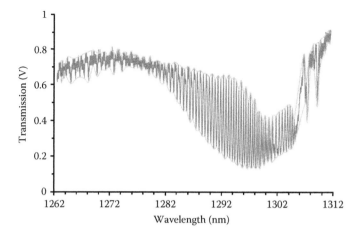

FIGURE 7.14 The transmission spectrum of a TFBG fabricated at 3° tilt angle, when the TFBG is surrounded by air. Envelopes are fitted to the spectrum to define an area occupied by cladding mode resonances. The TFBG is fabricated in SM fiber (PS1250, Fibercore).

resonances, but the ghost mode and the Bragg peak are unaffected. The degree of attenuation in the cladding mode resonances depends on the magnitude of the refractive index of the surrounding resin. The cladding mode resonances are permanently attenuated when the TFBG is surrounded by resins having a refractive index greater than that of the optical fiber cladding; thus, the TFBG cannot monitor the cure of such resins. The attenuation in the cladding mode resonances evolves continuously when a TFBG is surrounded by resins having refractive indices smaller than that of the optical fiber cladding, and the TFBG can be used to monitor cure for such resins. Figure 7.14 shows envelopes fitted to the TFBG transmission spectrum, defining an area occupied by cladding mode resonances. The magnitude of the enclosed area evolves as the cladding mode resonances attenuate with the progress of the cure of the resin, and the area can be computed online to provide cure monitoring, as has been demonstrated previously [12].

7.4 MEASUREMENT APPLICATIONS: LABORATORY MANUFACTURING

Experimental examples of the monitoring of laboratory composite manufacturing processes will be provided in this section. The section will cover the laying up of fiber-reinforced preform and the monitoring of resin infusion, temperature, resin flow, and resin cure.

7.4.1 RESIN INFUSION AND FLOW MONITORING

The need to determine the presence of resin at critical locations within a component is often a key requirement during the production of composite components, rendering the monitoring of the infusion a necessity. Existence of resin voids and of excess

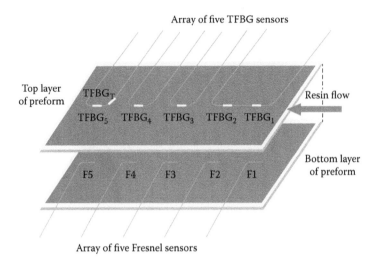

FIGURE 7.15 Experimental configuration for the infusion of resin, showing five embedded TFBG sensors laid serially near the top layer of the composite preform and five Fresnel sensors embedded into the middle layer aligned with the TFBGs. RTM6 resin is infused in-plane, directed from the right to the left (see arrow). The sixth $TFBG_T$ was embedded for temperature monitoring.

resin within the fabric will compromise the quality and thus the structural integrity of a composite part. Resin flow in complex shaped components and/or in complex woven fiber-reinforced preform is often complicated and unpredictable; thus, online experimental monitoring systems will be invaluable. Figure 7.15 shows the experimental layup that was implemented for monitoring resin infusion into eight plies of carbon fiber–reinforced preform of length 600 mm and width 100 mm using the vacuum bagging system. Five TFBG sensors were embedded and laid serially near the top layer of the composite preform. Five Fresnel sensors were embedded into the middle layer of the same preform such that they were aligned directly below the TFBG sensors. The preform was subsequently infused with RTM6 resin system, in which the flow was directed in-plane and along the length of the preform (Figure 7.15).

A refractometer instrument system was developed, as described in Section 7.3.1 (Figure 7.7), and used to interrogate the five Fresnel sensors simultaneously. The output from a superluminescent diode (Covega 1005), centered at the wavelength of 1550 nm within a 100 nm bandwidth, had its amplitude modulated at 777 kHz from a function generator (Stanford Research Systems, DS345) before being coupled into the refractometer. Multi-channel lock-in amplifier electronic boards (NI PXI-4462) were installed in an external chassis (NI PXI-1033) and were used to demodulate the signals reflected from the five Fresnel sensors that were embedded into the composite preform and from a reference Fresnel sensor, simultaneously. A LabView™ program was developed in order to facilitate real-time monitoring of the infusion and the degree of cure of the resin via the refractive index measurement of the resin. An interrogation system based on a swept laser source (Santec, HSL 2000), shown

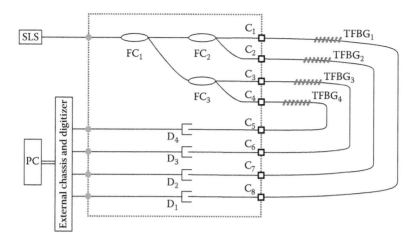

FIGURE 7.16 Schematic diagram of the experimental layout showing the interrogation system for multiplexed TFBG sensors. SLS: scanning laser source (Santec Ltd); FC_1–FC_3: SM fiber couplers (3 dB split ratio); D_1–D_4: photodetectors (New Focus, 2011-FC); C_1–C_8: fiber connector blocks; PC: personal computer; $TFBG_1$–$TFBG_4$: tilted fiber Bragg grating.

in Figure 7.16, was developed for monitoring simultaneously the transmission of six multiplexed TFBG sensors embedded into the preform. The laser scans a wavelength bandwidth of 49 nm, centered at 1287 nm, at a scan rate of 2.5 kHz. The signals transmitted through the six TFBG sensors were acquired using two digitizer boards (NI PXI-5152 and NI PXI-5122), which were installed in an external chassis (NI PXIe-1073). The two boards allowed a total of four channels to be interrogated, and as there were a total of six TFBG sensors to be interrogated, two channels consisted of two multiplexed TFBG sensors, while each of the remaining two channels was connected to a single TFBG sensor. A sampling rate of 100 MS/s was used, such that a full transmission spectrum was acquired every second from each of the four channels simultaneously. The TFBG and Fresnel sensor systems were synchronized such that they both acquired data every second, simultaneously. Figure 7.17a shows the carbon fiber–reinforced preform containing the sensors, laid onto a hot plate, vacuum bagged, sealed, and ready for the infusion, while Figure 7.17b shows the component after the infusion and cure. Pen markings on Figure 7.17b indicate the time and spatial instances of the resin front recorded visually during the progress of the infusion. Figure 7.18 shows the transmission spectra of all six TFBG sensors that were embedded into the preform. Figure 7.18a shows a transmission spectrum consisting of two multiplexed TFBG sensors ($TFBG_1$ and $TFBG_2$) fabricated at 3° tilt angles, while each of Figures 7.18b and c shows a spectrum for a TFBG ($TFBG_3$ and $TFBG_4$) fabricated at 3° tilt angle. The spectrum shown in Figure 7.18d is from two multiplexed TFBG sensors: $TFBG_5$ was fabricated at 3° tilt angle, while $TFBG_T$ was inscribed at 1.5° tilt angle. There were a total of five TFBG sensors for infusion monitoring, all fabricated at a tilt angle of 3°. The TFBG sensor at 1.5° tilt angle was to be used for temperature monitoring.

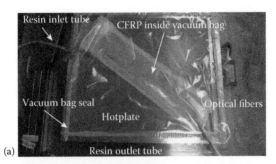

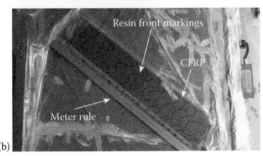

FIGURE 7.17   (a) Carbon fiber–reinforced preform, vacuum bagged and ready for infusion, and (b) the composite after cure, showing pen marks to indicate the resin front during the progress of infusion. CFRP: carbon fiber–reinforced preform.

Figure 7.19 shows the instances when each of the embedded TFBG and Fresnel sensors detected the arrival of resin, shown as significant attenuation in the signals. The arrival of resin at a location is detected because when the surrounding medium changes from air to resin, there is a concomitant significant increase in refractive index, which attenuates the sensor signals, as discussed in Sections 7.3.1 and 7.3.3. The infusion was also monitored visually, with markings being drawn by pen onto the top of the preform to indicate the time and spatial instances of the resin front as the infusion progressed. Table 7.2 shows good agreement between the Fresnel and TFBG sensors for resin flow detection during the infusion process, and the measurements compare very well with the visual monitoring results shown in Table 7.3. Figure 7.19a shows that Fresnel sensor F1 did not detect resin arrival at its location because the signal did not experience significant attenuation. The embedded optical fiber containing $TFBG_5$ and $TFBG_T$ was broken when vacuum pressure was applied to the preform, and the sensors were unable to continue providing measurements. Temperature monitoring by TFBG sensors was performed in another experiment (see Section 7.4.2).

A separate experiment was set up using an optical fiber consisting of a serial array of three multiplexed tapered regions embedded into the middle layer of an eight-ply carbon fiber–reinforced preform and aligned along the middle of the preform. The preform was 390 mm long and 390 mm wide. Optical fiber Taper 1 was

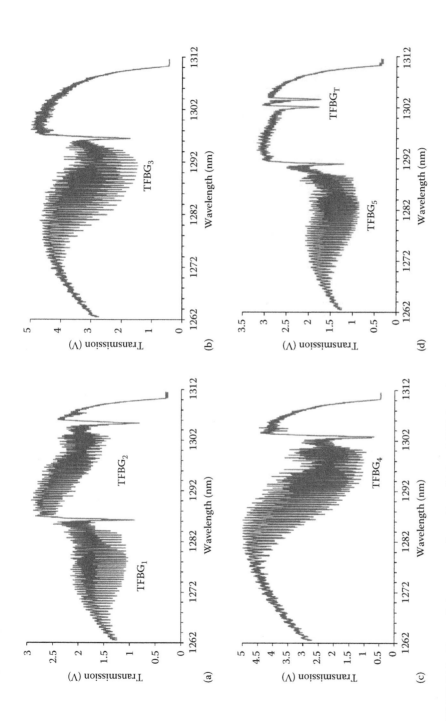

FIGURE 7.18 Transmission spectra from (a) multiplexed $TFBG_1$ and $TFBG_2$, (b) $TFBG_3$, (c) $TFBG4$, and (d) multiplexed $TFBG_5$ and $TFBG_T$. The sensors were embedded into a carbon fiber–reinforced preform described in Figure 7.15. $TFBG_T$, fabricated at a tilt angle of 1.5°, was for temperature monitoring. The rest of the TFBG sensors, fabricated at tilt angles of 3°, were for resin infusion and flow monitoring.

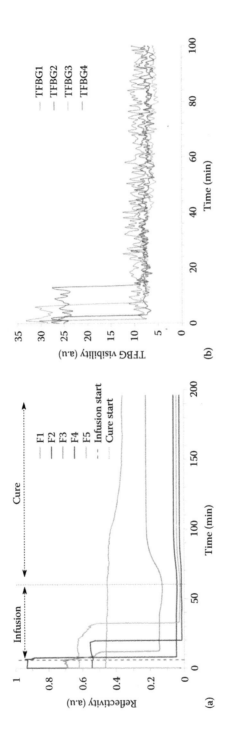

FIGURE 7.19 Significant attenuation (a) in the reflection intensity from Fresnel sensors and (b) in the cladding mode resonance visibility from the TFBG sensors, showing resin arrival at the various sensor locations within a carbon fiber–reinforced composite panel. The TFBGs were laid serially near the top layer of the composite panel and were aligned directly above the Fresnel sensors that were embedded in the middle layer of the panel (Figure 7.15).

TABLE 7.2

Summary of the Resin Arrival Times at Fresnel and TFBG Sensor Locations within a Carbon Fiber–Reinforced Composite Preform (Figures 7.15 and 7.19) during Infusion by RTM6 Resin

Sensor Location (mm)	570	440	330	190	55
Time of resin at Fresnel sensors (min)	28	15	6	3	
Time of resin at TFBG sensors (min)		14	7.5	3	1

Note: Sensor locations are measured from the end of the preform containing the resin inlet. Resin flow direction is from right to left in the table.

TABLE 7.3

Visual Monitoring of the Resin Front during the Infusion Process, Observed from the Top of the Panel (Figure 7.17b)

Resin-Front Distance (mm)	532	505	492	466	442	412	388	355	340	328	277	191
Visual time (min)	23	20	19	17	15	12	10	8	7	6	4	2

Note: Distance is measured from the end of the preform containing the resin inlet. Resin flow direction is from right to left in the table.

embedded 70 mm from one edge of the preform, Taper 3 was 70 mm from the other edge, and Taper 2 was located in the center of the preform (Figure 7.20a). RTM6 resin was infused through the thickness of the preform using a vacuum bagging system. Figure 7.21a shows the amplitude of the Rayleigh backscatter reflection from the embedded optical fiber containing three tapered sensors obtained by OFDR (OBR 4400, Luna Technologies), while Figure 7.21b shows the instances when the resin arrived at each of the sensor locations, which is indicated by the significant change in the amplitude of reflected signal. Resin arrival times were 9, 9.6, and 10.3 min, respectively, for Taper 3, Taper 1 and Taper 2. This result indicated that the center of the carbon fiber–reinforced preform was the last to be filled with the resin during the infusion, which is consistent with resin flow for through-the-thickness infusion.

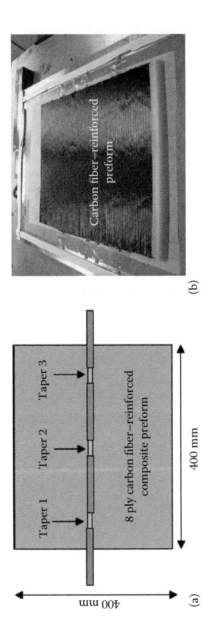

FIGURE 7.20 (a) Schematic diagram showing the configuration of a serial array of three multiplexed tapered optical fiber sensors embedded into the middle layer of an eight-ply carbon fiber–reinforced preform and (b) preform being laid onto a hot plate.

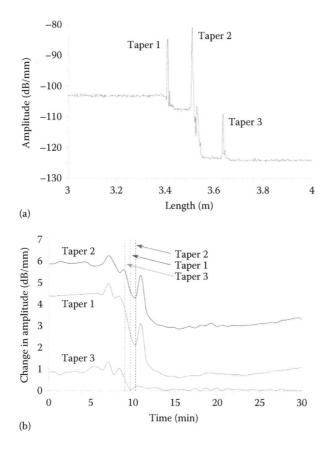

(a)

(b)

FIGURE 7.21   (a) Backscatter reflection amplitude obtained using OFDR (OBR 4400, Luna Technologies) from the embedded optical fiber containing a serial array of three multiplexed tapered sensors, and (b) the corresponding change in amplitude during the infusion of RTM6 resin through the thickness of the preform. Resin arrival times, indicated by dashed lines, were 9, 9.6, and 10.3 min, respectively, for Tapers 3, 1, and 2.

7.4.2 Process Temperature Monitoring

The quality of a manufactured composite part is largely influenced by the thermal cure profile used during production. Maintaining a precisely defined thermal profile throughout the composite preform during manufacture is thus a vital requirement. Sensors that have the capability to be embedded into composites without compromising their structural integrity will be ideal for providing temperature monitoring *in situ* and thus facilitate active temperature control during production. Thermocouple sensors can be embedded within composite parts intended for testing, but they are not suitable for embedding into components that are for real applications, as their relatively bulky electrical cabling will have to be embedded, too. Optical fiber sensors offer unrivaled capability to be embedded into composites and monitor temperature because they have many suitable properties, including their small dimensions and inert properties, making them compatible with composite material applications. A common hindrance in monitoring temperature by optical fiber sensors is their simultaneous sensitivity to both temperature and strain, and methods that have been tried to discriminate between the two parameters are often complex or cumbersome [18].

This chapter exploits the differential in temperature sensitivities between a strongly coupled cladding mode resonance (ghost mode) and a Bragg peak from two TFBG sensors, multiplexed in one optical fiber, to monitor the temperature cycle used during the cure of RTM6 resin infused into carbon fiber–reinforced preform. The optical fiber containing the two TFBG sensors was embedded in the middle layer of an eight-ply carbon fiber–reinforced preform, 600 mm long and 100 mm wide. Figure 7.22 is a schematic of the experimental layup, showing the two embedded TFBG sensors alongside a HiBi optical fiber containing FBG sensors used for transverse strain monitoring, to be discussed in Section 7.4.4. The production of the composite used a vacuum bagging system and a hotplate.

The two TFBG sensors were fabricated in SM fiber (PS1250, Fibercore) at a wavelength of around 1300 nm with tilt angles of 1.5° and 3°, producing the transmission spectrum shown in Figure 7.23. The tilt angles of $TFBG_T$ and $TFBG_3$ are 1.5° and 3°, respectively (Figure 7.23). Experiments were performed to characterize the response of the TFBG sensors with respect to axial strain and temperature independently, to determine the sensitivity coefficients, which are summarized in Table 7.4.

A total of six K-matrices (see Equation 7.20) from the sensitivity coefficients in Table 7.4 can be constructed between the two TFBG sensors, by using different combinations of ghost modes and Bragg peaks. The conditioning numbers of the six K-matrices are given in Table 7.5.

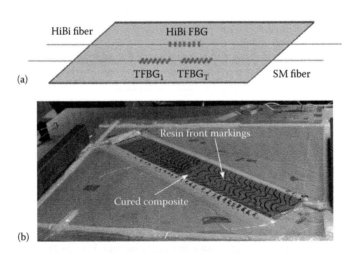

FIGURE 7.22 (a) Schematic of carbon fiber–reinforced preform containing an SM optical fiber with multiplexed $TFBG_1$ and $TFBG_T$ sensors and a HiBi optical fiber inscribed with an FBG. (b) Composite panel after cure. The preform was 600 mm long and 100 mm wide. The sensors were embedded in the middle of the eight-ply preform.

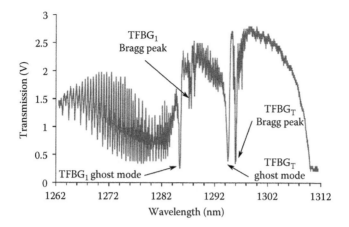

FIGURE 7.23 Spectrum from two TFBGs (Figure 7.22) multiplexed in a single optical fiber that was embedded into carbon fiber preform for temperature monitoring. $TFBG_T$ and $TFBG_1$ were fabricated in SM optical fiber (PS1250, Fibercore) at tilt angles of 1.5° and 3°, respectively.

TABLE 7.4

Summary of the Sensitivity Coefficients of the Multiplexed TFBG Sensors That Are Shown in Figure 7.23

TFBG Sensor	Strain Coefficient (pm/µε)		Temperature Coefficient (pm/°C)	
	Ghost Mode	Bragg Peak	Ghost Mode	Bragg Peak
$TFBG_1$	0.999	1.008	3.890	3.227
$TFBG_T$	1.053	1.070	5.816	6.090

TABLE 7.5

Summary of the Six Conditioning Numbers from the Various Combinations of Bragg Peaks and Ghost Modes of the Two Multiplexed TFBG Sensors, $TFBG_T$ and $TFBG_1$

		$TFBG_T$		
		Ghost Mode	Bragg Peak	385.8
$TFBG_1$	Ghost mode	29.8	28.3	
	Bragg peak	18.8	18.4	
	39.5			

Note: Calculated from the corresponding K-matrices (see Equation 7.20) by using the sensitivities in Table 7.4.

The conditioning numbers 18.4, obtained by combining the two Bragg peaks, and 18.8, obtained by combining the Bragg peak of $TFBG_1$ and the ghost mode of $TFBG_T$, are the smallest in Table 7.5. The K-matrices from which these conditioning numbers are derived, along with the K-matrix from which the conditioning number 28.3 (from the combination of the ghost mode of $TFBG_1$ and the Bragg peak of $TFBG_T$) is obtained, are used here for discriminating temperature from longitudinal strain. It was stated in Section 7.3.4 that reliable temperature and strain discrimination will require a well-conditioned K-matrix, that is, a small conditioning number [18].

The sensitivity coefficients (Table 7.4) from which the conditioning numbers 28.3, 18.8, and 18.4 are defined have been used in Equation 7.20 to provide temperature monitoring of the cure cycle during the cure of RTM6 resin in the carbon fiber–reinforced preform described in Figure 7.22. Figures 7.24 through 7.26

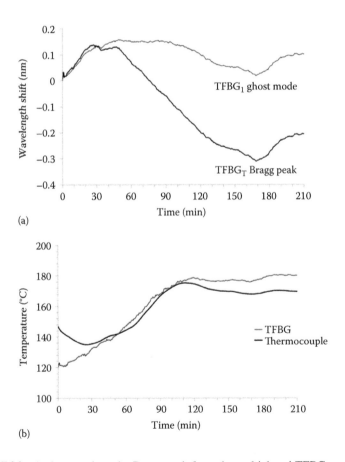

(a)

(b)

FIGURE 7.24 A ghost mode and a Bragg peak from the multiplexed $TFBG_1$ and $TFBG_T$, respectively (conditioning number = 28.3), showing (a) their wavelength shifts and (b) temperature monitored by the two TFBGs and by a thermocouple, during cure of a carbon fiber–reinforced composite described in Figure 7.22.

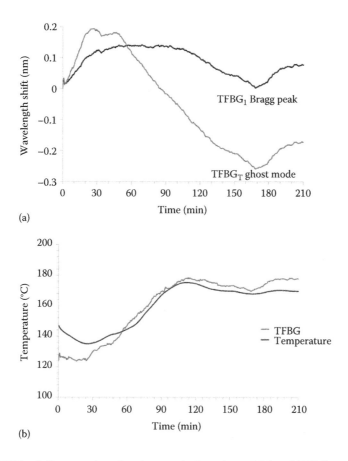

(a)

(b)

FIGURE 7.25 A Bragg peak and a ghost mode from the multiplexed TFBG$_1$ and TFBG$_T$, respectively (conditioning number = 18.8), showing (a) their wavelength shifts and (b) temperature monitored by the two TFBGs and by a thermocouple, during cure of a carbon fiber–reinforced composite described in Figure 7.22.

provide the wavelength shifts and the temperature measurements obtained by multiplexed TFBG$_1$ and TFBG$_T$ and by a thermocouple. Figures 7.24 through 7.26 show that the TFBG temperature measurement compares well with the measurement from a thermocouple sensor, thus demonstrating the capability of this sensor for temperature monitoring. Better comparison is obtained when the matrix conditioning number is smaller, as shown by Figure 7.26 for a conditioning number of 18.4. Discrepancy between the TFBG and thermocouple measurements is apparent early on in the experiment, before the temperature ramp is started. The signal processing used for the TFBG temperature measurements did not consider transverse strains that are likely to develop in the composite preform during cure, and this could account for some discrepancy between the measurements obtained by the two methods.

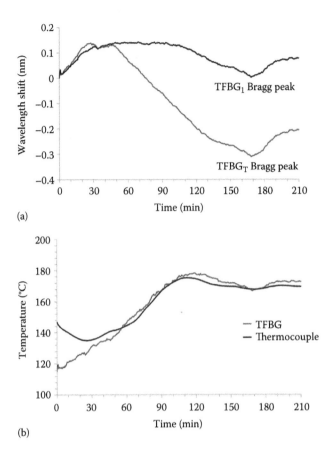

(a)

(b)

FIGURE 7.26 Two Bragg peaks from the multiplexed $TFBG_T$ and $TFBG_1$ in Figure 7.22 (conditioning number = 18.4), showing (a) their wavelength shifts and (b) temperature monitored by the two TFBGs and by a thermocouple, during cure of a carbon fiber–reinforced composite described in Figure 7.22.

7.4.3 CURE MONITORING OF FIBER-REINFORCED PREFORM: REFRACTIVE INDEX OF RESIN

The progress of a chemical cure reaction, also known as the degree of cure or conversion, is an essential parameter to monitor *in situ* and experimentally during composite manufacture. The degree of cure will serve as a quality indicator for the composite, while the information can also be used to improve or optimize the cure cycle. The refractive index of a material is determined by the types of molecular bonds it contains and by its density (Section 7.3.1). The curing of a resin involves the conversion of the matrix resin from a low–molecular weight viscous liquid or semisolid to a highly cross-linked rigid structure, accompanied by the refractive index evolution of the resin (Section 7.3.1). An optical fiber Fresnel sensor was used here for monitoring the refractive index evolution of four different resin systems

(Hexcel) during cure. The Fresnel sensor gives reflected intensity, which is converted to refractive index using Equation 7.9. The response of the Fresnel sensor to refractive index evolution as a function of the degree of cure was calibrated for each resin system prior to applying the Fresnel sensor for monitoring composite production. The degree of cure was measured using temperature-modulated DSC [36,37], while the refractive index evolution of the resin was measured using an optical fiber Fresnel sensor in a separate calibration experiment performed using identical cure cycles. Cure of resin was performed at an isotherm of 180°C after a temperature ramp of 10°C/min. Refractive index measurement, corresponding to the isothermal period of the cure cycle, was correlated with the degree of cure to provide the required correlation function.

The refractive index and degree of cure calibration from the optical fiber Fresnel sensor and the DSC, respectively, are plotted in Figure 7.27a for RTM6 resin. The decrease in refractive index during the temperature ramp, in the first ~18 min, is due to the thermo-optical property of the optical fiber. Once the temperature isotherm is

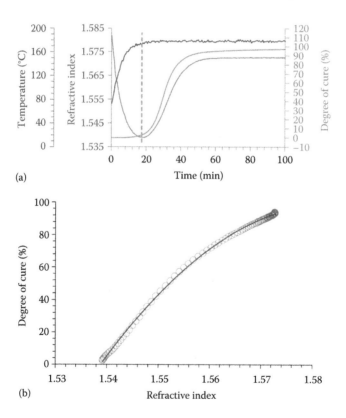

(a)

(b)

FIGURE 7.27 (a) Calibration of RTM6 resin with respect to the refractive index (blue line) measured by the Fresnel sensor and the degree of cure (green line) measured by the DSC, for the given temperature (red line) profile and (b) correlation of the refractive index with the degree of cure corresponding to the isotherm in (a); the red line is a quadratic fit to the data ($R^2 = 0.9994$).

reached, after ~18 min, the refractive index evolution shows strong correlation with the degree of cure, as the cure chemical reaction progresses. The refractive index and degree of cure measurements, corresponding to the isothermal period of the cure cycle, in Figure 7.27a are plotted against each other in Figure 7.27b, giving a quadratic correlation function of Equation 7.21 ($R^2 = 0.9994$).

$$\text{DoC} = -4.87 \times 10^4 * (RI)^2 + 1.54 \times 10^5 * (RI) - 1.22 \times 10^5 \qquad (7.21)$$

where:
  DoC is degree of cure (%)
  RI   is refractive index

The reflected intensity monitored by Fresnel sensor F2 in the experiment shown in Figure 7.15 and plotted in Figure 7.19a was converted into a refractive index measurement using Equation 7.9 in Section 7.3.1 and subsequently into a measurement of the degree of cure using Equation 7.21. The refractive index monitored by the Fresnel sensor is plotted in Figure 7.28a, while a measurement of the degree of cure, corresponding to the isothermal period of the cure cycle, is shown in Figure 7.28b. Temperature in Figures 7.27 and 7.28 was measured by a thermocouple sensor. A data acquisition and signal processing program was developed in LabView™ to facilitate real-time measurement of the degree of cure of the resin.

More experimental calibrations were performed to investigate whether good correlation was obtainable between refractive index measured by the Fresnel sensor and degree of cure measured by the DSC analysis, for different resin systems. The additional resin systems studied were EP2400, RTM890, and PR520, all manufactured by Hexcel. The calibration followed the same procedure as that performed for RTM6 resin. Results obtained from calibration experiments are plotted in Figure 7.29 for

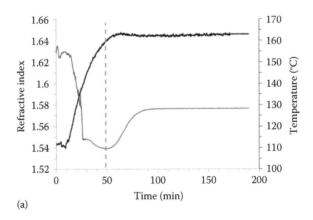

(a)

FIGURE 7.28 (a) Online monitoring of the refractive index (light gray line) of RTM6 resin using Fresnel sensor F2 in the experiment described in Figure 7.15.

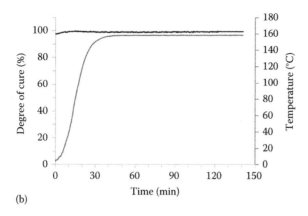

(b)

FIGURE 7.28 (Continued) (b) Direct conversion of the refractive index into degree of cure using Equation 7.21 for the period corresponding to the isotherm in (a). Temperature (black line) was measured by a thermocouple sensor embedded into the composite preform. The dashed line in (a) indicates the time the isotherm started.

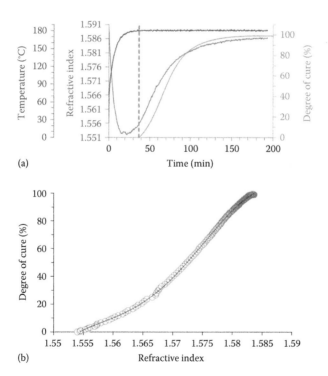

FIGURE 7.29 (a) The calibration of EP2400 resin with respect to the refractive index (blue line) measured by the Fresnel sensor and the degree of cure (green line) measured by the DSC, for the given temperature (red line) profile. (b) The correlation of the refractive index measurement with the measured degree of cure corresponding to the isotherm in (a); the red line is a quadratic fit to the data ($R^2 = 0.9998$).

TABLE 7.6

Correlation Functions between Refractive Index and Degree of Cure for the Different Resin Systems

Resin System	Degree of Cure, Doc (%), as Function of Refractive Index, *RI*	R^2
RTM6	Doc $= -4.87 \times 10^4 * (RI)^2 + 1.54 \times 10^5 * (RI) - 1.22 \times 10^5$	0.9994
EP2400	Doc $= -3.31 \times 10^8 * (RI)^4 + 2.07 \times 10^9 *$	0.9998
	$(RI)^3 - 4.88 \times 10^9 * (RI)^2 + 5.09 \times 10^9 * (RI) - 1.99 \times 10^9$	
RTM890	Doc $= 2.59 \times 10^4 * (RI)^2 - 7.90 \times 10^4 * (RI) + 6.02 \times 10^4$	0.9995
PR520	Doc $= 6.22 \times 10^3 * (RI) - 9.69 \times 10^3$	0.9948

EP2400 resin. The refractive index was found to be well correlated with the degree of cure for each of the resin systems. Table 7.6 summarizes the correlation functions and the coefficients of regression obtained for the different resins, including RTM6. It is clear from Table 7.6 that it is always possible to obtain good correlation between refractive index and degree of cure, but the correlation function is specific to the resin type. Linear correlation was exhibited for PR520, quadratic for RTM890, and polynomial of degree 4 for EP2400 resin.

Further experiments were performed to monitor directly the degree of cure for each of the three additional resin systems using Fresnel sensors, by making use of the correlation functions in Table 7.6. For each of the three experiments, an optical fiber Fresnel sensor was embedded into the middle layer of an eight-ply carbon fiber–reinforced preform having dimensions of 390 mm in length and width. Each preform was infused, through the thickness, by one of the three resins using a vacuum bagging system. The degree of cure monitored by the Fresnel sensor, online and in real time, is plotted in Figure 7.30 for EP2400 resin.

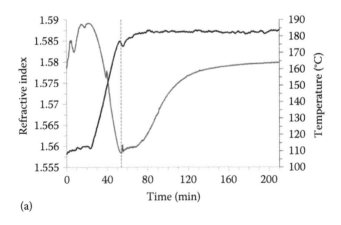

(a)

FIGURE 7.30   (a) Online monitoring of the refractive index (light gray line) of EP2400 resin using a Fresnel sensor embedded in carbon fiber–reinforced preform.

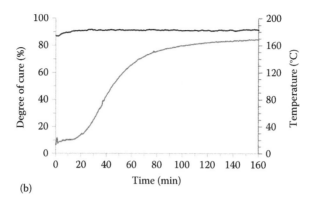

(b)

FIGURE 7.30 (Continued) (b) Direct conversion of the refractive index into degree of cure using the correlation function in Table 7.6 for the period corresponding to the isotherm in (a). Temperature (black line) was measured by a thermocouple sensor embedded into the composite preform. The dashed line in (a) indicates the time the isotherm started.

7.4.4 CURE MONITORING OF FIBER-REINFORCED PREFORM: STRAIN DEVELOPMENT

Fiber-reinforced composites undergo regimes of expansion and shrinkage during production, caused by thermo-mechanical and chemical reactions, as discussed in Sections 7.1 and 7.2.5. The quality of a manufactured composite is thus strongly influenced by its state of internal stress or strain developed during production. In this chapter, a HiBi FBG sensor embedded in the middle layer of an eight-ply carbon fiber–reinforced composite preform was used to monitor effective transverse strain developed in the composite during the cure of RTM6 resin. A schematic of the experimental layup is described in Section 7.4.2 and shown in Figure 7.22. The HiBi FBG was embedded with the slow axis aligned in-plane while the fast axis was oriented out-of-plane of the composite preform. The FBG, which was fabricated at the wavelength of 1552 nm (spectrum in Figure 7.11c) in a HiBi optical fiber (HB1500, Fibercore Ltd), was interrogated using an interrogation system based on a scanning FFP interferometer coupled to a HiBi optical fiber (Figure 7.11b), described in Section 7.3.3. The system interrogates the wavelengths reflected by the HiBi FBG along the two orthogonal polarization modes simultaneously and independently [56]. The composite preform was infused with RTM6 resin and cured as described in Section 7.4.2.

The two wavelength shifts reflected by the HiBi FBG along the two orthogonal polarization modes are plotted in Figure 7.31a along with the degree of cure of the resin, which was determined from a nonparametric cure kinetics model in conjunction with DSC measurements [36,37]. Figure 7.31b shows the effective transverse strain developed during the cure process, calculated from the wavelength shifts in Figure 7.31a using Equation 7.17, and the degree of cure. In Figure 7.31b, effective decrease in transverse strain of ~712 µε was obtained during cure, indicating shrinkage in the through-the-thickness direction of the composite. The final degree of cure attained at the end of the cure was ~93% (Figure 7.31b).

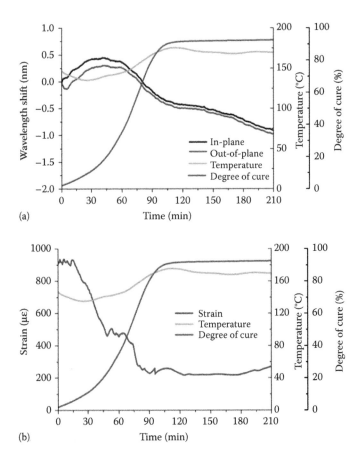

FIGURE 7.31 (a) Out-of-plane and in-plane FBG wavelength shifts, temperature, and degree of cure vs. time and (b) effective transverse strain, temperature, and degree of cure vs. time, for the strain-monitoring experiment using a HiBi FBG sensor.

7.5 MEASUREMENT APPLICATIONS: INDUSTRIAL MANUFACTURING

7.5.1 Carbon Fiber–Composite Aircraft Tailcone Assembly

Multiplexed optical fiber sensors have been embedded into carbon fiber–reinforced preform (CFRP) during the industrial production of a full-sized, one-piece tailcone assembly for a regional jet aircraft. The industrial manufacturing processes that were employed included automatic dry carbon-fiber placement using a robotic system, through-thickness reinforcement by tufting using a robotic machine, infusion and cure monitoring in an industrial oven, and industrial postproduction mechanical loading tests. The embedded sensors monitored the manufacturing process and thereafter the postproduction mechanical loading tests of the tailcone. Optical fiber

Fresnel sensors monitored the infusion and the degree of cure of the resin. The curing of the resin was also monitored by FBG sensors via the development of strains in the tailcone preform. Effective transverse strains in the through-thickness direction of the tailcone were monitored by HiBi FBG sensors, while the measurement signals from the SM FBG sensors were interpreted as longitudinal strains in the tailcone. For a complete report on the tailcone manufacturing and the experimental results obtained, readers are directed elsewhere [14,45,46].

7.5.1.1 Carbon Fiber Preform and Sensor Layup

The tailcone structure consisted of a total of seven plies of dry carbon fiber laid (Figure 7.32a) using automated dry fiber placement (ADFP) implemented by an industrial robotic system [23]. The orientation of the laid plies was 45/−45/0/90/0/−45/45, from the underside to the top. Optical fiber sensors were embedded in the tailcone preform in four different zones, that is, F1–F4 (Figure 7.32c). Zones F1 and F2 coincide with the shortest (2.3 m) and longest (2.5 m) dimensions, respectively, on the tailcone. The sensors were embedded after the first three plies by laying them onto the 0° oriented ply, after which four more plies were added (Figure 7.32e). Each of the zones F1–F4 consisted of three embedded polyimide-coated optical fibers, one SM optical fiber (Fibertronix, FX-PI-01-01-01) with a Fresnel sensor at its end, one SM optical fiber (Fibertronix, FX-PI-01-01-01) containing five equally spaced FBG sensors, and one HiBi optical fiber (Fibercore, HB1250P) containing five equally spaced FBG sensors. The five FBG sensors inscribed in a single length of optical fiber were 325 mm apart in Zone F1, 395 mm apart in Zone F2, and 335 mm apart in Zones F3 and F4 of the tailcone. The first FBG sensors in each optical fiber, toward the wider end of the tailcone, were also at these respective distances from the edge. Polyimide-coated fibers were used to ensure that the buffer coating was able to withstand the temperatures used when curing the composite material. The SM and HiBi optical fiber diameters for the core, cladding, and polyimide coating are provided in Table 7.7. The polyimide coating was removed from each of the 2 mm long sections of fiber to be exposed to the UV radiation prior to FBG inscription, and the FBG sensors were not recoated. The FBG sensors along the length of the optical fibers are marked from S1 to S5 (Figure 7.32). The end of each optical fiber containing the SM FBG sensors was also cleaved and used as a Fresnel sensor.

Hand-laying of the optical fiber sensors between the third and fourth plies was necessary to allow the precise positioning of the sensors in the tailcone and for the precise orienting of the HiBi FBG polarization Eigen-axes with the principal directions (in-plane and out-of-plane) of the tailcone. The measurements made by the HiBi FBG sensors were thus of transverse strains that are in-plane or out-of-plane [54,55] of the tailcone calculated using Equation 7.17, while the measurements from the SM FBG sensors were interpreted as longitudinal strain (Equation 7.12). The tailcone preform was reinforced in the through-thickness direction, after layup, by means of tufting at the feet of the stringers. There were 10 stringers in total, all in the interior of the tailcone (Figure 7.32). The online monitoring of the strains experienced by the needle during the tufting process was also performed using multiplexed SM FBG sensors, which were surface bonded to the needle. The results of these measurements have been published elsewhere [14,25].

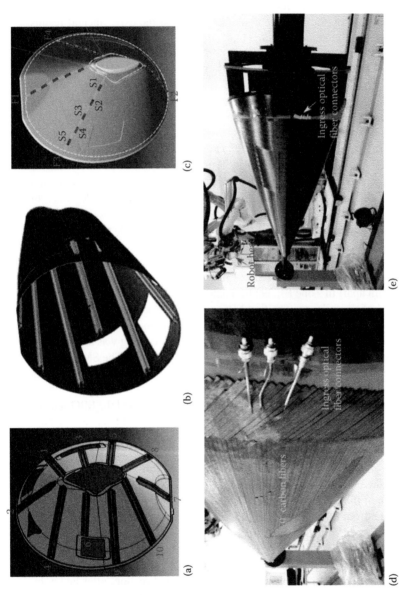

FIGURE 7.32 3D drawing of the tailcone showing (a) the locations of the stringers numbered 1–10, (b) the interior with the stringers, and (c) rear view of the tailcone exterior, showing optical fiber sensor locations S1–S5 within each of the zones F1–F4 of the tailcone. (d) Optical fiber sensors laid onto the third ply such that they were oriented parallel to the 0° carbon fibers, and (e) an industrial robot system laying a 90° ply of dry carbon fiber after the optical fiber sensors had been laid. The ingress optical fiber connectors are shown. The length and diameter of the tailcone are approximately 2.5 and 1.6 m, respectively.

TABLE 7.7

Specification of the Types of the Optical Fibers in Which the Sensors Were Fabricated

Fiber Type	Manufacturer	Fiber Identification Number	Coating Type	Core Diameter (μm)	Cladding Diameter (μm)	Total Diameter (μm)
SM	Fibertronix	FX-PI-01-01-01	Polyimide	9	125	135
HiBi	Fibercore	HB1250P	Polyimide	6	125	145

The ingress of the embedded optical fibers was located on the edge of the preform, at the wider end of the tailcone. Figure 7.33 shows the design of the ingress connections. A stainless steel hypodermic tube was used to protect the ingress of the optical fiber. The hypodermic tube withstands pressure loads in excess of 2.5 bar, which was greater than the pressure exerted by the robot head during the ADFP process. Additional protection

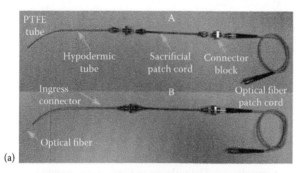

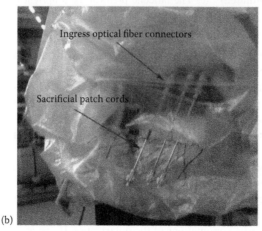

FIGURE 7.33 (a) Photograph of the ingress connector and the sacrificial patch cord for connection to the embedded optical fiber sensors and to the interrogating instruments, (b) the sacrificial patch cords connected to ingress connectors allowing the installation of vacuum bags.

Tubes for resin infusion

Optical fiber patch cords to external instrumentation

(c)

FIGURE 7.33 (Continued)   (c) The tailcone preform fixed upright in the oven ready for resin infusion and cure.

to the optical fiber inside the hypodermic tube was provided by PTFE tubing. The length of the hypodermic tubing used at the ingress was 70 mm, of which 30 mm was embedded in the edge of the preform. An optical fiber of length 150 mm, protected by PTFE tubing and a hypodermic tube, was made with connectors on both ends to form a sacrificial fiber patch cord or connection link. The sacrificial patch cord provided an agile link between the embedded sensors and the interrogating instruments, which were located externally to the oven, and was designed to ease the process of installing the vacuum bags over the tailcone preform for implementing the infusion and cure processes. The connection system ensured that the required vacuum pressure for the infusion of resin was not compromised. Two connector blocks were used to connect the sacrificial patch cord to the fiber sensors at one end and to the instrumentation at the other end. The sacrificial patch cords were made from SM and HiBi optical fiber in which the appropriate orientations of the polarization Eigen-axes were configured.

7.5.1.2 Infusion Monitoring

The installation of the vacuum bags over the tailcone preform in preparation for the resin infusion and cure processes is shown in Figure 7.33. The role of the sacrificial patch cords in enabling vacuum bag installation and in connecting the sensors to the external instrumentation is also shown in the figure. The tailcone was placed upright in the oven, with the exhaust end located at the top, during the infusion and cure processes. A portable Fresnel sensing instrument, configured to interrogate eight multiplexed Fresnel sensors and an air-reference Fresnel sensor simultaneously (described in Section 7.3.1), was used for monitoring the infusion of the tailcone with RTM6 resin. Details on the Fresnel sensing principles, data acquisition, and signal analysis are given in Section 7.3.1. The significant attenuation in the reflected signal that occurs when the RTM6 resin (refractive index $= 1.59$) covers the Fresnel sensor was used to identify the arrival of resin at a sensor during the infusion process.

Figure 7.34 shows three Fresnel sensor signals for the infusion of resin into the tailcone preform. The Fresnel-F3 sensor, embedded halfway up the length of the tailcone in Zone F3, detected the resin 24 min after the infusion process was started, before the resin was detected 1.5 min later by the Fresnel-F1 sensor, which was located close to the exhaust end of the tailcone in Zone F1 (Figure 7.34). The arrival of resin at the middle of the tailcone before the exhaust end was expected, as the resin inlet was located at the bottom of the tailcone (Figure 7.33c). The Fresnel-F4 sensor, embedded halfway up the tailcone in Zone F4, detected the resin 15 min after the infusion process was stopped, which was after the cure phase was started (Figure 7.34). This suggests that the resin had not infused into every part of the preform by the time the infusion process was terminated. The viscosity of the resin decreases when the temperature is increased, which consequently assists the flow and the spreading of the resin further into the preform. Each Fresnel sensor element

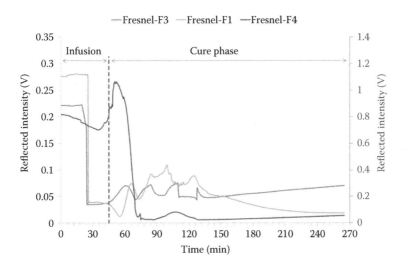

FIGURE 7.34 Resin infusion into the tailcone assembly detected by three Fresnel sensors located in different zones of the tailcone. The F3 and F4 sensors, embedded in Zones F3 and F4, respectively, were located halfway up the tailcone, while the F1 sensor, embedded in Zone F1, was located near the exhaust end. The resin arrival times at the three sensors were 24, 25.5, and 55 min from the start of the infusion process.

was 9 μm in diameter, allowing the detection of pockets of resin of microscopic size. The cure cycle (i.e., the temperature cycle) that was used for the tailcone production is not given here as it is confidential to the industrial partners. For this reason, the degree of cure monitored by the Fresnel sensors is also not given in this chapter.

7.5.1.3 Cure Monitoring: Development of Transverse Strain

The cure process was monitored by the embedded HiBi FBG and SM FBG sensors, which measured the transverse and longitudinal strains, respectively, developed during the chemical cure reaction. The HiBi FBG and SM FBG sensors were interrogated simultaneously using separate instrumentation systems.

A purpose-built instrument (Figure 7.35) based on a swept laser source (Santec, HSL 2000), with a 49 nm wavelength bandwidth centered at 1287 nm and a scan rate of 2.5 kHz, and HiBi optical fiber components, was used to interrogate the HiBi

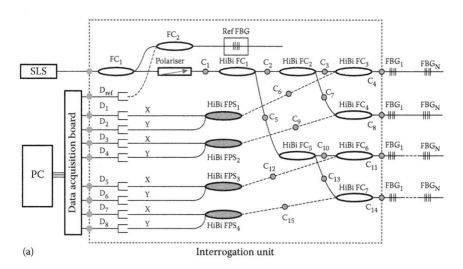

(a) Interrogation unit

(b)

FIGURE 7.35 (a) Schematic of the interrogation instrument for HiBi FBG sensors and (b) a photograph of the instrument. SLS: swept laser source (1300 nm wavelength); FC: SM fiber coupler; HiBi FC: HiBi fiber coupler, FBG: HiBi FBG, C_{1-15}: adjustable HiBi connectors; HiBi FPS: HiBi fiber polarization splitter; D: detector; X/Y: orthogonal polarization states; PC: personal computer; Ref FBG: reference SM FBG.

FBG sensors [45,46]. The instrument was configured to interrogate the reflections from each polarization Eigen-mode independently and simultaneously, using separate photodetectors. The data acquisition was configured to record a full spectrum from the five WDM HiBi FBG sensors at 15 s time intervals, as higher data rates were not necessary for monitoring the cure of resin. The polarization Eigen-axes of the HiBi FBG sensors were aligned with the in-plane and out-of-plane directions of the tailcone, making it possible to monitor the out-of-plane transverse strains developed during the curing process using Equation 7.17, as discussed in Section 7.3.3. Figure 7.36 shows the reflection spectra from five WDM HiBi FBG sensors fabricated in each of the four HiBi optical fibers embedded in the different zones of the tailcone. The HiBi FBG peaks correspond to the two orthogonal polarization Eigen-axes of the optical fiber. The temperature for the cure cycle, measured by thermocouple sensors attached to the surface of the tailcone preform, was used together with a prior calibration for the temperature responsivity of the Bragg peaks to compensate for the influence of temperature on the measured Bragg wavelengths before the effective transverse strain was calculated using Equation 7.17.

The cure process started immediately after infusion of the resin. Figure 7.37 shows the effective transverse strain measured by three HiBi FBG sensors embedded at locations S1, S3, and S5, from the exhaust to the wider end of the tailcone, in Zone F1 (identified in Figure 7.32c) of the tailcone. The information contained in Figure 7.37 is interpreted as follows. The positive strain corresponds to the combined thermal-expansion strain of all the components of the tailcone (i.e., the molding tool, carbon fiber, resin, etc.), while the decrease in the strain during the cure phase relates to shrinkage strain caused by the chemical cure reaction.

The sensor located at S1, closest to the exhaust end of the tailcone, exhibited thermal-expansion strain and shrinkage strain in the through-the-thickness (i.e., out-of-plane) direction of 109 and −1139 με respectively (Figure 7.37a). Similarly, the sensor located at S5, closest to the wider end of the tailcone, detected 516 and −1252 με for the

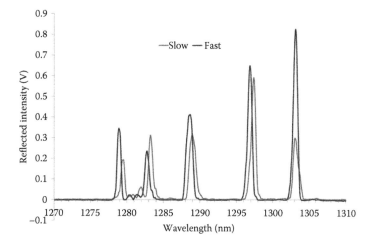

FIGURE 7.36 Five wavelength-division-multiplexed FBG sensors fabricated in polyimide-coated HiBi optical fiber (Fibercore, HB1250P).

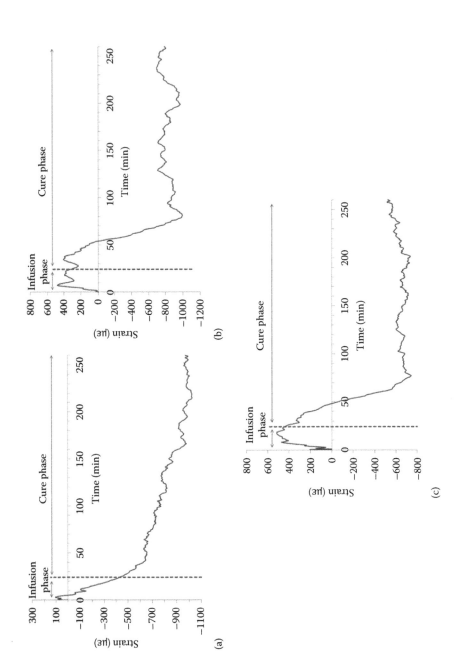

FIGURE 7.37 Effective transverse strain developed through the thickness of the tailcone during the curing process measured by HiBi FBG sensors embedded at sensor positions (a) S1, (b) S3, and (c) S5 in Zone F1 of the tailcone (Figure 7.32c). The numbering of the sensors, i.e., S1 to S5, is from the exhaust end to the wider end of the tailcone.

thermal-expansion strain and shrinkage strain, respectively (Figure 7.37c). The measured strains arise from a combination of various sources, all of which originate from the manufacturing process. Thus, it is appropriate to refer to these strains as process-induced strains. Process-induced strains originate from thermal strain due to the differential expansion between the reinforcement fibers and the resin; strain in the composite due to the resin shrinkage caused by the chemical cure reaction; and strain arising from mold–composite interactions caused by the differential in thermo-elastic response (Sections 7.1 and 7.2.5). The magnitude of the process-induced strain can also be influenced by the type of mold, by the structural design of the composite, including the orientation of the reinforcement fibers, and by the existence of a pressure gradient through the thickness of the composite during the cure process. The magnitudes of the effective transverse strains measured at the different locations of the tailcone were in good agreement.

7.5.1.4 Cure Monitoring: Development of Longitudinal Strain

The embedded SM FBG sensors monitored the longitudinal strains developed in the tailcone preform during the curing of the RTM6 resin. A purpose-built instrument (Figure 7.38) based on a swept laser source (Santec, HSL 2000), with a 110 nm wavelength bandwidth centered at 1318 nm and a scan rate of 20 kHz, and SM optical fiber components, was used to interrogate the WDM SM FBG sensors. A data recording rate of one full spectrum of the five WDM SM FBG sensors every 15 s was used, as for the HiBi FBG sensors described in Section 7.5.1.3. The instrument is shown schematically in Figure 7.38. Also shown is the input from the BBS, at a center wavelength of 1550 nm and

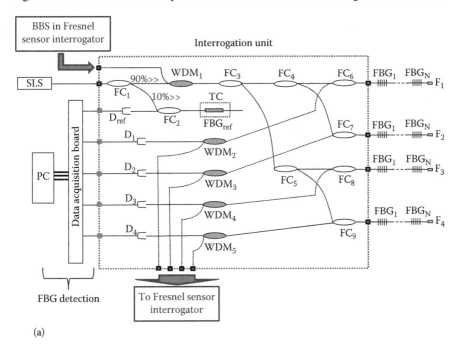

(a)

FIGURE 7.38 (a) Schematic of the interrogation instrument for SM FBG sensors and the Fresnel sensors at the distal ends of the optical fibers.

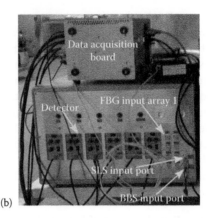

(b)

FIGURE 7.38 (Continued)   (b) Photograph of the instrument. SLS: swept laser source at 1300 nm; FC: SM fiber coupler (50/50 split ratio); BBS: broadband sources at 1550 nm; FBG: SM FBG; D: detector; F: Fresnel sensor at distal end; PC: personal computer; Ref FBG: reference SM FBG; TC: temperature controlled; WDM: wavelength-division multiplexer for wavelengths of 1300 (SM FBG sensors) and 1550 nm (Fresnel sensors).

100 nm bandwidth, which was used to interrogate the optical fiber Fresnel sensors that were located at the distal ends of the optical fibers containing the SM FBG sensors. The WDM coupler was used to separate the light from the Fresnel and SM FBG sensors. The strain measured by the SM FBG sensors was interpreted as longitudinal strain by using Equation 7.12 (Section 7.3.3), in which the FBG response to temperature was compensated for as discussed in Section 7.5.1.3. Figure 7.39 shows the reflection spectra from five WDM SM FBG sensors fabricated in each of the four SM optical fibers embedded in the different zones of the tailcone. The SM FBG sensors and HiBi FBG sensors described in Section 7.3.3 were interrogated simultaneously using separate instrumentation.

Figure 7.40 shows the longitudinal strains measured by three SM FBG sensors embedded at the locations S1, S3, and S5, from the exhaust to the wider end of

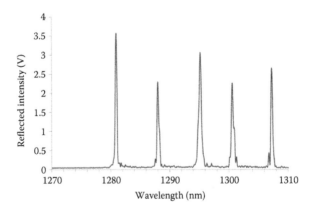

FIGURE 7.39   Five wavelength-division-multiplexed FBG sensors fabricated in polyimide-coated SM optical fiber (Fibertronix, FX-PI-01-01-01).

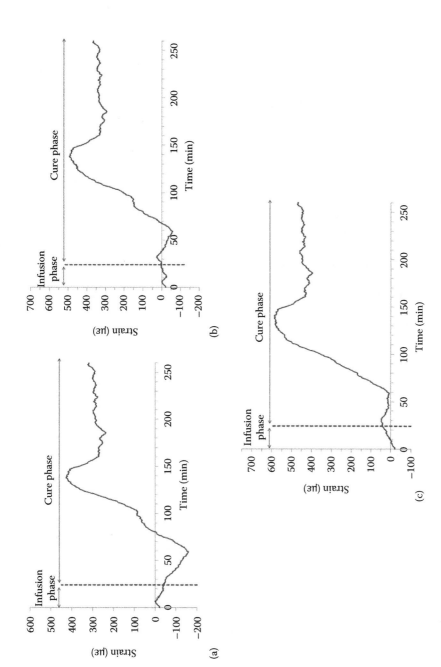

FIGURE 7.40 Longitudinal strain developed lengthwise of the tailcone during the curing process, monitored by SM FBG sensors embedded at sensor positions (a) S1, (b) S3, and (c) S5 in Zone F1 of the tailcone (Figure 7.31c). The numbering of the sensors, i.e., S1–S5, is from the exhaust end to the wider end of the tailcone.

the tailcone, in Zone F1 (identified in Figure 7.32c) of the tailcone. The thermal-expansion strains and shrinkage strains in Figure 7.40 are identified as discussed in Section 7.5.1.3. The sensor located at S1, closest to the exhaust end of the tailcone, detected 428 and −153 με for the thermal-expansion strain and shrinkage strain, respectively, monitored in the longitudinal direction of the tailcone (Figure 7.40a). The corresponding values measured by the SM FBG sensor located at S5, closest to the wider end of the tailcone, for the thermal-expansion strain and shrinkage strain were 588 and −158 με respectively (Figure 7.40c). Figure 7.40 shows that shrinkage strain in the longitudinal direction of the tailcone was detected ~140 min into the cure phase. The shrinkage strain in the through-thickness direction was detected ~115 min earlier, as discussed in Section 7.5.1.3 (Figure 7.37). The capability to detect shrinkage strain by HiBi FBG sensors significantly earlier in the through-the-thickness direction of the composite, compared with shrinkage strain detected by SM FBG sensors in the longitudinal direction, was previously reported [54]. The magnitudes of the longitudinal strains measured at the different locations of the tailcone were in good agreement.

7.5.1.5   Post-Fabrication Mechanical Testing: Strain Monitoring

The embedded SM FBG and HiBi FBG sensors used to monitor the processes involved in the manufacture of the tailcone were used subsequently, along with surfaced-mounted RFSG rosettes, to monitor structural strains during postfabrication mechanical testing (Figure 7.41). After the tailcone was manufactured, the RFSG sensors were bonded to the outer surface of the tailcone at locations close to where the FBG sensors were embedded. The RFSG rosette sensor at each location was configured to measure longitudinal, shear, and circumferential strains (Figure 7.41b). Figure 7.41b describes the strain components measured by the FBG and RFSG sensors. The longitudinal strain measured by the SM FBG sensors was compared with that measured by the RFSG sensors. Figure 7.41b shows that the in-plane transverse strain measured by the HiBi FBG sensors is oriented in the same direction as the circumferential strain measured by the RFSG sensors, which allowed the comparison of the strains measured by the two types of sensors. The in-plane transverse strain was determined from a rearranged Equation 7.17 in which the slow and fast Eigen-axes parameters were interchanged. The interior surface of the tailcone was lined with vacuum bags to allow the tailcone to be subjected to vacuum pressure loading. Two experimental configurations were implemented, the first using thin aluminum foil to seal the window and door areas while the thick aluminum plates were used in the second (Figure 7.41).

In-plane transverse strains and circumferential strains measured by the embedded HiBi FBG and surfaced-mounted RFSG sensors, respectively, are compared in Figure 7.42. The window and door areas were sealed with thin aluminum foil during the pressure loading test. The in-plane transverse strains and the circumferential strains monitored midway along the length of the tailcone at sensor location S3 were compared round the tailcone in Zones F1, F2, F3, and F4 (Figure 7.42).

The general trend and the sign of the strains measured by the embedded HiBi FBG and surfaced-mounted RFSG sensors were in good agreement. Figure 7.42 also shows that the trend in the measured strain varies from one sensor position to

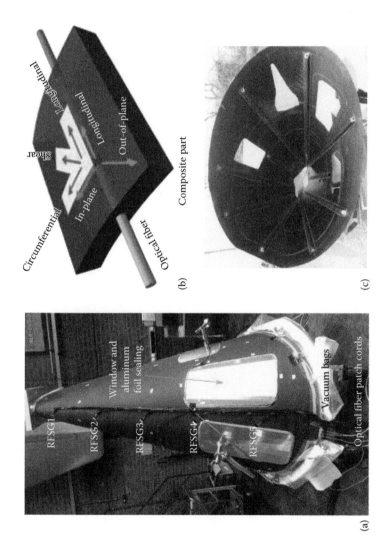

FIGURE 7.41 (a) Tailcone configuration for the vacuum loading tests in which aluminum foils and plates were used to seal the window and door areas. (b) Strain components measured by the embedded FBG (light gray arrows) sensors and surface-mounted RFSG (dark gray arrows) rosette sensors. (c) The interior of the tailcone after demolding. The in-plane and out-of-plane transverse strains were measured by the HiBi FBG sensor, while the longitudinal strain was measured by the SM FBG sensor embedded at the same location.

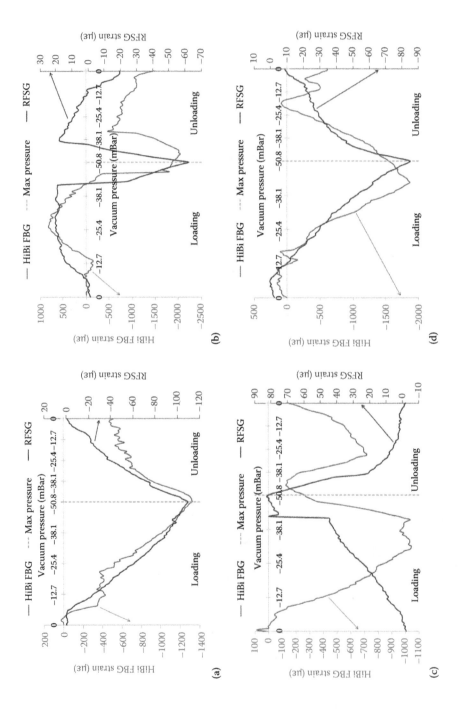

FIGURE 7.42 Comparison of the embedded HiBi FBG and the surface-mounted RFSG sensors for in-plane transverse and circumferential strains, respectively, measured at sensor position S3 in (a) Zone F1, (b) Zone F2, (c) Zone F3, and (d) Zone F4 of the tailcone (Figure 7.32c). The window and door areas were covered with thin aluminum foil.

the next, round the midway of the tailcone. Discrepancies in the measurements by the two sensor types were likely caused by the differences in the loadings experienced by the *embedded* HiBi FBG sensors and the *surface*-bonded RFSG sensors. This difference in the magnitude of the strains between the measurements from the embedded and surface-bonded sensors was expected due to the complex structure of the tailcone, which consisted of a distribution of 10 stringers, tufting zones, different ply-orientation architecture, windows, doors, and nonuniform dimensions.

Figure 7.43 shows that the SM FBG sensor embedded at location S5 in Zone F1 of the tailcone measured longitudinal strains that closely matched the longitudinal strains measured by the RFSG rosette sensor. The measurement exhibited a maximum of ~70 µε in longitudinal strain in the tailcone for a maximum vacuum pressure of −51 mbar when thin aluminum foil was used to seal window and door areas (Figure 7.43a). When the thin aluminum foil was replaced by thick aluminum plate covers, the maximum longitudinal strains recorded were ~56 and 65 µε at the

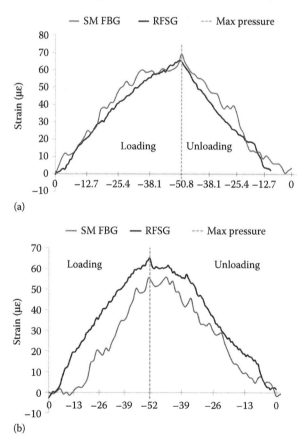

(a)

(b)

FIGURE 7.43 Comparison of the embedded SM FBG and surface-mounted RFSG sensors' longitudinal strain measurements at the sensor location S5 in Zone F1 of the tailcone when the window and door areas were sealed with (a) thin aluminum foil and (b) thick aluminum plates, during vacuum pressure loading tests.

vacuum pressure of -52 με for the embedded SM FBG and surface-mounted RFSG sensors, respectively (Figure 7.43b).

7.5.2 SUPERCONDUCTING MAGNET COIL: POSTFABRICATION TEST MONITORING

The commissioning of a superconducting magnet coil manufactured at Oxford Instruments (Oxford, UK) was investigated using HiBi FBG sensors that were embedded within the magnet structure. The magnet system consisted of steel over-bind reinforcement. The reinforced magnet exhibits high compression, of the order of 100 MPa, when compared with the unreinforced magnet, which experiences significantly less compression. The development of multi-strain components in the superconducting magnet coil was monitored during energization and when the magnet underwent quenching, which are the normal testing and training schedules for commissioning new superconducting magnets. Two WDM HiBi FBG sensors were installed in the superconducting magnet. The FBG sensors were fabricated in HiBi optical fiber (Fibercore HB1500) with core and cladding diameters of 8 and 125 μm, respectively, while the total diameter of the optical fiber including the protective polyacrylate buffer was 242 μm. The FBG sections (5 mm long) of the optical fiber were stripped prior to FBG inscription, and the FBG sensors were not recoated. The wavelengths reflected by the HiBi FBG sensors in each polarization Eigen-mode were measured independently and simultaneously using the custom-designed interrogation system based on a scanning FFP interferometer coupled to HiBi optical fiber described in Section 7.3.3 (Figure 7.11b).

The internal diameter and length of the prototype superconducting magnet that was investigated were 120 mm and 60 mm, respectively. Superconducting wire was wound onto a metallic former using a machine that maintained constant strain and ensured tight binding from turn to turn and from layer to layer. Glass fiber cloth was then wound on top of the coil (Figures 7.44a and b). The HiBi FBG sensors were placed onto the surface of the glass fiber cloth according to the schematic diagram of Figure 7.44c. This procedure ensured that there was no direct contact between the superconducting wire and the bare fiber sensors, thereby preventing potential mechanical damage to the sensors. It was ensured that, at the FBG sensor locations, the HiBi fiber was laid such that the slow axes lay in-plane (i.e., coil surface) while the fast axes lay out-of-plane (i.e., radially) to the surface of the coil. The sensors were all laid halfway between the end plates, such that one HiBi FBG lay parallel to the axis of the magnet (FBGa), referred to as the axial direction, and one along the circumference (FBGh) of the magnet, referred to here as the hoop direction (Figure 7.44c). Glass fiber cloth was then wound onto the magnet to fill up the region of the magnet between the two end plates/caps of the metallic former. The completed magnet was vacuum impregnated with MY750 resin and was cured in an autoclave that was maintained at a temperature of 80°C for 12 h to bind the coils, the glass fiber cloth, and the sensors together. The surface of the magnet was, after potting, machined and fitted with steel over-bind for reinforcement (Figure 7.44a). Figure 7.44d shows the magnet after potting. The superconducting magnet was mounted concentrically inside another magnet that was required for providing a background magnetic field to the superconducting coil. The magnet system was placed in a cryostat and cooled

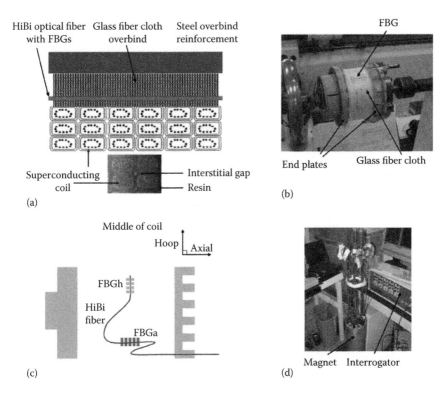

FIGURE 7.44 (a) Cross-section of the magnet, (b) picture of the magnet showing sensors laid on the glass fiber cloth, (c) schematic diagram of the layout of the FBG sensors on a horizontal magnet as in (b), and (d) picture of the magnet after potting. The slow axes of the HiBi FBG sensors were laid in-plane while the fast axes lay out-of-plane to the surface of the coil. FBGa is along the axis of the coil, while FBGh is along the circumference of the coil. The two FBGs are 9 cm apart.

down from room temperature to 4.2 K using liquid helium. The sensors were connected to the interrogation system via a feed-through at the top of the cryostat.

The outer magnet was energized and maintained at a current of 110 A, after which the inner superconducting coil was energized to 100 A. A forced quench was then induced via a trigger that allowed current to pass through a resistive element built into the outer coil, leading to local heating in the coil. Figure 7.45a shows the two orthogonal components of the wavelength response from HiBi FBGh, which was aligned with the hoop direction of the superconducting coil. The profile of the energizing electrical current is also plotted in the figure. Figure 7.45b is a similar representation for HiBi FBGa, which was aligned with the axial direction of the superconducting coil. The onset of the quench is characterized by a sudden rise in the electrical current at approximately 2 s, and this is illustrated in more detail in Figures 7.45c and d over a period of 5 s. The wavelength responses of the FBG sensors are well correlated with the electrical current. The wavelength responses after the quench are characterized by oscillations whose period gradually lengthens.

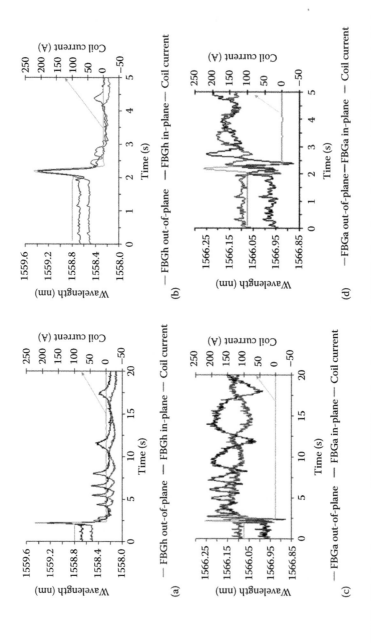

FIGURE 7.45 Wavelength response of the HiBi FBG to a forced quench for (a) FBGh aligned with the hoop and (b) FBGa aligned with the axial directions of the coil; (c) and (d) are magnifications of (a) and (b), respectively.

The in-plane wavelength component is in antiphase with the out-of-plane wavelength component for any given HiBi FBG (Figures 7.45a and b). FBGh, which was aligned with the hoop direction of the coil, exhibited a greater sensitivity to the quench than that of FBGa, which was aligned with the axial direction of the coil. The sudden increase in the wavelengths, followed by a sharp decrease during the quench (after 2 s), as illustrated by Figure 7.45a, is opposite to the behavior of Figure 7.45b. The evolution of the strain components of the sensor embedded aligned with the hoop is therefore opposite to that aligned with the axial direction of the coil.

Figure 7.46 shows the effective transverse strains recorded by FBGh and FBGa, which were aligned with the hoop and axial directions of the coil, respectively. The effective transverse strain evaluated for FBGh, which was aligned with the hoop direction of the coil, is largely in antiphase with the transverse strain exhibited by FBGa, which was aligned with the axial direction of the coil. The transverse strains are not exactly in antiphase, because a traveling wave emanating from the quench zone will excite each of the two sensors at slightly different times, as the sensors are in different spatial locations (Figure 7.44c). The transverse strain exhibits oscillations, which suggests that the coil, either locally at the location of the sensors between the energized windings and the over-bind, or over the bulk of the structure, is set into mechanical oscillations after the quench process. The period of the oscillations lengthens gradually with time. It may be inferred from the strain oscillations that when the coil expands along the hoop direction, it simultaneously contracts along the axial direction. Also plotted in Figure 7.46 are the average hoop and axial strains measured by two RFSGs that were attached to the outer surface of the magnet, one aligned with the hoop and the other aligned with the axial direction of the coil. The hoop and axial strains from the RFSGs increased to a maximum of 2000 and 1250 µε, respectively, at the onset of the quench before the strains decayed rapidly to approximately −480 and −290 µε, respectively. The RFSG responses imply that the coil will finally settle at negative states of strain, thus indicating

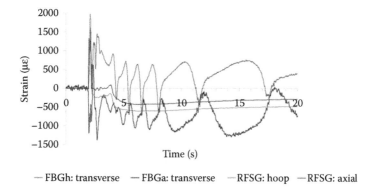

FIGURE 7.46 Effective transverse strains calculated from the wavelength shifts of Figures 7.45a and b for FBGh and FBGa, respectively. The hoop and axial strains measured by RFSGs are also plotted.

that the magnet has dissipated energy as a result of the quench. The maximum transverse strain excursion from the HiBi FBG sensors is about 1500 µε, which is comparable to the maximum strains measured by the RFSGs. The RFSGs were, however, unable to record the oscillations in strain that were measured by the HiBi FBG sensors.

7.6 CONCLUSIONS, CHALLENGES, AND FUTURE REQUIREMENTS

This chapter has shown the capability and current state of the optical fiber sensors in monitoring key manufacturing processes of composite materials. The monitoring of the following processes and parameters has been presented: tufting strains experienced by the needle during the through-thickness reinforcement of preform, resin infusion and flow, process temperature, degree of cure via the evolution of the refractive index of the resin, resin cure via the development of longitudinal and transverse strains, and structural strains during postfabrication testing. In summary, this chapter has demonstrated that optical fiber sensors embedded into the fiber-reinforced preform during layup can be configured to monitor all processes of manufacture through to the monitoring of postfabrication testing, thus providing multifunctionality as well as multiplexability. Laboratory productions of planar carbon fiber–reinforced (and glass fiber–reinforced) composites were monitored online using embedded optical fiber sensors. A full-scale one-piece all-carbon fiber–reinforced aircraft tailcone assembly was manufactured in industrial plants, where optical fiber sensors embedded into the tailcone were used to monitor the entire manufacturing processes as well as monitoring the postproduction mechanical testing. High-technology industrial application involved embedding optical fiber sensors into a superconducting magnet coil and monitoring the production process as well as the postfabrication commissioning process. The applications presented in this chapter are thus complex and challenging, which serves to highlight the capability of optical fiber sensors to provide online quality control in monitoring composite production.

Optical fiber sensors will require optimization when they are used in embedded scenarios, as strain transfer from the composite to the sensors is dependent on the ply architecture of the reinforcement fibers, and it was also shown previously that strain transfer depends on the relative orientations of the optical fiber sensor and the reinforcement fibers in its vicinity [54]. Lack of this optimization is currently responsible for the discrepancy between strains measured by embedded optical fiber sensors (which are usually an order of magnitude lower) and strains predicted by modeling/simulations. Other challenges still to be overcome include the development of suitable ingress–egress connectors for the optical fiber sensors and the need to produce and use smaller-diameter optical fibers that can be embedded with minimal intrusion. Automated optical fiber sensor placement into fiber-reinforced preform, simultaneously with the automated placement of the preform, should be the future of research if industrial mass production of composites is to be uninterrupted. Future research will also need to focus on performing online quality control and employing adaptive composite manufacturing procedures by exploiting the optical fiber sensors embedded into the composites.

REFERENCES

1. Konstantopoulos, S., Fauster, E., and Schledjewski, R. 2014. Monitoring the production of FRP composites: A review of in-line sensing methods. *Express Polymer Letters* **8** 823–840.
2. Skordos, A. A., Karkanas, P. I., and Partridge, I. K. 2000. A dielectric sensor for measuring flow in resin transfer moulding. *Measurement Science and Technology* **11** 25–31.
3. Hegg, M. C., Ogale, A., Mescher, A., Mamishev, A., and Minaie, B. 2005. Remote monitoring of resin transfer molding processes by distributed dielectric sensors. *Journal of Composite Materials* **39** 1519–1539.
4. Garschke, C., Weimer, C., Parlevliet, P. P., and Fox, B. L. 2012. Out-of-autoclave cure cycle study of a resin film infusion process using *in situ* process monitoring. *Composites Part A: Applied Science and Manufacturing* **43** 935–944.
5. Schmachtenberg, E., Schulte, Z. H. J., and Topker, J. 2005. Application of ultrasonics for the process control of resin transfer moulding (RTM). *Polymer Testing* **24** 330–338.
6. Tuncol, G., Danisman, M., Kaynar, A., and Sozer, E. M. 2007. Constraints on monitoring resin flow in the resin transfer molding (RTM) process by using thermocouple sensors. *Composites Part A: Applied Science and Manufacturing* **38** 1363–1386.
7. Xin, C., Gu, Y., Li, M., Li, Y., and Zhang, Z. 2011. Online monitoring and analysis of resin pressure inside composite laminate during zero-bleeding autoclave process. *Polymer Composites* **32** 314–323.
8. Lekakou, C., Cook, S., Deng, Y., Ang, T. W., and Reed, G. T. 2006. Optical fibre sensor for monitoring flow and resin curing in composites manufacturing. *Composites Part A: Applied Science and Manufacturing* **37** 934–938.
9. Jarzebinska, R., Chehura, E., James, S. W., and Tatam, R. P. 2012. Multiplexing a serial array of tapered optical fibre sensors using coherent optical frequency domain reflectometry. *Measurement Science and Technology* **23** 105203 (8 pp.).
10. Gupta, N. and Sundaram, R. 2009. Fiber optic sensors for monitoring flow in vacuum enhanced resin infusion technology (VERITy) process. *Composites Part A: Applied Science and Manufacturing* **40** 1065–1070.
11. Kim, C. B. and Su, C. B. 2004. Measurement of the refractive index of liquids at 1.3 and 1.5 micron using a fibre optic Fresnel ratio meter. *Measurement Science and Technology* **15** 1683–1686.
12. Buggy, S. J., Chehura, E., James, S. W., and Tatam, R. P. 2007. Optical fibre grating refractometers for resin cure monitoring. *Journal of Optics A: Pure and Applied Optics* **9** S60–S65.
13. Wang, P., Molimard, J., Drapier, S., Vautrin, A., and Minni, J. C. 2012. Monitoring the resin infusion manufacturing process under industrial environment using distributed sensors *Journal of Composite Materials* **46** 691–706.
14. Dell'Anno, G., Partridge, I., Cartie, D., Hamlyn, A., Chehura, E., James, S., and Tatam, R. 2012. Automated manufacture of 3D reinforced aerospace composite structure. *Structural Integrity* **3** 22–40.
15. Archer, E., Broderick, J., and McIlhagger, A. T. 2014. Internal strain measurement and cure monitoring of 3D angle interlock woven carbon fibre composites. *Composites Part B: Engineering* **56** 424–430.
16. Chehura, E., Jarzebinska, R., Da Costa, E. F. R., Skordos, A. A., James, S. W., Partridge, I. K., and Tatam, R. P. 2013. Multiplexed fibre-optic sensors for monitoring resin infusion, flow, and cure in composite material processing. *Proceedings of SPIE* **8693** 86930F.
17. Wong, R. Y., Chehura, E., James, S. W., and Tatam, R. P. 2013. A chirped long-period grating sensor for monitoring flow direction and cure of a resin. *Proceedings of SPIE* **8693** 86930E.

18. Chehura, E., James, S. W., and Tatam, R. P. 2007. Temperature and strain discrimination using a single tilted fibre Bragg grating. *Optics Communications* **275** 344–347.
19. Wang, Q., Xia, J., Zhao, Y., Liu, X., Wang, P., Zhang, Y.-N., Sang, C., and Wang, L. 2014. Simultaneous measurement of strain and temperature with polarization maintaining fiber Bragg grating loop mirror. *Instrumentation Science and Technology* **42** 298–307.
20. Chehura, E., Jarzebinska, R., Da Costa, E. F. R., Skordos, A. A., James, S. W., Partridge, I. K., and Tatam, R. P. 2015. Monitoring the processing of composite materials using optical fibre sensors. *Measurement Science and Technology Journal.* Submitted.
21. Toure, S. M. 2015. *Manufacture and Characterisation of Carbon Fibre Prepreg Stacks Containing Resin Rich and Resin Starved Slip Layers.* Engineering Doctorate Thesis. University of Manchester, UK.
22. Campbell, F. C. 2010. *Structural Composite Materials.* Materials Park, OH: ASM International.
23. Coriolis Composites, Fiber placement. http://www.coriolis-composites.com/products/fiber-placement-process.html.
24. ADVITAC, The Advitac Project. http://www.advitac.eu/.
25. Chehura, E., Dell'Anno, G., Huet, T., Staines, S., James, S. W., Partridge, I. K., and Tatam, R. P. 2014. On-line monitoring of multi-component strain development in a tufting needle using optical fibre Bragg grating sensors. *Smart Materials and Structures* **23** 075001 (9 pp.).
26. Rudd, C. D., Long, A. C., Kendall, K. N., and Mangin, C. 1997. *Liquid Moulding Technologies: Resin Transfer Moulding, Structural Reaction Injection Moulding and Related Processing Techniques.* Sawston, Cambridge: Woodhead Publishing.
27. Lim, S. T. and Lee, W. I. 2000. An analysis of the three-dimensional resin-transfer mold filling process *Composite Science and Technology* **60** 961–975.
28. Khoun, L., Oliveira, R.-D., Michaud, V., and Hubert, P. 2011. Investigation of process-induced strains development by fibre Bragg grating sensors in resin transfer moulded composites. *Composites Part A* **42** 274–282.
29. Vila, J., Gonzalez, C., and Llorca, J. 2014. A level set approach for the analysis of flow and compaction during resin infusion in composite materials. *Composites Part A* **67** 299–307.
30. Hardis, R., Jessop, J. L. P., Peters, F. E., and Kessler, M. R. 2013. Cure kinetics characterization and monitoring of an epoxy resin using DSC, Raman spectroscopy, and DEA. *Composites Part A* **49** 100–108.
31. Marin, E., Robert, L., Triollet, S., and Ouerdane, Y. 2012. Liquid resin infusion process monitoring with superimposed fibre Bragg grating sensor. *Polymer Testing* **31** 1045–1050.
32. Harsch, M., Karger-Kocsis, J., and Herzog, F. 2008. Monitoring of cure-induced strain of an epoxy resin by fiber Bragg grating sensor. *Journal of Applied Polymer Science* **107** 719–725.
33. Boeing Australia, Boeing Australia Component repairs. http://www.boeing.com.au/boeing-in-australia/subsidiaries/boeing-defence-australia/boeing-australia-component-repair-gallery.page?.
34. Cranfield University, Composites. https://www.cranfield.ac.uk/About/Cranfield/Themes/Register-Online-Now/Composites.
35. TA Instruments, A review of DSC kinetics methods. http://www.tainstruments.com/pdf/literature/TA073.pdf.
36. Bandara, U. 1986. A systematic solution to the problem of sample background correction in DSC curves. *Thermal Analysis* **31** 1063–1071.

37. Skordos, A. A. and Partridge, I. K. 2001. Cure kinetics modelling of epoxy resins using a non-parametric numerical procedure. *Polymer Engineering and Science* **41** 793–805.

38. Gabbott, P. 2007. *Principles and Applications of Thermal Analysis.* Hoboken, New Jersey: Wiley-Blackwell.

39. Karbhari, V. M. 2013. *Non-Destructive Evaluation (NDE) of Polymer Matrix Composites: Techniques and Applications.* Sawston, Cambridge: Woodhead.

40. Chen, J.-Y., Hoa, S. V., Jen, C.-K., and Wang, H. 1999. Fiber-optic and ultrasonic measurements for *in-situ* cure monitoring of graphite/epoxy composites. *Journal of Composite Materials* **33** 1860–1881.

41. Ye, C.-C., James, S. W., and Tatam, R. P. 2000. Simultaneous temperature and bend sensing using long period fibre gratings. *Optics Letters* **25** 1007–1009.

42. Buggy, S. J. 2008. *Composite Material Process Monitoring using Optical Fibre Grating Sensors.* PhD Thesis, Cranfield University, UK.

43. Robert, L. and Dusserre, G. 2014. Assessment of thermoset cure-induced strains by fiber Bragg grating sensor. *Polymer Engineering and Science* **54** 1585–1594.

44. Born, M. and Wolf, E. 1999. *Principles of Optics: 7th (Expanded) Edition.* Cambridge: Cambridge University Press.

45. Chehura, E., James, S. W., Staines, S., Groenendijk, C., Cartie, D., Portet, S., Hugon, M., and Tatam, R. 2015. Monitoring a full-sized all-carbon-fibre composite aircraft tailcone assembly during production and post-production mechanical testing using embedded optical fibre sensors. *Smart Materials and Structures Journal.* Submitted.

46. Chehura, E., James, S. W., Staines, S., Groenendijk, C., Cartie, D., Portet, S., Hugon, M., and Tatam, R. 2015. Monitoring the manufacturing process and the mechanical performance of a full-sized all carbon fibre composite aircraft tailcone assembly using embedded fibre optic sensors. *20th International Conference on Composite Materials (ICCM20,)* Copenhagen (on-line Proc. P-ID4220–3).

47. Paul, P. H. and Kychakoff, G. 1987. Fiber-optic evanescent field absorption sensor. *Applied Physics Letters* **51** 12–14.

48. Kumar, A., Subrahmanyam, T. V. B., Sharma, A. D., Thyagarajan, K., Pal, B. P., and Goyal, I. C. 1984. Novel refractometer using a tapered optical fibre. *Electronics Letters* **20** 534–535.

49. Jarzebinska, R., Cheung, C. S., James, S. W., and Tatam, R. P. 2009. Response of the transmission spectrum of tapered optical fibres to the deposition of a nanostructured coating. *Measurement Science and Technology* **20**, 034001.

50. Dockney, M. L., James, S. W., and Tatam, R. P. 1996. Fibre Bragg gratings fabricated using a wavelength tuneable laser source and a phase mask based interferometer. *Measurement Science and Technology* **7** 445–448.

51. Meltz, G., Morey, W. W., and Glenn, W. H. 1989. Formation of Bragg gratings in optical fibres by a transverse holographic method. *Optics Letters* **14** 823–825.

52. James, S. W., Dockney, M. L., and Tatam, R. P. 1996. Simultaneous independent temperature and strain measurement using in-fibre Bragg grating sensors. *Electronics Letters* **32** 1133–1134.

53. Udd, E., Schulz, W., Seim, J., Haugse, E., Trego, A., Johnson, P., Bennett, T. E., Nelson, D., and Makino, A. 2000. Multidimensional strain field measurements using fibre optic grating sensors. *Proceedings of SPIE* **3986** 254–262.

54. Chehura, E., Skordos, A. A., Ye, C.-C., James, S. W., Partridge, I. K., and Tatam, R. P. 2005. Strain development in curing epoxy resin and glass fibre/epoxy composites monitored by fibre Bragg grating sensors in birefringent optical fibre. *Smart Materials and Structures* **14** 354–362.

55. Chehura, E., Buggy, S. J., James, S. W., Johnstone, A., Lakrimi, M., Domptail, F., Twin, A., and Tatam, R. P. 2011. Multi-component strain development in superconducting magnet coils monitored using fibre Bragg grating sensors fabricated in highly linearly birefringent fibre. *Smart Materials and Structures* **20** 125004 (10 pp.).

56. Ye, C.-C., Staines, S. E., James, S. W., and Tatam, R. P. 2002. A polarization-maintaining fibre Bragg grating interrogation system for multi-axis strain sensing. *Measurement Science and Technology* **13** 1446–1449.

57. Lawrence, C. M., Nelson, D. V., Udd, E., and Bennett, T. 1999. A fibre optic sensor for transverse strain measurement. *Experimental Mechanics* **39** 202–209.

58. Laffont, G. and Ferdinand, P. 2001. Tilted short-period fibre-Bragg-grating-induced coupling to cladding modes for accurate refractometry. *Measurement Science and Technology* **12** 765–770.

59. Erdogan, T. and Sipe, J. E. 1996. Tilted fiber phase gratings. *Journal of the Optical Society of America A* **13** 296–313.

8 FBG-Based Structural Performance Monitoring and Safety Evaluation of a Composite Arch Bridge

Xiao-Wei Ye, Tan Liu, Chuan-Zhi Dong, and Bin Chen

CONTENTS

8.1 Introduction 269
8.2 Structural Health Monitoring System of the Composite Arch Bridge 271
    8.2.1 Structural Features of the Composite Arch Bridge 271
    8.2.2 Design and Implementation of the Structural Health Monitoring System 272
    8.2.3 Description of FBG-Based Structural Monitoring System 273
8.3 Structural Stress Monitoring Using FBG-Based Strain Sensors 274
    8.3.1 Fundamentals of FBG-Based Strain Sensors 274
    8.3.2 Deployment of FBG-Based Strain Sensors 275
    8.3.3 Analysis of Long-Term Strain Monitoring Data 278
        8.3.3.1 Strain Monitoring Data from Selected FBG Sensors 278
        8.3.3.2 Separating Temperature Effect Using Wavelet Multiresolution Analysis 282
8.4 Cable Force Monitoring Using FBG-Based Stress Ring Sensors 288
    8.4.1 Fundamentals of FBG-Based Stress Ring Sensors 288
    8.4.2 Deployment of FBG-Based Stress Ring Sensors 289
    8.4.3 Analysis of Long-Term Cable Force Monitoring Data 290
8.5 Concluding Remarks 295
Acknowledgments 295
References 295

8.1 INTRODUCTION

As the key components of transportation systems, bridges play a pivotal role in social and economic development. However, they are constantly exposed to harsh environments and suffer complex external loadings (e.g., highway traffic, railway traffic, wind, waves, earthquakes, and ship collisions). This induces damage to the structural elements of bridges and degrades their mechanical performances. As a result, structural failures or catastrophic accidents may happen when structural damage reaches a critical state. To assure the safety of bridge structures, long-term continuous monitoring of environmental factors and operational loadings is essential to promptly grasp the realistic structural conditions and to proactively formulate early warning strategies.

In the past three decades, the technology of structural health monitoring (SHM) has been broadly employed to track structural behaviors (e.g., acceleration, strain, and displacement) online in a continuous and real-time manner and to ensure that engineering structures are operated in the specified tolerance and safe range (Ni et al., 2010, 2012a,b; Ye et al., 2012, 2013, 2014, 2015). In addition, with the advances of innovative sensing and information processing technologies, SHM systems for various types of civil infrastructures have emerged to identify the extent of structural damage, evaluate the structural safety condition, and estimate the remaining service life. Up to now, a considerable number of SHM systems have been installed on large-scale engineering structures all over the world (Fuhr et al., 1999; Tennyson et al., 2001; Chan et al., 2006; Li et al., 2006; Jiang et al., 2010). With the aid of these instrumented SHM systems, multifarious monitoring parameters reflecting the global or local structural responses of target structures can be obtained to facilitate the assessment of structural health conditions and the establishment of inspection and maintenance schemes.

Among the newly developed cutting-edge sensing approaches, optical fiber sensing technology has gained increasing attention from researchers and engineers in the field of civil engineering (Casas and Cruz, 2003; Mehrani et al., 2009; Xiong et al., 2012). The monitoring items can be readily related to optical characteristics, such as changes of light intensity, interferometry, fiber Bragg grating (FBG), absorption, time-domain reflectometry, and frequency-domain reflectometry. The instrumentation of bridges using diversified optical fiber sensors is widely reported in the literature (Zhang et al., 2006; Kister et al., 2007a,b; Barbosa et al., 2008; Costa and Figueiras, 2012; Rodrigues et al., 2012; Surre et al., 2013). Fuhr and Huston (1998) measured the chloride penetration for corrosion detection in reinforced concrete bridges via optical fiber sensors. Lin et al. (2005a,b) realized the real-time monitoring of local scour by the use of FBG sensors. Talebinejad et al. (2009) proposed a vibration measurement method using serially multiplexed FBG accelerometers for the structural monitoring of bridges. Mokhtar et al. (2012) developed a temperature-compensated optical fiber-strain sensor system for arch bridge condition monitoring. Yau et al. (2013) measured the static vertical displacement of bridges using FBG sensors. The measurement of cable forces (Bronnimann et al., 1998; He et al., 2013), the damage evaluation of arch bridge suspenders (Li et al., 2012), and the estimation of structural prestress losses (Lan et al., 2014) could also be realized by the use of optical fiber sensors.

In this chapter, an SHM system instrumented on a composite arch bridge (Jiubao Bridge) located in Hangzhou, China, is introduced. The arrangement of the FBG-based sensing system is highlighted. The long-term monitoring data from the FBG-based strain sensors and the FBG-based stress ring sensors installed on selected sections are acquired. The structural safety condition of the composite arch bridge is evaluated by the use of the monitoring data and defined condition index.

8.2 STRUCTURAL HEALTH MONITORING SYSTEM OF THE COMPOSITE ARCH BRIDGE

8.2.1 STRUCTURAL FEATURES OF THE COMPOSITE ARCH BRIDGE

The Jiubao Bridge, with a total length of 1855 m, is a two-way, six-lane composite arch bridge located in Hangzhou, China, as illustrated in Figure 8.1. This bridge consists of three main parts: the south approach spans, the main navigation channel spans, and the north approach spans. The south approach spans (90 m + 9 × 85 m + 55 m) and the north approach spans (55 m + 2 × 85 m + 90 m) are made up of reinforced constant cross-sectional continuous concrete box girders. The main navigation channel spans (210 m + 210 m + 210 m) contain three spans of steel arch composite structures. The main spans are composed of constant cross section continuous steel box beams and concrete deck slabs.

As shown in Figure 8.2, the arch rib system of the Jiubao composite arch bridge is formed by three pairs of arch ribs. Each arch structure consists of a main arch rib, a deputy arch rib, and transverse connection rods connecting the main arch rib and the deputy arch rib. The span length of the main arch rib is 188 m, the inclination angle of the main arch rib is 12°, and the height of the main arch rib is 43.78 m. The axis

FIGURE 8.1   Jiubao composite arch bridge.

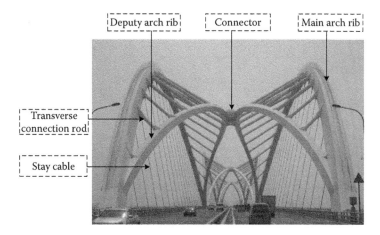

FIGURE 8.2   Arch rib system of the Jiubao Bridge.

of the deputy arch rib is a spatial curve and the height of the deputy arch rib is 33 m. The transverse links between the main arch rib and the deputy arch rib are circular steel tubes and the spacing between the two transverse links is 8.5 m. Thirty-eight stay cables are installed between the main arch rib and the bridge deck, and the distance between the two stay cables is 8.5 m. In total, there are 114 stay cables on the Jiubao Bridge.

8.2.2   DESIGN AND IMPLEMENTATION OF THE STRUCTURAL HEALTH MONITORING SYSTEM

The SHM system of the Jiubao composite arch bridge is designed and implemented to monitor the structural responses and assess the structural safety under the action of dead load, live load (highway traffic and wind), accidental load (earthquake and ship collision), and environmental load (temperature). As shown in Figure 8.3, the sensor subsystem is the basic ingredient of the SHM system that plays the role of collecting original monitoring data. It detects and acquires the structural responses and the input signals (e.g., displacement, strain, acceleration, and temperature) of the bridge under all kinds of excitation and environmental conditions. The arrangement points of the sensors are determined by finite element analysis.

The data collection and transmission subsystem consists of data acquisition stations, server clusters, and optical fiber signal transmission networks. It plays the role

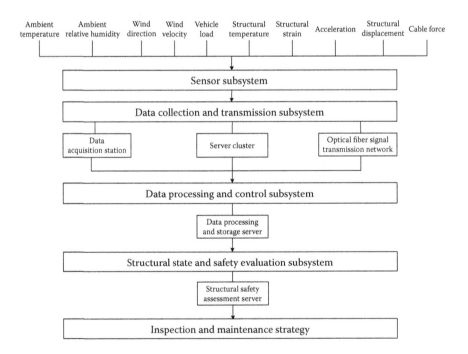

FIGURE 8.3   Architecture of the SHM system instrumented on the Jiubao Bridge.

of data transportation from the sensor subsystem. It transfers the signals measured by the sensors to the data processing and control subsystem. In the data processing and control subsystem, the pretreated signals are processed to obtain the effective information or mechanics indices. Then, they are delivered to the structural state and safety evaluation subsystem. The data processing and control subsystem also realizes the function of the data query by the bridge manager.

8.2.3 DESCRIPTION OF FBG-BASED STRUCTURAL MONITORING SYSTEM

In this section, the FBG-based monitoring system of the Jiubao Bridge will be described. As shown in Figure 8.4, the FBG sensor system consists of three types of sensors: the FBG-based strain sensors, the FBG-based temperature sensors, and the FBG-based stress ring sensors. The FBG-based strain and temperature sensors are installed to monitor the stress change and fatigue damage in the steel box girders and arch ribs. It should be noticed that the FBG-based temperature sensors are deployed adjacent to the FBG-based strain sensors to eliminate the effect of temperature. The FBG-based stress ring sensors are utilized to monitor the force change of cables. In the data acquisition and transmission subsystem, the wavelength data collected by the FBG-based sensors are classified and transferred to the data processing and

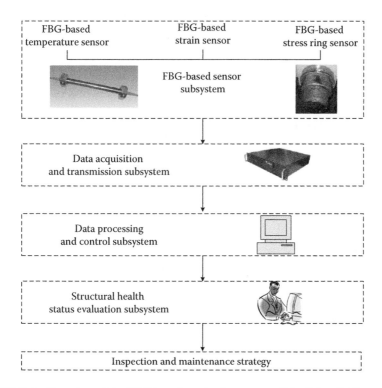

FIGURE 8.4 Flowchart of the FBG-based monitoring system of the Jiubao Bridge.

control subsystem. In the data processing and control subsystem, the stress state of the structural components and the force change of cables are analyzed, and then moved to the structural health status evaluation subsystem. In this subsystem, the evaluation of the structural condition is carried out to help the bridge managers make the optimal inspection and maintenance plans.

8.3 STRUCTURAL STRESS MONITORING USING FBG-BASED STRAIN SENSORS

8.3.1 FUNDAMENTALS OF FBG-BASED STRAIN SENSORS

The FBG sensor can be regarded as a type of optical fiber sensor with periodic changes of the refractive index that are formed when exposed to an intense ultraviolet (UV) interference pattern in the core of the optical fiber. According to Bragg's law, when a beam of white light generated from the broadband source is put into the FBG sensor, the Bragg wavelength, which is related to the grating period, is reflected, as illustrated in Figure 8.5.

FBG sensing technology realizes the absolute measurement of the strain and the temperature by monitoring the Bragg wavelength shift of reflection spectrum or transmission spectrum. The Bragg wavelength, λ_B, can be expressed by (Ye et al., 2014)

$$\lambda_B = 2n_{eff}\Lambda \tag{8.1}$$

where:

n_{eff} is the effective index of refraction
Λ is the grating period

Changes in either the strain or temperature could induce the shift of the Bragg wavelength. The shift of wavelength caused by the strain can be obtained by

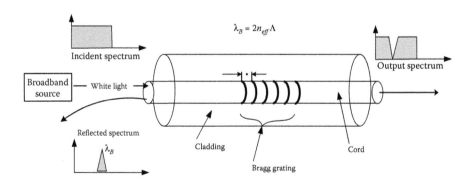

FIGURE 8.5 Fundamentals of an FBG sensor.

$$\Delta\lambda_{B\varepsilon} = \lambda_B\left(1-p_\varepsilon\right)\Delta\varepsilon \tag{8.2}$$

and the shift of wavelength caused by the temperature can be obtained by

$$\Delta\lambda_{BT} = \lambda_B\left(\alpha+\zeta\right)\Delta T \tag{8.3}$$

Generally, it is assumed that the strain and the temperature, respectively, cause the shift of the Bragg wavelength. The Bragg wavelength shift, $\Delta\lambda_B$, with respect to changes in axial strain, $\Delta\varepsilon$, and temperature, ΔT, is given as

$$\Delta\lambda_B = \lambda_B\left[(1-p_\varepsilon)\Delta\varepsilon+\left(\alpha+\zeta\right)\Delta T\right] \tag{8.4}$$

where:
  p_ε is the photo-elastic coefficient of the fiber
  α is the thermal expansion coefficient of the fiber material
  ζ is the thermo-optic coefficient of the fiber material

Since both the axial strain and the temperature cause the shift of Bragg wavelength, the temperature compensation must be done when using FBG strain sensors to monitor the change of structural strain/stress.

8.3.2 Deployment of FBG-Based Strain Sensors

For structural strain monitoring, 166 measurement points in total are determined by FEA. Table 8.1 lists the deployment of the strain monitoring points on the Jiubao composite arch bridge, also illustrated in Figures 8.6 and 8.7. The FBG-based temperature sensors are arranged near the FBG-based strain sensors for temperature compensation. Figure 8.8 shows a photo of the FBG-based strain and temperature sensors. Table 8.2 and Table 8.3 list the technical parameters of the FBG-based strain sensors and temperature sensors.

TABLE 8.1
Overview of Strain Monitoring Points

Structure	Section Number	Monitoring Point	Total
Arch rib	12	4	48
Steel box girder	7	8	56
Pier	2	2	4
Approach span	5	8/8/12/14/16	58

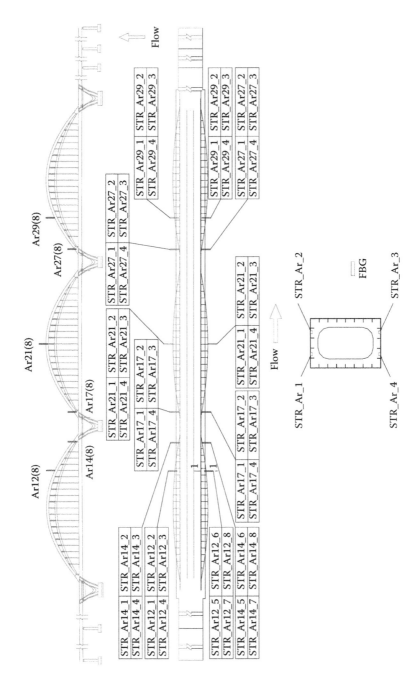

FIGURE 8.6 Deployment of FBG-based strain sensors on arch ribs.

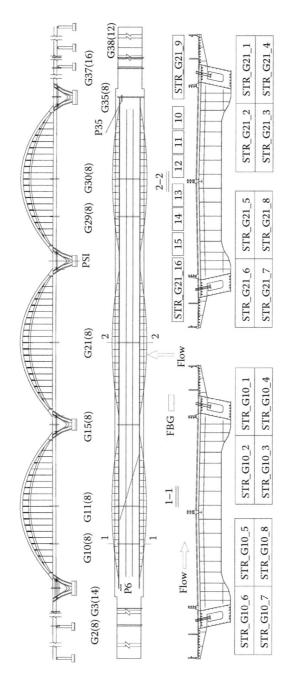

FIGURE 8.7 Deployment of FBG-based strain sensors on girders and piers.

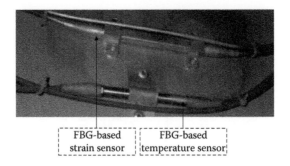

FBG-based strain sensor   FBG-based temperature sensor

FIGURE 8.8   FBG-based strain and temperature sensors.

TABLE 8.2
Technical Parameters of FBG-Based Strain Sensor

Item	Value
Measurement range	±1500 με
Sampling frequency	5 Hz
Resolution ratio	±1 με
Precision	±2–3 με
Strain sensitivity	1.18–1.22 pm/με
Temperature range	−30°C–60°C
Durability	More than 25 years

TABLE 8.3
Technical Parameters of FBG-Based Temperature Sensor

Item	Value
Sampling frequency	5 Hz
Measurement range	−30°C–+180°C
Resolution ratio	0.01°C
Precision	±0.2°C
Connection type	FC/PC or welding

8.3.3   Analysis of Long-Term Strain Monitoring Data

8.3.3.1   Strain Monitoring Data from Selected FBG Sensors

In this section, the strain monitoring data obtained from the selected FBG sensors are examined. The relative variation of structural stress during one week was analyzed to illustrate the general stress state of the Jiubao composite arch bridge. Strain monitoring data during the period from March 21, 2015 to March 27, 2015 were extracted for further analysis. The strain signals collected at midnight on March 21, 2015 were taken as the baseline data. The steel grade was Q345 and the elasticity modulus was 206 GPa. Two sections of the steel box girder were chosen to analyze the stress variation and, in total, eight FBG-based strain sensors were deployed in these sections, as shown in Figure 8.9.

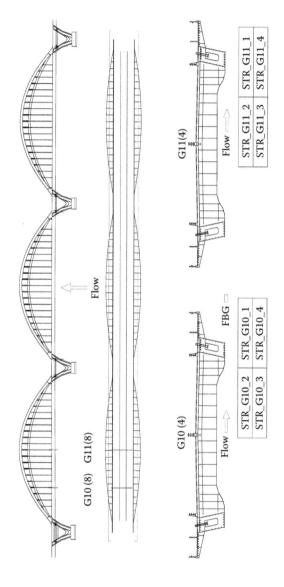

FIGURE 8.9 Deployment of selected FBG-based strain sensors.

The stress–time histories were derived from the acquired strain data by simply multiplying the elasticity modulus of the steel, as illustrated in Figures 8.10 through 8.17. It is observed in Figures 8.10 through 8.17 that the stress–time histories of the selected sensors have a periodic variation characteristic within a one-day period. These multicomponent stress–time histories are composed of low-frequency signals and high-frequency signals that may have been caused by different kinds of factors (e.g., highway traffic, wind, and temperature).

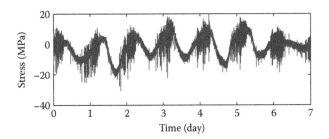

FIGURE 8.10   Stress–time history of STR_G10_1.

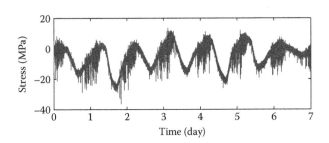

FIGURE 8.11   Stress–time history of STR_G10_2.

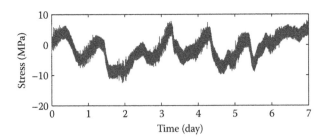

FIGURE 8.12   Stress–time history of STR_G10_3.

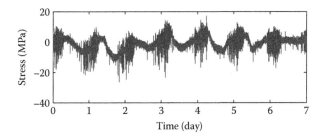

FIGURE 8.13 Stress–time history of STR_G10_4.

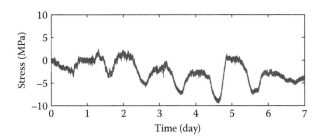

FIGURE 8.14 Stress–time history of STR_G11_1.

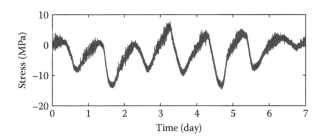

FIGURE 8.15 Stress–time history of STR_G11_2.

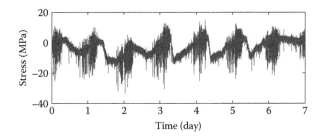

FIGURE 8.16 Stress–time history of STR_G11_3.

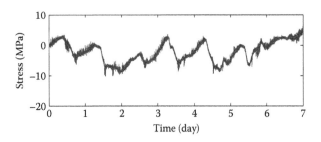

FIGURE 8.17   Stress–time history of STR_G11_4.

8.3.3.2   Separating Temperature Effect Using Wavelet Multiresolution Analysis

The in-service strain monitoring data acquired from the sensors deployed on the girders and arch ribs were mainly induced from three effects: highway traffic, wind, and temperature. (The static strain attributable to the initial dead load could not be obtained because the FBG sensors were installed after the completion of construction.) Strain components due to different effects play different roles in shaping strain quantities. As illustrated in Figures 8.10 through 8.17, the strain responses of the girder and arch rib were acquired simultaneously and continuously under operational conditions. It is seen from Figures 8.10 through 8.17 that there are trend ingredients (low-frequency components) in all the stress–time histories that can be attributed to the daily cycle effect of temperature variation (Ni et al., 2012a,b).

As reflected by the strain monitoring data, under the temperature effect, the bridge girder mainly behaves by expanding or contracting along the longitudinal direction. In contrast, under highway traffic load, the bridge deck undergoes flexural bending. These two types of distinct responses are mixed in the strain monitoring data. This mixture phenomenon implies that one of the effects may be contaminated or distorted by the others. It is desirable to characterize them separately when each effect on the structural behavior requires qualification. The extraction of a specific effect is not easy when the measured signals are nonstationary and non-Gaussian noisy in nature. In this section, a wavelet-based nonparametric approach is used to decompose the strain ingredients by multiresolution analysis.

Transform-domain processing of a signal involves mapping it from the signal space to the transform space using a set of basis functions. For the wavelet transform, a particular function is chosen as the mother wavelet and a family of daughter wavelets is defined by scaling and shifting, serving as a complete set of basis functions. In wavelet-based multiresolution analysis, the same function can be adopted and repeated with different scale and shift parameters. Multiresolution analysis is a process of choosing a set of basis functions that originate from the same mother wavelet. From a practical point of view, wavelet multiresolution analysis allows a decomposition of the signal into various resolution scales: the data with coarse resolution contain the information about low-frequency components, and the data with fine resolution contain the information about high-frequency components (Xia et al., 2012).

Using a selected mother wavelet function, $\Psi(t)$, the continuous wavelet transform of a signal is defined as

$$W_\Psi f(a,b) = \langle f, \Psi_{a,b} \rangle = \frac{1}{|a|^{1/2}} \int_{-\infty}^{\infty} f(t) \overline{\Psi} \left(\frac{t-b}{a} \right) dt, \quad a > 0 \tag{8.5}$$

where:

$\Psi(t)$ is the mother wavelet
a is a scale parameter
b is a time parameter

The overbar represents a complex conjugation. It is known that the function, $f(t)$, can be reconstructed from $W_\Psi f(a,b)$ by the double-integral representation as represented by

$$f(t) = \frac{1}{C_\Psi} \int_{-\infty}^{\infty} \int_{-\infty}^{\infty} W_\Psi f(a,b) \Psi \left(\frac{t-b}{a} \right) \frac{1}{a^2} da\, db \tag{8.6}$$

In practical signal processing, a discrete version of the wavelet transform is often employed by discretizing the scale parameter, a, and the time parameter, b. In general, the procedure becomes much more efficient if dyadic values of a and b are used. That is,

$$a = 2^j, b = 2^j k, j, k \in Z \tag{8.7}$$

where Z is a set of positive integers. With some special choices of $\Psi(t)$, the corresponding discretized wavelets, $\{\Psi_{j,k}\}$, are expressed by

$$\Psi_{j,k}(t) = 2^{j/2} \Psi \left(2^j t - k \right) \tag{8.8}$$

This constitutes an orthonormal basis for $L_2(R)$. With this orthonormal basis, the wavelet expansion of a function, $f(t)$, can be obtained as

$$f(t) = \sum_j \sum_k \alpha_{j,k} \Psi_{j,k}(t) \tag{8.9}$$

where

$$\alpha_{j,k} = \int_{-\infty}^{\infty} f(t) \Psi_{j,k} dt \tag{8.10}$$

In discrete wavelet analysis, a signal can be represented by its approximations and details. The detail at level, j, is defined as

$$D_j = \sum_{k \in Z} \alpha_{j,k} \Psi_{j,k}(t) \tag{8.11}$$

and the approximation at level J is defined as

$$A_J = \sum_{j>J} D_j \tag{8.12}$$

It follows that

$$f(t) = A_J + \sum_{j \leq J} D_j \tag{8.13}$$

Equation 8.12 provides a tree-structure decomposition of a signal and a reconstruction procedure, as well. By selecting different dyadic scales, a signal can be broken down into numerous low-resolution components.

Utilizing the above-described procedure, the measured strain signals are decomposed into approximations and details (i.e., high-frequency and low-frequency components) at various levels. For each level, the high-frequency part (details) is separated, and the remaining low-frequency part (approximations) is transferred into the next level of decomposition. Through wavelet-based multiresolution analysis, the strain component attributable to the temperature effect can be obtained from the lowest-frequency part in the wavelet transform domain.

Figures 8.18 through 8.25 illustrate the wavelet-based decomposed stress–time histories for the eight selected FBG-based strain sensors. In these time histories, the low-frequency parts of the 12-level decomposition of stress data represent the reconstructed temperature-induced stress. Comparison analysis between the initial

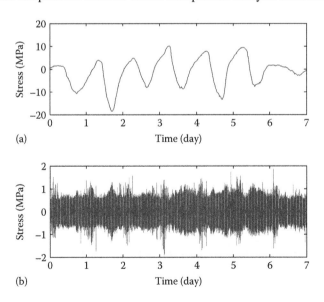

(a)

(b)

FIGURE 8.18 Wavelet-based decomposed stress data of STR_G10_1 with 12-level decomposition: (a) low-frequency part and (b) high-frequency part.

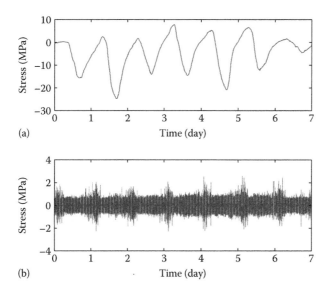

(a)

(b)

FIGURE 8.19   Wavelet-based decomposed stress data of STR_G10_2 with 12-level decomposition: (a) low-frequency part and (b) high-frequency part.

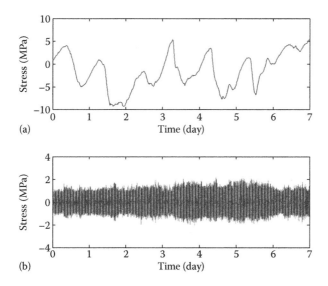

(a)

(b)

FIGURE 8.20   Wavelet-based decomposed stress data of STR_G10_3 with 12-level decomposition: (a) low-frequency part and (b) high-frequency part.

stress and the temperature-induced stress indicates that the temperature-induced stress–time histories have the same variation tendency and periodic characteristics in one day. The temperature-induced stress takes a major part of the total stress, which means that the main stress change during the bridge operational status is caused by the temperature effect. Also, Figures 8.18 through 8.25 show the high-frequency parts

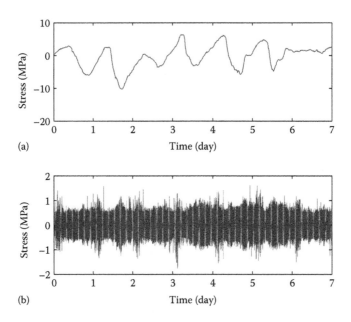

FIGURE 8.21 Wavelet-based decomposed stress data of STR_G10_4 with 12-level decomposition: (a) low-frequency part and (b) high-frequency part.

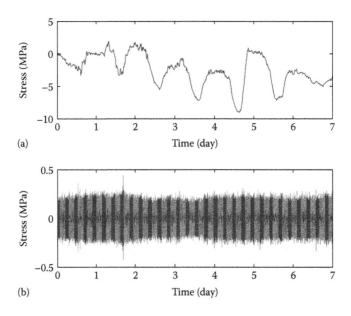

FIGURE 8.22 Wavelet-based decomposed stress data of STR_G11_1 with 12-level decomposition: (a) low-frequency part and (b) high-frequency part.

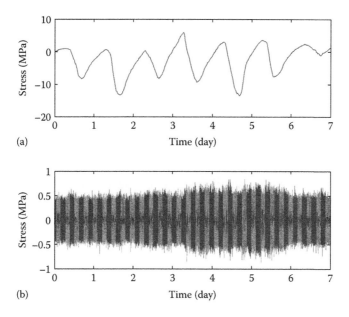

(a)

(b)

FIGURE 8.23 Wavelet-based decomposed stress data of STR_G11_2 with 12-level decomposition: (a) low-frequency part and (b) high-frequency part.

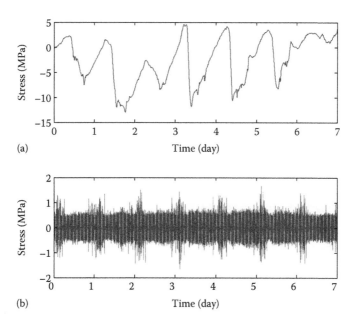

(a)

(b)

FIGURE 8.24 Wavelet-based decomposed stress data of STR_G11_3 with 12-level decomposition: (a) low-frequency part and (b) high-frequency part.

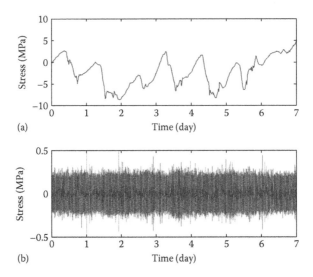

(a)

(b)

FIGURE 8.25 Wavelet-based decomposed stress data of STR_G11_4 with 12-level decomposition: (a) low-frequency part and (b) high-frequency part.

of the 12-level decomposition of stress data that consists of live load (highway traffic and wind)–induced stress. The stress amplitudes vary approximately within 2 MPa, indicating that the live load–induced stress is taking a minor part of the total stress.

8.4 CABLE FORCE MONITORING USING FBG-BASED STRESS RING SENSORS

8.4.1 FUNDAMENTALS OF FBG-BASED STRESS RING SENSORS

As illustrated in Figure 8.26, the FBG-based stress ring sensor is utilized to monitor the tension force of the bridge cables. It is mainly composed of an elastomeric element, FBG strain sensors, and a protective envelope. The operational principle of the FBG-based stress ring sensor can be described as follows. The FBG-based stress ring sensor is tightly attached between the nut and the anchor plate during its operational status. According to the knowledge of structural mechanics, the tension force of the cable, F, is equal to the pressure of the elastomeric element, N. The elastomeric element is deformed

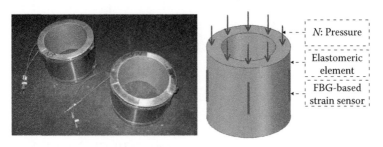

FIGURE 8.26 FBG-based stress ring sensor for cable force monitoring.

with an elastic strain, ε_e, that can be measured by the FBG strain sensor deployed on the surface of the elastomeric element. With Young's elastic modulus, E_e and the cross-sectional area of the elastomeric element, A_e, the cable force, F, can be calculated by

$$F = N = \varepsilon_e E_e A_e \tag{8.14}$$

In reality, both the mechanical properties of the elastomeric element and the measurement accuracy of the FBG strain sensor will affect the precision of the FBG-based stress ring sensor. For the FBG-based stress ring sensor in this project, there were a total of four FBG strain sensors mounted on the surface of the elastomeric element, and, thus, the strain of the elastomeric element was derived by averaging the strain values obtained from the four FBG strain sensors. The FBG strain sensors were protected by the use of a special packaging technique. The temperature effect was eliminated by taking the measure of temperature compensation. In order to meet the requirement of long-term cable force monitoring, the elastomeric element and the FBG strain sensors were wrapped by a protective envelope.

8.4.2 DEPLOYMENT OF FBG-BASED STRESS RING SENSORS

As shown in Figures 8.27 and 8.28, 22 FBG-based stress ring sensors for cable force monitoring were installed, including 10 sensors for force measurement

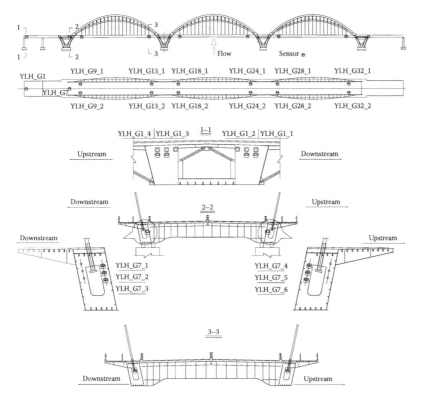

FIGURE 8.27 Deployment of FBG-based stress ring sensors.

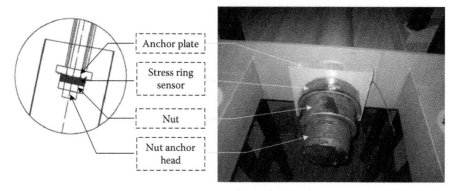

FIGURE 8.28 Site photo of FBG-based stress ring sensor.

of the externally prestressed tendons (see sections 1–1 and 2–2 in Figure 8.27), and 12 sensors for tension force measurement of the cables (see section 3–3 in Figure 8.27).

8.4.3 Analysis of Long-Term Cable Force Monitoring Data

For this section, four FBG-based stress ring sensors installed on four short cables on the north arch of the main channel bridge were selected to analyze the tension force change of the cables. Figure 8.29 shows the deployment locations of the four selected FBG-based stress ring sensors. As described earlier, the monitoring cable force data covered the period from March 21, 2015 to March 27, 2015. The data collected at midnight on March 21, 2015 were regarded as the baseline data for analysis of the change of the cable force. The time histories of the cable force change during one week for the selected four cables are illustrated in Figures 8.30 through 8.33. It is seen from Figures 8.30 through 8.33 that there are periodic peaks and valleys that may suffer from the temperature effect. In this connection, the upper and lower threshold values and the confidence intervals of cable force change can be preset as the early alarming ranges for structural damage detection.

Table 8.4 lists the final tension forces of the cables when the construction of the bridge was completed. Figures 8.34 through 8.37 show the change of the cable forces of the four selected cables during one week (from 21 March, 2015 to 27 March, 2015). Figures 8.38 through 8.41 illustrate the percentage between the cable force change and the final cable force for the four selected cables during one week, which indicates that the daily change of cable force was not significant in comparison with the final cable force.

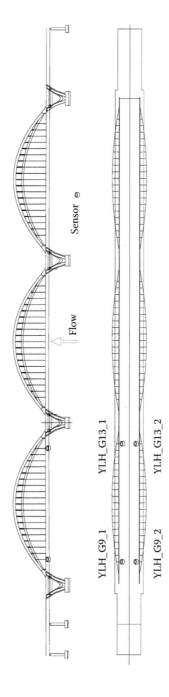

FIGURE 8.29 Deployment of selected four FBG-based stress ring sensors.

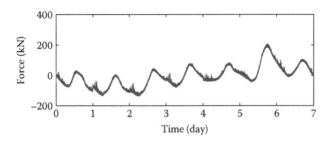

FIGURE 8.30   Time history of cable force change of YLH-G9-1.

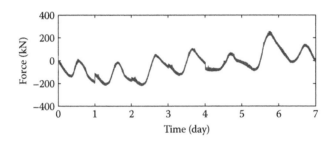

FIGURE 8.31   Time history of cable force change of YLH-G9-2.

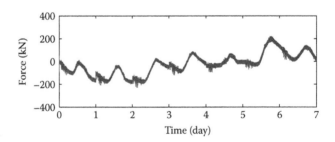

FIGURE 8.32   Time history of cable force change of YLH-G13-1.

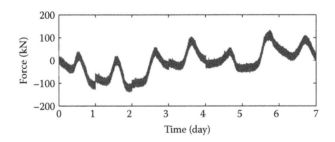

FIGURE 8.33   Time history of cable force change of YLH-G13-2.

TABLE 8.4
Final Cable Forces of Selected Cables

Sensor Number	Final Cable Force (kN)
YLH-G9-1	1453.91
YLH-G9-2	1449.06
YLH-G13-1	1310.14
YLH-G13-2	1720.85

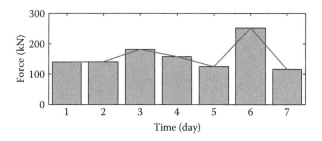

FIGURE 8.34 Change of tension force of YLH-G9-1 during one week.

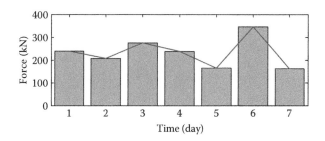

FIGURE 8.35 Change of tension force of YLH-G9-2 during one week.

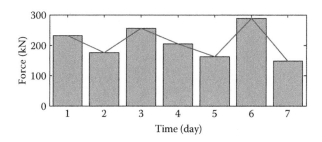

FIGURE 8.36 Change of tension force of YLH-G13-1 during one week.

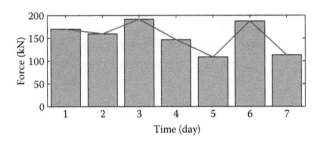

FIGURE 8.37   Change of tension force of YLH-G13-2 during one week.

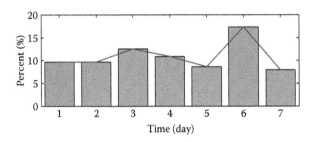

FIGURE 8.38   Percentage between cable force change and final cable force of YLH-G9-1.

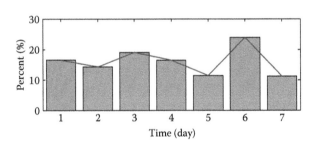

FIGURE 8.39   Percentage between cable force change and final cable force of YLH-G9-2.

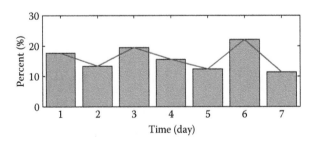

FIGURE 8.40   Percentage between cable force change and final cable force of YLH-G13-1.

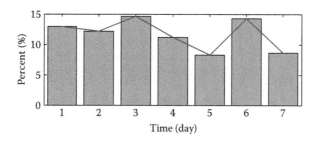

FIGURE 8.41 Percentage between cable force change and final cable force of YLH-G13-2.

8.5 CONCLUDING REMARKS

In this chapter, the optical fiber sensing technology applied in the field of bridge engineering was addressed. A composite arch bridge (Jiubao Bridge) instrumented with an SHM system was introduced, emphatically elaborating the FBG-based monitoring system. Eight FBG-based strain sensors and four FBG-based stress ring sensors were selected to examine the stress status and the cable tension force at their deployed locations. The analysis results revealed that the stress status and the cable tension force at the selected locations exhibited apparent periodic characteristics. Another observation was that temperature had a major effect on the change of their amplitudes in comparison with live load (highway traffic and wind). The FBG-based monitoring system demonstrated strong stability and reliability in data acquisition, and the data acquired by the system can supply effective information for structural safety condition evaluation.

ACKNOWLEDGMENTS

The work described in this chapter was jointly supported by the National Science Foundation of China (Grant No. 51308493) and the Research Fund for the Doctoral Program of Higher Education of China (Grant No. 20130101120080).

REFERENCES

Barbosa, C., Costa, N., Ferreira, L.A., Araujo, F.M., Varum, H., Costa, A., Fernandes, C., and Rodrigues, H. (2008), Weldable fibre Bragg grating sensors for steel bridge monitoring, *Measurement Science and Technology*, 19(12): 1–10.

Bronnimann, R., Nellen, P.M., and Sennhauser, U. (1998), Application and reliability of a fiber optical surveillance system for a stay cable bridge, *Smart Materials and Structures*, 7(2): 229–236.

Casas, J.R., and Cruz, P.J.S. (2003), Fiber optic sensors for bridge monitoring, *Journal of Bridge Engineering*, ASCE, 8(6): 362–373.

Chan, T.H.T., Yu, L., Tam, H.Y., Ni, Y.Q., Liu, S.Y., Chung, W.H., and Cheng, L.K. (2006), Fiber Bragg grating sensors for structural health monitoring of Tsing Ma bridge: Background and experimental observation, *Engineering Structures*, 28: 648–659.

Costa, B.J.A., and Figueiras, J.A. (2012), Fiber optical based monitoring system applied to a centenary metallic arch bridge: Design and installation, *Engineering Structures*, 44: 271–280.

Fuhr, P.L., and Huston, D.R. (1998), Corrosion detection in reinforced concrete roadways and bridges via embedded fiber optic sensors, *Smart Materials and Structures*, 7(2): 217–228.

Fuhr, P.L., Huston, D.R., Nelson, M., Nelson, O., Hu, J., and Mowat, E. (1999), Fiber optic sensing of a bridge in Waterbury, Vermont, *Journal of Intelligent Material Systems and Structures*, 10(4): 293–303.

He, J.P., Zhou, Z., and Ou, J.P. (2013), Optic fiber sensor-based smart bridge cable with functionality of self-sensing, *Mechanical Systems and Signal Processing*, 35: 84–94.

Jiang, G.L., Dawood, M., Peters, K., and Rizkalla, S. (2010), Global and local fiber optic sensors for health monitoring of civil engineering infrastructure retrofit with FRP materials, *Structural Health Monitoring*, 9(4): 309–322.

Kister, G., Badcock, R.A., Gebremichael, Y.M., Boyle, W.J.O., Grattan, K.T.V., Fernando, G.F., and Canning, L. (2007a), Monitoring of an all-composite bridge using Bragg grating sensors, *Construction and Building Materials*, 21(7): 1599–1604.

Kister, G., Winter, D., Badcock, R.A., Gebremichael, Y.M., Boyle, W.J.O., Meggitt, B.T., Grattan, K.T.V., and Fernando, G.F. (2007b), Structural health monitoring of a composite bridge using Bragg grating sensors. Part 1: Evaluation of adhesives and protection systems for the optical sensors, *Engineering Structures*, 29(3): 440–448.

Lan, C.G., Zhou, Z., and Ou, J.P. (2014), Monitoring of structural prestress loss in RC beams by inner distributed Brillouin and fiber Bragg grating sensors on a single optical fiber, *Structural Control and Health Monitoring*, 21(3): 317–330.

Li, D.S., Zhou, Z., and Ou, J.P. (2012), Dynamic behavior monitoring and damage evaluation for arch bridge suspender using GFRP optical fiber Bragg grating sensors, *Optics and Laser Technology*, 44(4): 1031–1038.

Li, H., Ou, J.P., Zhao, X.F., Zhou, W.S., Li, H.W., and Zhou, Z. (2006), Structural health monitoring system for the Shandong Binzhou Yellow River Highway Bridge, *Computer-Aided Civil and Infrastructure Engineering*, 21(4): 306–317.

Lin, Y.B., Chen, J.C., Chang, K.C., Chern, J.C., and Lai, J.S. (2005a), Real-time monitoring of local scour by using fiber Bragg grating sensors, *Smart Materials and Structures*, 14(4): 664–670.

Lin, Y.B., Pan, C.L., Kuo, Y.H., Chang, K.C., and Chern, J.C. (2005b), Online monitoring of highway bridge construction using fiber Bragg grating sensors, *Smart Materials and Structures*, 14(5): 1075–1082.

Mehrani, E., Ayoub, A., and Ayoub, A. (2009), Evaluation of fiber optic sensors for remote health monitoring of bridge structures, *Materials and Structures*, 42(2): 183–199.

Mokhtar, M.R., Owens, K., Kwasny, J., Taylor, S.E., Basheer, P.A.M., Cleland, D., Bai, Y., Sonebi, M., Davis, G., Gupta, A., Hogg, I., Bell, B., Doherty, W., McKeague, S., Moore, D., Greeves, K., Sun, T., and Grattan, K.T.V. (2012), Fiber-optic strain sensor system with temperature compensation for arch bridge condition monitoring, *IEEE Sensors Journal*, 12(5): 1470–1476.

Ni, Y.Q., Xia, H.W., Wong, K.Y., and Ko, J.M. (2012a), In-service condition assessment of bridge deck using long-term monitoring data of strain response, *Journal of Bridge Engineering*, ASCE, 17(6): 876–885.

Ni, Y.Q., Ye, X.W., and Ko, J.M. (2010), Monitoring-based fatigue reliability assessment of steel bridges: Analytical model and application, *Journal of Structural Engineering*, ASCE, 136(12): 1563–1573.

Ni, Y.Q., Ye, X.W., and Ko, J.M. (2012b), Modeling of stress spectrum using long-term monitoring data and finite mixture distributions, *Journal of Engineering Mechanics*, ASCE, 138(2): 175–183.

Rodrigues, C., Cavadas, F., Felix, C., and Figueiras, J. (2012), FBG based strain monitoring in the rehabilitation of a centenary metallic bridge, *Engineering Structures*, 44: 281–290.

Surre, F., Sun, T., and Grattan, K.T. (2013), Fiber optic strain monitoring for long-term evaluation of a concrete footbridge under extended test conditions, *IEEE Sensors Journal*, 13(3): 1036–1043.

Talebinejad, I., Fischer, C., and Ansari, F. (2009), Serially multiplexed FBG accelerometer for structural health monitoring of bridges, *Smart Structures and Systems*, 5(4): 345–355.

Tennyson, R.C., Mufti, A.A., Rizkalla, S., Tadros, G., and Benmokrane, B. (2001), Structural health monitoring of innovative bridges in Canada with fiber optic sensors, *Smart Materials and Structures*, 10(3): 560–573.

Xia, H.W., Ni, Y.Q., Wong, K.Y., and Ko, J.M. (2012), Reliability-based condition assessment of in-service bridges using mixture distribution models, *Computers and Structures*, 106–107: 204–213.

Xiong, W., Cai, C.S., and Kong, X. (2012), Instrumentation design for bridge scour monitoring using fiber Bragg grating sensors, *Applied Optics*, 51(5): 547–557.

Yau, M.H., Chan, T.H.T., Thambiratnam, D., and Tam, H.Y. (2013), Static vertical displacement measurement of bridges using fiber Bragg grating (FBG) sensors, *Advances in Structural Engineering*, 16(1): 165–176.

Ye, X.W., Ni, Y.Q., Wai, T.T., Wong, K.Y., Zhang, X.M., and Xu, F. (2013), A vision-based system for dynamic displacement measurement of long-span bridges: Algorithm and verification, *Smart Structures and Systems*, 12(3–4): 363–379.

Ye, X.W., Ni, Y.Q., Wong, K.Y., and Ko, J.M. (2012), Statistical analysis of stress spectra for fatigue life assessment of steel bridges with structural health monitoring data, *Engineering Structures*, 45: 166–176.

Ye, X.W., Su, Y.H., and Han, J.P. (2014), Structural health monitoring of civil infrastructure using optical fiber sensing technology: A comprehensive review, *The Scientific World Journal*, 2014: 1–11.

Ye, X.W., Yi, T.H., Dong, C.Z., Liu, T., and Bai, H. (2015), Multi-point displacement monitoring of bridges using a vision-based approach, *Wind and Structures*, 20(2): 315–326.

Zhang, W., Gao, J.Q., Shi, B., Cui, H.L., and Zhu, H. (2006), Health monitoring of rehabilitated concrete bridges using distributed optical fiber sensing, *Computer-Aided Civil and Infrastructure Engineering*, 21(6): 411–424.

9 Smart Composite Textiles with Embedded Optical Fibers

Tamer Hamouda and Abdel-Fattah M. Seyam

CONTENTS

9.1 Introduction 299
9.2 Optical Fiber Embedded into Textile Preform for Composite Applications...300
    9.2.1 Optical Fiber Embedded into Woven Preforms................................300
    9.2.2 Optical Fiber Embedded into 3D Braid Preforms 307
9.3 Bending Loss of Optical Fiber Integrated in Textile Preforms 310
    9.3.1 Macrobending..310
    9.3.2 Microbending...311
    9.3.3 Critical Bend Radius...312
9.4 Optical Fiber Side-Emitting Properties 313
9.5 Application of Optical-Fiber Sensors in Smart Composites: Cure
    Monitoring 315
9.6 Application of Optical Fiber Sensors in Smart Composites: Monitoring
    Composite Mechanical Properties...323
Conclusion 333
References...334

9.1 INTRODUCTION

Sensors can be attached or embedded into composite materials to create a *smart skin* or *smart structure*. So, the main goal is to create a structural element that can monitor the hidden internal conditions of the component and determine their present state throughout the use of the structures. To acquire a robust composite structure with a reliable health-monitoring system, there are requirements that need to be fulfilled: minimizing the influence of the structures' components on the sensors' sensing integrity and at the same time ensuring that the structure integrity will not be degraded substantially as a result of attaching or embedding the sensors. Therefore, selection of the right sensors plays a major role in acquiring a smart and reliable composite structure.

For decades, strain gages, accelerometers, and displacement sensors have been used for health monitoring of structures [1–4]. These sensors are usually attached to the structure's surface and are thus negatively affected by the adverse weather conditions. They are also characterized by their large size, which prevents their use as integrated sensors to monitor the internal structure damage. In addition, they are negatively affected by ambient electromagnetic interference, and their corrosion resistance is low. Complicated systems are required to triangulate the fault location, because they cannot be multiplexed. On the other hand, optical fiber sensors such as silica optical fibers (SOF) and polymer optical fibers (POF) have many advantages over conventional sensors: small size, high strength, low attenuation, immunity to electromagnetic interference, multiplexing capability, resistance to corrosion, ease of embedding into structures, high sensitivity, and high temperature resistance. Optical fiber sensors can be used to sense pressure, temperature, strain, quasi-distributed strain and displacement, acceleration, vibration, humidity, rotation, position, acoustic waves, sensing chemicals, and biomolecules [5]. For these reasons, numerous researchers have used SOF as embedded and surface sensors [6–9]. While SOF sensors possess these advantages, they are very brittle and require very precise positioning with respect to the connectors as well as the light source, which demands very expensive special tools and a long time for system development [10].

On the other hand, POF sensors are characterized by high numerical aperture and high flexibility; they are easy to handle, inexpensive, and easy to splice and cleave, do not require a protective layer, and can handle high tensile strain and high fracture toughness [11,12]. High numerical aperture of the POF provides high system power for transmitters with high broad angle emission. Additionally, high numerical aperture allows an easy assembly and optical fiber coupling with low-cost connectors. Studies have been conducted on the behavior of POF sensors, their ability to be used as sensors embedded into geotextile, civil engineering, and composite structures, and their ability to localize fault conditions while monitoring a large area [13–18].

9.2 OPTICAL FIBER EMBEDDED INTO TEXTILE PREFORM FOR COMPOSITE APPLICATIONS

Due to the good, uniform distribution of the matrix and reinforcing fibers, textile preforms are considered to be the backbone structure of textile-reinforced composite materials [19]. The ongoing major developments in textile technology facilitate the ability to produce textile preform using high-performance fibers, such as carbon, glass, and aramid. Several techniques have been developed to produce textile preforms, such as nonwoven preforms, two-dimensional (2D) and three-dimensional (3D) woven preforms, 2D and 3D braided preforms, knitted preforms, and filament ending preforms.

9.2.1 Optical Fiber Embedded into Woven Preforms

There are two main types of woven preforms: 2D and 3D. 2D woven preform is produced by interlacing of warp yarns (0°) and weft (filling) yarns (90°). This interlacing follows different weave styles according to the required performance. 2D woven preforms possess numerous bending points along the yarns as a result of weave interlacing, as can be seen from Figure 9.1. Bending of optic fibers is a major concern due to optical signal attenuation. The integration of optical fiber in 2D preforms along

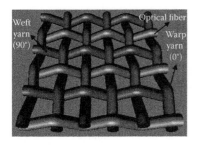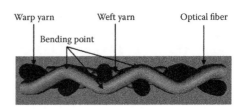

FIGURE 9.1 Schematic of 2D woven preform with embedded optical fiber.

with the nature of continuous insertion techniques of the optical fibers in woven preforms cause macrobending in the integrated optical fiber at the interlacing points. Therefore, studying the bending loss of optical fibers is essential to identify the range of applicability and sensitivity of such sensors and the stability of their integration in 2D and 3D preforms for composite applications.

3D preforms are formed using different technologies, such as stacks of 2D woven preforms, 3D orthogonal preforms, and stitched woven preforms. 3D orthogonal woven preforms are characterized by their integral bonded structure. This integrated structure provides the final composite with higher resistance to crack propagation, eliminates delamination, provides faster resin transfer, and could be formed with very low to very high fiber volume fraction [20]. 3D orthogonal woven preforms (Figure 9.2) are integrated structures with three different yarn systems that are combined without interlacing (crimp) [21]. Since SOF and POF sensors are characterized by their small diameters, they can be integrated between the layers of stacked 2D fabrics, or embedded into woven preform, whether 2D or 3D, during the weaving process to become part of the preform [22]. Optical fibers can be integrated into 3D orthogonal woven preforms in three directions: x-, y-, and/or z-direction. Integration of optical fibers in the x- and y-directions ensures the straightness of the optical fibers inside the 3D orthogonal woven preform. Figure 9.2 depicts optical fibers embedded into a 3D orthogonal woven preform in the x-direction [23]. To ensure that a large area of the preform is covered, the optical fiber may bend at each side of the preform (Figure 9.3).

Rothmaier et al. [24] investigated the use of embedded plastic optical fibers in woven textile as pressure sensors. Thermoplastic silicone optical fiber made

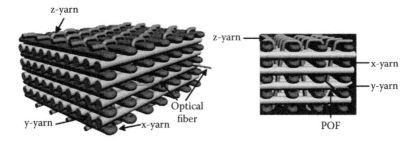

FIGURE 9.2 Schematic of 3D orthogonal woven structure with embedded optical fiber.

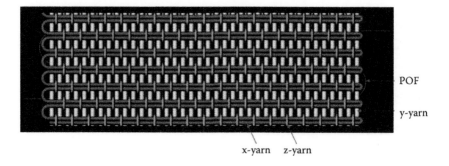

POF

y-yarn

x-yarn   z-yarn

FIGURE 9.3   Large area of 3D preform with embedded optical fiber.

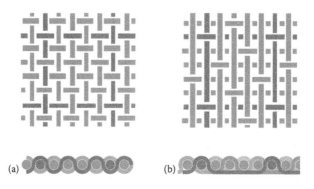

(a)   (b)

FIGURE 9.4   Embedded optical fiber into plain weave (a) and atlas weave (b). (From Rothmaier, M. et al., *Sensors*, 8, 4318–4329, 2008.)

of polysiloxane–urea–copolymer with a polysiloxane content of >90% was used. Optical fibers with diameters of 0.51 and 0.98 mm were integrated into two different textile woven samples (plain weave and atlas 1/4 weave) in warp and weft directions to form a pressure-sensitive matrix. Plain weave means that at any one unit repeat, each optical fiber will move over and under the weft yarn, as shown in Figure 9.4a; atlas 1/4 weave means that at any one unit repeat, the optical fiber will move under one weft warp and over for weft yarn, as shown in Figure 9.4b. Where the optical fiber moves over or under the other yarns, a microbending effect will occur. Therefore, the atlas weave is preferable, as the interlacing between the optical fiber and the other fiber will be minimal, and thus the light attenuation caused by the microbending will be low.

Optical fiber ends were connected to four light-emitting diode (LED) light sources and four light detectors, as shown in Figure 9.5. A force of 20 N was applied at nine different fabric locations, four of them at POF intersections (1, 3, 7, and 9), four at non-optical fiber intersection locations (2, 4, 6, and 8), and the last at a location (5) that is free of optical fiber, which is neglected. The results of applied force versus sensor signal showed that optical fibers with smaller diameter of 0.51 mm have a very sensitive response to small forces compared with 0.98 mm diameter optical fibers and can detect forces of 0.5 N. However, small-diameter optical fibers failed

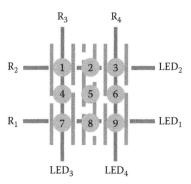

FIGURE 9.5   Optical fibers connected to LED light source and light signal detectors (R). (From Rothmaier, M. et al., *Sensors*, 8, 4318–4329, 2008.)

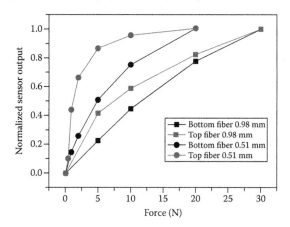

FIGURE 9.6   Applied force vs. sensor signal. (From Rothmaier, M. et al., *Sensors*, 8, 4318–4329, 2008.)

when applied forces reached 20 N (Figure 9.6). Thicker optical fibers (0.98 mm diameter) withstand forces of up to 30 N. Figure 9.6 shows that both optical fibers (0.51 and 0.98 mm) located on the top have a higher response to the applied forces than the optical fibers located at the bottom. This may be due to the fact that the top sensors bend and squeeze before the bottom sensors when forces are applied.

Castellucci et al. [25] used optical fibers embedded in 3D woven composite structures as three-axis-distributed strain sensors. A 3D carbon fiber cantilever beam was fabricated with embedded optical fiber sensors. Low–bend loss optical fiber of 155 μm diameter with polyamide coating was used. During the weaving process of the textile preform, optical sensors were embedded in three directions: warp, fill, and z-directions. Warp and filling optical sensors are located 1 mm below the surface. Embedded optical sensors in the warp direction were fully straight, while the embedded optical sensors in the fill direction turned around at the edges with 50 mm curvature. z-Direction optical sensors ran from the top surface to the bottom surface, and vice versa, through the preform thickness (Figure 9.7).

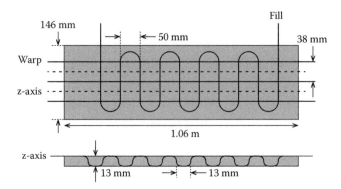

FIGURE 9.7 Embedded optical fiber in warp, fill, and z-directions. (From Castellucci, M. et al., *Proceedings of the Industrial and Commercial Applications of Smart Structures Technologies*, Smart Structures/NDE, San Diego, CA, 2013.)

The final composite samples were prepared by vacuum-assisted resin transfer molding (VARTM) followed by curing in an autoclave oven at 50 psi and 350 °F. Different static loads located 0.9 m from the clamps were applied to the composite sample as a cantilever and in torsion, as shown in Figure 9.8. Since the warp and filling optical sensors were 1 mm below the surface, all measurements were recorded with the embedded sensors under tension. The cantilever beam was flipped to record the measurement of the optical fiber on the compression side. The applied static load increased from 4.5 kg to 9.0, 11.3, and 13.6 kg (10, 20, 25, and

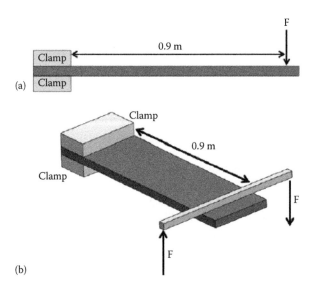

FIGURE 9.8 Cantilever loading configuration (a), and torsional loading configuration (b). (From Castellucci, M. et al., *Proceedings of the Industrial and Commercial Applications of Smart Structures Technologies*, Smart Structures/NDE, San Diego, CA, 2013.)

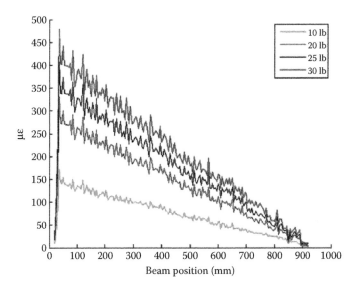

FIGURE 9.9 Strain profile for one load cycle measured by embedded warp-direction fiber along the length of the entire beam. (From Castellucci, M. et al., *Proceedings of the Industrial and Commercial Applications of Smart Structures Technologies*, Smart Structures/NDE, San Diego, CA, 2013.)

30 lb). The three embedded warp optical sensors were spliced to form a single long sensor.

A commercially available optical frequency-domain reflectometry (OFDR) instrument was used to measure the static distributed strain. Figure 9.9 shows the strain profile that was measured by the embedded warp-direction optical sensors under different static loads. The measured strain is linear, with higher strain near to the clamp and decreasing as it approaches the tip of the cantilever beam. The results also show that as the static load increased, the strain magnitude increased, and the strain profile remained the same under all different loads. The strain measured by fill-direction optical sensors under torsion test is shown in Figure 9.10. The horizontal blue line indicate the filling region, while tension and compression are measured in the turn-around regions. The fill-direction sensing region and tension–compression in the turn-around area are shown in Figure 9.11. Unlike the warp-direction and fill-direction sensors, the z-direction sensor experiences a transition from tension to compressive strain as the fiber travels from one side (surface) of the beam surface to the other side (bottom). Figure 9.12 illustrates the strain profile measured by an embedded z-direction fiber sensor. The figure shows the strain under tension for the part of the sensor that is located on the top surface along with the strain under compression for the region of the sensor that is located on the bottom surface. It also shows that as the applied load increases, the strain magnitude increases.

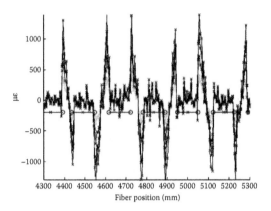

FIGURE 9.10 Strain profile measured by fill optical sensors under torsion test. (From Castellucci, M. et al., *Proceedings of the Industrial and Commercial Applications of Smart Structures Technologies*, Smart Structures/NDE, San Diego, CA, 2013.)

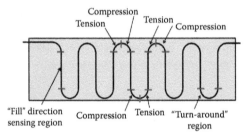

FIGURE 9.11 Schematic diagram of fill-direction sensing region and tension and compression region at turn-around area. (From Castellucci, M. et al., *Proceedings of the Industrial and Commercial Applications of Smart Structures Technologies*, Smart Structures/NDE, San Diego, CA, 2013.)

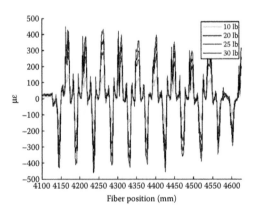

FIGURE 9.12 Strain profile measured by the z-direction sensing fiber. (From Castellucci, M. et al., *Proceedings of the Industrial and Commercial Applications of Smart Structures Technologies*, Smart Structures/NDE, San Diego, CA, 2013.)

9.2.2 OPTICAL FIBER EMBEDDED INTO 3D BRAID PREFORMS

The braiding process is simply defined as a process of intertwining three or more parallel yarns to fabricate a continuous, seamless textile structure with nonorthogonal fiber orientation.

3D braided preforms are characterized by their ability to form several complex preform shapes. Thus, all the processes of cutting, overlaps, and splices needed to acquire the complex shapes using regular 2D preform are eliminated with 3D braiding. 3D braided preforms exhibit high torsional stability and structural integrity [26].

Much research has been conducted using optical sensors embedded into braided structures for structural health monitoring. The internal strain and curing behavior of 3D braided composites has been investigated using embedded fiber Bragg grating (FBG) sensors [27]. FBGs were embedded along the curved braiding yarns and in the machine direction to investigate the internal strain and the local deformation. A vertical-type circular four-steps braiding machine was used to fabricate the seven-layer 3D circular braided preform. Only 336 carriers (7 rows and 48 columns of carriers) out of the total of 2016 carriers (14 rows and 144 columns of carriers in the radial and circumferential directions) were used. S2 glass yarns were connected to the 336 carriers, while 48 Kevlar yarns were connected to stationary carriers to form the exterior surface layer. FBGs were not directly embedded into the 3D braided preform during the fabrication process. Instead of direct embedding, the authors of this study [27] used co-braided nylon fibers in some of the carriers, and these nylon fibers were replaced by FBGs after the braiding process. This technique was used to eliminate the probability of optical sensors breaking under the action of the beating motion during the braiding fabrication. After the braiding process finished, the end of the nylon fiber was glued to the FBG sensor, and the nylon fiber was pulled out and replaced by the FBG. Figure 9.13 shows the location and orientation of FBGs in 3D braided preform. F1 was embedded in the machine direction in layer 1, and F2 was embedded along the axial yarn in layer 7. F3 was attached to the interior surfaces of the preform in the machine direction. F5 was embedded along a curved braiding yarn (S2 glass yarn). F4 and F6 were attached to the exterior surface of the tubular preform; F4 was oriented in the machine direction, while F6 was oriented with the same orientation angle as F5.

Yuan et al. [28] investigated the internal strain under tension, bending, and thermal environments in 3D braiding composite using an embedded co-braided

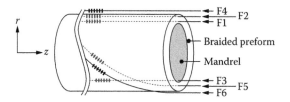

FIGURE 9.13 Schematic diagram of the orientations and positions of embedded FBGs in tubular braided preform. (From Jung, K. and Kang, T.J., *Journal of Composite Materials*, 41, 1499–1518, 2007.)

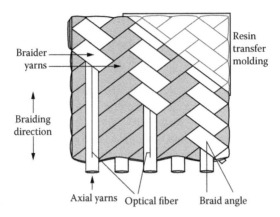

FIGURE 9.14 Embedded optical fibers in braided composites. (From Yuan, S. et al., *Journal of Materials Science & Technology*, 20, 199–202, 2004.)

Fabry–Perot (F-P) fiber-optic sensor and polarimetric fiber-optic sensors. The two types of optical sensor were integrated directly into the braided preform as an axial yarn during the manufacturing process. The embedded optical sensors were straight and parallel to the braiding direction. Figure 9.14 shows the location and method of embedding optical fiber in 3D braid preform. It also shows that the structural yarns are braided around the optical fibers; the optical sensors are straight, and no large residual stress appears to have happened during the manufacturing process. An F-P cavity is maintained at the center of the specimen to measure the internal strain in the local area, while polarimetric fiber-optic sensors are used to measure the global strain along the braided length.

To validate the efficiency and performance of the embedded optical sensors, a strain gage was mounted on the surface of the composite samples. Tensile testing was conducted on the braided composite to measure the composite's strain. A load–strain curve using the strain gage and optical sensors is shown in Figure 9.15. The results indicate that the load–strain curves obtained from the strain gage and the F-P sensor are similar. Thus, an embedded F-P optical sensor is effective in measuring the local internal strain of 3D braided composites.

Real-time monitoring of 3D braided composite with embedded optical fiber sensors using a spatial modulation technique was investigated by El-sherif and Ko [29]. Graphite/epoxy braided composite was fabricated with embedded optical fiber. A rectangular-vertical laboratory braiding machine was used to fabricate the preform. Step index SOF with core and cladding diameter of 100 and 125 μm, respectively, was embedded into the 3D braid preform during the fabrication process. Optical fibers were integrated into the 3D braided preform in one of three different configurations: as a co-braiding with the carbon fibers, parallel to the axis and near to the surface, or into the center axis. Composite specimens were fabricated using a compression molding technique. Spatial modulation of an optical signal propagating through a multimode optical fiber has been used successfully for real-time monitoring [30]. Modal power distribution (MPD) as a result of changing the boundary condition of the optical fiber is used for sensing the perturbation in the spatial modulation of the

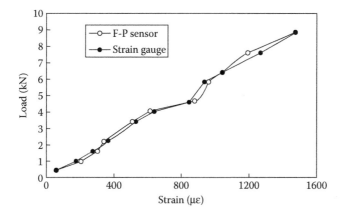

FIGURE 9.15 Load–strain curve using strain gage and F-P optic sensor. (From Yuan, S. et al., *Journal of Materials Science & Technology*, 20, 199–202, 2004.)

multimode optical fibers. This technique depends on light source conditions and the geometry and optical properties of the optical fiber.

Figure 9.16 shows the experimental setup and the testing protocol of the 3D braided composite with embedded optical fibers. Each of the three samples was tested using a point compression load on the top surface. MPD as a result of applied load was measured using three separate charge-coupled device (CCD) cameras. The result of normalized output intensity versus the output scanning angle is illustrated in Figure 9.17. The highest intensity was recorded at zero load, which indicates that no change in the MPD was detected (curve a). When the load increased and the applied stress reached 25 psi, the induced change in MPD was detected (curve b). As

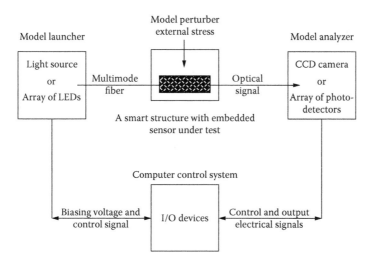

FIGURE 9.16 Experimental setup and testing protocol. (From El-sherif, M. A. and Ko, F. K., *Technical Digest of the First European Conference on Smart Structures and Materials*, CRC Press, Glasgow, pp. 85–88, 1992.)

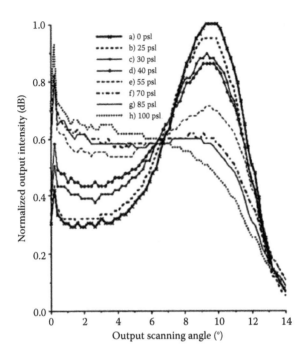

FIGURE 9.17 Normalized output intensity vs. output scanning angle. (From El-sherif, M. A. and Ko, F. K., *Technical Digest of the First European Conference on Smart Structures and Materials*, CRC Press, Glasgow, pp. 85–88, 1992.)

applied stress increases, the MPD increases and the output intensity decreases. The two specimens with embedded optical fibers oriented near to the surface showed the same results. However, the other sample showed less sensitivity.

9.3 BENDING LOSS OF OPTICAL FIBER INTEGRATED IN TEXTILE PREFORMS

Optical fiber is made of two optically dissimilar materials: core and cladding. Cladding is made of low–refractive index material, while the core is made of high–refractive index material [31]. This difference in refractive index between the core and the cladding causes the passing light to propagate along the length of the optical fiber by total internal reflection. Optical fibers' material imperfections, absorption, scattering, and bending will cause signal attenuation to occur as light propagates along the optical fiber. Macrobending and microbending are two types of bending that cause loss of the power signal.

9.3.1 MACROBENDING

Macrobending of an optical fiber is the attenuation associated with a bending radius that is much greater than the optical fiber radius. Macrobending causes light to leak out, and thus signal loss, which reduces the performance of the optical fiber as a sensor.

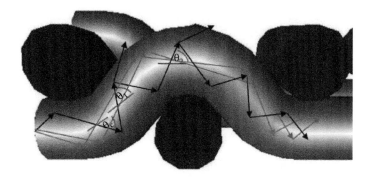

FIGURE 9.18 Optical fiber macrobending loss.

Light propagates from one end of the fiber to the other end through the total internal reflection at the interface between the core and the cladding. This internal reflection occurs at a critical angle that keeps the light reflected along the axis of the optical fiber. Macrobending of an optical fiber causes the incident light to be reflected at an angle greater than the critical angle. Macrobending can occur when an optical fiber is integrated into a 2D woven preform, in which the interlacing between the optical fiber and the warp yarn causes the optical fiber to bend with a bending radius greater than the optical fiber diameter (Figure 9.18). A schematic diagram of light traveling inside an optical fiber interlaced into a 2D woven preform is shown in Figure 9.18. A trace of light rays is indicated by the black arrows reflected at angles Θ_1 and Θ_2 as a result of weaving the optical fiber in a 2D woven structure, for example. These angles are greater than the critical angle Θ_c, which causes some of the light to refract out of the optical fiber core through the cladding, hence causing signal attenuation. Figure 9.18 also shows light rays, indicated by blue arrows, that follow the total internal reflection and reflect with an angle smaller than the critical angle.

9.3.2 MICROBENDING

Microbending (bending with small curvature) is defined as microscopic curvature or deformation that creates local axial displacement with an amplitude of micrometers or less over lengths of millimeter level. In other words, it is a bend in the optical fiber with a radius equal to or smaller than the critical bending radius, causing the traveling light in the core to penetrate into the cladding and leak out from the optical fiber. Microbending loss is caused by imperfections and irregularities of the optical fiber material during the manufacturing process [32]. These irregularities or imperfections of the core or cladding will cause deviation from perfect total internal reflection. Figure 9.19 shows a light ray reflecting with angles Θ_3 and Θ_4; these angles are greater than the critical angle Θ_c and cause the light to refract out of the optical fiber core through the cladding, resulting in signal loss [33]. Microbending may also occur at the crossover points between warp and weft yarns if the weave structure is too tight and requires high beat-up force. This issue must be considered when designing optical fiber sensors to avoid signal loss.

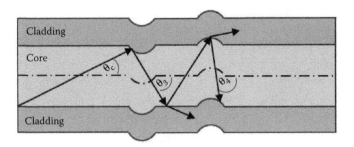

FIGURE 9.19 Optical fiber microbending loss.

9.3.3 CRITICAL BEND RADIUS

Critical radius is the minimum bending radius at which the optical fiber exhibits no signal loss. Optical fiber is optically unaffected by bending above some threshold radius (critical bend radius), and this radius varies according to the specific fiber's design. As a bending curvature is reduced to a critical value, though, some portion of the traveling light at the core/cladding interface cannot be refracted back to the core. This portion of light leaks out of the optic fiber and is measured as insertion loss (IL). Thus, it is essential to find the critical radius associated with each optical fiber and to consider this value when constructing and designing sensors embedded into textile preform. As stated in Section 9.3.2, integration of optical fiber into textile preform, whether 2D or 3D, will induce microbending at the crossover point (the interlacing point between fibers in the warp and weft directions) and will also create macrobending at the preforms edges during continuous insertion.

Figure 9.20 shows the effect of bending radius on induced signal attenuation of the three optical fibers investigated [34]. Several rods with different diameter were used to create the data shown in the figure. Each optical fiber was wrapped a half turn around each rod (180° wrap angle), as seen in Figure 9.21, which simulates the bending at the preform edges of a composite structure. The results show that as the rod radius increases, the induced attenuation decreases. This indicates that a small radius is associated with more light reflection beyond the critical angle, such that more light rays leak

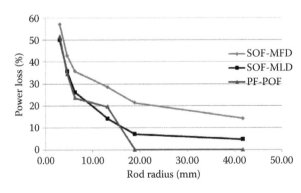

FIGURE 9.20 Optical fibers' critical bend radius. (From Seyam, A. F. M. and Hamouda, T., *Journal of the Textile Institute*, 104, 892–899, 2013.)

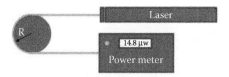

FIGURE 9.21 Critical bend radius experimental setup. (From Seyam, A. F. M. and Hamouda, T., *Journal of the Textile Institute*, 104, 892–899, 2013.)

from the optical fiber core through the cladding, causing signal attenuation. The results also indicate that no attenuation occurs in the POF when the bending radius is 19 mm or greater. Figure 9.21 also shows the results of signal loss at a wrap angle of 180° for two types of SOF (SOF-MFD and SOF-MLD). For these fibers, the power loss at radius 19 mm was 15% and 7%, respectively. The power loss data for SOF suggest that the SOF will continue to exhibit attenuation even with rod radii higher than those used in this experiment. However, the loss is very small for a bending radius of 40 mm and higher.

There are several factors that influence the bending loss and critical bend radius, such as the radius of curvature, the refractive index difference between the core and the cladding, the coating material refractive index, and the wavelength. The higher the cladding and coating refractive index, the higher is the bending loss and the greater is the critical radius.

9.4   OPTICAL FIBER SIDE-EMITTING PROPERTIES

As optical fibers experience any type of bending, whether macrobending or micro-bending, the side-emission effect is created as a result of some light leaking from the fiber's core to its cladding and further, to the surrounding medium. Nevertheless, side emission can be achieved in several ways, such as by adding fluorescent additives or specific scatterers to the fiber core or cladding material, creating asymmetries in the fiber core/cladding geometry, increasing the refractive index of the fiber cladding material over that of the core material, or leaving diffusive cavities. The relationship between side-emitting intensity and bending radius of POF was investigated via theoretical derivation, simulation experiments, and POF fabric experiments [35]. Weave structure was simulated by using metal rods fixed between two metal slabs. These metal rods represent the warp yarn with 1 mm diameter. POF was used as a weft, whereby it interlaced with the rods in the form of a plain weave structure, as shown in Figure 9.22. The POF's bending radius was calculated by the following equation:

$$r = \frac{h}{4} + \frac{p^2}{4h} \tag{9.1}$$

where:
   h is the height of the crimp weave of the POF
   p is the distance between two metal rods

Woven samples were fabricated using polyester yarn as a warp (0°) at 67 yarn/cm and POF as weft yarn (90°) at 20 weft/cm. POFs with diameters of 250 and 500 μm were used. Five different weave structures were used: plain, 1/2 twill,

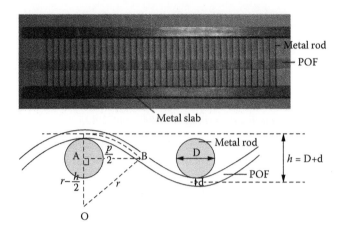

FIGURE 9.22 Module consists of metal rods as warp and POFs as weft. (From Wang, J. et al., *Textile Research Journal*, 38, 1170–1180, 2012.)

5-thread sateen, 8-thread sateen, and 16-thread sateen. A green LED with a diameter of 5 mm was used as a light source and was connected to one end of the POF bundle while a luminance meter was used to measure the side-emitting intensity of bent POF. Simulation results showed that as the optical fiber diameter increases, the side-emitting intensity increases under the same bending radius (Figure 9.23). The results also indicated that side-emitting intensity of the bent optical fiber increased with the decrease of bending radius. The side-emitting intensity results for the woven fabrics with different weave structures show that sateen weave fabrics exhibit the highest side-emitting intensity compared with twill and plain weave (Figure 9.24). In spite of the fact that the plain weave structure possessed the highest interlacing ratio between warp and weft yarns, the optical fiber remained

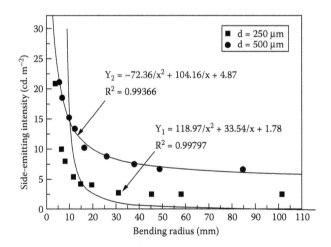

FIGURE 9.23 Side-emitting intensity vs. bending radius for 250 and 500 μm POFs. (From Wang, J. et al., *Textile Research Journal*, 38, 1170–1180, 2012.)

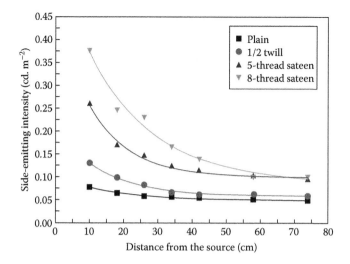

FIGURE 9.24 Side-emitting intensities of woven fabrics with different weave structures. (From Wang, J. et al., *Textile Research Journal*, 38, 1170–1180, 2012.)

straight and did not bend. This is due to the high bending rigidity of the optical fiber and the insufficient stress of the warp yarn over the interlaced optical fiber. The results for the sateen weave structure show that the optical fiber became easy to bend as the weave repeat size increased and the float length became longer; therefore, the highest side-emitting intensity was recorded for the sateen weave.

9.5 APPLICATION OF OPTICAL-FIBER SENSORS IN SMART COMPOSITES: CURE MONITORING

Curing process techniques and conditions have a noticeable effect on final composite structure properties. Changes in composite mechanical properties were observed with changes in curing cycle, curing speed, and curing temperature [36–39]. Therefore, monitoring the curing process using embedded optical sensors during composite fabrication has been investigated by many researchers. Aktas et al. [40] monitored the curing process of unsaturated polyester/Plyophen and unsaturated polyester/Durez resin blends using embedded FBGs. Two different types of FBG were used: FBG-1 (with a wavelength of 1545 nm) detects both temperature and strain, while FBG-2 (with wavelength of 1560 nm) is affected by only the change in temperature (Figure 9.25). FBGs were embedded between the seventh and eighth layers of 15 layers of noncrimp triaxial glass fabric. A thermocouple was used for temperature validation. A vacuum infusion technique was used, and the changes in the wavelength of the FBGs during the compaction, infusion, anduring cycle were recorded.

Figures 9.26 and 9.27 show the FBGs' recorded wavelength versus time and temperature. The results indicated that FGB-2 was sensitive to temperature changes, while FBG-1 detected changes in both temperature and strain. As the temperature increased (curing cycle), the wavelength changed for both FBGs. After the full curing, when the temperature dropped to room temperature, the

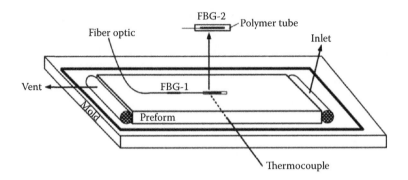

FIGURE 9.25 Experimental setup.

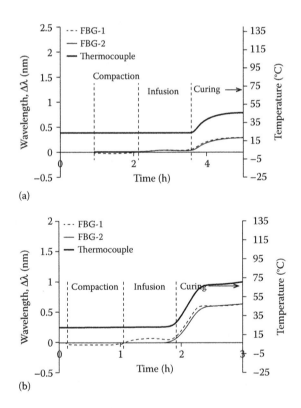

FIGURE 9.26 FBG results before and after the infusion process and at the beginning of the curing cycle of (a) unsaturated polyester/Plyophen, (b) unsaturated polyester/Durez. (From Aktas, A. et al., *Journal of Composite Materials,* submitted 0021998314568165, 2015.)

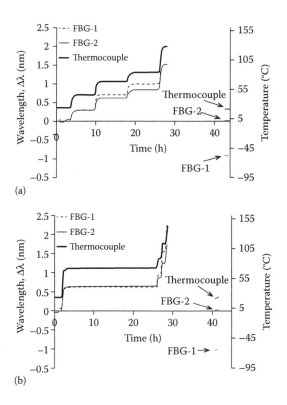

FIGURE 9.27 Cure monitoring results: (a) unsaturated polyester/Plyophen, (b) unsaturated polyester/Durez. (From Aktas, A. et al., *Journal of Composite Materials,* submitted, 0021998314568165, 2015.)

FBG-2 wavelength returned to the original status, as indicated by the arrows in Figure 9.27. Converting the wavelength data of FBG-2 to temperature indicated that the FBG-2 temperature results were in good agreement with the thermocouple temperature data.

Hamouda et al. [23] investigated the capability of POF to monitor a composite fabrication process, starting from the 3D weaving preform fabrication and the infusion process and under bending loads and impact energy. Optical time domain reflectometry (OTDR) was used to collect the changes in the backscattering level and signal attenuation during the fabrication process. Low-attenuation and high-bandwidth graded-index (GI) POF was used. The POF's core and cladding were made of a perfluorinated polymer (polyperfluoro-butenyvinyleether) with core diameter of 62.5 μm and total diameter of 750 μm. 3D woven preform samples were fabricated using a 3D machine located at NCSU. POFs were embedded in the filling direction during the fabrication process (Figure 9.28). Preform samples ranged from two warp layers and 1.57 x-yarns/cm/layer to four warp layers and 4.72 x-yarns/cm/layer.

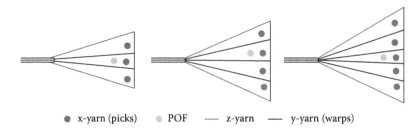

FIGURE 9.28 Embedded POF location and orientation in 3D woven preform. (From Hamouda, T. et al., *Smart Materials and Structures*, 24, 1–11, 2015.)

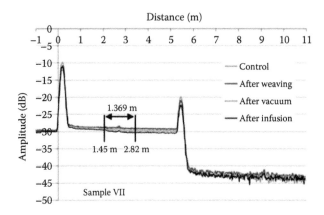

FIGURE 9.29 Backscattering of embedded POF into four y-yarn layers and 1.57 x-yarns/cm/layer in different fabrication processes. (From Hamouda, T. et al., *Smart Materials and Structures*, 24, 1–11, 2015.)

The backscattering level was measured for original POF as a control measurement. The backscattering level was then measured for the embedded POF after preform formation, after applying the vacuum of resin infusion, and after 24 h of resin treatment, measured from the time of the start of infusion. Figure 9.29 illustrates typical OTDR traces of sample VII (four layers and 1.57 x-yarns/cm/layer) at the different fabrication processes. The recorded distributed loss of the control POF before and of the POF after weaving was 177 dB/km, which indicated that weaving the POF into the preform did not cause any damage to the POF. After applying 100 KPa vacuum pressure to the preform (preparing for the vacuum infusion), a noticeable drop in the backscattering level at a distance of 2.13 m was observed. This drop in the backscattering results from a macrobending loss due to the applied pressure on the embedded POF. After 24 h of infusion and curing, the drop in the backscattering was negligible as compared with that immediately after the vacuum pressure.

Figure 9.30 illustrates the signal attenuation after each fabrication step for all nine composite samples. The results revealed that all embedded POFs in two-warp-layer samples showed a significant attenuation after the weaving process (0.14–0.22 dB),

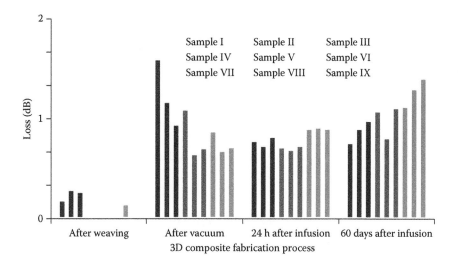

FIGURE 9.30 Signal attenuation of embedded POF after each process of fabricating 3D woven composite. (From Hamouda, T. et al., *Smart Materials and Structures*, 24, 1–11, 2015.)

while the POFs in three- and four-layer samples did not exhibit significant signal attenuation.

This attenuation may be due to the open and flexible structure of two-warp-layer samples; embedded POF may be easily distorted and suffer from macrobending at the contact points with preform yarns and during sample handling. The same observation may be made for the samples with the lowest x-yarn density. However, by 24 h after resin infusion, the signal attenuation of all samples was about the same, which is evidence of distortion recovery of the POF in the cases where the attenuation was the highest. To investigate the effect of resin cross-linking, signal attenuations of the POFs in all nine composite samples were also collected 60 days after resin infusion (Figure 9.30). The highest signal loss after 60 days was 1.0 dB for sample IX (four warp layers and high filling density) after resin infusion.

Sampath et al. [41] used FBG sensors and Fresnel reflection measurement for in-situ cure monitoring of a large-scale wind turbine blade. Their technique for monitoring the curing process relied on measuring the Fresnel reflectivity at the interface between the optical fiber and epoxy resin (Figure 9.31). This was based on the fact that the refractive index of epoxy resin changes throughout the curing stages (resin polymerization). An FBG sensor was used to separate the temperature-induced cross effects.

The Fresnel reflection coefficients at the fiber end/resin interface for the parallel and perpendicular polarizations are expressed as

$$r_p = \frac{n_m \cos\theta_1 - n_f \cos\theta_2}{n_m \cos\theta_1 + n_f \cos\theta_2} \tag{9.2}$$

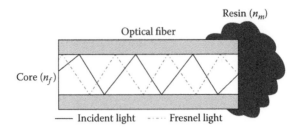

FIGURE 9.31 Fresnel reflection at fiber end/resin medium. (From Sampath, H. et al., *Sensors*, 15, 18229–18238, 2015.)

$$r_n = \frac{n_f \cos\Theta_1 - n_m \cos\Theta_2}{n_f \cos\Theta_1 + n_m \cos\Theta_2} \tag{9.3}$$

where:

n_f and n_m are the refractive indices of the fiber and the resin, respectively
Θ_1 and Θ_2 are the angles of incidence and reflection, respectively

Assuming the paraxial paths in standard single-mode fiber, that is, $\Theta_1 \approx \Theta_2 \approx 0$, the Fresnel reflectivity R can be expressed as

$$R = |r_f|^2 = |r_m|^2 = \left| \frac{n_f - n_m}{n_f + n_m} \right|^2 \tag{9.4}$$

Measuring the optical signal system uses two light sources: a broadband source (1550 nm) for the embedded FBGs and a laser source with wavelength of 1310 nm for Fresnel reflectivity sensing. The FBG sensors are connected to the broadband source by two 50:50 single-mode couplers, and Bragg wavelengths are demodulated by a spectrometer. Fresnel reflection output is referenced by the tapped output power of the 1310 nm laser source. The sensor output is calculated by

$$P_{cal} = \frac{P_{out}}{P_{ref}} \tag{9.5}$$

A curing test was carried out at three different temperatures: 80, 90, and 100°C. Figure 9.32 shows the collected Fresnel reflection output data. It is clearly seen that at the beginning, all sensors show a decrease in recorded optical signal, as the density of the resin is very low. As the cross-linking reaction between the polymer resin's chains is initiated, the viscosity of the resin starts to increase, and the resin's refractive index also increases, so the optical signal start to increase as well. Once the resin reaches the consolidation state, no further changes occur to either the viscosity or the refractive index; thus, no changes were recorded for the Fresnel reflection.

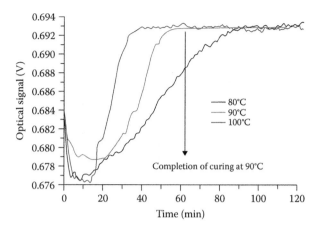

FIGURE 9.32 Sensor Fresnel output at 80, 90, and 100°C. (From Sampath, H. et al., *Sensors*, 15, 18229–18238, 2015.)

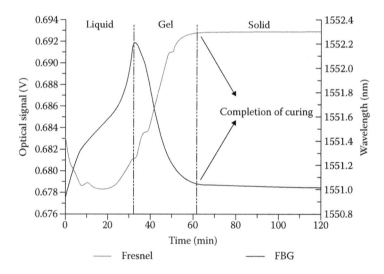

FIGURE 9.33 Fresnel and FBG signal measurements at 90°C. (From Sampath, H. et al., *Sensors*, 15, 18229–18238, 2015.)

The results of FBG and Fresnel reflection are shown in Figure 9.33. Both sensors show a constant measurement once the resin reaches the solid state, which is after 63 min of the curing reaction. At the gel point, which is the point when the resin's cross-linking reaction starts to take place, a noticeable drop in FBG wave length is induced, as the resin density increases when the cross-linking reaction between the resin's chains starts.

Chehura et al. [42] used the same measurement principle, using multiplexed fiber-optic sensors for monitoring resin infusion and cure in composite material processing. A series of FBG and Fresnel sensors were embedded near to

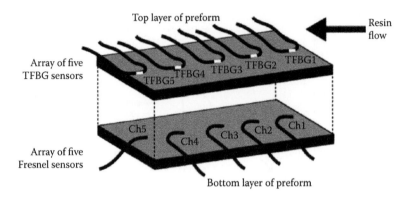

FIGURE 9.34 Tilted FBG sensors were laid near the top layer of the composite and aligned directly above Fresnel sensors. (From Chehura, E. et al., *Smart Sensor Phenomena, Technology, Networks, and Systems Integration*, 8693, 86930–86937, 2013.)

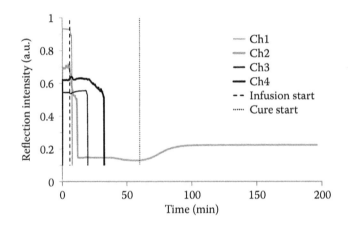

FIGURE 9.35 Fresnel sensors' reflection measurement. (From Chehura, E. et al., *Smart Sensor Phenomena, Technology, Networks, and Systems Integration*, 8693, 86930–86937, 2013.)

the top layer and in the middle layer of the composite material, respectively, as shown in Figure 9.34. Resin transfer molding using a vacuum bagging system was used to fabricate the composite panels, while the resin infused through the thickness. The reflection intensity of the Fresnel sensors was measured using OFDR. Figure 9.35 illustrates the measured reflective intensity of each Fresnel sensor during the resin flow. Sudden attenuation in the reflective intensity was observed as the infused resin reached the sensors, while the viscosity of the infused resin was low. When resin starts to cure and the cross-linking reaction begins, the reflective index of the resin increases due to the increase in its density. The measured Fresnel reflective index against the degree of curing is shown in Figure 9.36.

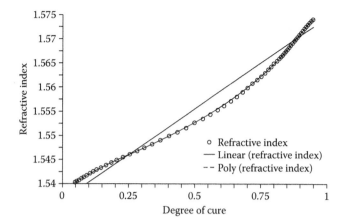

FIGURE 9.36 Refractive index of the resin measured using an optical fiber Fresnel refractometer. (From Chehura, E. et al., *Smart Sensor Phenomena, Technology, Networks, and Systems Integration*, 8693, 86930–86937, 2013.)

9.6 APPLICATION OF OPTICAL FIBER SENSORS IN SMART COMPOSITES: MONITORING COMPOSITE MECHANICAL PROPERTIES

Because of their light weight, ease of building, long life, low need for maintenance, and high strength, textile composite materials are excellent candidates for replacing traditional materials in infrastructure, such as steel and concrete. These advantages of textile composites have led to numerous applications, such as aircraft, automobiles, military uses, and wind turbine blades. Composite materials have a lifetime, which depends on age, usage, location, surrounding atmosphere (earthquakes, tornadoes, etc.) and also frequent maintenance needed to keep them safe for use. Thus, there is an increased need for structural health monitoring (SHM) systems that are able to detect any change in the structure's mechanical properties that may indicate damage or failure.

Kim et al. [43] used metal-coated FBG sensors to monitor the strain of composite plates used under quasi-static indentation. Four FBG sensors were used in this experiment: two were mounted on the top surface of the carbon fiber–reinforced polymer (CFRP) plate (FBG 1 and FBG2), and the other two (FBG 3 and FBG 4) were embedded into the middle layer of the CFRP plate and under the neutral axis of the composite. FBG 1 and FBG 3 sensors are metal coated with aluminum (Figure 9.37); the metal-coated optical sensors produce permanent residual strains due to the elasto-plastic characteristics of the metal coating, while normal optical fiber sensors barely yield residual strains when host structures are strained and released [44]. CFRP samples were subjected to eight increased quasi-static loads using a hemispherical steel indentor with a diameter of 1.27 mm, as shown in Figure 9.37. The behavior of the composite sample under the second loading cycle is shown in Figure 9.38. Strain measurement results show that the top surface-mounted FBGs (FBG 1 and FBG 3)

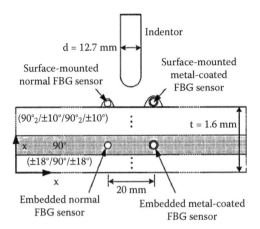

FIGURE 9.37 FBGs' location and direction. (From Kim, S. et al., *Composites: Part B*, 66, 36–45, 2014.)

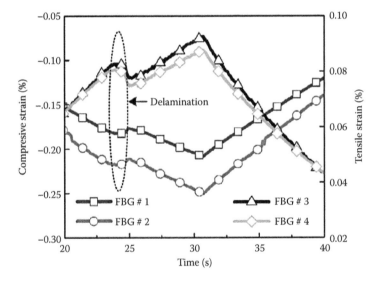

FIGURE 9.38 Strains from FBG sensors at second loading. (From Kim, S. et al., *Composites: Part B*, 66, 36–45, 2014.)

have a negative strain, while the embedded FBGs (FBG 2 and FBG 4) showed a positive strain. A decrease in the strain measured by embedded FBG sensors was observed after 25 s of loading. This decrease is due to an observed delamination near to the front surface. This delamination reduced the tensile stress, which resulted in a reduction in the strain measured by the top-mounted FBG sensors due to the effects of Poisson's ratio at this moment. This delamination caused the neutral axis to be shifted downward. Thus, the embedded FBGs were consequently compressed due to Poisson's ratio of the plate because of a temporary extension of the internal laminates near the original neutral axis.

Thermal residual stress has a negative effect on the fatigue life and dimension stability of composite structures. This residual stress occurs during the manufacturing process of composite materials due to the changing temperature during the curing cycle [45–47]. Several studies have been conducted to reduce the thermal residual effect through monitoring the mechanical performance of the composite structure during the curing cycle [48]. Kim et al. [49] used an embedded FBG between the fourth and fifth layers to monitor the strain growth and curing reaction of composite laminates to reduce the thermal residual stress. Another FBG temperature- and strain-free sensor was attached to the top surface of the composite laminate. This sensor was used to eliminate the temperature effect on the wavelength shift measured by the embedded FBG sensor. The evolution of strain was calculated using the following equation:

$$\Delta_\epsilon = \frac{\lambda_B - K_T \Delta T}{K_\epsilon} \tag{9.6}$$

where:

$\Delta\epsilon$ is the strain
$\Delta\lambda_B$ is the wavelength shift
ΔT is the temperature change of the FBG
K_T and K_ϵ are the sensitivity coefficients of the strain and the temperature, respectively

Thermal residual strain was monitored when samples were subjected to two different curing cycles: a conventional curing cycle and a modified curing cycle. The conventional curing cycle consists of four regions: the heating region (25°C–125°C), the gelation point, the vitrification region, and the final region. Figure 9.39 shows the measured strain during the curing cycle. Strain decreased to a negative value as the temperature increased from 25°C to 125°C. Increasing the temperature liquefies the epoxy resin; thus, the vacuum infusion process causes hydrostatic compaction of the liquid resin around the FBG sensor, which may cause a negative strain [50]. When the resin started to gel (Region 2) and the cross-linking reaction was initiated, positive strain was observed. This conversion from negative compressive strain to positive tensile strain may have been due to the resin contracting in

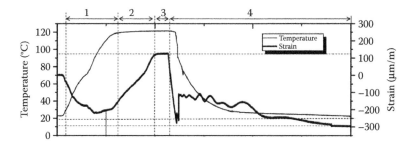

FIGURE 9.39 Residual strain after conventional curing cycle. (From Kim, H. S. et al., *Composites: Part B*, 44, 446–452, 2013.)

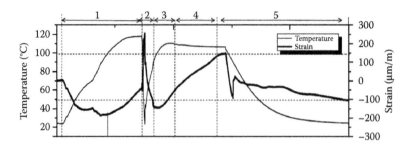

FIGURE 9.40 Residual strain after modified curing cycle. (From Kim, H. S. et al., *Composites: Part B*, 44, 446–452, 2013.)

thickness due to the applied negative vacuum pressure (−0.1 MPa), whereby the contraction may have squeezed the resin through the carbon fibers, and positive in-plane strains due to the high Poisson's ratio [51,52]. In Region 3, the resin was fully cross-linked and the strain remained constant at 150 µm/m, as no change in dimension occurred. After the full consolidation of the resin, the temperature dropped from 125°C to room temperature (25°C) (Region 4). The composite laminate shrank according to laminate thermal expansion coefficient and residual strain of −300 µm/m was recorded.

In the modified curing cycle (Figure 9.40), a sudden drop in temperature was involved before the gelation point (Region 2), whereby the temperature decreased from 125°C to 50°C. During this abrupt cooling, a decrease in strain was observed as a result of thermal contraction. Two hours of postcuring at 110°C followed after the abrupt cooling (Region 4). After the full curing and cooling of the composite laminate to room temperature (Region 5), the recorded residual thermal strain was −100 µm/m. This reduction in the residual strain in the modified curing cycle may be attributed to the abrupt cooling (Region 2) causing the resin to dissipate the thermal strain, in addition to the highly visco-elastic property of the epoxy resin [53]. This decreased the bonding temperature between the composite laminates and reduced the thermal residual stress. It is worth mentioning that tensile strength and tensile modulus are the same for both curing cycles.

Shin et al. [54] investigated the use of embedded FBG sensors for monitoring the postimpact fatigue damage. FBG sensors were embedded under the two outer 0° laminae layers of graphite/epoxy prepreg and along the axial loading direction. Both FBG sensors were 3 mm away from the center line of the specimen. One end of each embedded FBG protruded from the specimen, while the other end was located inside the specimen (Figure 9.41). A drop weight impact test with a falling weight of 260 g and 140 cm height was conducted, and impact locations are indicated in Figure 9.41 by A and B. Fatigue loading from 0.5 to 5 kN at a frequency of 5 Hz was performed on impacted samples.

The original FBG sensors had peak wavelengths between 1550 and 1551 nm. Figure 9.42 shows the wavelength shift occurring under cycle fatigue test without impact. A slight shift in peak wavelength in the original FBG was observed due to the residual stresses induced during the fabrication processes (embedding, curing,

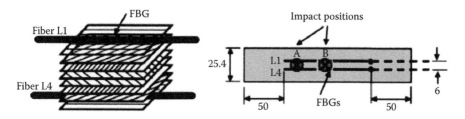

FIGURE 9.41 Location and direction of embedded FBG sensors. (From Webb, S. et al., *23rd International Conference on Optical Fibre Sensors* (*Proceedings of SPIE-the International Society for Optical Engineering*, 9157), 1–4, 2015.)

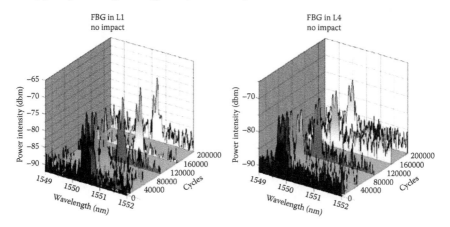

FIGURE 9.42 Developed spectra from embedded FBGs under fatigue cycles without impact.

and cutting into testing specimens), which can be seen in Figure 9.42. Under different fatigue cycles, a wavelength shift was observed, which indicates the good response of the FBG to strain and stress. Wavelength shifting results also indicated that the FBG was not damaged by the fatigue test, as the wavelength shifted to a longer wavelength and retained its original shape and intensity after 200,000 cycles. Other samples were subjected to impact test prior to the fatigue cycle test, and the results are shown in Figure 9.43. It can be seen that the upper embedded FBG spectrum peaks continued to widen with increasing intensity throughout the fatigue cycles (Figure 9.43b). The other FBG sensor, embedded at the bottom, showed different behavior, whereby the intensity of the wavelength surrounding the original peak continued to rise as the fatigue cycles increased. The indistinguishable peaks at the end of 200,000 fatigue cycles indicate that impact damage continued to develop and spread.

Webb et al. [55] monitored the behavior of a composite structure under repeated low-velocity impact with embedded FBG at an impact energy of 12.1 J. The spectral response of the FBG sensor after each of the five individual strikes is shown in Figure 9.44. Different features were observed, such as peak splitting and oscillating behaviors, along with changes from compressive strain to tensile strain throughout the impact events. An increase in the impact's event duration was noticed, which is an indication of matrix relaxation and growth of damage in the composite. The

(a)　　　　　　　　　　　　　　　　　　　　(b)

FIGURE 9.43 Developed spectra from embedded FBGs under fatigue cycle test after impact. (From Webb, S. et al., *23rd International Conference on Optical Fibre Sensors* (*Proceedings of SPIE-the International Society for Optical Engineering, 9157*), 1–4, 2015.)

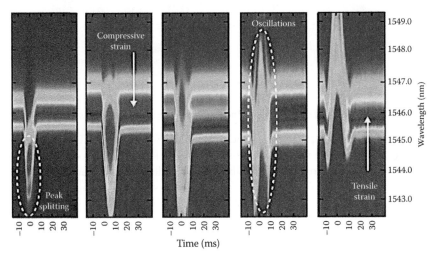

FIGURE 9.44 FBG spectra during repeated impacts of composite laminated plate. Color scale corresponds to reflected intensity, with red as maximum intensity. (From Potluri, P. et al., *Applied Composite Materials*, 19, 799–812, 2012.)

changes from compressive strain to tensile strain indicated that the natural axis moved due to induced failure. This information could be used to define the remaining lifetime prognostics for composite structures.

Hamouda et al. [23] have extended their work to monitor the mechanical properties of 3D composites with embedded POF. POFs were integrated in the middle layer (between the second and third layers of y-yarns in the case of a four-layer composite) in the filling direction (x-direction), as shown in Figure 9.2. To cover a large area

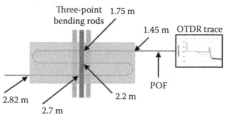

FIGURE 9.45   Bending test setup using OTDR. (From Hamouda, T. et al., *Smart Materials and Structures*, 24, 1–11, 2015.)

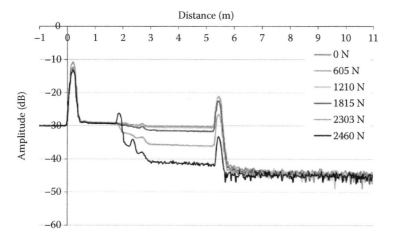

FIGURE 9.46   Backscattering of POF embedded into composite sample VII with four yarn layers and 1.57 x-yarn/cm/layer at different bending loads.

of the 3D composite (Figure 9.3), the POFs were bent at both selvedges of the 3D preform with a 1.9 cm radius (critical radius) [34]. Nine different samples were fabricated, with two, three, and four layers and with filling densities of 1.57, 3.14, and 4.72. The prepared samples were tested for bending and impact. OTDR was used to monitor the backscattering level, signal attenuation, and location of the failure after each test. Figure 9.45 shows the bending test setup. OTDR was connected to one end of the embedded POF.

A bending load was applied to the required level; then the test was paused, and the backscattering signal was collected. The load was then released and the signal was again measured. The backscattered signal was also measured for all samples at failure and after releasing the final load. Figure 9.46 shows the OTDR signal trace of embedded POF in a composite of four y-yarn layers with x-yarn density 1.57 (Sample VII) under different bending loads. No changes in the backscattering level were recorded under a bending load of 1210 N. When the applied load reached 1815 N, a noticeable drop in the POF backscattering level associated with signal loss was observed at three different locations: 1.75, 2.24, and 2.70 m. These locations are the three contact points of the composite sample with the upper rod. This signal loss is due to induced macrobending and flattening of the POF, whereby the change in cross section shape leads to a change in the refractive index of POF, which results in a reduced level of backscattering at these locations. The POF signal loss was substantial when the applied load reached 2460 N. At this load, the OTDR trace showed both a loss of 10.43 dB and reflection, which is an indication of permanent structural changes in the POF due to the high bending stress. To investigate the recovery performance of embedded POF, the backscattering level was recorded after releasing each load (Figure 9.47). The results showed that the backscattering returned to its original level after release of loads up to 2303 N. However, a signal loss of 2.9 dB remained after release of the 2460 N load, indicating permanent damage to the POF. Figure 9.48 shows the relation between applied bending loads and recorded signal loss for nine 3D orthogonal composites. The results revealed that as the bending load increases, the signal loss also increases.

The effect of repeated low-velocity impact on the signal loss of embedded POF in 3D orthogonal composite was also investigated [23]. 3D woven composites are particularly resistant to impact loading and are therefore often used in impact-critical applications [56,57]. Impact testing was terminated when the POF reached the maximum signal loss, meaning that the POF could no longer be used for measurements. However, samples that were structured with the highest x-yarn density of 4.72 and three and four y-yarn layers were repeat impacted 30 times, and the POF maintained dynamic ranges of 9 and 11 dB, respectively. The POFs embedded in the other samples, which were fabricated with low filling densities, lost their dynamic range completely due to their low resistance to impact.

Figure 9.49 shows OTDR traces after each impact for 3D composite samples with four warp layers and filling density of 1.57. The impacts were repeated 14 times, till

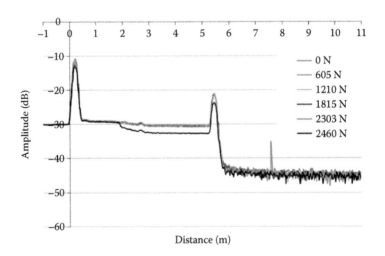

FIGURE 9.47 Backscattering of POF embedded into composite sample with four yarn layers and 1.57 x-yarn/cm/layer after unloading. Key shows maximum bending load prior to unloading.

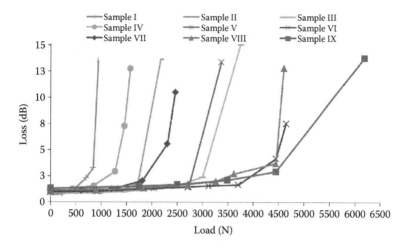

FIGURE 9.48 Signal loss vs. applied load for different 3D composite configurations under different bending loads.

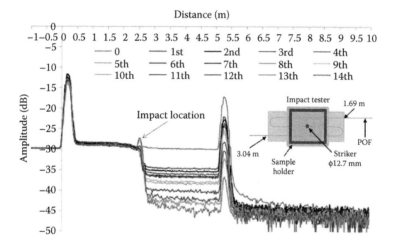

FIGURE 9.49 OTDR traces of embedded POF in four layers and 1.57 filling density 3D composite after repeated impact test.

the POF signal loss reached the maximum value and the dynamic range dropped to zero. Results show that as the sample is subjected to repeated impact, a noticeable drop in the backscattering level is observed. The recorded backscattering level before starting the impact test (at zero impact) was −28.18 dB. After the first impact, the backscattering level showed a significant drop, with signal loss of 4.87 dB at 2.48 m. When the repeated impact reached the fourth strike, the total signal loss reached 6.34 dB, indicating that the stress and deformation on the POF increased up to this impact level. The backscattering level continued to drop with the repeated impact; when the number of repeated impacts reached the 14th strike, the POF signal loss reached a maximum. A small reflection was observed at the 14th impact due to partial fracture or slight crimping in the POF. A visual examination of the extent of the damaged area with increasing numbers of repeated impacts is shown in Figure 9.50. These images indicate that the increase in POF signal loss corresponds to the increase in the damaged region.

Figure 9.51 shows backscattering levels after repeated impact of a 3D orthogonal composite sample made of three warp layers and filling density of 4.72 filling/cm. After the first impact, a drop in the backscattering level at 2.5 m with a signal loss of 1.67 dB was observed. When the impact test was repeated 30 times, the dynamic range gradually dropped to a final signal loss of 4.0 dB. Figure 9.52 shows some images of the impacted sample along with the back surface after 30 impact strikes. It

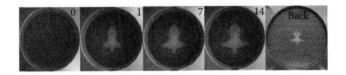

FIGURE 9.50   Impact surface before and after repeated impacts (1, 7, and 14 strikes) and back surface of the sample after final impact.

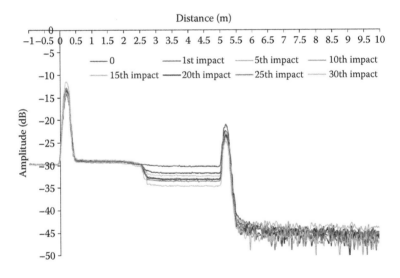

FIGURE 9.51   OTDR trace of POF in 3D orthogonal composite sample made of three warp layers and filling density of 4.72 filling/cm after repeated impact.

FIGURE 9.52   Selected images of impact surface before and after repeated impacts.

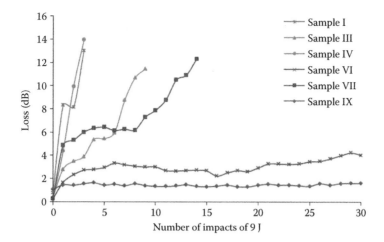

FIGURE 9.53 POF signal loss vs. repeated impacts.

can be seen that after the 15th strike, the area of damage in the composite was almost unchanged. The negligible change in damaged area after the 15th strike is consistent with the small changes in POF backscattering level.

The compiled results of signal loss versus the number of impacts for different sample configurations are shown in Figure 9.53. Samples made of two and three layers with filling densities of 1.57 and 3.14, represented by I, III, IV, and VII, respectively, showed an increase in signal loss as the number of impacts increased till backscattering levels dropped to the full dynamic range. The POFs in Samples VI and IX, three and four warp layers with filling densities of 4.72 filling/cm, respectively, showed an increase in signal loss after the first impact and then insignificant losses after further impacts. When the repeated impact reached the 30th strike, the damage propagated significantly (Figure 9.52); however, the POF signal did not reflect that level of damage. This may be attributed to the position of the POF, which was between the thick samples VI and IX. Therefore, in such samples, the POF sensor needs to be configured differently.

CONCLUSION

This chapter presents the importance of SHM systems for real-time monitoring of global and local response of civil engineering and composite structures. The behavior of composite structures during the fabrication process and under static and dynamic loads needs to be monitored to detect any sign of failure before it occurs, which both saves lives and reduces maintenance costs. Sudden collapse of structures may occur due to internal damage, while the visual observation of structures shows no damage on the structure's surface. The small size of optical fiber sensors allows them to be embedded into composite structures during the fabrication process with no negative effect on the structure's integrity. Using textile preforms such as 2D or

3D woven preform permits the optical fiber to be easily integrated into textile pre-forms during the preform fabrication process. Due to the high sensitivity of optical fiber, the critical bending radius needs to be defined correctly prior to the integration process to properly embed the optical fiber and to avoid any negative effects on its signal integrity.

Wavelength shift, microbending, and macrobending loss in embedded optical fibers are used as indications of thermal residual strain and internal stress and strain that the structure experiences during the fabrication process (preform fabrication and curing process) and also under static and dynamic loads.

REFERENCES

1. Sirohi, J. and Chopra, I. Fundamental understanding of piezoelectric strain sensors. *Journal of Intelligent Material Systems and Structures*, 11 (2000), 246–257.
2. Ide, H., Abdi, F., Miraj, R., Dang, C., Takahashi, T., Sauer, B. Wireless-Zigbee strain gage sensor system for structural health monitoring. *Photonics in the Transportation Industry: Auto to Aerospace II*, 7314, 12 (2009), 731403–731403.
3. Rao, M., Bhat, M., Murthy, C. Structural health monitoring (SHM) using strain gauges, PVDF film and fiber Bragg grating (FBG) sensors: A comparative study. *Proceedings of National Seminar on Non-Destructive Evaluation* (2006), 333–337.
4. Rumsey, M., Paquette, J., White, J., Werlink, R., Beattie, A., Pitchford, C., Dam, J. Experimental results of structural health monitoring of wind turbine blades. *American Institute of Aeronautics and Astronautics* (2008), 1–14.
5. Lee, D. G., Mitrovic, M., Stewart, A., Garman, G. Characterization of fiber optic sensors for structural health monitoring. *Journal of Composite Materials*, 36, 11 (2002), 1349–1366.
6. Zhan-Feng, G., Yang-Liang, D., Bao-Chen, S., Xiu-Mei, J. Strain monitoring of railway bridges using optic fiber sensors. *Journal of Quality in Maintenance Engineering*, 13, 2 (2007), 186–197.
7. Leng, J. and Hao, J. Damage assessment of composite laminate by using fiber optic AE sensors based on fused-tapered coupler. *Proceedings of SPIE*, 6531 (2007), 1021–1025.
8. Rippert, L., Papy, J., Wevers, M., Huffel, S. Fibre optic sensor for continuous health monitoring in CFRP composite materials. *Smart Structures and Materials*, 4693 (2002), 312–323.
9. Shaaf, K., Cook, B., Ghezzo, F., Starr, A., Nemat-Nasser, S. Mechanical properties of composite materials with integrated embedded sensor networks. *Smart Structures and Materials*, 5765 (2005), 781–785.
10. Ishigure, T., Nihei, E., Koike, Y. Optimum refractive-index profile of the graded-index polymer optical Fiber, toward gigabit data links. *Applied Optics*, 35, 12 (1996), 2048–2053.
11. Peters, K. Fiber optic sensors in concrete structures: A review. *Smart Materials and Structures*, 20 (2011), 1–17.
12. Ziemann, O., Krauser, J., Zamzow, P., Daum, W. *POF Handbook: Optical Short Range Transmission Systems*, Berlin: Springer (2008).
13. Kiesel, S., Van Vickle, P., Peters, K., Abdi, O., Hassan, T., Kowalsky, M. Polymer optical fiber sensors for the civil infrastructure. *Mechanical and Aerospace Systems, SPIE Sensors and Smart Structures,* SPIE 6174 (2006), 1–12.
14. Nakamura, K., Husdi, I., Ueha, S. Memory effect of POF distributed strain sensor. In *Second European Workshop on Optical Fibre Sensors, Proceedings of SPIE*, 5502 (2004), 144–147.

15. Liehr, S., Lenke, P., Krebber, K., Seeger, M., Thiele, E., Metschies, H., Gebreselassie, B., Munich, J., Stempniewski, L. Distributed strain measurement with polymer optical fibers integrated into multifunctional geotextiles. *Optical Sensors, Proceedings of SPIE*, 7003 (2008).

16. Liehr, S., Lenke, P., Wendt, M., Krebber, K., Seeger, M., Thiele, E., Metschies, H., Gebreselassie, B., Münich, J. Polymer optical fiber sensors for distributed strain measurement and application in structural health monitoring. *IEEE Sensors Journal*, 9, 11 (2009), 1330–1338.

17. Kuang, K., Akmaluddin, Cantwell, W. J., Thomas, C. Crack detection and vertical deflection monitoring in concrete beams using plastic optical fibre sensors. *Measurement Science and Technology*, 14, 2 (2003), 205–216.

18. Kuang, K., Quek, S., Koh, C., Cantwell, W., Scully, P. Plastic optical fibre sensors for structural health monitoring: A review of recent progress. *Journal of Sensors*, 2009 (2009), 13.

19. Ogale, V. and Alagirusamy, R. Textile preforms for advanced composites. *Indian Journal of Fiber & Textile Research*, 29 (2004), 366–375.

20. Yi, H. and Ding, X. Conventional approach on manufacturing 3D woven preforms used for composite. *Journal of Industrial Textiles*, 34, 1 (2004), 34–39.

21. Mohamed, T. S., Johnson, C., Rizkalla, S. Behavior of three-dimensionally woven glass fiber reinforced polymeric bridge deck. *Composites Research Journal*, 1 (2007), 27–42.

22. Grillet, A., Kinet, D., Witt, J., Schukar, M., Krebber, K., Pirotte, F., Depre, A. Optical fiber sensors embedded into medical textiles for healthcare monitoring. *IEEE Sensors Journal*, 8, 7 (2008), 1215–1222.

23. Hamouda, T., Seyam, A. M., Peters, K. Polymer optical fibers integrated directly into 3D orthogonal woven composites for sensing. *Smart Materials and Structures*, 24, 2 (2015), 1–11.

24. Rothmaier, M., Phi Luong, M., Clemens, F. Textile pressure sensor made of flexible plastic optical fibers. *Sensors*, 8 (2008), 4318–4329.

25. Castellucci, M., Klute, S., Lally, E. M., Froggatt, M. E., Lowry, D. Three-axis distributed fiber optic strain measurement in 3D woven composite structures. *Proceedings of the Industrial and Commercial Applications of Smart Structures Technologies*. San Diego, CA: Smart Structures/NDE (2013).

26. Du, G.-W. and Ko, F. K. Unit cell geometry of 3D braided structures. *Journal of Reinforced Plastics and Composites*, 12, 7 (1993), 752–768.

27. Jung, K. and Kang, T. J. Cure monitoring and internal strain measurement of 3-D hybrid braided composites using fiber Bragg grating sensor. *Journal of Composite Materials*, 41, 12 (2007), 1499–1518.

28. Yuan, S., Huang, R., Rao, Y. Internal strain measurement in 3D braided composites using co-braided optical fiber sensors. *Journal of Materials Science & Technology*, 20, 2 (2004), 199–202.

29. El-sherif, M. A. and Ko, F. K. Spatial modulation within embedded fiber-optic sensors for smart structures characterization. *Technical Digest of the First European Conference on Smart Structures and Materials*. Glasgow: CRC Press (1992), pp. 85–88.

30. Herczfeld, P. R., El-Sherif, M. A., Kawase, L. R., Ko, F., Bobb, L. Embedded fiber-optic sensor utilizing the modal power distribution technique. *Journal of Optics Letters*, 15, 21 (1990), 1242–1244.

31. Casas, J. R. and Cruz, P. J. S. Fiber optic sensors for bridge monitoring. *Journal of Bridge Engineering*, 8, 6 (2003), 362–372.

32. Zendehnam, A., Mirzaei, M., Farashiani, A., Farahani, L. Investigation of bending loss in a single-mode optical fiber. *Pramana-Journal of Physics*, 74, 4 (2010), 591–603.

33. Gupta, S. C. *Textbook on Optical Fiber Communication and Its Applications*. New Delhi, India: PHI Learning (2004), 40–47.
34. Seyam, A. F. M. and Hamouda, T. Smart textiles: Evaluation of optical fibres as embedded sensors for structure health monitoring of fibre reinforced composites. *Journal of the Textile Institute*, 104, 8 (2013), 892–899.
35. Wang, J., Huang, B., Yang, B. Effect of weave structure on the sideemitting properties of polymer optical fiber jacquard fabrics. *Textile Research Journal*, 38, 11 (2012), 1170–1180.
36. Zhang, K., Gu, Y., Zhang, Z. Effect of rapid curing process on the properties of carbon fiber/epoxy composite fabricated using vacuum assisted resin infusion molding. *Materials & Design*, 54 (2014), 624–631.
37. Liu, L., Zhang, B. M., Wang, D. F., Wu, Z. J. Effects of cure cycles on void content and mechanical properties of composite laminates. *Composite Structures* (2006), 303–309.
38. Khattab, A. Cure cycle effect on high-temperature polymer composite structures molded by VARTM. *Journal of Composites*, 2013 (2013), 1–6.
39. Davies, L. W., Day, R. J., Bond, D., Nesbitt, A., Ellis, J., Gardon, E. Effect of cure cycle heat transfer rates on the physical and mechanical properties of an epoxy matrix composite. *Composites Science and Technology*, 67, 9 (2007), 1892–1899.
40. Aktas, A., Boyd, S. W., Shenoi, R. A. Cure and strain monitoring of novel unsaturated polyester/phenolic resin blends in the vacuum infusion process using fibre Bragg gratings. *Journal of Composite Materials,* 49, 29 (2015), 3599–3608.
41. Sampath, H., Kim, H., Kim, D., Kim, Y., Song, M. *In-situ* cure monitoring of wind turbine blades by using fiber Bragg grating sensors and Fresnel reflection measurement. *Sensors*, 15 (2015), 18229–18238.
42. Chehura, E., Jarzebinska, R., Da Costa, E. F., Skordos, A. A., James, S. W., Partridge, I. K., Tatam, R. P. Multiplexed fibre optic sensors for monitoring resin infusion, flow, and cure in composite material processing. *Smart Sensor Phenomena, Technology, Networks, and Systems Integration*, 8693 (2013), 86930–86937.
43. Kim, S., Cha, M., Lee, I., Kim, E., Kwon, I., Hwang, T. Damage evaluation and strain monitoring of composite plates using metal-coated FBG sensors under quasi-static indentation. *Composites: Part B*, 66 (2014), 36–45.
44. Kim, S. W., Jeong, M. S., Lee, I., Kim, E. H., Kwon, I. B., Hwang, T. K. Determination of the maximum stresses experienced by composite structures using metal coated optical fiber sensors. *Composite Science and Technology*, 78 (2013), 48–55.
45. Hosseini-Toudeshky, H. and Mohammadi, B. Thermal residual stresses effects on fatigue crack growth of repaired panels bounded with various composite materials. *Composite Structures*, 89, 2 (2009), 216–223.
46. Hodges, J., Yates, B., Darby, M. I., Wostenholm, G. H., Clement, J. F., Keates, T. F. Residual stresses and the optimum cure cycle for an epoxy resin. *Journal of Materials Science*, 24, 6 (1989), 1984–1990.
47. Kim, H. S., Park, S. W., Lee, D. G. Smart cure cycle with cooling and reheating for co-cure bonded steel/carbon epoxy composite hybrid structures for reducing. *Composites Part (A)*, 37, 10 (2006), 1708–1721.
48. Kim, S. S., Murayama, H., Kageyama, K., Uzawa, K., Kanai, M. Study on the curing process for carbon/epoxy composites to reduce thermal residual stress. *Composites Part A: Applied Science and Manufacturing*, 43, 8 (2012), 1197–1202.
49. Kim, H. S., Yoo, S. H., Chang, S. H. *In situ* monitoring of the strain evolution and curing reaction of composite laminates to reduce the thermal residual stress using FBG sensor and dielectrometry. *Composites: Part B*, 44, 1 (2013), 446–452.

50. Mulle, M., Collombet, F., Olivier, P., Grunevald, Y. H. Assessment of cure residual strains through the thickness of carbon–epoxy laminates using FBGs, Part I: Elementary specimen. *Composites Part A: Applied Science and Manufacturing*, 40, 1 (2009), 94–104.

51. Plepys, A. R. and Farris, R. J. Evolution of residual stresses in three-dimensionally constrained epoxy resins. *Polymer*, 31, 10 (1990), 1932–1936.

52. Park, S. J., Seo, M. K., Lee, J. R. Relationship between viscoelastic properties and gelation in the epoxy/phenol-novolac blend system with N-Benzylpyrazinium salt as a latent thermal catalyst. *Journal of Applied Polymer Science*, 79, 12 (2001), 2299–2308.

53. Kim, H. S. and Lee, D. G. Avoidance of the fabrication thermal residual stress of co-cure bonded metal/composite hybrid structures. *Journal of Adhesion Science and Technology*, 20, 9 (2006), 959–979.

54. Shin, C. S., Liaw, S. K., and Yany, S. W. Post-impact fatigue damage monitoring using fiber bragg grating sensors. *Sensors*, 14 (2014), 4144–4153.

55. Webb, S., Peters, K., Zikry, M., Stan, N., Chadderdon, S., Selfridge, R., Schultz, S. Fiber Bragg grating spectral features for structural health monitoring of composite structures. In *23rd International Conference on Optical Fibre Sensors (Proceedings of SPIE-the International Society for Optical Engineering, 9157)* (2015), 1–4.

56. Potluri, P., Hogg, P., Arshad, M., Jetavat, D., Jamshidi, P. Influence of fibre architecture on impact damage tolerance in 3D woven composites. *Applied Composite Materials*, 19, 5 (2012), 799–812.

57. Chen, F. and Hodgkinson, J. M. Impact behaviour of composites with different fibre architecture. *Journal of Aerospace Engineering*, 223, 7 (2009), 1009–1017.

10 Smart Aerospace Composite Structures

Gayan Kahandawa and Jayantha Ananda Epaarachchi

CONTENTS

10.1 Introduction: Advance Composite Structures...339
    10.1.1 Composite Structural Damage...342
10.2 Fiber-Optic Sensors 345
    10.2.1 FBG Sensors 345
        10.2.1.1 Evolution of FBG Sensors...346
        10.2.1.2 Embedding FBG Sensors in FRP Structures....................348
    10.2.2 PPP-BOTDA Sensors...349
        10.2.2.1 Brillouin Scattering-Based Fiber-Optic Measuring
                Technique 350
    10.2.3 Smart Structures and Materials...351
    10.2.4 Fiber Optics in Smart Aerospace Composite Structures.................352
        10.2.4.1 Damage Detection Technologies 352
        10.2.4.2 Signal Processing Technologies for Fiber-Optic Sensors 353
        10.2.4.3 Damage Prediction...356
        10.2.4.4 Decoding Distorted FBG Sensor Spectra..........................356
        10.2.4.5 Development of ANN-Based Signal Processing
                Techniques 360
        10.2.4.6 ANN-Based Damage Detection...361
        10.2.4.7 Neural Networks 361
        10.2.4.8 Postprocessing of FBG Data Using ANN..........................365
        10.2.4.9 The ANN Model...365
Summary 367
References..367

10.1 INTRODUCTION: ADVANCE COMPOSITE STRUCTURES

Fiber-reinforced polymer (FRP) composites have been used as an engineering material for more than six decades. The main attraction of FRP is its superior strength-to-weight ratio. The aerospace industry has been spending billions of dollars on investment in these composites to produce lightweight subsonic and supersonic aircrafts. Its other desirable properties, such as the ease of fabrication of complex shapes and the ability to tailor desirable properties to suit different engineering applications, are enviable for an advanced material. Since research and development in the aircraft industry and space exploration agencies have been focused on FRPs for many years, most of the advanced fiber composites available today have, in one way or another, their origins in these fields.

The weight-save or positive weight spiral in the aerospace industry is directly translated to the enhancement of the load-carrying capacity of civil aircraft, while for the fighters it will be translated to performance enhancement (mainly fuel-carrying capacity vs. flying speed). As composites are partially made from polymer-based materials, they possess very good damping and fatigue-resistance properties compared with traditional metallic materials.

The commercial aircraft industry is gradually replacing metallic parts with FRP composites as much as possible. Hence, FRP composites are frequently applied to primary load-bearing structures in newly developed aircraft, such as Boeing 787 and Airbus 380. However, the main disadvantages of using FRP composites in the aircraft industry are their difficulty for repair, anisotropic behavior, high initial setup cost, and, most importantly, their complex failure criteria. Because of these undesirable properties, FRP composite structures in aircraft need to be closely monitored to prevent unexpected failure.

These structures can include stress-concentrated regions such as pin-loaded holes and other cutouts. These stress concentrations easily induce damage that includes concurrent splitting, transverse cracking, and delamination (Chang and Chang, 1987; Kamiya and Sekine, 1996; Kortscho and Beaumont, 1990). Unlike metals, the failure modes of composites are very difficult to predict. Therefore, there are no established standards for composite materials. For this reason, it is essential to monitor advance composites regularly. In view of the aforementioned issues, a structural health monitoring (SHM) technique has recently been developed for these composite structures (Chang, 2003; Zhou and Sim, 2002).

Monitoring of composites began with damage detection technologies such as vibration and damping methods (Adams et al., 1978). Then, sophisticated off-line nondestructive test (NDT) methods were developed for the safe operation of composite structures. However, with the increasing complexity of structures, the off-line NDT was insufficient and development of online SHM systems became vital.

The process of implementing the damage detection and characterization strategy for engineering structures is referred to as SHM. Here, the indicator for damage is defined as changes to the material properties or changes to the structural response of the structure. The SHM process involves the observation of a system over time using periodically sampled dynamic response measurements from an array of sensors.

To address complex failure modes of FRP composites, an SHM system must be efficient, robust, and accurate. Due to recent developments in the aerospace industry, the utilization of FRP composites for primary aircraft structures, such as wing leading edge surfaces and fuselage sections, has increased. This has led to a rapid growth in the field of SHM. Impact, vibration, and loading can cause damage, such as delamination and matrix cracking, to the FRP composite structures. Moreover, internal material damage can be undetectable using conventional techniques, making inspection of the structures for damage and clear insight into the structural integrity difficult using currently available evaluation methods.

An SHM system developed to monitor aircraft and space structures must be capable of identifying multiple failure criteria of FRP composites (Reveley et al., 2010).

Since the behavior of composites is anisotropic, multiple numbers of sensors must be in service to monitor these structures under multidirectional complex loading conditions. The layered structure of the composites makes it difficult to predict the structural behavior by using only surface sensors. To address this issue, embedded sensors must be used, and the sensors used must possess sufficient durability, as it is not possible to replace embedded sensors after fabrication of the parts.

The fiber Bragg grating (FBG) sensor is one of the most suitable sensors for the SHM of aircraft structures. The diminutive FBG sensors can be embedded in FRP composites during the manufacturing of the composite part with no effect on the strength of the part. This sensor has a narrowband response with a wide operating range; hence, it can be highly multiplexed. This nonconductive sensor can operate in electromagnetically noisy environments without any interference. The FBG sensor is made up of glass possessing a long lifetime comparable to that of FRP composites. Because of its low transmission loss, the sensor signal can be monitored from longer distances, making it suitable for remote sensing (Hill and Meltz, 1997; Kersey et al., 1997).

The FBG sensor's capability of detecting strain gradients along its length can be used to identify the strain variations at a critical part in the FRP composite's structure through an optical phenomenon called *chirp*, in the reflected spectra of the FBG sensor (Hill and Eggleton, 1994; Kersey et al., 1997). This phenomenon has been used for decades of research to detect damage in composite structures (Okabe et al., 2000; Takeda et al., 2002). But there are numerous cases reported where the chirp or the distortion of the FBG spectrum was not limited to stress concentrations (Wang et al., 2008). There are other causes of chirp, and it is necessary to eliminate such effects to identify damage accurately.

As such, there are many remaining unresolved problems and engineering challenges in FBG-based SHM systems. An extraction of important data from FBG spectra is a significant problem that remains unsolved after many decades of research work. It is clear that multiple causes lead to distortion to the FBG response spectra. Most of the effects cannot be eliminated in advanced aerospace applications. In order to identify damage from the distortions to the FBG response spectra, the individual contribution from each effect needs to be identified and eliminated. To unambiguously identify the effects from the damage, extensive computational power is required for postprocessing of the spectral data. Figure 10.1 shows FBG response spectra from an FBG embedded near a damaged location with the part under complex multidirectional loading.

With the complex layout of composite structures, it is difficult to embed two sensors identically in two similar structures. The sensitivity of an FBG is in the nanometer range and a small shift of location makes the two environments different to the two sensors. As a result, it is difficult to compare fine spectral data of two spectra from two FBG sensors. Consequently, it is impossible to have universal decision-making algorithms work with FBG sensors. This complication is a major unsolved problem in the FBG-based SHM. The scarcity of adaptive decision-making systems needs to be addressed.

The manufacturing of advanced composite components with embedded sensors is another significant engineering challenge associated with FBG-based SHM systems.

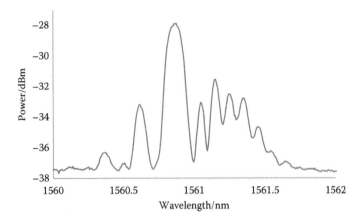

FIGURE 10.1   Distorted FBG spectra due to multiple effects.

The number of sensors must be optimized, and there should be backup systems in place for any redundant sensors after manufacture. Further, the embedded FBG network must be robust enough for compensating situations, such as redundant sensors, during operations. These unresolved critical problems have caused barriers to the development of SHM systems.

10.1.1   COMPOSITE STRUCTURAL DAMAGE

Damage in FRP structures can be defined as the *changes to material properties* and/ or *changes to the structural response* of the structure. Damage can be in both the matrix and/or fiber.

Many reinforced plastics consist of brittle fibers, such as glass or carbon, in a weak, brittle polymer matrix, such as epoxy or polyester resin. An important characteristic of these composites, however, is that they are surprisingly tough, largely as a result of their heterogeneous nature and construction. During deformation, microstructural damage is widespread throughout the composite, but much damage can be sustained before the load-bearing ability is impaired. Beyond some critical level of damage, failure may occur by the propagation of a crack, which usually has a much more complex nature than cracks in homogeneous materials. Crack growth is inhibited by the presence of interfaces, both at the microstructural level between fibers and the matrix, and at the macroscopic level as planes of weakness between separate laminae in a multiply laminate. The fracturing of a composite, therefore, involves not only the breaking of the load-bearing fibers and the weak matrix, but a complex combination of crack deviations along these weak interfaces.

Composite materials can be analyzed in different levels, as shown in Figure 10.2. The analysis of damage in composite structures starts from micromechanics to macromechanics. The failure of composites begins with microcracks in the matrix. The microcracks increase in number under increasing deformation. Due to a mismatch in the elastic module between the neighboring layers, there exists an onset of delamination. As the critical value of the strain energy release rate is reached,

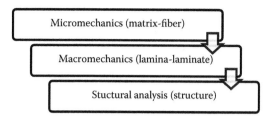

FIGURE 10.2   Levels of analysis for damage in composites.

the delamination starts growing rapidly. As a result, under the condition that the matrix crack density is very high and that the angle plies can no longer take tensile load, the entire tensile load is transferred through the 0° fibers. With further tensile loading, the 0° fibers break. This is the final stage of failure. Under a compressive load, also, the laminates start buckling at the delaminated zone and may fail instantaneously.

The evolution of a matrix crack as the initial stage of damage, followed by delamination, is also true for composites subjected to impact load (Marshall et al., 1985). Most of the damage models reported in literature are transverse matrix cracks, splitting, delamination, and so on, and a combination of those are considered. The degradations of effective elastic modules of damaged laminates are used as the damage parameters. Once estimated through online health monitoring, such parameters can further be correlated to damage states.

Numerous failure theories have been proposed and are available to the composite structural designer (Daniel and Ishai, 2005). The failure theories have been classified into three groups: limit or noninteractive theories (maximum stress, maximum strain); interactive theories (Tsai and Wu, 1971); and partially interactive or failure mode-based theories (Hashin and Rotem, 1973; Puck and Schürmann, 1998). The validity and applicability of a given theory depend on the convenience of application and agreement with experimental results.

Sun (2000) reviewed six failure theories and showed comparisons of theoretical predictions with experimental results for six different composite material systems under various loading conditions. He included uniaxial (normal and shear) loading, off-axis loading, and biaxial (normal and shear) loading later. It was found, as observed previously, that most theories differed little from each other in the first quadrant (tension–tension). The biggest differences among theories occurred under combined transverse compression and shear. In this case, predictions of the Tsai-Wu interactive theory were in better agreement with experimental results than other theories.

Hinton et al. (2004) experimentally evaluated the predictive capabilities of current failure theories (Hinton et al., 2004). One observation of this exercise was that, even for the unidirectional lamina, predictions of the various theories differed by up to 200%–300% from each other. The difficulty in evaluating failure theories is much greater in the case of a multidirectional laminate. Since the failure modes greatly depend on material properties and type of loading, different failure theories work well with different applications.

Delamination, which is the failure of the interface between two plies, is one of the main types of damage in aerospace composite structures, and is known as the silent killer of composite structures. It is caused by normal and shear tractions acting on the interface, which may be attributed to transverse loading, free-edge effect, ply drop-off, or local load introduction. Delamination can significantly reduce structural stiffness and load-carrying capacity and, therefore, is considered as one of the critical failure modes in laminated composites.

A laminate is constructed from groups of individual unidirectional plies that are laid at various angles, depending upon design requirements, or from layers of woven cloth laid at various angles to the main stress axes. The tension/shear coupling effects cause shear stresses to develop in the plane of the laminate, especially near free edges, when the material is stressed. As interlaminar planes in nonwoven composites are always planes of weakness, the interlaminar shear stresses may easily become large enough to disrupt or delaminate a composite well before its overall tensile strength is reached. A crack traveling through a given ply may, therefore, find it energetically favorable to deviate along an interlaminar plane. Under such circumstances, considerations of fiber debonding and pullout are of little significance.

For predictions of onset of interface damage, numerical methods have been developed based on the interface strength. A general, strength-based first ply failure (FPF) criterion suitable for the assessment of interface damage has been proposed by Puck and Scharmann (1998). For predicting the formation of an initial delamination in an intact interface, a strength/energy approach is proposed (Wimmer et al., 2009) that combines strength criteria with fracture mechanics. For the simulation of delamination growth, fracture mechanics are frequently used. It is widely accepted that conventional fiber–reinforced epoxy resins show brittle fracture behavior. Consequently, local material nonlinearity in the vicinity of the delamination front is neglected and linear elastic fracture mechanics (LEFM) can be used. Several techniques based on LEFM have been developed and are utilized successfully within the framework of the finite element method (FEM) for the simulation of delamination growth, such as crack-tip elements (Jha and Charalambides, 1998), the virtual crack extension technique, and the virtual crack-closure technique (VCCT) (Liu et al., 2011). Wimmer and Pettermann (2008) suggested a numerically efficient semianalytical approach for the prediction of delamination growth and its stability was also proposed. An alternative method that takes into account the nonlinear interface behavior at the delamination front introduces a cohesive zone (Barenblatt et al., 1962). Based on this idea, cohesive zone elements (CZE) have been developed within the FEM (Alfano and Crisfield, 2001) for the simulation of delamination.

Among the failure modes of composite structures, delamination is one of the most dangerous failures. In real applications, it is difficult to predict the behavior of delamination. The hidden nature of the growth of delamination can result in the sudden failure of composite structures. Consequently, for safer operation of composite structures, it is vital to have SHM systems on board to identify delamination both qualitatively and quantitatively.

10.2 FIBER-OPTIC SENSORS

Fiber-optic sensors (FOSs) are one of the most suitable sensors for the SHM of aircraft FRP structures. The sensors can be embedded into FRP composites during the manufacture of the composite part with no adverse effect on the strength of the part, as the sensor is diminutive in size. Among fiber-optical sensors, FBG sensors are well established in the field of SHM. An FBG sensor is suitable for networking as it has a narrowband response with wide wavelength operating range; hence, it can be highly multiplexed. As it is a nonconductive sensor, it can also operate in electromagnetically noisy environments without any interference. FOSs are made up of glass that is environmentally more stable and has a long lifetime similar to that of FRP composites. Because of its low transmission loss, the sensor signal can be monitored from longer distances, making it suitable for remote sensing (Hill and Meltz, 1997; Kersey et al., 1997). The sensor's capability of detecting stress gradients along its length can be used to identify the stress variations in the FRP composites (Hill and Eggleton, 1994); (Le Blanc et al., 1994). This can be used to detect damage in the composite structures (Okabe et al., 2000; Takeda et al., 2002).

The most recent development in the FOS field is the pulse-prepump Brillouin optical time domain analysis (PPP-BOTDA) (Che-Hsien et al., 2008). The PPP-BOTDA is capable of achieving a 2 cm spatial resolution for strain measurements. The PPP-BOTDA based system has been successfully used in various industrial applications; however, PPP-BOTDA is, so far, only able to measure static or quasi-static strain.

10.2.1 FBG SENSORS

FGBs are formed by constructing periodic changes in the index of refraction in the core of a single-mode optical fiber. This periodic change in the index of refraction is typically created by exposing the fiber core to an intense interference pattern of UV radiation. The formation of permanent grating structures in optical fiber was first demonstrated by Hill et al. in 1978 at the Canadian Communications Research Centre (CRC) in Ottawa, Ontario (Hill et al., 1978). In groundbreaking work, they launched a high intensity argon-ion laser radiation into germanium-doped fiber and observed an increase in reflected light intensity. After exposing the fiber for a period of time, it was found that the reflected light had a particular frequency. After the exposure, spectral measurements were taken, and these measurements confirmed that a permanent narrowband Bragg grating filter had been created in the area of exposure. This was the beginning of a revolution in communications and sensor technology using FBG devices.

The Bragg grating is named for William Lawrence Bragg, who formulated the conditions for x-ray diffraction (Bragg's Law). These concepts, which won him the Nobel Prize in 1915, related energy spectra to reflection spacing. In the case of FGBs, the Bragg condition is satisfied by the abovementioned area of the modulated index of refraction in two possible ways based on the grating's structure. The first is the Bragg reflection grating that is used as a narrow optical filter or reflector. The second is the Bragg diffraction grating that is used in wavelength-division multiplexing and demultiplexing of communication signals.

The gratings first written at the CRC, initially referred to as "Hill gratings," were actually a result of research on the nonlinear properties of germanium-doped silica fiber. It established, at the time, a previously unknown photosensitivity of germanium-doped optical fiber that led to further studies resulting in the formation of gratings, Bragg reflection, and an understanding of its dependence on the wavelength of the light used to form the gratings. Studies of the day suggested a two-photon process, with the grating strength increasing as a square of the light intensity (Lam and Garside, 1981). At this early stage, gratings were not written from the "side" (external to the fiber) as commonly practiced now, but were written by creating a standing wave of radiation (visible) interference within the fiber core introduced from the fiber's end.

After their appearance in the late 1970s, FBG sensors were used for SHM of composite materials efficiently for more than two decades. Recent advances in FBG sensor technologies have provided great opportunities to develop more sophisticated *in situ* SHM systems. There have been a large number of research efforts on the health monitoring of composite structures using FBG sensors. The ability to embed them inside FRP material between different layers provides a closer look at defects. The attractive properties such as small size, immunity to electromagnetic fields, and multiplexing ability are some of the advantages of FBG sensors. The lifetime of an FBG sensor is well above the lifetime of the FRP structures and it also provides the measuring of multiple parameters such as load/strain, vibration, and temperature (Kashyap, 1999).

10.2.1.1 Evolution of FBG Sensors

The Hill's gratings were made in the fiber core by a standing wave of 488 nm argon laser light (Hill et al., 1978). The grating exposure in this case was shown to be a two-photon process. Hill et al. (1978) found a 4% back reflection due to variation in refractive index. The field did not progress until Meltz et al. (1989), of United Technologies, proposed that fiber gratings could be formed by exposure through the cladding glass of two interfering beams of coherent UV light, thus exciting the 240 nm band directly by one-photon absorption.

Reliable fabrication of Bragg gratings depends on a detailed knowledge of the underlying mechanisms of photo-induced index changes. The basis of all proposed mechanisms is the ionization of GeO_2 deficiency centers that exhibit an absorption band centered at 240 nm. Hand and Russell (1990) suggested that the photo-induced index changes in optical fibers originated from the bleaching of the 240 nm band and the creation of two bands at 281 and 213 nm (Hand and Russell, 1990). Their work explained a photo-induced index change of almost 10% at visible wavelengths. However, experimental measurements show much larger index changes.

To explain the observed large index changes, Sceats et al. (1993) proposed thermo-elastic stress relaxation of the glass network caused by the formation of regions of low density around broken Ge-Si bonds (stress relief model). Alternatively, Bernardin and Lawandy (1990) suggested that UV irradiation may induce rearrangement of the molecular structure, leading to a compaction of the glass matrix; this is referred to as the *compaction model*. The fabrication of FBG

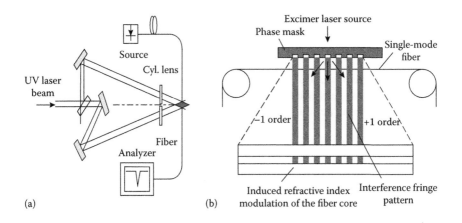

(a)

(b) Induced refractive index Interference fringe
modulation of the fiber core pattern

FIGURE 10.3 (a) Split beam interferometer. (b) Phase mask technique.

sensors has been developed to commercial level and the main techniques used are split beam interferometer, phase mask technique, and point-by-point method (Figure 10.3).

As described previously, FBG sensors are fabricated in the core region of specially fabricated single mode low-loss germanium-doped silicate optical fibers. The grating is the laser-inscribed region that has a periodically varying refractive index. This region reflects only a narrow band of light corresponding to the Bragg wavelength, λ_B, whichis related to the grating period Λ_o:

$$\lambda_B = \frac{2n_o \, \Lambda_o}{k} \qquad (10.1)$$

where:

    k is the order of the grating

    n_o is the initial refractive index of the core material prior to any applied strain

Due to the applied strain, ε, there is a change in the wavelength, $\Delta\lambda_B$, for the isothermal condition,

$$\frac{\Delta\lambda_B}{\lambda_B} = \varepsilon \, P_e \qquad (10.2)$$

where P_e is the strain-optic coefficient and is calculated as 0.793. The Bragg wavelength also changes with the reflective index. Any physical change in the fiber profile will cause variation of the reflective index. The variation of the Bragg wave length, λ_B, as a function of change in the refractive index $\Delta\delta n$, and the grating period, $\delta\Delta_0$, is given as

$$\delta\lambda_{Bragg} = 2\Lambda_o\eta\Delta\delta n + 2n_{eff}\delta\Lambda_o \tag{10.3}$$

where:

    η is the core overlap factor of about 0.9 time the shift of the Bragg wavelength
    n_{eff} is the mean refractive index change
    Λ_o is the grating period

For the Gaussian fit, the sensor reflectivity can be expressed as

$$s(\lambda,\lambda_s) = y_0 + S_0 \exp\left[-\alpha_s(\lambda-\lambda_s)^2\right] \tag{10.4}$$

where:

    y_0 is the added offset to represent the dark noise
    α_s is a parameter related to the full width at half maximum (FWHM)
    λ is the wavelength
    λ_s is the central wavelength
    S_0 is the initial reflectivity of the fiber

A major advantage of using FRP composites is the possibility of deciding on the number of layers and layup orientation based on the required structural behavior. In an FRP composite aerospace structure there are a number of layers with multiple orientations. The layers are placed one on top of the other; hence, it is possible to embed FBG sensors in any layer during the manufacturing of the structure.

10.2.1.2   Embedding FBG Sensors in FRP Structures

The process of embedding FBG sensors in FRP composites is quite complicated. The level of difficulty is largely dependent on the geometry of the part, layup configuration, and embedding location of the sensors in the part. In general, FBG sensors will be placed closer to critical sections of the structure where high-stress concentrations are predicted. However, in reality, locating FBG sensors in predicted locations is not always possible. On the other hand, reliance on a single sensor is not recommended, as it is not possible to replace failed embedded sensors after manufacturing. As a result, many FBG sensors need to be embedded in the surrounding area closer to the critical locations of the structure in order to capture strain levels reliably.

As such, multiplexed FBG sensors play a critical role in the SHM of aerospace structures. Normally in FRP, the damage starts from stress concentrations. In the process of implementing SHM systems, identification of the locations that have potential for damage is essential. Finite element analysis (FEA) techniques are being widely used to identify stress concentrations and, hence, to locate FBG sensors. It is less likely that FBG sensors are placed in simple planer structures in real applications, apart from where the requirement is mere strain rather than damage detection.

Difficulties associated with the manufacturing of composite structures with embedded FBG sensors are the main problem with placing FBG sensors in a complicated location. The aerospace industry's advanced manufacturing technologies (such

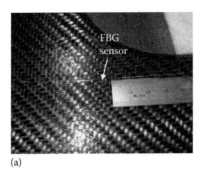
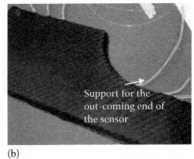

(a)　　　　　　　　　　　　　　　　　(b)

FIGURE 10.4 (a) Embedding FBG sensors and the support for the out-coming end before sending to the autoclave. (b) Cured sample from the autoclave.

as prepreg and autoclave process) create hazardous environments for brittle sensors. Every precaution needs to be taken to not apply loads on the sensor in the noncured resin matrix during the manufacturing process. With applied pressures as high as 700 kPa, even the egress ends of the sensors need to be supported to avoid breakage. It is essential to develop methods to protect FBG sensors during the FRP composite manufacturing processes. Since there is no way of replacing damaged FBG sensors after manufacturing of the component, a strict set of procedures must be developed to follow during manufacture.

Figure 10.4b shows a support given to the egress end of the sensor. Sometimes it is helpful to have an extra protective layer of rubber applied to the fiber to maximize the handling of samples without damage to the sensors.

10.2.2 PPP-BOTDA Sensors

As a novel and advanced sensing technique, FOSs have been extensively researched for SHM purposes due to their distributed sensing, rapid data transmission, small dimension, ease of installation, and immunity from electromagnetic influence. Among them, Brillouin-based sensing is one of the most useful and important sensing techniques (Horiguchi et al., 1989; Horiguchi and Tateda, 1989; Gayan et al., 2012). Investigations on the application of optic fiber sensing techniques for SHM have been widely carried out and significant progress has been made recently.

In addition to the aforementioned advantages, Brillouin scattering-based distributed FOSs show some potential merits over general FOSs for SHM that can be characterized by the following advantages. Optical fiber used as strain sensors can be installed in several and even scores of kilometers. Along the optic fiber, the sampling point can be taken every 5 cm, and thus, the strain distribution along the entire length of optical fiber can be obtained in a continuous and distributed way. The sufficient measurement can solve the problem of instability, including nonuniqueness and discontinuity of solutions that many damage detection algorithms have encountered. Through the point fixation (PF) installation of the optical fiber, there will be formed a uniform strain distribution between two bonding points, the interval between these two bonding points can vary from scores of centimeters to several meters, and the measured strain is the uniform strain between these two bonding points. Thus, the measured stain is less susceptible to local stress/strain concentration and hence more representative of the entire structural member, which is in favor of monitoring the global behavior of structures. One of the most applicable approaches for the Brillouin scattering-based distributed optical fiber sensing technology method is the Brillouin optical time domain reflectometry (BOTDR) based technique as demonstrated by Horiguchi et al. and Horiguchi and Tateda (Horiguchi et al., 1989; Horiguchi and Tateda, 1989).

Some investigations on the application of BOTDA techniques for SHM have been carried out and significant progress has been made. However, the spatial resolution of the BOTDR technique is 1 m with a strain measuring accuracy of 50 µε. Consequently, the BOTDA sensing technique is difficult to directly monitor the localized deformation (fine crack) of civil structures. In our previous investigation, loop installation of optical fiber is a good countermeasure, but this method is still not convenient for monitoring the practical structure. Recently, the newly developed PPP-BOTDA sensing technique has improved largely the spatial resolution (10 cm), and some applications of PPP-BOTDA-based distributed fiber optic sensors (DFOSs) for SHM have been carried out (Hao and Zhishen, 2008; Zhang and Wu, 2012). Meng et al. (2015) reported the detection of microcracks using this technique successfully.

10.2.2.1 Brillouin Scattering-Based Fiber-Optic Measuring Technique

As a continuous distributed sensing method, spatial resolution and strain measurement accuracy are the main concerns of the BOTDR technique. A light wave transmitted through a fiber is scattered by a nonlinear interaction with acoustic waves, which is called Brillouin scattering. Because phonons decay exponentially, the Brillouin scattered light spectrum is a Lorentzian function in form. Figure 10.5a shows an illustrative 3D Brillouin scattering spectrum describing the Brillouin scattering power (intensity) distribution at different frequencies along the fiber. Figure 10.5b shows the frequency shift of the Brillouin backscattering light at a certain location due to its corresponding strain, and Figure 10.5c shows the Brillouin scattering power (intensity) spectrum at a certain frequency along the fiber (Hao and Zhishen, 2008; Zhang and Wu, 2012).

The scattered Brillouin light frequency is shifted from the incident light frequency due to the velocity of the acoustic wave. This frequency is called the Brillouin frequency shift and it is given by the following equation:

$$v_B = \frac{2n v_A}{\lambda} \tag{10.5}$$

where:
    n is the refractive index
    v_A is the acoustic wave velocity
    λ is the wavelength of incident light

If there is longitudinal strain or temperature changes along the optical fiber, the Brillouin frequency shift v_B changes in proportion to that variety of strain or temperature.

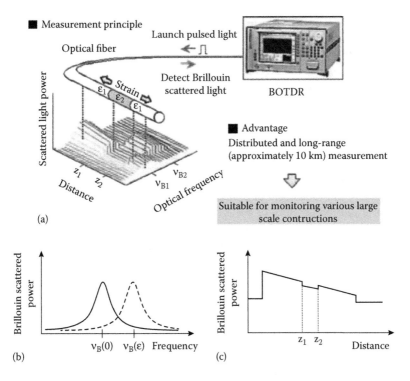

FIGURE 10.5 Measuring principle of Brillouin scattering technique. (a) Brillouin scattering power distribution at different frequencies along the fiber. (b) Frequency shift of the Brillouin backscattering light at a certain location due to its corresponding strain. (c) Brillouin scattering power (intensity) spectrum at a certain frequency along the fiber.

10.2.3 Smart Structures and Materials

A smart structure is a system containing multifunctional parts that can perform sensing, control, and actuation; it is a primitive analogue of a biological body (Cao et al., 1999). Smart materials are used to construct these smart structures that can perform both sensing and actuation functions. The "IQ" of smart materials is measured in terms of their "responsiveness" to environmental stimuli and their "agility." The first criterion requires a large amplitude change, whereas the second assigns faster response materials with higher "IQ" Commonly encountered smart materials and structures can be categorized into three different levels: (1) single-phase materials, (2) composite materials, and (3) smart structures. Many ferric materials, and those with one or more large anomalies associated with phase-transition phenomena, belong to the first category. Functional composites are generally designed to use nonfunctional materials to enhance functional materials or to combine several functional materials to make a multifunctional composite. The third category is an integration of sensors, actuators, and a control system that mimics the biological body in performing many desirable functions, such as synchronization with environmental changes, self-repair of damages, and so on. These three levels cover the general definition of smart materials and structures.

10.2.4 FIBER OPTICS IN SMART AEROSPACE COMPOSITE STRUCTURES

10.2.4.1 Damage Detection Technologies

The majority of research work on FBG sensors in the SHM of composite structures has focused on investigation of the spectra of FBG sensors embedded in the vicinity of damage. Observations of the distorted sensor spectra due to stress concentrations caused by delamination and cracks have been used to estimate the damage conditions. Many researchers have investigated purposely damaged axially loaded specimens, and the changes of FBG spectra were attributed to the damage, and successfully identified the damage (Takeda et al., 2008). In real-life situations, the applied loads are not limited to uniaxial loads; hence, the performance of FBGs in multiaxial loading situation needs to be investigated for a complete understanding of damage status. The FBG spectral response is significantly complicated under multiaxial loading conditions (Gayan et al., 2007). The distortion of FBG spectra is not only due to the accumulated damage, but also the loading types. In the previous section, it was shown that embedding FBGs between nonparallel fiber layers and the application of torque caused substantial distortions to the FBG spectra.

Even though the distortion of the response spectra of an FBG sensor has been widely used in SHM applications to detect structural integrity, unfortunately, there is still no definite method available to quantify the distortion. This has been a significant drawback in the development of SHM systems using embedded FBG sensors for decades. The quantification of distortion of the FBG sensor will allow referencing and comparison to monitor for the progressive damage status of a structure. An explicit method to quantify distortion to the sensor including self-distortion needs to be developed.

It is clear that multiple causes lead to the distortion to the FBG response spectra. Most of the effects, such as the embedment of FBG sensor only between parallel fiber laminates, parallel to the fiber orientation, and multidirectional loading on FBG sensor, cannot be eliminated in advanced aerospace applications. In order to identify damage from the distortions to the FBG response spectra, the individual effect from each effect needs to be identified and eliminated. To identify the pure effects from the damage, distinguished from the other effects, extensive computational power is required for postprocessing of the spectral data. Figure 10.6 shows FBG response spectra from an FBG embedded near a damaged location with the part under complex multidirectional loading.

As a consequence, in the laboratory environment, it is possible to discuss and interrelate the FBG response spectra with the damage by creating an artificial damage and observing the spectrum of an FBG that is embedded closer to the damage location. But in real applications, if such a spectrum is observed, it is very difficult to interpret the spectrum in order to identify the damage. The one directional accuracy, which is if there is a known damage in the structure, response spectra (distortion) of embedded FBG can be explained, but if a distorted response spectrum is observed,

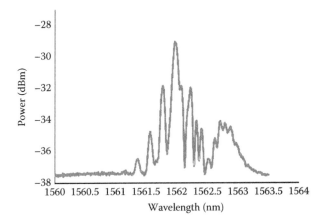

FIGURE 10.6 Distorted FBG spectra due to multiple effects.

it is not possible to identify it as a presence of damage. This incongruity made some of the SHM researchers disappointed and discouraged. There was a huge demand for an out-of-the-box approach to overcome this discrepancy.

The system discussed in the next section is a novel approach to overcome the complications previously mentioned. The approach is to develop a system to adapt to the initial conditions of the structure and to identify new conditions by comparing them with the initial condition. The response of the FBG during the undamaged states of the structure is recorded, and these recorded data are used as a "reference." Therefore, the isolation of possible "reference" data from a distorted spectrum of any embedded FBG sensor will definitely provide the subsequent distortions to the spectra caused by accumulated damage.

The main difficulty for this approach is to develop a system to reference the FBG response spectra. Historically, statistical methods such as artificial neural networks (ANNs) have been used to analyze such complicated data associated with a large number of random variables. The main advantage is the ability to train an ANN with undamaged data, and, subsequently, the trained ANN can be used to distinguish any new spectral variation. The following section introduces a signal processing technique used for damage detection in smart aerospace composite structures.

10.2.4.2 Signal Processing Technologies for Fiber-Optic Sensors

FBG sensors are very good in strain measurements, and the linear unidirectional sensitivity in the axial direction of the sensor is desirable for accurate and reliable strain readings. In such applications, the FBG sensor undergoes pure elongation or contraction, and hence, the cross section always remains in a circular shape. In multidirectional loading cases, an FBG sensor may be subjected to torsional deformations other than linear elongation or contraction. For example, when a torque is applied to a composite sample that has an embedded FBG sensor, it undergoes a twist that may cause changes to its cross section.

Another possibility of changed cross section of FBG sensors under torsional loading is due to the microbending of the grating. The embedded sensor is not always laid on the matrix and there is a possibility of laying an FBG between reinforced fibers. In the case of structure under a lateral pressure, the fiber sitting on the FBG sensor will press the FBG sensor against the fibers, causing the sensor to experience microbends. These changes of the cross section of the FBG lead to changes in the refractive index of the core material of the sensor. Since the changes are not uniform along the grating length, the refractive index of the sensor unevenly varies along the grating length of the sensor, causing distortion in the FBG spectra.

It is obvious that the distortion of FBG sensors depends on the type of loading. The effect of the twist and microbending of FBG sensors under multiaxial loading has been identified as the causes for this discrepancy. The change of section geometry of the FBG sensor due to microbending and twisting leads to a variation of the refractive index of the FBG core material that causes distortion of the FBG response spectra. Figure 10.7 illustrates a distorted FBG sensor response due to tension and torsion combined loading on the FRP panel in which the FBG is embedded (Kahandawa et al., 2010a,b; 2011; 2013b).

Even though the distortion of the response spectra of an FBG sensor has been widely used in SHM applications to detect structural integrity, unfortunately, there is still no definite method available to quantify the distortion. This has been a significant drawback in the development of SHM systems using embedded FBG sensors for decades. Quantification of distortion of the FBG sensor will allow referencing and comparison to monitor for the progressive damage status of a structure. An explicit method to quantify distortion to the sensor including self-distortion needs to be developed.

Kahandawa et al. (2013a) has proposed a "distortion index" to quantify distortion in the FBG spectrum. The distortion index is calculated related to the original spectrum before the presence of any damage (Kahandawa et al., 2013a). Distortion (Ds) is defined using the FWHM value of the FBG spectra and the maximum power of the FBG response spectrum. FWHM is an expression of the extent of a function

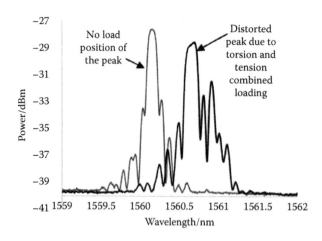

FIGURE 10.7 Distortion of the peak due to applied torsion and tension combined loading.

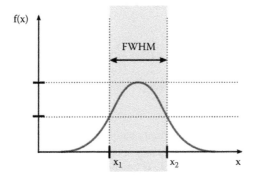

FIGURE 10.8 Full width at half maximum.

given by the difference between the two extreme values of the independent variable at which the dependent variable is equal to half of its maximum power (Figure 10.8).

Generally for a FBG response spectrum, the FWHM value increases with the distortion while the peak value decreases: $FWHM = (x_2 - x_1)$.

For the comparison, the distortion, Ds, and the distortion index, DI, need to be calculated for the same load case. Distortion of the peak at a particular load case, D_s, can be expressed as

$$D_s = \frac{FWHM}{P} \tag{10.6}$$

where P is the peak strength of the FBG sensor reflection spectra in dBm, as shown in Figure 10.9. From the calculated distortion value, the DI can be calculated.

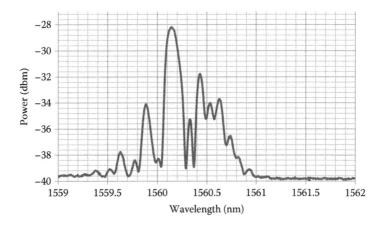

FIGURE 10.9 Peak value and the FWHM of FBG response spectra.

The DI at the same load case is

$$DI = \frac{D_{si}}{D_{s0}}$$ (10.7)

where:
   D_{si} is the current distortion
   D_{s0} is the distortion at the original condition (no damage)

If the structure is undamaged, D_{si} is equal to D_{s0} for the same load case. In that case, the DI is equal to unity. With the presence of damage, the response spectrum of an FBG sensor broadens (FWHM increases), while the peak power of the spectrum decreases. As a result, D_{si} increases, making the DI value above unity. This phenomenon can be used to identify the presence of damage in a structure.

To verify the DI, it is better to use several load cases to calculate distortion and the corresponding distortion indices. At the initial stage, the structure can be subjected to several known load cases, and the corresponding distortion, D_{s0} can be recorded. In future operations, those load cases can be used to calculate the DI in order to identify damage. In the following experiment, the DI has been calculated to investigate the relationship of DI to a growing defect.

10.2.4.3   Damage Prediction

In a laboratory condition, a crack can be simulated and the related FBG spectra can be obtained easily. In these purposely created experimental situations, the response spectra (distortion) of embedded FBG can be explained. However, the distorted response spectrum from an embedded FBG sensor is not easy to convert to the exact damage condition of the structure. This incongruity disappointed some of the SHM researchers.

10.2.4.4   Decoding Distorted FBG Sensor Spectra

During the past decade, many systems for decoding FBG spectra using fixed FBG filters have been developed. Figure 10.10 illustrates a general arrangement of a fixed FBG filter system. The system consists of a tunable laser source (TLS), fixed FBG filter, optical couplers (CPs), and a photodetector (PD). A high frequency data acquisition system (DAQ) has been used to acquire the PD voltage values.

Figure 10.10 illustrates the simplest form of this system, using only one FBG filter that is the building block of the complete decoding system. A tunable laser light, A, is transmitted to the FBG sensor and the reflected light, B, of the FBG sensor is fed to the FBG filter through an optical coupler. The intersection of the wavelengths reflected from the sensor and the wavelengths' reflected light, C, by the filter (λ), or conversely, the wavelengths that are not transmitted through the filter, are reflected to the photodetector. L_1 and L_2 are the light transmitted through the FBG sensor and filter, respectively.

While the sensor receives the total wavelength range from the tunable laser source, the filter only receives the wavelengths reflected by the sensor. Hence, the

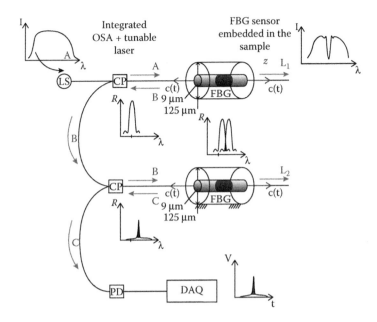

FIGURE 10.10 FBG spectrum decoding system.

filter can only reflect light (to the photodetector) if the wavelength from the sensor is within the filter's grating range (λ). The reflected light of the filter is captured using the photodetector, converted to a voltage, and recorded in the DAQ system. A system can be implemented with multiple FBG filters with λ_n wavelengths as required for a specific application.

Figure 10.11 shows the reflected spectra of the FBG sensor and the filter. The filter can only reflect light if the received wavelength from the sensor reflection is within the filter's grating range. Thus, the filter reflects the intersection as shown in Figure 10.11.

The reflected light of the FBG filter was captured using a PD and the voltage was recorded using a DAQ. Figure 10.12 shows the PD voltage in the time domain corresponding to the intersection of the spectra shown in Figure 10.11. The tunable laser sweeping frequency allows transformation of voltage reading to time domain. Since the filter spectrum is fixed, the intersection of the two spectra depends only on the sensor spectrum position. Variation of the intersection can be used to identify the location of the peak, the strain at sensor, and the damage status of the structure. Any distortion to the spectrum is visible from the PD voltage–time plot (Figure 10.12). By matching the tunable laser swept frequency with the DAQ sampling frequency, it is possible to transform voltages to their respective wavelength values accurately. More filter readings will increase the accuracy, the operating range, and the robustness of the system.

There were several attempts to fit the FBG spectra using mathematical functions such as the commonly used Gaussian curve fit (Nunes et al., 2004). Sensor reflectivity can be expressed as

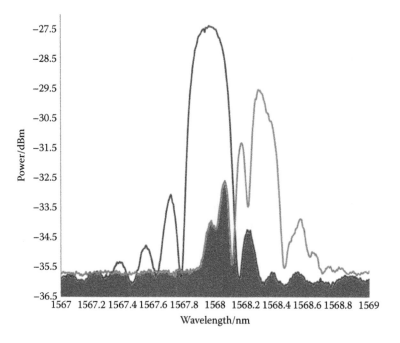

FIGURE 10.11   Intersection of the FBG spectra.

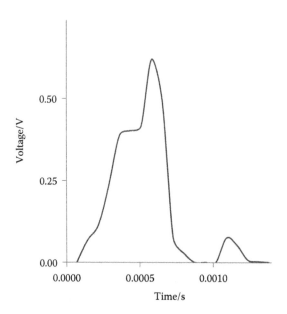

FIGURE 10.12   PD reading due to the intersection of the FBG spectra.

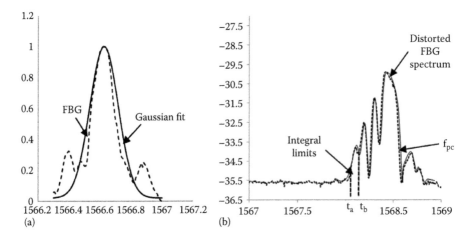

FIGURE 10.13   (a) Gaussian fit. (b) The piecewise continuous function.

$$S(\lambda,\lambda_s) = y_o + S_o \exp\left[-\alpha_s(\lambda - \lambda_s)^2\right]$$

(10.8)

where:

y_o is the added offset to represent the dark noise
α is a parameter related to FWHM
λ is the wavelength

Unfortunately, a Gaussian fit always gives an error for a distorted spectrum, as shown in Figure 10.13a. Realistically, a distorted spectrum must be considered as a piecewise continuous function, f_{pc}, in order to capture the distortion (Figure 10.13b). Consequently, the optical power, P, of the distorted signal can be obtained using the following integral:

$$P = \beta \int_{t_a}^{t_b} f_{pc} dt$$

(10.9)

where:

β is a constant dependent on the power of the source
t_a and t_b are the integral limits in the time domain in Figure 10.13b

The power integral at each point can be used to estimate the strain in the sensor by using an ANN. The sensitivity of the integrated data depends on the integral limits; larger integration limits reduce the sensitivity and very small limits cause data to scatter. Both cases make the algorithms inefficient. Optimum limit values have to be set to achieve better results.

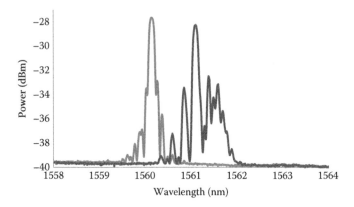

FIGURE 10.14   A typical distortion of FBG spectra.

10.2.4.5   Development of ANN-Based Signal Processing Techniques

The SHM systems used in damage detection in FRP composites must be capable of identifying the complex failure modes of composite materials. The damage accumulation in each layer of a composite laminate is primarily dependent on the properties of the particular layer (McCartney, 1998; 2002) and the loads that are imposed onto the layer. As such, the layered structure of the composite laminates makes it difficult to predict the structural behavior using only surface-attached sensors. Over the past few years, this issue has been critically investigated by many researchers using embedded FBG sensors (Takeda et al., 2002, 2003, 2008; Lee et al., 1999).

The majority of the research works were focused on the investigation of the spectra of FBG sensors embedded in the vicinity of damage loaded with unidirectional loading. However, in real-life situations, the applied loads are not limited to uniaxial loads, and hence the performance of FBGs in multiaxial loading situation needs to be investigated for comprehensive damage characterization.

The FBG spectral response is significantly complicated by multiaxial loading conditions (Sorensen et al., 2007), fiber orientation, and the type of damage present in the structure (Kahandawa et al., 2010a,b). It has been shown that FBGs embedded between nonparallel fiber layers and subjected to torque create significant distortions in the spectra (Figure 10.14).

It is clear that the cause of the distortion of FBG spectra depends, not only on the consequences of accumulated damage, but also on loading types and fiber orientation. Embedding FBGs in between nonparallel fiber layers and the application of torsional loading to the component have caused substantial distortions to FBG spectra. In order to identify damage using the response of the FBG sensor, the other effects imbedded in the response need to be identified and eliminated. The introduction of a referencing technique for the FBG spectrum using fixed wavelength FBG filters provides the capability of identifying the variations to the FBG spectrum and distinguishing the other effects causing distortions. Consequently, elimination of distortions caused by other effects will permit identification of distortions of FBG spectra caused by the damage. The proposed system is used to capture the distortions

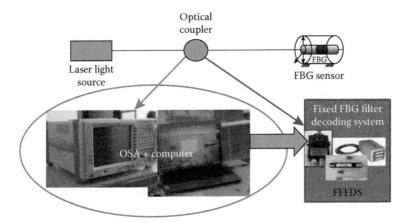

FIGURE 10.15 Replacement of OSA with fixed filter decoding system (FFFDS).

of reflected spectra of an embedded FBG sensor inside a composite laminate, thus enabling a quantitative estimate of the damage size in the vicinity of the sensor.

The aforementioned effects on the FBG spectrum along with the accumulation of damage make the response of the FBG highly nonlinear. The nonlinearity of the response varies from structure to structure; hence, the estimation of transfer functions is extremely difficult. In this scenario, statistical methods provide promising results for data processing. Among the methods available, the ANN has provided proven results for nonlinear systems with high accuracy. Application of ANNs is an efficient method for modeling the nonlinear characteristics of physical parameters and creates a system that is sensitive to wide ranges of noise.

10.2.4.6 ANN-Based Damage Detection

With the complex damage modes of composite materials and complex spectral responses of FBG sensors under complex operational loading, the damage detection in composite materials using FBG sensors becomes extremely difficult. Incorporating multiple sensor readings is also a challenging task. Further, extraction of important data and the elimination of valueless data imbedded in the response spectra of an FBG is a challenging task. Even though it is possible to avoid these complications in the laboratory environment, in real applications these are not avoidable. To overcome these difficulties, the introduced novel FFFDS with an ANN is being used (Figure 10.15).

The FFFDS uses the desirable characteristics of ANN to work and train with complications, to identify the real working environment. As a result of considering the working environment as the base (reference), the system's sensitivity to changes, such as damage, was remarkably improved. The following section introduces the field of ANN and its characteristics.

10.2.4.7 Neural Networks

ANNs are commonly referred to as "neural networks." This concept emerged while scientists were looking for a solution to replicate the human brain. In some cases, like identification and prediction, the human brain tracks the problem more efficiently than other controllers. That is because the human brain computes in an entirely different way from a conventional digital computer.

The human brain is a highly complex, nonlinear and parallel computer (information processing system). It has the capacity to organize its structure constituents, known as neurons, so as to perform certain computations many times faster than the fastest digital computer available today.

For example, in human vision, the human routinely accomplishes perceptual recognition tasks such as recognizing a familiar face embedded in a familiar scene in approximately 100–200 ms, whereas tasks of much lesser complexity may take hours on a conventional computer (Freeman and Skapura, 2007). Hence, we can say that the brain processes information super-quickly and super-accurately. It can also be trained to recognize patterns and to identify incomplete patterns. Moreover, the trained network works efficiently even if certain neurons (inputs) fail. The attraction of ANN, as an information processing system, is due to the desirable characteristics presented in the next section.

Fixed wavelength filters and a data capturing system are used to decode spectral data from an FBG sensor to a form that can be fed into an ANN in order to estimate strain and/or damage in a composite structure.

Figure 10.16 illustrates a data flow diagram of the structural strain and damage assessment process. First, reflected spectral data from the FBG sensor mounted on the structure is entered into an FBG filter. The reflected spectral data from the FBG filter (representing the spectral intersection of the reflected FBG sensor data and the inherent filter characteristics) is entered into a photodetector. The voltage time domain output of the photodetector is entered into

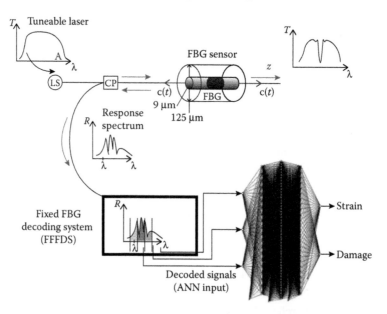

FIGURE 10.16 Decoding FBG spectrum using fixed FBGs and use of ANN.

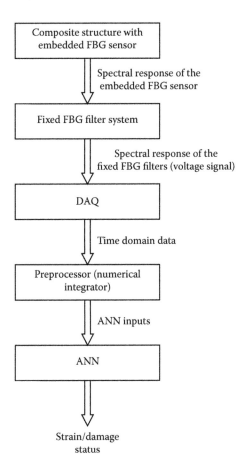

FIGURE 10.17 Flowchart of the process for a fixed FBG filter system.

a data acquisition unit. The data are then processed and entered into a neural network. Finally, the neural network output is the state of strain and/or damage (Figure 10.17).

Figure 10.18 depicts a general arrangement of the proposed system. The system consists of a tunable laser (TLS), three fixed FBG filters (filters 1–3), optical couplers (CP), and three photodetectors (PD). A high frequency data acquisition system (DAQ) was used to record the photodetector output voltages that are subsequently fed into the processor to determine the state of strain and damage.

Three filters are used to increase the accuracy, operating range, and robustness of the system, and the composite signal created by the photodetectors represents a unique signature of the sensor spectrum. Furthermore, the output of the photodetectors contains information relative to the sensor spectra in a form that can be used with an ANN. The number of filters and FBG sensors is not limited

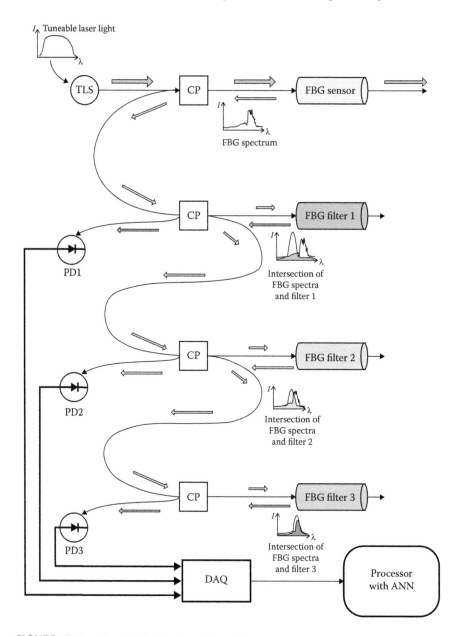

FIGURE 10.18 Fixed FBG filter decoding system.

to the example shown in this figure. Any one filter in the system is capable of covering an approximate range of 500 microstrain/1 nm movement of the peak of the sensor spectrum. More filters can be used with the same system to cover a wider operating range of the embedded FBG sensors and to obtain more precise data.

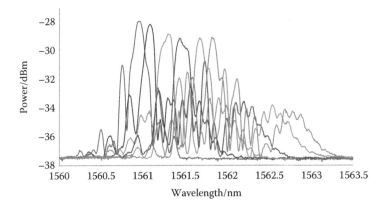

FIGURE 10.19   Distorted FBG response.

10.2.4.8   Postprocessing of FBG Data Using ANN

As discussed in earlier sections, under complex load conditions, the spectrum of the FBG distorts. Figure 10.19 shows the FBG sensor response of an FBG embedded in a composite structure during operation under several load cases. The complicated response makes it extremely difficult to model using mathematical transfer functions in order to estimate strain as well as damage. In such cases, an ANN provides promising functional approximations to the system.

One of the commonly used neural network architectures for function approximation is the multilayer perceptron (MLP). Back-propagation (BP) algorithms are used in a wide range of applications to design a MLP successfully (Lopes and Ribeiro, 2001).

10.2.4.9   The ANN Model

Figure 10.20 shows a general arrangement of the ANN used in this study. The ANN consists of three input neurons that accommodate three FBG-fixed filters, three hidden layers, and an output layer.

The neurons of the hidden layers are with Gaussian activation functions, and the k value used is 1, as shown in Figure 10.21a. It takes a parameter that determines the center (mean) value of the function used as a desired value. The neuron of the output layer is with sigmoid activation function with k = 1, as shown in Figure 10.21b. This function is especially advantageous for use in neural networks trained by back-propagation algorithms because it is easy to distinguish, and can minimize the computational capacity of training (Karlik and Olgac, 2010). Initial weights (at the start of the training process) of the neurons were randomly placed between −1 and 1.

Three different composite specimens, with embedded FBG sensors, were investigated using the developed system in order to evaluate the system performance for estimation of strain and/or damage.

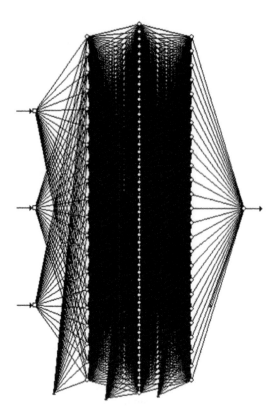

FIGURE 10.20   ANN developed to estimate strain.

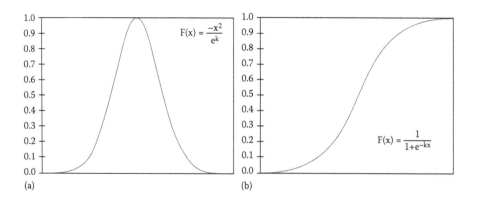

FIGURE 10.21   Activation functions of the neurons of an ANN. (a) Hidden layer neurons.
(b) Output neuron.

SUMMARY

Optical fiber sensors have a great potential for health monitoring of advanced composite structures. Recent years have shown many research and laboratory developments in this area. The superior performances and the unique advantages of the optical sensors have strongly established their place for the SHM of FRP composite structures. At this stage, the success of the SHM with optical sensors has expanded from the laboratory environment to real structures. However, to make this technology reliable and readily accessible, further research is warranted. The embedding technology, robustness of the sensors, and signal processing techniques must be critically addressed. The postprocessing of data needs to be developed with the recent advancements of statistical data analysis algorithms.

REFERENCES

Adams, R. D., Cawley, P., Pye, C. J., and Stone, B. J. (1978). A vibration technique for non-destructively assessing the integrity of structures. *Journal of Mechanical Engineering Science*, *20*, 93–100.

Alfano, G., and Crisfield, M. A. (2001). Finite element interface models for the delamination analysis of laminated composites: Mechanical and computational issues. *International Journal for Numerical Methods in Engineering*, *50*(7), 1701–1736.

Barenblatt, G. I. (1962). The mathematical theory of equilibrium cracks in brittle fracture. In H. L., Dryden, T. V., Karman, G., Kuerti, F. H. V. D., Dungen, and, L. Howarth (eds.), *Advances in Applied Mechanics*. Vol. 7, pp. 55–129, Philadelphia, PA: Elsevier.

Bernardin, J. P., and Lawandy, N. M. (1990). Dynamics of the formation of Bragg gratings in germanosilicate optical fibers. *Optics Communications*, *79*(3–4), 194–199.

Cao, W., Cudney, H. H., and Waser, R. (1999). Smart materials and structures. *Proceedings of the National Academy of Sciences*, *96*(15), 8330–8331.

Chang, F., and Chang, K. (1987). A progressive damage model for laminated composites containing stress concentrations. *Journal of Composite Materials*, *21*, 834–855.

Chang, F. K. (2003). *Structural Health Monitoring*. Lancaster, PA: DESTechnol Publications.

Che-Hsien, L. I., Nishiguti, K., and Miyatake, M. (2008). PPP-BOTDA method to achieve 2 cm spatial resolution in Brillouin distributed measuring technique. The Institute of Electronics, Information and Communication Engineers (IEICE) Japan Technical Report OFT2008-13 (2008-05).

Daniel, I. M., and Ishai, O. (2005). *Engineering Mechanics of Composite Materials*. Vol. 13. Oxford, UK: Oxford University Press.

Freeman, J. A., and Skapura, D. M. (2007). *Neural Networks, Algorithms, Applications, and Programming Techniques*. Pearson Education.

Hand, D. P., and Russell, P. S. J. (1990). Photoinduced refractive-index changes in germanosilicate fibers. *Optics Letters*, *15*(2), 102–104.

Hao, Z., and Zhishen, W. (2008). Performance evaluation of BOTDR-based distributed fiber optic sensors for crack monitoring. *Structural Health Monitoring*, *7*(2), 143–156.

Hashin, Z., and Rotem, A. (1973). A fatigue failure criterion for fiber reinforced materials. *Journal of Composite Materials*, *7*, 448–464.

Hill, K. O., Fujii, Y., Johnson, D. C., and Kawasaki, B. S. (1978). Photosensitivity in optical fiber waveguides: Application to reflection filter fabrication. *Applied Physics Letters*, *32*(10), 647–649.

Hill, K. O., and Meltz, G. (1997). Fiber Bragg grating technology fundamentals and overview. *Journal of Lightwave Technology*, *15*(8), 1263–1276.

Hill, P. C., and Eggleton, B. J. (1994). Strain gradient chirp of fiber Bragg gratings. *Electronics Letters*, *30*(14), 1172–1174.

Hinton, M. J., Soden, P. D., and Kaddour, A. S. (2004). Failure criteria in fiber-reinforced-polymer composites. *Composite Science and Technology*, *64*, 321–327.

Horiguchi, T., Kurashima, T., and Tateda, M. (1989). Tensile strain dependence of Brillouin frequency shift in silica optical fibers. *IEEE Photonics Technology Letters*, *1*(5), 107–108.

Horiguchi, T., and Tateda, M. (1989). BOTDA-nondestructive measurement of single-mode optical fiber attenuation characteristics using Brillouin interaction: Theory. *Journal of Lightwave Technology*, *7*(8), 1170–1176.

Jha, M., and Charalambides, P. G. (1998). Crack-tip micro mechanical fields in layered elastic composites: Crack parallel to the interfaces. *International Journal of Solids and Structures*, *35*(1–2), 149–179.

Kahandawa, G., Epaarachchi, J., and Wang, H. (2011). Identification of distortions to FBG spectrum using FBG fixed filters. *Proceedings of the 18th International Conference on Composite Materials (ICCM 2011)*, 1–6.

Kahandawa, G., Epaarachchi, J., Wang, H., and Lau, K. T. (2012). Use of FBG sensors for SHM in aerospace structures. *Photonic Sensors*, *2*(3), 203–214.

Kahandawa, G. C., Epaarachchi, J. A., and Lau, K. T. (2013a). Indexing damage using distortion of embedded FBG sensor response spectra. 87930K-87930K-6.

Kahandawa, G. C., Epaarachchi, J. A., Wang, H., and Canning, J. (2010a). Effects of the self distortions of embedded FBG sensors on spectral response due to torsional and combined loads. *APWSHM3*, Tokyo, Japan.

Kahandawa, G. C., Epaarachchi, J. A., Wang, H., Followell, D., and Birt, P. (2013b). Use of fixed wavelength fiber Bragg grating (FBG) filters to capture time domain data from the distorted spectrum of an embedded FBG sensor to estimate strain with an Artificial Neural Network. *Sensors and Actuators A: Physical*, *194*, 1–7.

Kahandawa, G. C., Epaarachchi, J. A., Wang, H., Followell, D., Birt, P., Canning, J., and Stevenson, M. (2010b). An investigation of spectral response of embedded fiber Bragg grating (FBG) sensors in a hollow composite cylindrical beam under pure torsion and combined loading. *ACAM 6*, Perth, Australia.

Kamiya, S., and Sekine, H. (1996). Prediction of the fracture strength of notched continuous fiber-reinforced laminates by interlaminar crack extension analysis. *Composites Science and Technology*, *56*(1), 11–21.

Karlik, B., and Olgac, A. V. (2010). Performance analysis of various activation functions in generalized MLP architectures of neural networks. *International Journal of Artificial Intelligence and Expert Systems (IJAE)*, *1*, 111–122.

Kashyap, R. (1999). *Fiber Bragg Gratings*. Vol. 1. San Diego, CA: Acadamic Press.

Kersey, A. D., Davis, M. A., Patrick, H. J., LeBlanc, M., Koo, K. P., Askins, C. G., and Friebele, E. J. (1997). Fiber grating sensors. *Journal of Lightwave Technology*, *15*(8), 1442–1463.

Kortscho, M. T., and Beaumont, P. W. R. (1990). Damage mechanics of composite materials: I—Measurements of damage and strength. *Composites Science and Technology*, *39*(4), 289–301.

Lam, D. K. W., and Garside, B. K. (1981). Characterization of single-mode optical fiber filters. *Applied Optics*, *20*(3), 440–445.

Le Blanc, M., Huang, S. Y., Ohn, M. M., and Measures, R. M. (1994). Tunable chirping of a fiber Bragg grating using a tapered cantilever beam. *Electronics Letters*, *30*(25), 2163–2165.

Liu, P. F., Hou, S. J., Chu, J. K., Hu, X. Y., Zhou, C. L., Liu, Y. L., and Yan, L. (2011). Finite element analysis of postbuckling and delamination of composite laminates using virtual crack closure technique. *Composite Structures*, *93*(6), 1549–1560.

Lopes, N., and Ribeiro, B. (2001). Hybrid learning in a multi-neural network architecture. *Neural Networks, 2001. Proceedings. IJCNN '01. International Joint Conference on, 2001. 4*, 2788–2793.

Marshall, D. B., Cox, B. N., and Evans, A. G. (1985). The mechanics of matrix cracking in brittle-matrix fiber composites. *Acta Metallurgica, 33*(11), 2013–2021.

Mccartney, L. N. (1998). Prediction transverse crack formation in cross-ply laminates. *Composites Science and Technology, 58*, 1069–1081.

Mccartney, L. N. (2002). Prediction of ply crack formation and failure in laminates. *Composites Science and Technology, 62*, 1619–1631.

Meltz, G., Morey, W. W., and Glenn, W. H. (1989). Formation of Bragg gratings in optical fibers by a transverse holographic method. *Optics Letters, 14*(15), 823–825.

Meng, D., Ansari, F., and Feng, X. (2015). Detection and monitoring of surface micro-cracks by PPP-BOTDA. *Applied Optics, 54*(16), 4972–4978.

Nunes, L. C. S., Valente, L. C. G., and Braga, A. M. B. (2004). Analysis of a demodulation system for fiber Bragg grating sensors using two fixed filters. *Optics and Lasers in Engineering, 42*, 529–542.

Okabe, Y., Yashiro, S., Kosaka, T., and Takeda, N. (2000). Detection of transverse cracks in CFRP composites using embedded fiber Bragg grating sensors. *Smart Materials and Structures, 9*(6), 832.

Puck, A., and Scharmann, H. (1998). Failure analysis of FRP laminates by means of physically based phenomenological models. *Composites Science and Technology, 58*(7), 1045–1067.

Reveley, M. S., Kurtoglu, T., Leone, K. M., Briggs, J. L., and Withrow, C. A. (2010). Assessment of the state of the art of integrated vehicle health management technologies as applicable to damage conditions. NASA/TM-2010–216911. Cleveland, OH.

Sceats, M. G., Atkins, G. R., and Poole, S. B. (1993). Photolytic index changes in optical fibers. *Annual Review of Materials Science, 23*, 381–410.

Sorensen, L., Botsis, J., Gmur, T., and Cugnoni, J. (2007). Delamination detection and characterisation of bridging tractions using long FBG optical sensors. *Composites Part A: Applied Science and Manufacturing, 38*(10), 2087–2096.

Sun, C. T. (Ed.). (2000). *Strength Analysis of Unidirectional Composites and Laminates.* Oxford, UK: Elsevier.

Takeda, S., Okabe, Y., and Takeda, N. (2002). *Detection of Delamination in Composite Laminates Using Small-Diameter FBG Sensors.* San Diego, CA: Acadamic Press.

Takeda, S., Okabe, Y., and Takeda, N. (2003). Application of chirped FBG sensors for detection of local delamination in composite laminates, *Proceedings of SPIE.* San Diego, CA, United states, 171–178.

Takeda, S., Okabe, Y., and Takeda, N. (2008). Monitoring of delamination growth in CFRP laminates using chirped FBG sensors. *Journal of Intelligent Material Systems and Structures, 19*(4), 437–444.

Tsai, S. W., and Wu, E. M. (1971). A general theory of strength for anisotropic materials. *Journal of Composite Materials, 5*, 58–80.

Wang, Y., Bartelt, H., Ecke, W., Willsch, R., Kobelke, J., Kautz, M., and Rothhardt, M. (2008). Fiber Bragg gratings in small-core Ge-doped photonic crystal fibers. Paper presented at the Optical Fiber Sensors Conference, 2008. APOS '08. 1st Asia-Pacific.

Wimmer, G., and Pettermann, H. E. (2008). A semi-analytical model for the simulation of delamination in laminated composites. *Composites Science and Technology, 68*(12), 2332–2339.

Wimmer, G., Schuecker, C., and Pettermann, H. E. (2009). Numerical simulation of delamination in laminated composite components: A combination of a strength criterion and fracture mechanics. *Composites Part B: Engineering, 40*(2), 158–165.

Zhang, H., and Wu, Z. (2012). Performance evaluation of PPP-BOTDA-based distributed
    optical fiber sensors. *International Journal of Distributed Sensor Networks*, *2012*, 12,
    414692.
Zhou, G., and Sim, L. M. (2002). Damage detection and assessment in fiber-reinforced
    composite structures with embedded fiber optic sensors-review. *Smart Materials and
    Structures*, *11*(6), 925–939.

11 Advances in FRP-Based Smart Components and Structures

Zhi Zhou, Yung William Sasy Chan, and Jinping Ou

CONTENTS

11.1 Introduction 372
11.2 Principles of FRP-Based Smart Components and Structures 373
11.3 Existing FRP-Based Smart Components.. 375
    11.3.1 Smart FRP-Based Rebars/Bars 375
        11.3.1.1 Manufacturing 375
        11.3.1.2 Properties of the Smart FRP Rebar (Mechanic,
                 Sensing, and Microstructure)... 376
    11.3.2 Smart CFRP-Based Prestressed Plate 380
        11.3.2.1 Manufacturing 380
        11.3.2.2 Sensing Performance Evaluation 380
    11.3.3 Smart FRP-Based Tube for Concrete Confinement........................382
        11.3.3.1 Manufacturing of the Smart FRP Tube 383
        11.3.3.2 Mechanical and Sensing Performance Evaluation 383
    11.3.4 Smart Bridge Cables System and Suspender................................384
        11.3.4.1 Concept and Fabrication of Smart Cable-Based
                 Components 385
        11.3.4.2 Operating Principles of Smart FRP-Based Cable.............386
        11.3.4.3 Temperature Self-Compensation of the Smart Cable 388
        11.3.4.4 Mechanical and Sensing Performance of the Smart
                 Cable..389
        11.3.4.5 Some Existing Smart Bridge Cables................................390
    11.3.5 Smart Prestressed Strand Based on FRP...390
        11.3.5.1 Monitoring Principle of the Smart Strand 392
        11.3.5.2 Sensing Properties Characterization of the Smart
                 Strand...394
    11.3.6 Smart Anchorage System 394
        11.3.6.1 FBG-Based Smart Anchorage System..............................394
        11.3.6.2 Distributed-Based Smart Anchor Rod..............................397

11.4 Applications of FRP-Based Smart Components to Structures......................400
      11.4.1 Prestress Loss Monitoring of Beam Members Based on Novel
              Smart Prestressed Strand..400
              11.4.1.1 Description of the Beam Structure and
                        Experiment Setup 401
              11.4.1.2 Experimental Results...402
              11.4.1.3 Health Monitoring of Bridge Cable in
                        Yanghe River, China..404
              11.4.1.4 Monitoring of Suspenders in Erbian Dadu River
                        Arch Bridge, China...404
              11.4.1.5 Monitoring of Tianjin Yanghe Bridge, China....................406
      11.4.2 Experimental Application of the Smart Anchor Rod 408
11.5 Discussion..409
11.6 Conclusion 411
Acknowledgments.. 411
References... 412

11.1 INTRODUCTION

Nowadays, the number of civil engineering structures (buildings, bridges, etc.) is growing significantly due to increased demand from users. Moreover, changes in external conditions (including dead loads, live loads, and weather changes), unspecified decreases in material properties, as well as the mechanism of aging effects, could lead to damage (small or large) to structural components. One of the greatest problems encountered in any civil construction is the damage induced by the "corrosion mechanism" of certain conventional structural materials, such as the use of steel reinforcement in concrete. Corrosion of steel can lead to the destruction of its surrounding environment, leading to the reduction of material properties. One of the habitual facts is the degradation of reinforced concrete (RC) structures due to reinforced steel corrosion spreading into the concrete area. Fiber–reinforced polymer (FRP) composites, including carbon, aramid, and glass fiber, have been introduced in aerospace for aircraft production. A few decades later, the use of FRP material as an upgrade for steel corrosion problems took place in civil engineering applications. FRP composites have great advantages over other traditional materials, including very lightweight, good corrosion resistance, high strength in the fiber direction, and so on. Therefore, the use of FRP composites became worldwide in civil engineering to replace certain reinforcement materials or to upgrade deteriorated structures [1]. Additionally, countries, including China, are seeking a novel construction based on a prefabrication method. FRP composites are very suitable for this type of structure to reflect the "green construction" requirement. Moreover, FRP material can be manufactured in different profiles by the pultrusion process or other methods. Like other structural materials, FRP composites present some drawbacks, limiting their applications in some civil engineering structures. These disadvantages are related to different parameters, as cited by Hollaway [2]. Thus, the self-sensing in real-time approach could solve the problems of the applications of FRP as a replacement for conventional structural components for modern construction purposes.

On the other hand, structural health monitoring (SHM) is a tool for ensuring the health state of structures. SHM implements the use of different sensor systems by installing the sensors in structures in different ways for the purpose of damage detection. However, the difficulties of installing sensors in some structures, due to lack of space and the large strain developed, have restricted the use of some conventional transducers. Additionally, some sensors, such as optical fiber sensors, are very fragile. Therefore, strong protection is needed to avoid any causes that could harm the lifespan of the detecting system. Some transducers are strong enough to resist a certain peak of strain when used in harsh environments; however, they are not economical or suitable for applications in confined spaces, and their measurement is less accurate. According to an investigation of optical fiber Bragg grating (FBG) sensors by Ou and Zhou [3], it has been demonstrated that FRP composites could be used instead of metal or epoxy for protective packaging of FBG sensors.

Based on the advantages of small size and low cost of fiber-optic sensors, and the benefits of using FRP materials, some researchers have developed smart components and structures to detect the health of the components, and of course, the stability state of the structure. On the basis of what has been mentioned in this section, researchers have accomplished and tried to improve the use of FRP composites as smart components and structures to replace conventional structural elements as well as to improve the detection of structural damage. This is a very important claim, because the impact of structural defects and the need for a structural monitoring method in real applications are increasing.

This chapter gives a basic overview of smart components and structures based on FRP and optical-fiber (OF) sensors, which will enable readers to understand the development and proposed uses for smart components based on FRP and OF sensors. Following this, some of the smart components existing at present, including smart rebars, intelligent prestressed carbon fiber–reinforced polymer (CFRP) plates, smart anchors, and so on, are explained in detail, and their sensing and mechanical performance is characterized. Moreover, some of the applications of the smart components, in real structures or experimentally, are also described briefly. Finally, a short and clear discussion section further develops the context of smart structures based on FRP and OF in civil engineering applications.

11.2 PRINCIPLES OF FRP-BASED SMART COMPONENTS AND STRUCTURES

The basic principles of FRP-based smart components and structures are introduced briefly in this section, followed by a concrete schematic process of the formation of smart components based on OF sensors and FRP composites. In general, all structures and components involving the integration of sensors and/or actuator systems are classified as "intelligent structures." These intelligent structures are able to perform self-SHM and/or provide an actuating response without human intervention. Therefore, the concepts of an intelligent structure are based on safety and economic improvement concerns, weight and time savings, as well as sensing and auto-control abilities. Since the appearance of SHM, many researchers have tried to discover and develop the possibilities of structures for self-sensing and self-healing abilities. For example, a smart aggregate-based active sensing system was introduced by Yan et al. [4] to evaluate the severity of damage in a concrete shear wall. Kuang and Ou [5] proposed a smart self-healing RC beam, made of superelastic shape memory alloys (SMA) and containing fiber-optic sensors for temporary crack repair. Nevertheless, there are still significant problems in the application of such smart structures to the existing innovations proposed by many authors. For example, the proposed smart RC beam requires electric or thermal power to activate the smart actuator's SMAs. The principles of smart components based on FRP and OF sensors rely on the principles and advantages of OF sensors and the high-performance properties of FRP. In addition, the potential use of FRP to replace existing key components in structures would contribute to the principles of smart structure based on FRP and OF transducers. Using smart components in structures results in smart structures with sensing performance based on the smart components.

OF sensors have been used since their discovery in the telecommunication industry. Fiber-optic sensors, especially FBGs, have also been recognized as reliable tools to detect strain and temperature changes of their surrounding environment or the structure to which they are bonded. The application of OF sensors to detect structural change could be achieved either by bonding optical sensors on the surface by using some special glue, or by packaging the FBG sensors before pasting or embedding the packaged sensor into the structure for damage detection. Many methods have been proposed by researchers to package optical sensors. Biswas et al. [6] investigated the use of FBG sensors bonded at the surface or integrated into concrete with stainless steel or epoxy resin encapsulation techniques. Moreover, Chan et al. [7] used proof acrylonitrile butadiene styrene (ABS) enclosures for the Tsing Ma Bridge in Hong Kong to protect the FBG sensors from moisture and dust. However, these methods seem to be inconvenient for durable health monitoring due to the fragility of glass fiber, the difficulties of sensor installation, and the corrosive effects of certain materials. On the other hand, OF sensors (distributed and FBGs) have been accepted as a sensing tool allowing local and global structural health monitoring. When local and global OF sensors are combined, a self-sensing network ability has been demonstrated by Zhou et al. [8] using different couplers and network schemes.

Therefore, the lack of corrosion and good mechanical properties of FRP, as well as the sensing ability of OF sensors result in the development of up-to-date smart components. The smart component solves the issues encountered in civil engineering construction and realizes the dreams of smart structures. The basic principles of the formation of the smart components mentioned so far are illustrated in Figure 11.1.

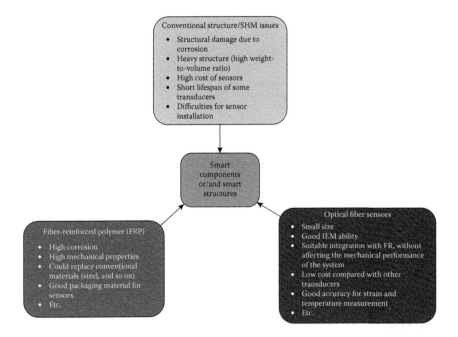

FIGURE 11.1 Principles of smart components/structures based on FRP and OF sensors. (Adapted from Sasy Chan, Y.W. et al. *Pacific Science Review*, 16: 1–7 (2014) [9].)

11.3 EXISTING FRP-BASED SMART COMPONENTS

11.3.1 Smart FRP-Based Rebars/Bars

Unidirectional FRP bars have been produced almost everywhere to replace the conventional steel bars used for RC structures. However, the disadvantages of low transversal strength of FRP rebars and the nonplastic behavior of the unidirectional composite limit the applications of these continuous fiber composites when subjected to high transversal deformation. Therefore, advanced smart FRP rebars were produced in the Harbin Institute of Technology (HIT) for structural safety improvement, as well as reinforcing materials. The smart FRP rebar could be a distributed or a local sensor, or even a combination for both local and global sensing.

11.3.1.1 Manufacturing

The smart FRP rebar system consists of a built-in OF sensor (FBG or distributed fiber) pultruded with raw fibers during the pultrusion process. Figure 11.2 shows a sketch of the pultrusion machine and the process for the production of the smart rebar. Optical fiber sensors and impregnated unidirectional roving fibers are pulled together through the pultruded machine, passing through a die mold to form the final product. During the manufacturing process, optical devices such as Brillouin optical time domain analysis (BOTDA), Brillouin optical time domain reflectometry (BOTDR), FBG demodulator, and optical time domain reflectometry (OTDR) are connected to the OF sensors being pulled out for the fabrication quality control

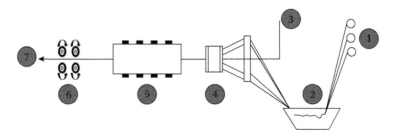

FIGURE 11.2 Schematic fabrication procedures of smart pultruded FRP bars. 1, Fibers; 2, resins; 3, optical fiber sensors and devices; 4, die; 5, heater; 6, pulling devices; 7, final smart FRP bar product.

process. The use of these optical devices attempts to control any defects of production induced by OF breaks during pultrusion. Other methods for the manufacturing of a self-sensing FRP bar involve using a protective sheath to cover the OF sensor before any pultrusion is done [10]. However, this method seems to be more complicated and could affect the strain transfer between the outside FRP and the OF sensor bar. Some OFBG-FRP-based smart rebars produced at HIT are shown in Figure 11.3a.

11.3.1.2   Properties of the Smart FRP Rebar (Mechanic, Sensing, and Microstructure)

The mechanical and sensing properties of OFBG-FRP bars are measured by a simple tensile test conducted with the OF bar and the OFBG-FRP bars simultaneously. The overall diameter of the smart bar is around 6 mm, of which the OF sensor bar represents only 2% of the FRP. The tensile test could be conducted using the universal tensile test machine MTS or using a hydraulic pressure system fixed at the steel shelf, by pulling out the smart FRP bar with one end fixed. Extensometers are fixed to the specimens to compare the results of strain obtained from the OF sensors being embedded into the FRP bars. The results of previous tests on smart FRP-based bars are shown in Figure 11.3b, which describes the strain sensitivity of each smart system. According to the output analysis of the data from the extensometers, good linearity and sensitivity could be obtained with a value similar to the OF bar, as well as the FBG sensor. The values of strain sensitivity are 1.21 and 1.19 pm/με for the CFRP-OFBG and glass fiber–reinforced polymer (GFRP)-OFBG bars, respectively. Additionally, the coefficients of correlation of basalt fiber reinforced polymer (BFRP)-OF-FBG, CFRP-OFBG, and GFRP-OFBG bars were determined to be 0.99973, 0.99968, and 0.99965, respectively. Therefore, from both coefficients of sensitivity and correlation values, we can conclude that the novel smart FRP composite bars have good linearity compared with the bare OFBG for strain monitoring.

The microstructure properties of the smart bar have been also studied for efficiency monitoring of the FRP bars with OFBG. Figure 11.4 shows the images from scanning electronic microscopy (SEM) of the microstructures of both GFRP and CFRP smart bars.

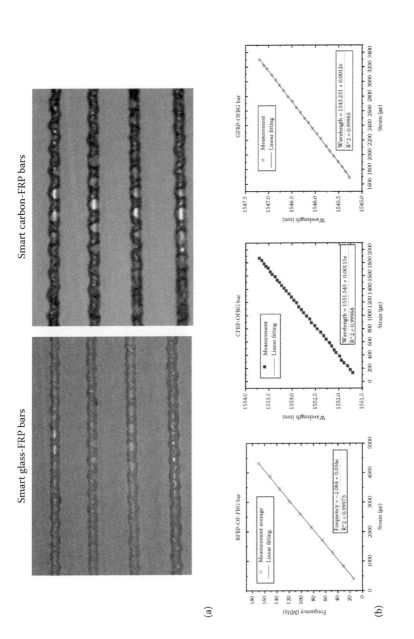

FIGURE 11.3 (a) Prototypes of some smart FRP reinforcement bars and (b) their strain sensitivity studies.

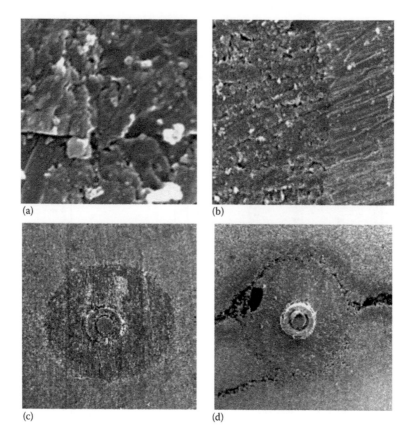

FIGURE 11.4 SEM of (a) bare FBG and GFRP; (b) bare FBG and CFRP; (c) uncracked interface between CFRP-OFBG surface and FRP; and (d) cracked interface between GFRP-OFBG and FRP.

As the wavelength shifts and the Brillouin frequency shifts of OF sensors (FBG and distributed OF) are sensitive to temperature change, temperature sensing tests of the novel smart bar were done in a low-temperature bath filled with alcohol and water. Based on the data from previous tests, the temperature sensitivity of each smart FRP-based bar is described in Figure 11.5. For the BFRP-OF-FBG, self-sensing is possible. According to the results displayed in Figure 11.5, the temperature sensitivities of the smart FRP bars are linear, with a linear fitting of 0.99643, 0.99944, and 0.99979 for each bar type, respectively.

The smart FRP bar is one of the most useful in civil engineering applications due to the high tensile strength of the unidirectional FRP composite. The applications of the smart bar are mainly in RC structures to replace conventional steel reinforcement. Moreover, the smart bar has the potential to detect strain and temperature in concrete structures. With the advances in smart FRP bars, the problems of corrosion and strain monitoring in RC are overcome. Actually, most contractors would prefer the use of smart FRP bars to conventional steel reinforcement due to their self-sensing ability and better performance in tensile strength.

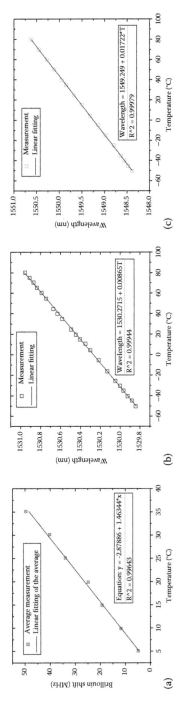

FIGURE 11.5 Temperature sensing of smart FRP-based bars: (a) BFRP-OF-FBG bar; (b) CFRP-OFBG bar; (c) GFRP-OFBG bar.

11.3.2 SMART CFRP-BASED PRESTRESSED PLATE

External strengthening of structures that have deteriorated as a result of earth-quakes and aging damage is one of the most interesting methods for rehabilitation and improving the strength of structures in civil engineering nowadays. Many external strengthening methods have been proposed by Mostofinejad et al. [11] and Wang et al. [12], including external bonded reinforcement (EBR), nonsurface-mounted (NSM), CFRP-based plates, and so on. However, CFRP-based plates are the most popular for external strengthening use due to their property of high-strength carbon in the longitudinal direction. The lightweight property of carbon fiber gives the attractive possibility of the structure being reinforced by such materials without changing the architecture of the structural members. Much research has been done on the behavior of structures being strengthened with this technique, as well as its advantages and drawbacks. Literature reviews and real examples reveal that structures strengthened with CFRP plates show a great improvement in bearing capacity. In addition, the technique is very inexpensive compared with a new replacement, very rapid, and simple to perform. However, the sudden failure property of carbon fiber (elongation at break ≤1.5%) and the damage of strengthened beams with carbon fiber need to be monitored in real time to avoid any major failure of strengthened structures and human losses. Therefore, a smart CFRP-based prestressed plate is proposed for self-sensing of damage in strengthened members by embedding OF sensors during the pultrusion process.

11.3.2.1 Manufacturing

To produce the smart CFRP-based plate, high-strength carbon-fiber and FBG sensors should follow the same production process as described in Figure 11.1, similar to the FRP rebar production using a pultrusion process. Only the shape of the mold used is different. FBG sensors are aligned at the middle of the plate along its length, with a 30–50 cm space between them. Raw carbon fiber Toray (T700-12 K) made in Japan, with a diameter ranging from 0 to 8 μm, is used. Figure 11.6 shows the industrial production steps and the final product prototype of the smart carbon plate made by ZhiXing FRP Reinforcement Nantong Co., Ltd, Nantong 226010, China.

11.3.2.2 Sensing Performance Evaluation

To investigate the sensing performance of the smart CFRP plate, calibration tests under loading and unloading of specimens were performed by Wang et al. at Nantong ZhiXing Laboratory in Nantong, China [13]. The characteristics of each specimen are summarized in Table 11.1. The strain readings of loading and unloading from optical device demodulators are collated to plot each calibration curve for each specimen. From these data, we are able to redraw each specimen calibration linear curve and determine the coefficient of correlation, as in Figure 11.7. This can be achieved using Origin 8.0 Software for comparative study. From linear fitting graphs, the coefficients for each specimen are listed in Table 11.1. Additionally, the repeatability, hysteresis, and linearity characteristics of the intelligent CFRP plate were also studied using the data from loading–unloading cycles. The calculated

Step 1: Processing of carbon fiber in the yarn

Step 2: Resin tank process of the carbon fiber

Step 3: Placing the FBG optical fiber sensor into the hole

Step 4: Preparation of the traction device

Step 5: Pulling process

The final prototype of the smart CFRP plate

FIGURE 11.6 Manufacturing process and prototype of the smart carbon plates from ZhiXing. Ltd, Nantong, China.

TABLE 11.1
Specimen Parameters

Specimen Name	Initial Wavelength of FBG (nm)	Specifications (Width × Thickness) (mm)	Load (kN)	Coefficient of Correlation (R^2)
SC-1	1524.727	50 × 3	100	0.99854
SC-2	1546.617	50 × 3	120	0.99835
SC-3	1547.565	100 × 1.4	150	0.99974
SC-4	1522.016	100 × 1.4	100	0.99865

values for each sensing performance parameter are listed in Table 11.2. From the sensing parameters of all specimens seen in Table 11.2, we observe that the sensitivity range value is 1.227–1.276 pm/µε, the linearity <1.5%, the repeatability <0.9%, the hysteresis <0.4%, and the overall accuracy <2%. According to these parameter values, an intelligent CFRP plate based on OF sensors and carbon fiber could be used as strengthening materials, as well as sensors for civil engineering applications. RC beams are well suited for the smart CFRP plate. Additionally, the novel intelligent plate could be also used prestressed due to the high strength of carbon fiber longitudinally.

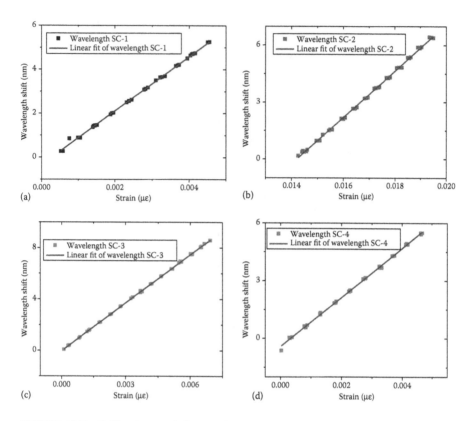

FIGURE 11.7 Calibration graph for each intelligent carbon specimen: specimens (a) SC-1, (b) SC-2, (c) SC-3, and (d) SC-4.

TABLE 11.2

Sensing Performance Parameters of the Intelligent CFRP Plate

Specimen Name	Coefficient of Correlation (R²)	Sensitivity (pm/με)	Linearity (%)	Repeatability (%)	Hysteresis (%)	Overall Accuracy (%)
SC-1	0.99854	1.230	0.318	0.622	0.197	0.726
SC-2	0.99835	1.227	1.290	0.716	0.380	1.700
SC-3	0.99974	1.250	0.708	0.894	0.192	1.160
SC-4	0.99865	1.276	0.897	0.742	0.208	1.183

11.3.3 SMART FRP-BASED TUBE FOR CONCRETE CONFINEMENT

FRP tubes based on pultrusion process are classified among the FRP profiles in the composite industry. FRP tubes have been used for modular construction based on the assembly method. In addition, FRP tubes are commonly used in composite structures such as concrete-filled tube columns, beams, and prefabricated deck panels. Tubes used for confinement concrete could be made using the pultrusion or wrapping (hand-layup) process. Recently, an FRP tube has been advantageously used to replace steel tube in confined columns due to the good corrosion properties of FRP. Moreover, structural members based on FRP tube confinement are well confined due to the fiber being oriented in the hoop direction. However, the well-known transversal weakness of FRP and its sudden failure property are crucial for key structural members such as columns. Therefore, to monitor the strain developed in the FRP tube used for confined concrete or for other applications, researchers have proposed the use of OF sensors, replacing the unreliable piezo-electric (PZT) sensors that have been used for many years. In this section, we will describe the manufacturing process of smart tube used in concrete confinement and its mechanical and sensing characteristics.

11.3.3.1 Manufacturing of the Smart FRP Tube

FRP tubes used for confinement concrete could be made in different ways according to the shape of the members and the ability of the producers. The preferable methods are wrapping cured concrete, filament winding, which requires a filament winding machine, and the pultrusion method. Most research has adopted the wrapping method, which consists of using impregnated carbon, glass, or aramid unidirectional sheets, orienting the fiber in the hoop direction. However, this method is more complicated to realize in practice. The second easiest way to produce a smart FRP tube is "filament-winding." The filament winding process can be done manually, automatically, and semiautomatically. Fibers are wound at 90° to the longitudinal axis of the tube around a steel mold. OF sensors are embedded between the second and third plies during fabrication. The position of the OF sensors (if they consist of FBG) is well calculated to fit at the right position where we want to monitor the hoop strain of the tube. The filament-wound smart FRP tube is passed through an oven at 80°C for 3 h for the curing and shaping steps. Finally, when the wound tube is sufficiently cured, the steel molds are removed at room temperature from the FRP tube. Hence, a final intelligent tube product is obtained, which could be used for concrete column confinement as well as to produce a self-sensing concrete-filled FRP tube.

11.3.3.2 Mechanical and Sensing Performance Evaluation

To evaluate the performance of the smart tube [14], the filament-wound FRP tube was used to manufacture self-sensing concrete-filled specimens. E-glass fibers were used for the smart tube, with five plies of 0.32 mm/ply thickness. An equivalent concrete mixed to ASTM C150/Type I with a water/cement ratio of 0.39 was used. During the concrete casting, the smart prefabricated FRP tubes are placed on a plate surface and could work as formwork. The specimens were cured at 20°C till concrete strength at 28 days was achieved. A mechanical test to evaluate the behaviors of the novel concrete-filled FRP tubes was set up using an MTS Systems Corporation test machine with a capacity of 2500 kN. An FBG interrogator device from Micron Optics Inc. was used to record the wavelength changes of the FBG sensors. Three electrical strain gages were bonded on the surface of the FRP tube at the same positions as the FBGs for comparative purposes.

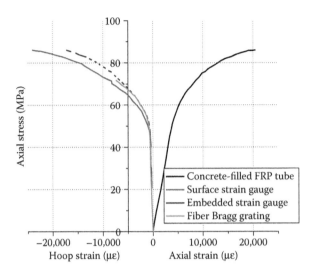

FIGURE 11.8 Stress–strain curve of concrete-filled FRP tube in compression.

The concrete-filled FRP tubes failed in brittle mode due to the rupture of the FRP tube at the middle height and in the hoop direction. All data from OF sensors and strain gages are used to plot the stress–strain curves of the concrete-filled FRP tube in compression, shown in Figure 11.8. From the curves, we could obviously confirm that the strength of the concrete-filled FRP tube was almost 90% higher (45.2–85.9 MPa) compared with the unconfined concrete. Moreover, the FBG sensors record the hoop strain of the concrete-filled FRP tube more reliably compared with the electric resistance strain gages. In addition, the maximum strain measured by the FBG sensor (7300 με), even smaller than the ES gages, does not exceed the strain normal range of the structures.

Therefore, the smart FRP tube was successfully used in confined concrete columns to evaluate the hoop strain of the specimen before the rupture of the FRP tube occurred.

11.3.4 SMART BRIDGE CABLES SYSTEM AND SUSPENDER

Bridge cables are the most important components used for suspension bridges, arch bridges, cable-stayed bridges, and other cable-based structures. In practice, cables are challenged by very harsh condition due to environmental factors such as rainfall and wind, provoking the corrosion of the cables. Additionally, fatigue, material aging, and stress redistribution might also harm the structural functionalities of bridge cables. Empirical formulas and conventional transducers (elasto-magnetic sensors, acoustic emission (AE) sensors, vibration sensors, and so on) have been used to measure the local forces and losses around the cable anchor area [15]. However, these methods are not suitable now that bridges are being built with increasing spans, which results in the use of longer cable. It would be necessary to install a reliable sensor to evaluate the evolution of stress–strain developed in the cable system along its length, preventing early damage and collapse or total failure of the structure. To replace conventional sensors, Zhou et al. [16] have proposed using a combination of the advantages of FBG sensors and the benefits of FRP materials to fabricate smart bridge cable components. The first FRP-OF-based bridge cable component was made of FRP-OFBG rebar mounted with either FRP-based cable or steel wire–based cable [16]. The smart bridge cables are then used to assess the strain at the bridge cables. However, the progress of long-span bridge construction makes the FRP-OFBG-based cable useless, requiring a large number of FBG sensors to evaluate the strain along the cable due to the FBGs being local point sensors. The performance based on strain and temperature measurement of the cable-based bridges along its length could be better evaluated by using a combination of distributed-measurement OF sensors (BOTDA/R) and a local FBG measurement for accuracy improvement of the smart FRP-based cable [17].

11.3.4.1 Concept and Fabrication of Smart Cable-Based Components

The most common technique used to fabricate smart cables to date is the smart FRP rebar (FRP-OFBG rebar or FRP-OF-FBG rebar) developed by HIT in 2003. Favorable methods for installation of a smart FRP-based bar are (1) replacing one or some of the FRP rebars forming the FRP-based cables, (2) adding or substituting some of the external wires of the steel-based cables with smart FRP rebar (the external choice avoids the crushing of the smart FRP bar induced by the pressure generated by the steel wire when the cable is twisted), or (3) installing the smart FRP bar externally along the cable length. The basic manufacturing procedures for smart cables are shown in Figure 11.9. The step of fabrication might start

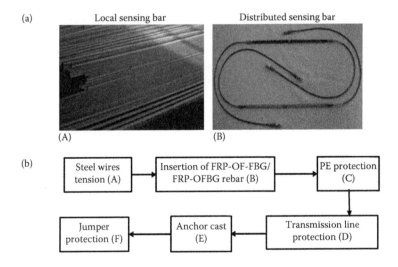

FIGURE 11.9 Smart cable bridges. (a) Existing types of FRP-OFBG bars installed in smart cables: (A) local sensing bar; (B) distributed sensing bar. (b) Diagram of the fabrication process for smart cable components.

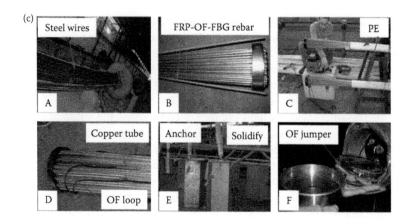

FIGURE 11.9 **(Continued)** Smart cable bridges. (c) Pictures of the fabrication process of the smart cable components.

with the manufacturing of the smart FRP rebar if the latter does not exist yet or when a specific case of sensor installation is needed. The strain developed in the FRP smart bar is basically assumed to be equal to the strain in each steel wire. The FRP-OFBG bars used to fabricate the smart bridge cable are shown in Figure 11.9.

11.3.4.2 Operating Principles of Smart FRP-Based Cable

The basic method of monitoring a cable bridge's performance consists of measuring the force being developed at the cable. Therefore, a set of equations is given here to briefly enable readers to understand the relationship between the cable force and the strains at the cable wires, as well as the strains at the OF sensors embedded in the FRP rebar.

For more understanding of the concept of the smart cable, a schematic illustration of the combined BOTDA/R and FBG smart-cable system is shown in Figure 11.10. From Figure 11.10, we can observe in more detail the sensing system incorporated

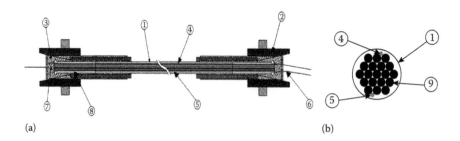

FIGURE 11.10 Schematic illustration of the smart FRP-OF-FBG cable. (a) Longitudinal view; (b) cross section of the cable. PE (1), anchor cup (2), mixture of epoxy-iron (3), FRP-OF-FBG1 (4), FRP-OF-FBG2 (5), OF jumper (6), fiber loop (7), copper tube (8), and steel wires (9).

into the cable, which enables both global and local measurement of strain developed along its length.

Basically, the force developed at the cable could be evaluated from the strains of the cable wires and the embedded FRP-OF rebar. The cable force is then given by the sum of the force in the steel wires added to the force of the smart FRP rebar, as seen in

$$F = F_s + F_{FRP} = n.E_s.\varepsilon_s.A_s + E_{FRP}.\varepsilon_{FRP}.A_{FRP} \tag{11.1}$$

where:

n	is the number of wires in the cable
F_s and F_{FRP}	are the forces on the steel wires and the FRP-OF-FBG rebar, respectively
E_s and E_{FRP}	are the Young modulus of steel and FRP, respectively
ε_s and ε_{FRP}	are the strain applied to steel and FRP-OF, respectively
A_s and A_{FRP}	are the cross-sectional areas of steel wire and FRP-OF-FBG rebar, respectively

The friction force and slippage between the cable wires and the FRP rebar are not considered in calculating the cable force. Therefore, the FRP rebar deforms coherently with the steel wires, which results in an equal value of strains seen in the cable wires and the FRP rebar ($\varepsilon_s = \varepsilon_{FRP}$). Considering the values of elastic modulus and the diameter of steel wire and smart FRP rebar in bridge cables as 210 GPa, 50 GPa, 7 mm, and 5 mm, respectively, we could write the relationship between the forces on cable wires and FRP rebars as

$$\frac{F_{FRP}}{F_s} = \frac{25E_{FRP}}{n.49E_s} = \frac{125}{1029n} \tag{11.2}$$

Equations 11.1 and 11.2 result in

$$F = F_s + F_{FRP} = \left(1 + \frac{125}{1029n}\right)E_s \varepsilon A_s n \tag{11.3}$$

in which $\varepsilon = \varepsilon_s = \varepsilon_{FRP}$.

Equation 11.3 can be simplified as

$$F = \varepsilon k \tag{11.4}$$

where $k = \left(1 + \dfrac{125}{1029n}\right)E_s A_s n$.

The cable force (F) can be determined by the strain increment method if small strain is assumed. Therefore, we have

$$F = F_0 + \Delta F = F_0 + k\Delta\varepsilon \tag{11.5}$$

where:
 F_0 is the initial force at the cable after installation
 $\Delta\varepsilon$ is the strain increment due to applied force after installation

Assuming that the cable is under constant temperature conditions, the increment values of wavelength and frequency shifts are proportional to the increment strain. By replacing the strain increment ($\Delta\varepsilon$) with the rapport of the changes in wavelength to the sensitivity coefficient of optical sensors, results in

$$F = F_0 + k\frac{\Delta\Psi}{K_\varepsilon} \tag{11.6}$$

where:
 $\Delta\Psi$ is the incremental wavelength or frequency shift
 K_ε is the sensitivity coefficient of FBG or OF

Thus, the cable force can be obtained from the following equations:

$$F = F_0 + \frac{\Delta\Psi}{c} \tag{11.7}$$

where c is the cable force sensitivity given by

$$c = \frac{K_\varepsilon}{k} = \frac{K_\varepsilon}{(1+125/1029n)E_sA_sn} \tag{11.8}$$

Equation 11.8 describes the sensitivity of the bridge cable with specified parameters (material properties of the FRP and steel wires used, as well as its diameter). Therefore, we can conclude from Equation 11.8 that the cable sensitivity depends on the size and material properties of the design.

To determine the force developed at the bridge cable based on smart FRP bars, Equations 11.1 through 11.8 can be used for either FRP-OFBG- or FRP-OF-FBG-based cable systems.

11.3.4.3 Temperature Self-Compensation of the Smart Cable

Generally, both FBG and OF sensors reveal strain and temperature information corresponding to wavelength and frequency shifts, as described in

$$\begin{pmatrix} \Delta\lambda_B \\ \Delta\nu_B \end{pmatrix} = \begin{pmatrix} K_\varepsilon^{FBG} & C_T^{FBG} \\ K_\varepsilon^B & C_T^B \end{pmatrix} \begin{pmatrix} \Delta\varepsilon \\ \Delta T \end{pmatrix} \tag{11.9}$$

where:

$\Delta\lambda_B$ and Δv_B are the changes of Bragg wavelength and Brillouin frequency

$\Delta\varepsilon$ and ΔT are the changes in strain and temperature

K and C are the strain and temperature sensitivity coefficients of OF sensors

From Equation 11.9, we can obtain the strain and temperature applied to the system, as described in

$$
\begin{pmatrix} \Delta\varepsilon \\ \Delta T \end{pmatrix} = \begin{pmatrix} K_\varepsilon^{FBG} & C_T^{FBG} \\ K_\varepsilon^{B} & C_T^{B} \end{pmatrix}^{-1} \begin{pmatrix} \Delta\lambda_B \\ \Delta v_B \end{pmatrix}
\tag{11.10}
$$

The temperature self-compensation ability of the smart FRP-OF-FBG cable system was demonstrated by a test in a tank filled with water [17]. The test results affirmed that a maximum error of 9% is found for strain calculated from absolute temperature compensation by OF. However, the strain measured by OF-FBG with temperature self-compensation agrees well with the reference strain sensor. Therefore, we can conclude that the *in situ* cable force can be accurately measured by FRP-OF-FBG rebar with temperature self-compensation. Moreover, the smart sensor is inexpensive compared with the conventional sensors used years ago.

11.3.4.4 Mechanical and Sensing Performance of the Smart Cable

A series of specimen tests under high stress were conducted by He et al. [17] to investigate the mechanical and sensing performance of the smart-cable bridge system. Thirty smart FRP-OF-FBG-based cables provided by Jili Group Inc., China were used to calibrate and check the sensing properties of the cables. The test setup is shown in Figure 11.11. An FBG interrogator, a BOTDA/R instrument, and an optical switch/coupler were used as an interrogation system to gather information on wavelength shifts and frequency shifts from the OF embedded into the cables. The FBG interrogator device has a sample frequency of 5 Hz and a measurement precision ranging from 1 to 2 µε, while the DiTeSt STA200 has a spatial resolution of 0.5 m with an accuracy of ±20 µε (for strain measurement) and 0.5°C (for temperature measurement). The test results demonstrated a coefficient of linearity of 0.99973 and 0.99954 for FBG and BOTDA/R cable, respectively. Additionally, a variation of strain measurement ranging from +15 µε to +25 µε was obtained from both sensors (FRP-OF-FBG 1 and FRP-OF-FBG2) when applying a load over an 11 h period. From the test results, the high stability of the sensors' measurement under high stress was revealed. The stability of measurement means that the FRP-OF-FBG bars embedded into the cables bridge deform consistently along with the cable.

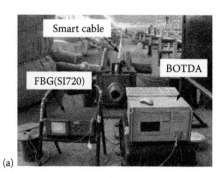

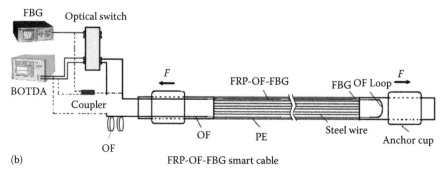

FIGURE 11.11 Test setup (a) and measurement system (b). (From He, et al. *Mechanical Systems and Signal Processing*, 35: 84–94 (2013).)

11.3.4.5 Some Existing Smart Bridge Cables

11.3.4.5.1 Smart CFRP Bridge Cable Based on FRP-OFBG

The smart CFRP bridge cable system consists of CFRP bars, some of which are replaced by smart GFRP bars. All the FRP bars together are wrapped by a protective layer of polyethylene (PE).

11.3.4.5.2 Smart Steel Wire Bridge Cable Based on FRP-OFBG

Cables based on steel wires are mostly used in cable-stayed bridges due to the easy way at the anchorage zone. An embedded FRP-OFBG rebar in the periphery of the cable makes it possible to detect the strain along the cable without crushing the FRP rebar, as well as the strain at the anchorage area.

11.3.5 SMART PRESTRESSED STRAND BASED ON FRP

Prestressing is a very well-known technology used since 1950 to develop lightweight and large spans of concrete structure. The main problem in prestressed structures is the lack of a reliable monitoring technique for prestress loss. Prestress loss might result from several causes, including steel relaxation, anchorage loss, and creep. Therefore, the real-time monitoring of prestress force loss started by using some conventional transducers, including electric resistance strain gages, vibrating-wire sensors, and magnetic flux transducers, installed in the prestressed strand. On the other hand, the use of FRP materials in construction due to their well-known high-strength property in the fiber direction makes it possible to design FRP-based components used as prestressed components. However, the transverse weakness of FRP, in addition to the high stress developed in prestressed members, limits the use of all-FRP prestressed strands. As result, Zhou et al. [18] have proposed a new method to monitor the prestress loss in prestressed strand by combining FRP, steel wire, and OF sensors. The details for the design of the smart prestressed strand are described in the following.

Basically, conventional prestressed strands are composed of seven wires, including one axial wire surrounded by six peripheral wires, to shore up the strength of the strand. To create the smart strand, the axial wire is replaced by a smart FRP rebar, as shown in Figure 11.12a. The first smart strand built was quasi-distributed; however, the use of numerous FBG sensors led to a high cost for the product. Moreover, strains along the entire length of the strand could not be obtained. Therefore, a final smart strand based on the mixing of local (FBG) and global (BOTDA) OF sensors in a single fiber is produced for cost reduction of the sensors, as well as enhancing the

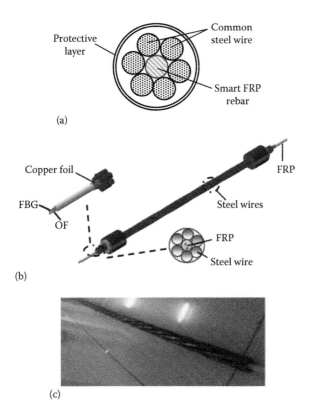

(a)

(b)

(c)

FIGURE 11.12 Details of the smart strand: (a) cross section, (b) configuration, and (c) prototype of smart strand. (a,b: From Zhou, et al. *Journal of Intelligent Material Systems and Structures*, 20(16): 1901–1912 (2013); c, from the authors of this chapter.)

sensitivity of the smart component. The manufacturing of the smart strand is done easily by twisting the six steel wires with a special rotating machine. The six steel wires and the FRP rebar are bundled into a protective layer to form the seven-wire strand. The configuration of the smart strand is shown in Figure 11.12b.

11.3.5.1 Monitoring Principle of the Smart Strand

Similarly to the smart cable, the smart FRP rebar is assumed to deform with the steel wires. Therefore, the strain in the strand could be obtained from the wavelength/frequency shifts of the optical fiber integrated into the smart FRP rebar. For FRP-OFBG-based smart strand, temperature compensation is required, as FBGs are sensitive to both temperature and strain changes. In contrast, the latest smart strand developed by Zhou et al. [19] has a self-temperature compensation ability. Additionally, global and local monitoring is feasible for the mixed BOTDA/R-FBG system. The first line of OF sensors with FBG measures the strain at each FBG sensor's position, which could be regarded as local monitoring. The second line detects the whole strain along the length of the strand, with a spatial resolution equivalent to the value specified in BOTDA/R, and could be regarded as global monitoring. Thus, both OF lines combined allow local and global monitoring of the strand simultaneously. Existing sensing configurations of the smart strand based on local and global monitoring (BOTDA/R-FBG system) are shown in Figure 11.13A.

The stress of the prestressed strand decreases as time increases. Without taking account of the relative deformation between the steel strands and the smart FRP rebar embedded into the smart strand, we could calculate the value of prestress loss during service by the following steps.

The stress of the smart strand at any time can be obtained from

$$\sigma(t) = \frac{\varepsilon\left(E_s.A_s + E_{frp}.A_{frp}\right)}{A_{frp} + A_s} \tag{11.11}$$

where:

E_{frp}, E_s, A_{frp}, and A_{frp} are the Young modulus and section of FRP and steel wires, respectively

ε is the strain of the smart strand

Writing

$$k = \frac{E_s.A_s + E_{frp}.A_{frp}}{\left(A_{frp} + A_s\right)E_s},$$

Equation 11.11 could be written as

$$\sigma(t) = k\varepsilon E_s \tag{11.12}$$

The loss of prestress at any time t is then obtained by subtracting the stress at time t ($\sigma(t)$) from the initial stress given by the hydraulic pressure $\sigma(0)$:

$$\Delta\sigma = \sigma(t) - \sigma(0) \tag{11.13}$$

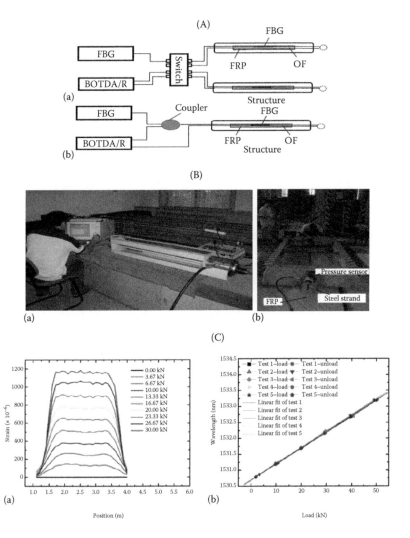

FIGURE 11.13 (A) Existing sensing configuration; (B) tensile test setup; and (C) test data results from Prof. Zhou's studies on smart prestress strand. (a) BOTDA distributed sensing; (b) FBG discrete sensing.

From Equations 11.12 and 11.13, and using the relationship between the changes in wavelength or frequency of OF and the change in strain, we can obtain the prestress loss of the smart strand as

$$\Delta\sigma = \frac{k.E_s}{\alpha_\varepsilon} \Delta\lambda_{B/FBG} \tag{11.14}$$

where:

α_ε is the strain sensitivity of the sensor

$\Delta\lambda_{B/FBG}$ is the change in wavelength or frequency read from optical devices

TABLE 11.3

Sensing Parameters for Smart Steel Strand and Components

Sensor Type	Parameter	FBG	BOTDA	BOTDA-FBG
Bare OF-FBG	Slope	0.999	1.031	0.993
	R^2	1.000	0.999	0.999
FRP-OF-FBG	Slope	0.937	0.957	0.914
	R^2	0.999	0.999	0.999
Smart steel strand	Slope	1.051	–	1.009
	R^2	1.000	–	0.999

11.3.5.2 Sensing Properties Characterization of the Smart Strand

To validate the sensing properties of the smart strand at each loading stage, a test setup in tensile is performed as illustrated in Figure 11.13B. The test setup is subdivided in two sets: the first is for the BOTDA monitoring characterization (Figure 11.13Ba), and the second is for the FBG measurement characterization (Figure 11.13Bb). The data from the BOTDA of the smart strand are similar to the strain distribution of the FRP-OF-FBG rebar and the OF-FBG bar after analysis (see Figure 11.13C). The coefficients of regression (R^2) from each sensor are listed in Table 11.3. The test results obviously demonstrate that the smart strand has very similar strain sensing properties to the smart FRP rebar or the bare fiber under tensile load.

11.3.6 SMART ANCHORAGE SYSTEM

Anchor rod is one of the key components for structural integrity improvement of a ground system. Anchor rod can be found in several structures, especially foundations, tunnels, and so on. The anchor system plays an important role due to the pulling load induced by the whole structure. Additionally, the high interface and axial stress developed at the anchorage zone characterize the anchor as a complex and useful system in construction. The stresses developed must be accurately measured in real time for damage prevention. The evaluation of the anchor state remained baffling till the breakthrough of the smart anchor rod system. Traditional methods, such as using load cells, have been used to evaluate the axial stress of the anchor. However, the transversal strain could not be detected, and the method proved costly. Anchors are fundamentally built with steel material. The use of FRP materials as anchors remains a big challenge due to the limited transversal strength of the FRP and its tendency to fail suddenly under lateral forces. Consequently, researchers have developed a novel component based on FRP and OF sensors to overcome these issues. The different components to date are described in the following section.

11.3.6.1 FBG-Based Smart Anchorage System

11.3.6.1.1 Manufacturing

The anchor system consists primarily of an FRP-FBG rod, a steel sleeve, and an epoxy resin grout adhesive. Additional located rings and sealing rings are mounted to keep the FRP in the middle of the steel sleeve and to prevent the epoxy resin from leakage during the fabrication process. The smart GFRP rod is inserted first between the steel sleeves during manufacturing of the smart anchor. Then, epoxy resins are filled into the void formed by the steel sleeve and the FRP rod. The smart anchor with built-in FBG sensor system is shown in Figure 11.14A. As seen in the figure, one FBG sensor is located outside the anchorage zone to evaluate the material properties, and seven other FBG sensors are quasi-distributed inside the anchorage zone for axial strain evaluation and interfacial mechanical behavior investigation.

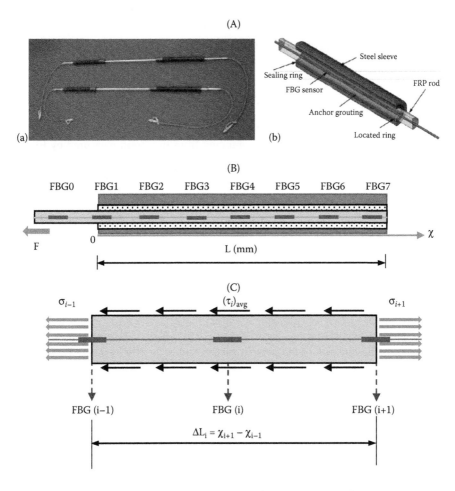

FIGURE 11.14 Smart FRP anchorage based on FBG. (A) Smart FRP-OFBG based anchor: (a) prototype of FRP-OFBG-based anchor; (b) Sketch of FRP-OFBG-based anchor. (B) FBG sensors layout. (C) Shear stress calculation.

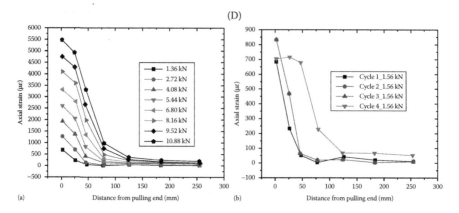

FIGURE 11.14 **(Continued)** Smart FRP anchorage based on FBG. (D) Axial strain characterization (a) with increasing load and distance, and (b) per cycle.

11.3.6.1.2 Mechanical and Sensing Characterization of the Smart Anchor–Based FRP-OFBG

11.3.6.1.2.1 Young Modulus of the FRP Rod

The Young modulus of the FRP rod embedded into the anchor can be obtained from the strain measured by FBG0 and the stress calculated from the pulling load test in cycle 1 [20]. Equation 11.15 gives the relationship between the pulling load and the stress of the FRP rod:

$$\sigma_{rod} = \frac{4P}{\pi D_{rod}^2} \tag{11.15}$$

where:

    P is the pulling load in cycle 1

    D_{rod} and σ_{rod} are the diameter and the stress of the FRP rod, respectively

The linear fitting of the stress–strain gives a coefficient of correlation equal to 0.999, which means the linearity of the strain measurement at the FBG0. Thus, the value of the Young modulus of the FRP rod, obtained from Hook's law, is equal to 51.8 GPa.

11.3.6.1.2.2 Axial Strain Variations

The axial strain–sensing characteristics of the smart anchor were determined from a pulling test under different loading cases. The pulling tests were repeated for four total cycles to determine the debonding characteristics from the pulling end. The test data gathered by Huang et al. (2010) are used to reproduce the graphs of axial strain versus distance from pulling end, as shown in Figure 11.14D. Based on these curves, the following conclusions are highlighted concerning the variations of axial strain for the system-based FBG: (1) the axial strain is very high near the pulling end (FBG1), and decreases as the distance from the pulling end increases, which is similar to the numerical simulation as reference; (2) the axial strain increases as the pulling load increases; (3) from the comparison of cyclic tests, gap measurement between FBG1 and FBG2 could be noticed; however, the difference decreased in cycle 3, while in cycle 4, the strains measured at FBG 1 and FBG2 are almost identical. In other words, the interface debonding occurs only at the FBG1 and FBG2 locations.

11.3.6.1.2.3 Interfacial Shear Stress Characterization From Figure 11.14C, the mechanical equilibrium of the smart anchor for an infinitesimal length ΔL_i between FBG $(i - 1)$ and $(i+1)$ can be written as

$$\sigma_{i-1}.A_{\text{rod}}+\left(\tau_i\right)_{\text{avg}}.A_\tau = \sigma_{i+1}.A_{\text{rod}} \ \ (i=1, 2, \ldots,n) \tag{11.16}$$

where:

σ_i and $(\tau_i)_{\text{avg}}$ are the axial stress and interfacial stress of the FRP rod at the ith FBG sensor, respectively

A_{rod} and $A\tau$ are the cross section of the FRP rod and the surface area between FBG $(i - 1)$ and FBG $(i+1)$, respectively

The average interfacial shear stress of the FRP rod can be calculated by Equation 11.17, using the data recorded from the FBG demodulator device.

$$\left(\tau_i\right)_{\text{avg}} = \frac{E_R.D_R\left(\varepsilon_{i+1} - \varepsilon_{i-1}\right)}{4\left(\chi_{i+1} - \chi_{i-1}\right)} \ \ (i=1,2,\ldots,n) \tag{11.17}$$

The interface properties of the anchor can be characterized by using the distribution of interfacial shear stress calculated from Equation 11.17. According to the analysis, there is no doubt that (1) the interfacial shear stress of the smart anchor changed along with the smart anchor; (2) the interfacial shear stress increased with the pulling load; (3) the interface started softening when the stress limit was reached; and (4) the interface started debonding and propagated into the far end.

11.3.6.2 Distributed-Based Smart Anchor Rod

11.3.6.2.1 Manufacturing and Operational Principle of the Self-Sensing Anchor

Distributed anchor rod, as the name implies, is made of a built-in single-mode distributed OF method, pulled along with the FRP rod to monitor its axial strains at full scale. The manufacturing process of the smart distributed anchor rod is illustrated in Figure 11.15a. The OF embedded into the FRP rod is sensitive to temperature and strain changes. Therefore, a BOTDA interrogator is used to interrogate the frequency shifts of the OF. Additionally, the FRP rod is inserted into a metallic anchor bolt, as shown in Figure 11.15a. Moreover, the front/back jumpers of the smart anchor rod are well prepared by fusing additional jacket fiber at both ends, which are connected to the BOTDA/R equipment. A prototype of the smart distributed FRP rod provided by ZhiXing Ltd. in Nantong can be seen in Figure 11.15b.

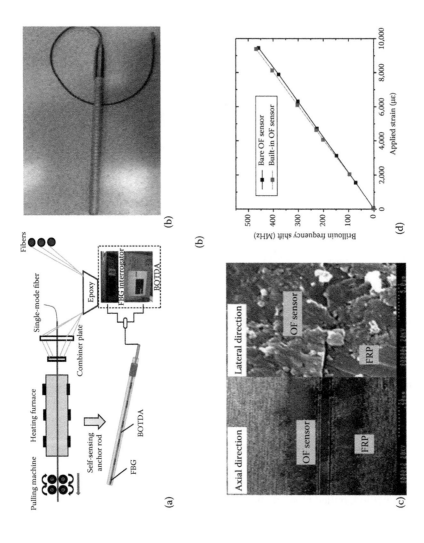

FIGURE 11.15 Smart distributed FRP anchor rod. (a) Manufacturing process of the smart distributed FRP anchor. (b) Prototype distributed anchor. (c) SEM of the distributed anchor rod. (d) Strain sensitivity comparison.

The operational principles of the self-sensing FRP anchor rod are described as follows Without considering the cross-coupling effect of temperature and strain, the Brillouin frequency shifts along the OF sensor are linearly proportional to the applied strains and the temperature changes, as given in

$$\Delta v_B\left(\Delta T,\Delta\varepsilon\right) = C_T\Delta T + C_\varepsilon\Delta\varepsilon \qquad (11.18)$$

where:

Δv_B is the Brillouin frequency shift

$\Delta\varepsilon$ and ΔT are the applied strain and temperature change

C_ε and C_T are the strain and temperature sensing coefficients of the OF sensor, respectively

In the unstrained OF condition, Equation 11.18 is written as

$$\Delta v_B\left(\Delta T,0\right) = C_T\Delta T \qquad (11.19)$$

The applied strain can be then obtained by subtracting Equation 11.19 from Equation 11.18:

$$\Delta\varepsilon = \left[\Delta v_B\left(\Delta T,\Delta\varepsilon\right) - \Delta v_B\left(\Delta T,0\right)\right]/C_\varepsilon \qquad (11.20)$$

11.3.6.2.2 Mechanical and Sensing Characteristics of the Smart Distributed Anchor Rod

A calibration test and SEM were carried out to distinguish the mechanical and sensing performance of the smart distributed anchor rod. From the results illustrated in Figure 11.15c,d, it is seen that the OF is well integrated into the FRP rod. Moreover, the strain sensitivity of the bare OF and the built-in OF sensor is almost identical, with a value of 0.048 and 0.050 MHz/με, respectively.

Additionally, a laboratory test setup was conducted by Huang et al. [21] to investigate the axial strain distribution, as well as the interfacial debonding characteristics of the proposed distributed anchor. The test was done by installing the novel anchor in a channel filled with sand. For sensing comparison purposes, electrical strain gages (ES) were fixed with the anchor rod, with a space of 0.5 m between them. Moreover, an OF temperature sensor was embedded along the sand channel for temperature compensation. The results of the test are shown in Figure 11.15c,d.

11.3.6.2.2.1 Axial Strain Variations

Axial strain along the distributed anchor rod could be calculated from the frequency shifts read from the BOTDA system by using Equation 11.20. According to the figure of Brillouin frequency shifts of the built-in OF sensor [19], the change in Brillouin frequency increases as the applied load is increased. Moreover, the axial strain decreases along the distance from the pulling end. Therefore, we can conclude that the axial strain variation of the smart novel distributed anchor rod is similar to that of the FBG-based anchor rod. Additionally, the strains measured by the built-in OF sensors are more stable and agree well with those measured by the ES gages.

11.3.6.2.2.2   Interfacial Debonding Characteristics of the OF Built-in Sensor
The interfacial shear stresses are calculated from the same principle as in the FBG-based smart anchor rod. Instead of the *i*th FBG measured, any *i*th measured location is taken, as BOTDA can measure all strain along the length of the anchor. According to [19], a maximum shear stress value is reached at the pulling end when a small pulling load (below 0.8 or 0.4 kN) is applied. The interfacial debonding occurs in sequence with the increasing of the pulling load, resulting in a sudden reduction of the related shear stress. Additionally, the built-in OF provides a platform ranging from 3 to 6 m, which is sufficiently bigger than the one provided by the ES. Therefore, the built-in OF can effectively predict the interfacial shear stress, as well as the interfacial damage progression along the anchor rod.

11.4   APPLICATIONS OF FRP-BASED SMART COMPONENTS TO STRUCTURES

The working principles and characteristics of the existing smart components have been described in Section 11.3. Accordingly, it is well known that these components are useful; however, the applications of the smart components based on FRP in real structures have not yet been described. Therefore, some of the existing successful practical applications of smart FRP-OF-based components will be covered in this section. The applications of smart components in structures results in the term *smart structures* due to the ability of the components to extract information on the real state of the structure.

11.4.1   Prestress Loss Monitoring of Beam Members Based on Novel Smart Prestressed Strand

The mechanical and sensing performance characteristics of the smart prestressed strand have been tested several times using calibration tests. From those studies, we can obviously say that the novel smart prestressed strand has good self-sensing properties and is cost effective compared with the conventional method for the determination of prestress loss forces. Additionally, accurate and real-time measurement of prestress loss is quite possible with the novel smart strand, which could be used as a sensor as well as a key component for concerned structures. In this section, the experimental applications of smart FRP-based prestressed strand in RC beams conducted by Lan et al. [22] are highlighted in detail. The smart strand can be applied for the purpose of checking its own performance, as well as distinguishing the different type of prestress force loss (time-dependent and instantaneous losses) occurring in the RC beam within a period of time. The experiments were done in two different cases: an RC beam with a single smart strand and an RC beam with five smart strands (bundle strand). The objective of the latter was to validate the combination of the theoretically predicted strand relaxation and the monitored prestress loss from the smart strand to eliminate the relative error of smart strand induced by strand relaxation found in a single-strand RC beam.

11.4.1.1   Description of the Beam Structure and Experiment Setup

RC beams made of C30- and C40-grade concrete were posttensioned with the novel smart strand of 15.34 mm diameter. The span and cross section of the beam were 2 m and 0.10 ×0.20 m, respectively. The apertures where the prestressed strands were inserted were 25 mm in diameter. The beam span was 2 and 4 m, respectively. Conventional steel reinforcements were also used to reinforce the beam transversally and longitudinally. Smart strands with a diameter of 15.24 mm and a standard strength (f_{ptk}) equal to 1660 MPa were used as the main tensile reinforcement of the beam. A hydraulic jack was mounted at one of the end of the beam to tension the prestressed strand at a given stress, which was controlled by a load cell mounted at the end of the beam setup. The test setup of the RC beam is shown in Figure 11.16. The control stress of each strand used to test the beam was obtained from $\sigma_{con} = \alpha.f_{ptk}$, where α was set to be between 0.7 and 0.75. Five FBG sensors were installed along each intelligent strand for quasi-distributed requirement measurement. The wavelengths of the FBG sensors range from 1525 to 1545 nm, separated by 5 nm from each other.

To collect data on strain and temperature from the smart strand, BOTDA/R and FBG demodulator were connected to the single FBG-OF sensors, separated by a light coupler. Additionally, a load cell was installed at one end of the specimen to compare the results obtained from the smart strand.

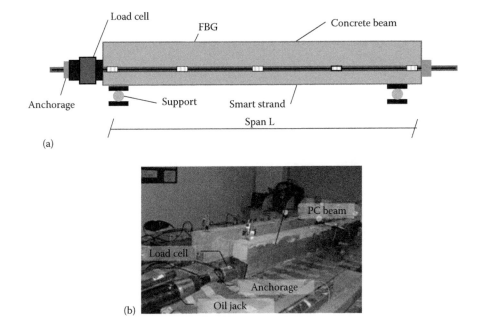

FIGURE 11.16   Sketch (a) and photo b) of the test setup of the prestressed RC beam.

11.4.1.2 Experimental Results

11.4.1.2.1 Beam with a Single Smart Strand

It has been demonstrated by Lan et al. [22] that time-independent (instantaneous) losses depend reliably on the control stress. When control stress is increased, instantaneous losses increase too. This phenomenon is normal for all prestressed members. The prestress losses measured from the novel strand and the load cell are recapitulated in Table 11.4 for each load. The values from the smart strand agree well with the load cell with a relative error less than 5%, induced by the strand relaxation. This means that the smart strand allows the prestress loss monitoring of the RC beam with good accuracy.

Time-dependent losses are caused by strand relaxation and concrete creep and shrinkage. The losses increased gradually with time. Using Origin 8.0 software and the data from [22], the time-dependent losses are given in Figure 11.17a.

TABLE 11.4

Instantaneous Loss of Prestressed Concrete Beam with One Smart Strand

Load (kN)	Smart Strand (MPa)	Load Cell (MPa)	Relative Error (%)
30	9	9	0
60	21	22	4.7
90	42	42	0
120	62	64	3.2
150	83	82	1.2
180	122	122	0

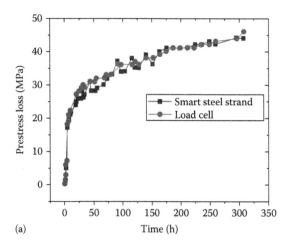

(a)

FIGURE 11.17 Time-dependent prestress loss. (a) RC beam with one strand.

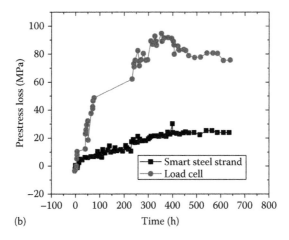

(b)

FIGURE 11.17   (**Continued**) Time-dependent prestress loss. (b) RC beam with multi-strand.

11.4.1.2.2   Beam with a Multi-Strand

The instantaneous loss measured from different sensors and theoretical predictions are recapitulated in Table 11.5. Prediction from smart strands agrees well with that from the load cell, with a relative error less than 7%. However, a large difference between theoretical values and sensors is found. Additionally, the time-dependent losses from the smart steel strand compared with the load cell during 30 days are shown in Figure 11.17b. Obviously, we can observe from Figure 11.17b the difference induced by strand relaxation effects. Therefore, a compensating method was proposed: introducing the theoretically predicted prestress loss resulting from strand relaxation into the monitored results. The modified results from smart strands and Chinese national code (GB50010-2002) are shown in Table 11.6. Compared with the load cell, the modified results agree well, with a relative error of 2.7%, which is less than the previous value without modification.

Therefore, we can conclude that the proposed intelligent strand can provide promising prediction of the instantaneous loss and the time-dependent prestress loss of prestressed concrete structure for control stress higher than half of the standard strength of the steel stand. From both RC beams, a buildup of prestress

TABLE 11.5

Instantaneous Loss of Prestressed Concrete Beam with Five Smart Strands

Load (kN)	Strand 1 (MPa)	Strand 2 (MPa)	Strand 3 (MPa)	Load Cell (MPa)	Theoretical Prediction (MPa)
Strand 1–140	203	−4	0	215	255
Strand 2–140	−2	169	3	177	255
Strand 3–140	1	4	209	223	255
Strand 2–160	−1	242	3	250	255

TABLE 11.6
Comparison of the Modified Results of Prestress Loss from Smart Strand to Load Cell

FBG (MPa)	Theoretical Strand Relaxation (MPa)	Combined Prediction of Prestress Loss (MPa)	Load Cell (MPa)	Relative Error (%)
39	34	73	75	2.7

loss could be seen. Successful results are assured by combining the theoretical prediction of strand relaxation and the monitored prestress loss from the smart strand.

11.4.1.3   Health Monitoring of Bridge Cable in Yanghe River, China

Yanghe Bridge and Erbian Bridge are both located in China. Both bridges presented some deficiencies due to the aging of infrastructures and the growing demand from users. The bridges were then rebuilt for structural and loading capacity improvement, as well as for the purpose of SHM. Therefore, the first applications of the smart cable based on FRP and OF techniques described in the previous section took place to replace conventional cable in the Yanghe River cable bridge, as well as the suspenders in the Erbian arch bridge. In this section, the application of smart cable in Erbian Bridge and Tianjing Yanghe Bridge are reviewed to characterize the durability and sensing performance of the smart components in real applications.

11.4.1.4   Monitoring of Suspenders in Erbian Dadu River Arch Bridge, China

Erbian Dadu River Arch Bridge across Dadu River was finished in 1995 in Erbian country. Figure 11.18a shows the structure of the Arch Bridge being constructed on site. The Arch Bridge was built to connect Sichuan province to other nearby provinces between Erbian country and Chengdu city. The bridge, with a main span of 138 m, was built with concrete-filled steel tube, and supported by eight suspenders and 50 tie cables. However, corrosion deterioration affected the state of the bridge and seriously damaged the bridge components, especially the cables and suspenders. Considering the aggressive environment at the bridge location and the difficulties of embedding reliable sensors, the owners have decided to replace some of the existing suspenders and cable ties of the old bridge by using smart FRP-OFBG bars, replacing one of the wires in the suspenders to improve the health monitoring of the bridge. Therefore, the stretching procedure of the suspenders was monitored during construction.

A floatation-tensile craft procedure was adopted for the stretching tensing construction of the suspender bridge. Up and down stream suspenders were simultaneously stretched. The reading from the strokes hoisting jack and the smart composite bars embedded into the suspender are combined for tensile change monitoring.

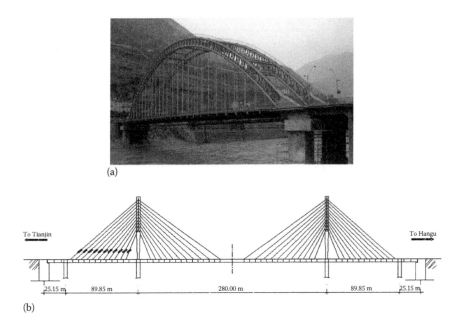

(a)

To Tianjin

To Hangu

25.15 m | 89.85 m | 280.00 m | 89.85 m | 25.15 m

(b)

FIGURE 11.18 Description of each bridge system. (a) Erbian Dadu River arch bridge; (b) Tianjin Yanghe cable-stayed bridge.

Figure 11.19 and Table 11.7 show the typical curve of the total monitoring procedure data from OF sensors and the comparison of different sensors [23], respectively. From the collected data, it is obvious that the smart GFRP-OFBG bar can track the whole stretching process perfectly. However, Table 11.7 reveals the existence of a certain error value between the actual tension and the designed force. The error

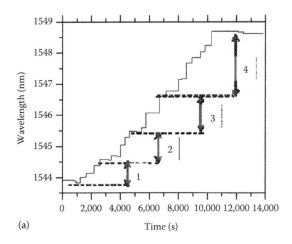

(a)

FIGURE 11.19 Whole-process monitoring of suspender during stretching. (a) GFRP-OFBG smart sensor wavelength change during suspender tension.

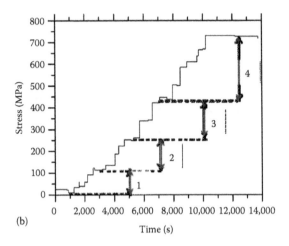

(b)

FIGURE 11.19   (**Continued**) Whole-process monitoring of suspender during stretching. (b) tension monitored by intelligent sensor.

TABLE 11.7

Comparison Result of Data from Different Sensors

Suspender Location	Tension Value before Anchoring (kN)	OFBG Sensors Monitoring Value after Anchoring (kN)	Design Value (kN)	Tension Error (%)
Upstream inner suspender	2908.79	2686.81	2893.54	7.69
Upstream outer suspender	3058.35	2844.27	2978.79	4.73
Downstream inner suspender	2922.39	2648.87	2893.54	9.24
Downstream outer suspender	3003.97	2663.02	2978.79	11.86

could be due to the fact that the smart bars used are not distributed in the suspenders, which might limit the reading during the anchoring. Additionally, Li et al. [24] used monitoring data from the smart suspenders to evaluate the behavior of the Erbian arch bridge under dynamic traffic load with good results.

Therefore, structural health monitoring of the bridge based on the strain data recorded by optical devices from smart sensors is provided by the application of smart FRP-OFBG-based sensors into cables or suspenders of Erbian Arch Bridge.

11.4.1.5   Monitoring of Tianjin Yanghe Bridge, China

Tianjin Yanghe Bridge is one of the earliest cable-stayed bridges built in mainland China. The bridge was opened to traffic in December 1987. The cable-stayed bridge has a main span of 260 m, comprising two side spans of 25.15+99.85 m each, as shown in Figure 11.18b. After 19 years of service, the damaged parts of the bridge were reinforced and repaired in 2007. The real-time monitoring of the cable bridge was done by replacing the existing bridge cables with the smart stay cables proposed in previous section.

Smart stay cables, when installed in a cable-stayed bridge, can record the time history of their own stress. As the bridge cables are subjected to cyclic traffic loads, accumulative fatigue damage, which is one of the main failure modes of cable-stayed bridges, needs to be assessed. This could be done by using the monitoring data of strain from the smart stay cables. Li et al. [25] conducted fatigue damage evaluation on two of the cables—C1 (38.185 m) and C11 (115.655 m)—in the Tianjin Yanghe Bridge. The monitoring system is composed of optical devices, smart cables, and a data acquisition system. The time history of strain from the FBG sensors embedded into the smart cables was recorded by using an FBG demodulator (Si-425) made by MOI. Inc., USA. The time history of stress during 1 h of each cable read from FBG sensors was plotted by Li et al. [25]. The peak values in Figure 11.20a indicate

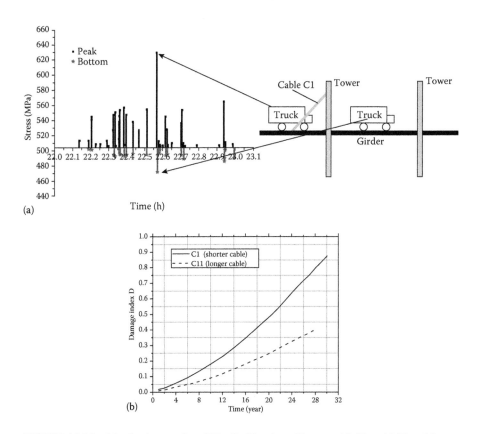

FIGURE 11.20 Monitoring results of Tianjin Yanghe cable-stayed bridge. (a) Time history of stress from OFBG of cable C1 in 1 h (screenshot figure). (b) Damage accumulation index comparison between cables C1 and C11.

that a vehicle is moving near the cable, and the successive values at the bottom are induced by the same vehicle moving near the symmetric position of the cable with respect to the tower. Using the Rainflow method and the fatigue accumulative damage index formulation proposed by Miner [26], the results for fatigue damage evaluation of cables C-1 and C-11 are represented in Figure 11.20b. According to the results shown in Figure 11.20b, the damage index of cable C1 is much larger than that of the longer cable C11. This shows that the shorter cable is exposed to more severe damage than the longer one.

Additional monitoring of the cable-stayed bridge has been conducted by the same group of researchers. For example, Lan et al. [27] monitored the cable force in the same bridge by using the strain data provided by the smart cables. Moreover, they demonstrated the use of local strain response from cable number 7 (C-7) in Yanghe Bridge to assess the force of the cable under random vehicle load. The results from the study revealed without doubt that the force of the cable was perfectly monitored. Hence, smart cables based on FRP-OF-FBG/OFBG are useful for the health monitoring of cable-based bridges.

11.4.2 EXPERIMENTAL APPLICATION OF THE SMART ANCHOR ROD

The anchor plays an important role in geotechnical engineering applications. The necessity to monitor the axial/transversal strength of the anchor and to detect the interfacial debonding of the anchor from the ground structure remains an issue in real applications. In the following, the experimental use of the smart anchor proposed in Section 11.3.6 inside a 12 m sand channel is described in more detail. The purpose is to demonstrate the feasibility of the novel anchor in a real application to monitor different parameters of structures where the anchor is being used.

The distributed anchor was placed into a 12 m long channel full of sand, as shown in Figure 11.21a. A load cell, ES gages, and OF temperature sensors were also mounted for comparative purposes. Axial strain and the interfacial damage evolution

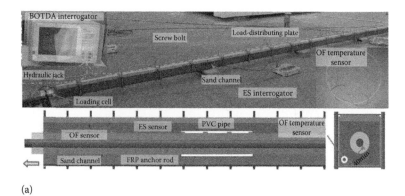

(a)

FIGURE 11.21 Application of the distributed anchor in 12 m of sand channel. (a) Experimental setup of the distributed anchor. (b) Displacements of the anchor at pulling end.

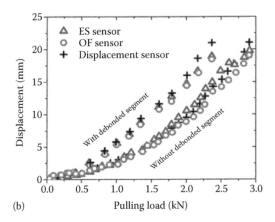

(b)

FIGURE 11.21 (Continued)

have been already demonstrated in Section 11.3.6.2.2. The investigation of displacements at the pulling end is discussed here. The displacements could be obtained by using Newton–Cotes rule integration, which is a function of the axial strain and the free length of the anchor. Moreover, the graph in Figure 11.21b shows the comparison of results from different sensors and the displacements with and without a debonded segment. It could be concluded that the distributed anchor embedded into the sand channel could reveal the displacement of the anchor from the pulling end. Additionally, the results obtained from the smart anchor agree well with the displacement sensors both with and without debonding segments. Therefore, a smart distributed anchor could be used in SHM of the ground system by describing different measurement parameters.

11.5 DISCUSSION

The existing smart FRP-based components and structures were successfully reviewed in the previous section. Therefore, according to the advantages and drawbacks of the novel smart systems, it is possible to make some remarks in this section. Moreover, feasible suggestions are also highlighted to further develop the existing smart components, as well as for a new design of intelligent components useful for civil engineering applications.

1. To date, a combination of distributed OF and FBG sensing techniques allows the successful monitoring of structures based on OF sensors. Based on the combined method, local monitoring at one point is ensured by the Bragg grating sensors and the global evaluation is ensured by the data from OF sensors embedded along the structure. The OF sensor networks are just separated by an optical splicer. Following the development of OF sensing equipment, it is an obvious step to research whether there is a new and reliable way to develop a novel, better network system for FRP monitoring

based on optical sensors without any evident change of the mechanical properties of the composites. Authors who are interested in smart FRP-based components should consider finding possible ways to design a new system, more reliable and accurate than the combined OF-FBG system, which could be embedded into FRP.

2. Intelligent FRP bars are the first structural component with self-sensing ability described in this chapter. According to the investigations of their mechanical and sensing properties in the previous section, we found that smart FRP bars have excellent mechanical strength in the longitu-dinal direction. Good linearity of strain sensing has been demonstrated. Additionally, smart FRP bars are easy to produce using a pultrusion process by pulling out fibers and optical sensors together with a pultrusion machine. To address the transversal weakness of the intelligent bar, methods such as winding the bar with similar or a different type of fiber at 90° to the longi-tudinal dimension is viable to increase the strength of the smart bar in the transversal direction.

3. As remarked, most smart FRP-based components described in this chapter are based on the use of the smart FRP rebar produced at HIT. Smart FRP rebar-based components include smart prestressed steel strand, smart dis-tributed anchor, and smart cable. However, the constraints applied to each component make the difference. This means that, although these compo-nents are based on the same design, researchers have investigated in more detail the monitoring principles of the smart components and how the strain and temperature data from the smart bars embedded into the smart compo-nents should be used.

4. The intelligent prestressed carbon plate proposed in Section 11.3.2 might resolve the problems of damage detection in external NSM for structural strengthening. The OF sensors are better protected than those bonded on the laminate surface. However, debonding between the plate and the RC members is still under investigation by a group of research-ers conducted by Professor Zhou at Dalian University of Technology, China.

5. Smart tube for concrete confinement, reviewed in this chapter, is also inter-esting, as confinement concrete could be classified among prefabricated structural members. Moreover, more contractors prefer to use confined-based members rather than conventional RC. According to our investiga-tion, the smart FRP tube could predict the strain of confined concrete up to a certain value of strain, which corresponds to the range of strain applied to the confined column. However, the smart tube reviewed in this chapter dates from 2007, although there has already been enormous progress in this type of composite structure. Therefore, many studies should be proposed to develop smart confinement members that can more reliably and accurately predict the behavior and damages of current composite members, which is one of the actual research directions of the author.

6. Nowadays, assembly of structures based on a prefabrication method is an emerging technology for construction. As components are transported to

the site for assemblage, it is advisable to design some lightweight smart structural elements. Therefore, the use of FRP such as smart bars and other FRP profiles would contribute to the development of smart FRP-based prefabricated components.

7. A smart FRP-based cable system was described, with successful applications in some bridges located in China. According to our investigations on the application of smart cables, we can report that the data from the smart cable could monitor the health of bridges in different situations. Additionally, the anchors of the smart cables were investigated.

8. Anchor based on FRP is meaningful, especially for geotechnical applications. More studies taking account of the effect of high shear strain are needed. This means that it is better to wind the anchor transversally with a smart fiber. Therefore, the transversal winding technique will strengthen the anchor as well as ease prediction of the shear strain.

11.6 CONCLUSION

This chapter has reviewed the most significant smart FRP-based components and their applications in civil engineering as sensors and structural key components. A variety of smart components such as reinforced bars, ground anchor, CFRP plate, and tube found to date have been described in detail. The basic principles of those intelligent structures/components were also given, based on the problems found in existing structures and the challenges encountered in SHM. The details of each smart system provided in the chapter will enlarge the knowledge of researchers who have an interest in smart reliable structures. Moreover, the analysis of the sensing performance and mechanical characteristics of each smart component based on FRP revealed that all smart components cited in the chapter are reliable and strong enough to detect and resist the desired deformation. Additionally, it has been demonstrated that components with combined distributed OF and FBG sensors are more attractive by providing local and global monitoring, as well as temperature self-compensation ability. Moreover, smart components based on FRP could gradually reduce the weight of a structure without reducing its strength. However, according to the remarks provided for each system, most research studies did not take account of the inconvenience of using FRP when the components are used in a very harsh environment, where large strain can happen. In spite of the progress, many discoveries and improvements are still ongoing in the research area of smart components and structures, including the design of new sensors based on FRP and OF, which are more robust and could endure very harsh conditions. Based on the developments in civil construction, more innovation is possible in smart components and structures based on FRP.

ACKNOWLEDGMENTS

The authors acknowledge the financial support of China Government Scholarship Council (CSC) under Grant No. 2012450T21, the project from National Natural Science Foundation of China (No. 61328501), and the High-tech support project

(2011BAK02B02). Special thanks are also addressed to ZhiXing Ltd Company in Nantong for providing necessary data for the realization of this work.

REFERENCES

1. Y. Xiao, Applications of FRP composites in concrete columns, *Advances in Structural Engineering*, 7(4): 335–343 (2004).
2. L.C. Hollaway, A review of the present and future utilization of FRP composites in the civil infrastructure with reference to their important in-service properties, *Construction and Building Materials*, 24(12): 2419–2445 (2010).
3. J. Ou and Z. Zhou, Optic fiber Bragg-grating-based sensing technologies and their applications in structural health monitoring, *Proceedings of the SPIE*, 6595: 659503 (2010).
4. S. Yan, et al., Health monitoring of reinforced concrete shear walls using smart aggregates, *Smart Materials and Structures*, 18(4): 047001 (2009).
5. Y. Kuang and J. Ou, Self-repairing performance of concrete beams strengthened using superelastic SMA wires in combination with adhesives released from hollow fiber, *Smart Materials and Structures*, 17(2): 025020 (2009).
6. P. Biswas, et al., Investigation on packages of fiber Bragg grating for use as embeddable strain sensor in concrete structure, *Sensors and Actuators A*, 157(1): 77–83 (2009).
7. T.H.T. Chan, et al., Fiber Bragg grating sensors for structural health monitoring of Tsing Ma bridge: Background and experimental observation, *Engineering Structures*, 28(5): 648e659 (2006).
8. Z. Zhou et al., A novel self-healing optical fiber network, *Applied Mechanics and Materials*, 330: 553–560 (2013).
9. Y.W. Sasy Chan and Z. Zhou, Advances of FRP-based smart components and structures, *Pacific Science Review*, 16: 1–7 (2014).
10. Y. Tang, et al., A new type of smart basalt fiber-reinforced polymer bars as both reinforcements and sensors for civil engineering application, *Smart Materials and Structures*, 19: 115001 1–15 (2010).
11. D. Mostofinejad, S. M. Shameli and A. Hosseini, EBROG and EBRIG methods for strengthening of RC beams by FRP sheets, *European Journal of Environmental and Civil Engineering*, 18(6): 652–668 (2014).
12. B. Wang, et al., Strain monitoring of RC members strengthened with smart NSM FRP bars, *Construction and Building Materials*, 23: 1698–1711 (2010).
13. Z. Wang and P. Ren, New smart carbon reinforced polymer plate for strengthening and its sensing performance test, *China Measurement & Test* 42(3): 113–117 (2016).
14. X. Yan and H. Li, Self-sensing concrete-filled FRP tube using FBG strain sensor, *Proceedings of the SPIE*, 6595: 659531 (2007).
15. J.M. Ko, et al., Dynamic monitoring of structural health in cable-supported bridges, *Proceedings of the SPIE*, 3671: 161–172 (1999).
16. Z. Zhou, et al., Applications of FRP-OFBG sensors on bridge cables, *Proceedings of the SPIE*, 5765: 668–677 (2005).
17. J.P. He, et al., Optical fiber sensor-based smart bridge cable with functionality of self-sensing, *Mechanical Systems and Signal Processing*, 35: 84–94 (2013).
18. Z. Zhou, et al., R&D of smart FRP-OFBG based steel strand and its application in monitoring of prestressing loss for RC, *Proceedings of the SPIE*, 6933(13): 1–9 (2013).
19. Z. Zhou, et al., A smart steel strand for the evaluation of prestress loss distribution in post-tensioned concrete structures, *Journal of Intelligent Material Systems and Structures*, 20 (16): 1901–1912 (2013).

20. M. Huang, et al., Smart fiber-reinforced polymer anchorage system with optical fiber Bragg grating sensors, *Proceedings of the SPIE Smart Materials and Structure*, 7648 (2010).
21. M. Huang, et al., A distributed self-sensing FRP anchor rod with built-in optical fiber sensor, *Measurement*, 46(4): 1363–1370 (2013).
22. C. Lan, et al., Monitoring of structural prestress loss in RC beams by inner distributed Brillouin and fiber Bragg grating sensors on a single optical fiber, *Structural Control and Health Monitoring*, 21(3): 317–330 (2014).
23. D. Li, et al., Development and sensing properties study of FRP-FBG smart stay cable for bridge health monitoring applications, *Measurement*, 44: 722–729 (2011).
24. D. Li, et al., Dynamic behavior monitoring and damage evaluation for arch bridge suspender using GFRP optical fiber Bragg grating sensors, *Optics and Laser Technology*, 44: 1031–1038 (2012).
25. H. Li, et al., Applications of optical fibre Bragg gratings sensing technology-based smart stay cables, *Optics and Lasers in Engineering*, 47: 1077–1084 (2009).
26. M.A. Miner, Cumulative damage in fatigue, *Journal of Applied Mechanics*, 12(3): 159–164 (1945).
27. C. Lan, et al., FBG based intelligent monitoring system of the Tianjin Yonghe Bridge, *Proceedings of the SPIE*, 6933(12): 1–9 (2008).

12 Fiber-Optic Structural Health Monitoring Systems for Marine Composite Structures

Raju and B. Gangadhara Prusty

CONTENTS

12.1 Curved Composites.. 417
12.2 Out-of-Plane Joints 418
    12.2.1 T-Joints... 418
    12.2.2 Top-Hat Stiffener 419
    12.2.3 Keels 420
        12.2.3.1 Types of Keel 420
        12.2.3.2 Full Keel 421
        12.2.3.3 Fin Keel.. 421
        12.2.3.5 Swing Keels 421
        12.2.3.6 Canting Keels... 421
        12.2.3.7 Keel–Hull Connection 421
12.3 Composite Failure..423
    12.3.1 Matrix Failure..424
    12.3.2 Fiber Failure...424
    12.3.3 Interface Failure...426
    12.3.4 Interlaminar Failure: Delamination...427
12.4 Nondestructive Evaluation of Composites...428
    12.4.1 Types of NDE 428
    12.4.2 Optical-Fiber Sensing 430
        12.4.2.1 Basic Principles...431
        12.4.2.2 Optical-Fiber Modulation Techniques 432
        12.4.2.3 Bragg Grating Sensors 434
        12.4.2.4 Multiplexing..436
        12.4.2.5 FBG Interrogation..436
12.5 Damage Characterization of Advanced Composites Using Optical-Fiber Sensors 437
    12.5.1 Matrix Cracking 437
    12.5.2 Debonding 439
    12.5.3 Crack Monitoring in a Bonded Repair System............................439

        12.5.4 Detection of Transverse Cracks 441
        12.5.5 Delamination 442
        12.5.6 Effects of Embedded Optical Fibers on Structural Integrity 445
        12.5.7 Summary of Optical-Fiber Sensors 446
12.6 Experimental Investigation of Top-Hat Stiffener 447
        12.6.1 VARTM Manufacturing Process 447
        12.6.2 VARTM Top-Hat Stiffener Results 448
                12.6.2.1 VARTM Top-Hat Stiffener Layup 1 Experimental Results ... 449
                12.6.2.2 VARTM Top-Hat Stiffener Layup 2 Experimental Results ... 450
                12.6.2.3 VARTM Top-Hat Stiffener Layup 3 Experimental Results ... 451
12.7 FBG Strain Sensing 452
        12.7.1 Fabrication of FBG Sensor 453
        12.7.2 Layup and Embedding Optical Fibers 456
        12.7.3 Optical Fiber Embedding Procedure with VARTM 459
        12.7.4 FBG Strain Measurement Setup 462
                12.7.4.1 Test Procedure 465
        12.7.5 FBG Results 465
                12.7.5.1 FBG Survival after Specimen Manufacture 465
                12.7.5.2 VARTM Top-Hat Stiffener Layup 1 Results 466
12.8 Conclusions 474
References 475

The use of composites in structural applications dates back at least 60 years. Due to limited research and availability of composites, early structural applications were limited; only military and defense projects could afford their high cost. More recently, reinforced polymer composite applications have become popular for marine, aerospace, and automobile structures, sporting goods, and biomedical components. Composite materials are preferred over metal structures for added benefits such as a high strength-to-weight ratio, lower maintenance, small electromagnetic signature, longer life, and customizable properties. Glass fiber–reinforced plastic (GFRP) composites are popular in the marine industry due to their low cost and ease of manufacture. Using composites in marine structures reduces the weight by 10%–30% compared with metal structures, increasing design efficiency and reducing fuel consumption. The absence of oxidation and rust formation reduces the need for continual maintenance. At the same time, disadvantages of GFRP include high initial cost, weaker matrix and low toughness, joining difficulty, and environmental degradation of the matrix.

Use of composite materials in aero/mechanical/marine structures presents a challenge in the continuous assessment of the strength and integrity of the structure. The various loads acting on a structure create a complex state of stresses in the structure. The process of implementing a damage identification strategy for aerospace, civil, and mechanical engineering infrastructure is referred to as structural health monitoring (SHM). This process involves the observation of a structure or mechanical system over time using periodically spaced measurements, the extraction of damage-sensitive features from these measurements, and the statistical analysis of these features to determine the current state of system health. Among the available SHM systems for composites, the use of embedded fiber-optic sensors for temperature, strain, pressure, acoustic emission, and so on is increasing due to their smaller size (~125 µm). This requires an understanding of the complex sensor architecture, the type of sensing required, the location of the sensor, and the ingress and egress of the optical fiber. In comparison with a flat panel, embedding and obtaining an egress of the optical fiber is a challenge for curved composites.

The issue of frequent keel failures is a mystery to many boat builders and designers. The marine industry has suffered a number of casualties and material losses during the past two decades. In the structural composite context, the initiation of cracks (or fractures) does not indicate a catastrophic failure. A stable crack propagation stage associated with a constant external load precedes a catastrophic failure. This is frequently seen in the structural application of composites. Nevertheless, due to the alignment of the fibers to the load path, some designs exhibit reserve strength after crack initiation has occurred. Understanding and assessing such designs requires adequate consideration of the strength and fracture behavior of the composite laminate. An accurate prediction of static collapse behavior for marine structures is essential for assessing the reserve strength and likelihood of damage.

12.1 CURVED COMPOSITES

Layered composites allow the manufacture of structures with complex shapes and sizes. These complex structures generally consist of curved segments, which may enhance efficiency but can involve complex failure mechanisms. Premature failures of curved composites in the past were predominantly due to a lack of appreciation of the interlaminar tensile strength, which may be less than 3% of the in-plane strength of a composite laminate. The primary failure mode in curved composites occurs when bends are opened or closed by external loading or pressure, as shown in Figure 12.1.

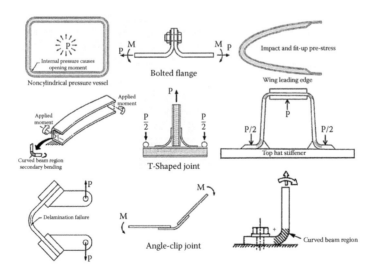

FIGURE 12.1 Curved structures subjected to opening moments inducing interlaminar tensile/compressive stress. (From Kedward, K.T. et al., *Composites*, 20(6), 527–536, 1989.)

TABLE 12.1

Effects of Geometry and Material Variations

Response Feature	In-Plane Stress in Laminate	Through-Thickness Stress in Laminate
Increasing thickness of laminate	Decreases	Increases
Increasing bend radius	Decreases	Decreases

Source: Shenoi, R.A. and Hawkins, G.L., *Composite Materials in Maritime Structures: Volume 2—Practical Considerations*, 1st edition, Cambridge University Press, New York, 1993.

12.2   OUT-OF-PLANE JOINTS

One of the challenges in designing a composite marine vessel is the high-strength joints. Often, the load needs to be transferred in an out-of-plane mode. The joints carrying such loads must be carefully designed. Out-of-plane joints are formed by placing additional composite layers on both sides and filling the gap with an appropriate filler material. Generally, the out-of-plane joints used are T-joints and flanges of top-hat stiffeners.

In the case of a hull–bulkhead joint (a T-joint), the load is transmitted between the bulkhead and the shell through the laminate in its in-plane direction. When the load is applied to the joint, high through-thickness stress develops in the corner of the joint, and it fails by delamination. As the load increases, the structure suffers multiple delaminations until the structure fails. Traditionally, one criterion considered in designing this joint is that the sum of the in-plane properties of the two laminates is equivalent to that of the weakest member being joined. In most applications, the material used in the laminate is the same as that used in the structure; the thickness of each laminate is specified as a minimum of half the thickness of the thinner structural element. Such joints are susceptible to bond-line porosity and localized disbonding due to either manufacturing defects or high operational loads.

Generally, the designer is faced with the challenge of opting for either strength (requiring a thicker laminate) or flexibility (requiring a thin laminate) (Shenoi and Wellicome, 1993). These opposing design requirements present different failure mechanisms based on in-plane and through-thickness stresses, as shown in Table 12.1.

12.2.1   T-JOINTS

A T-joint is constructed by connecting two orthogonal panels (web and flange) perpendicular to each other. This joint is formed by connecting the two plates with laminate around the joint. T-joints are used between the hull and the bulkhead, where tensile, shear, or combined loads are transferred. A typical T-joint is shown in Figure 12.2. The main failure mode with T-joints is delamination between the

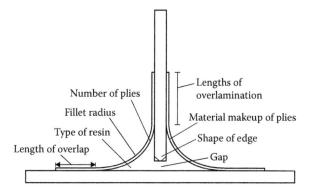

FIGURE 12.2   Typical composite T-joint. (From Junhou, P. and Shenoi, R.A., *Composites: Part A*, 27A, 89–103, 1996.)

overlap and the base panel when the vertical member is pulled against the base panel. Matrix cracking is typically the initial failure mechanism, which precedes delamination and fiber breakage.

12.2.2 TOP-HAT STIFFENER

Top-hat stiffeners are generally used in the fabrication of stiffened panels, such as decks, bulkheads, or the hull shell, joining the keel to the boat hull (Dodkins et al., 1994). The key benefits of using top-hat stiffeners are high bending stiffness and torsional resistance. As the name suggests, the stiffener has a hat shape with a flange, a web, and a crown, as shown in Figure 12.3. The top bend between the crown and the web is stiffened by adding more composite layers (preferably unidirectional) (Raju et al., 2010).

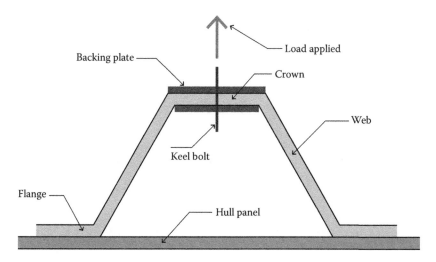

FIGURE 12.3   Typical top-hat stiffener configuration. (From Raju et al., *Ocean Engineering*, 37, 1180–1192, 2010.)

Due to the absence of through-thickness reinforcements and the lack of bond ductility, out-of-plane joints suffer from relatively low strength and fatigue resistance against out-of-plane tensile, bending, and shear loads (Smith, 1990; Shenoi and Hawkins, 1995). In addition, they are susceptible to failure by peel or delamination well before the ultimate in-plane laminate material stress is reached (Dodkins et al., 1994; Raju et al., 2010).

12.2.3 KEELS

The keel is the considered to be the basic structural member of the ship, which runs from bow to stern. It serves as the backbone of the structure; it is the major source of structural strength of the hull and provides greater directional control and stability. Traditionally, laying the keel was the first part of shipbuilding; however, modern ships are built using a prefabrication process whereby the keels are attached the hull at a later stage using keel bolts. In nonsailing vessels, the keel helps the hull to move forward, rather than slipping to the side. In the case of sailing ships, keels provide adequate lift to counteract the lateral force generated by the mast. Generally, a fair amount of weight is added to the keel to assist the righting effect of the hull's hydrodynamic forces. A keel serves as a dual-purpose hydrofoil. First, it counteracts any heeling movement; second, it minimizes sideways drift. The key factors in keel design include the root chord, span, taper, twist, longitudinal position at tip, cant angle, junction angle, junction fairing and section characteristics (Larsson and Eliasson, 1994). Earlier designs concentrated on lowering the center of gravity (CG) by flattening the wings, so that more weight could be added. However, in modern keels, the weight is placed on the lower part of the keel as a bulb.

12.2.3.1 Types of Keel

The two types of commonly used keels are fixed and nonfixed. Fixed keels include full keels, long keels, fin keels, winged keels, bulb keels, bilge keels, skeg keels, and centerboard keels, shown in Figure 12.4. Nonfixed keels are swing and canting keels.

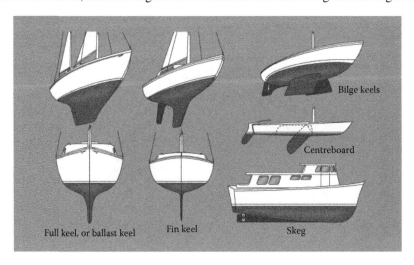

FIGURE 12.4   Major types of fixed keels. (From Encyclopaedia Britannica, 2010.)

12.2.3.2 Full Keel

This keel runs at least 50% of the length of the hull. Its main advantages are providing for safe grounding and improved directional stability, reducing the chance of fall-off as well as less swinging off course due to wind gusts and wave action. However, this type of keel makes it difficult for the vessel to make sharp turns. The large keel produces significant drag and the speed is lower than other keel types. There is a large contact area between the keel and the ship. This keel suits stability more than speed.

12.2.3.3 Fin Keel

The shape is similar to that of a shark fin, deep, flat and sharp edged; its length is typically less than 50% of the hull length. There is less wetted area with reduced drag; thus, speed is enhanced. The turning radius is small, and therefore, this is the keel generally used in racing boats. The reduced length provides less resistance to forces such as wind gusts and waves. Bulbs are a type of fin keel that have a ballast weight at the lower end of the fin, allowing a shallower design. A wing keel at the trailing edge of the keel provides additional hydrodynamic stability with reduced tip-vortex turbulence.

12.2.3.5 Swing Keels

These are weighted, narrow, fin-type keels providing both ballast and lateral stability. Centerboard keels are similar to swing keels but not weighted. Swing keels can be inserted in the water or retracted, depending on the depth of the water. A major drawback is the required maintenance. Swing keels are generally lighter than fixed-fin keels.

12.2.3.6 Canting Keels

A canting keel mounts on a hinge and is hydraulically rotated to windward when the boat heels. Rotating the keel away from the heeling direction generates force that both rights the boat and propels it forward. The drawbacks are the required maintenance and the complexity of the mechanism. Racers prefer canting keels for increased speed and stability.

12.2.3.7 Keel–Hull Connection

The keel is normally attached using a series of bolts that penetrate the hull (Figure 12.5). These bolts load the hull through washer plates (Figure 12.6a), which bear down upon the flanges of top-hat stiffeners called keel floors (Figure 12.6b).

With such high loadings (possibly a combination of tensile/compression, flexure, and torsional loads), the design of the supporting structure in this region is critical. However, because of the complex loading, this region is often overdesigned to prevent failure. Wind force on the sails causes a sailboat to heel. Resistance to heeling, called righting moment (RM), results from the lateral movement of the boat's center of buoyancy away from the CG. The RM is the heart of loads throughout the boat and its structure. Placing the weight as low as possible (often in a large bulb at the bottom of the keel) extracts the maximum RM from a given mass. However, increasing the RM can result in a rapid increase of power and, in turn, the structural

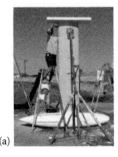

(a)   (b)

FIGURE 12.5   (a) Typical keel with keel bolts and (b) keel–hull mating. (Courtesy of EMP Composites, Sydney.)

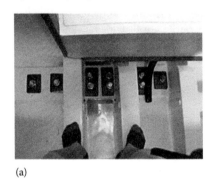
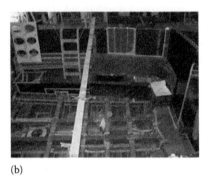

(a)   (b)

FIGURE 12.6   Keel bolt: (a) washer plate arrangement and (b) typical keel floor arrangement. (Courtesy of EMP Composites, Sydney.)

stresses, especially where the keel attaches to the hull. The parameters required to calculate the minimum diameter of the keel bolts are shown in Figure 12.7.

Figure 12.8 shows the structural configuration of a keel attached to the boat hull and the top-hat stiffener. Steel/titanium alloy bolts are used to attach the keel to the hull (Prusty and Raju, 2014).

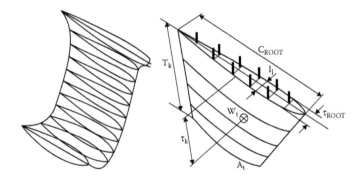

FIGURE 12.7   Typical keel showing the keel bolts and keel design. (From ABS, *Guide for Building and Classing Motor Pleasure Yachts*, Houston, TX, USA, 2000.)

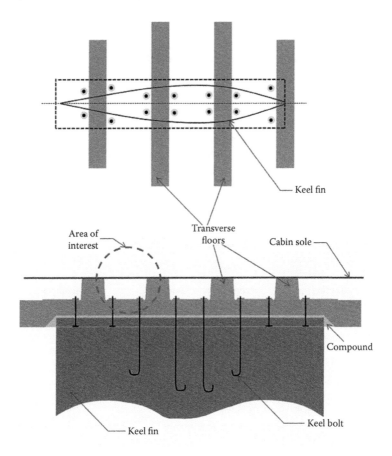

FIGURE 12.8   Keel–hull assembly. (From Raju et al., *Ocean Engineering*, 37, 1180–1192; 2010; Raju et al., *Proceedings of the Institution of Mechanical Engineers, Part M Journal of Engineering for the Maritime Environment*, 227(1), 61–80, 2013.)

12.3 COMPOSITE FAILURE

Despite their numerous advantages, composite structural applications may pose a serious risk if the load transfer and damage mechanisms are not properly analyzed. Structural components are subjected to multiple loadings such as tension, compression, shear, and flexure. The increase in the number of composite structure applications also demands in-depth study of the process of load transfer between the fiber and matrix at the microscopic level to large delamination failures at the macroscopic level. Stress distribution and failure modes in GFRP composites are complex in nature. Geometric and material nonlinearity complicates composite failure prediction; however, a fundamental feature of composite structures is that the failure is not usually a unique event but an accumulation of a gradual sequence of microcracking and delaminations leading to structural collapse (Hinton et al., 2004).

TABLE 12.2

Different Defects Found in Composite Materials

Matrix damage: cracking	Delamination: fastener holes	Overaged prepreg
Matrix damage: crazing	Delamination: fiber/matrix debond	Over-/undercured
Matrix damage: incorrect cure	Delamination: holes and penetration	Ply underlap or gap
Matrix damage: moisture pickup	Delamination: fills or fuzz balls	Prepreg variability
Matrix damage: porosity	Delamination: surface Swelling	Reworked areas
Fiber damage: broken	Barely visible impact damage (BVID)	Surface damage
Fiber damage: fiber/matrix debonding	Contamination of bonded surfaces	Surface oxidation
Fiber damage: misalignment	Corner cracking	Thermal stresses
Fiber damage: miscollination	Creep	Translaminar cracks
Fiber damage: uneven distribution	Crushing	Variation in density
Fiber damage: wrinkles or kinks	Cuts and scratches	Variation in thickness
Delamination: bearing surface damage	Dents	Variation in resin fraction
Delamination: blistering	Incorrect fiber volume ratio	Missing plies
Delamination: contamination	Incorrect materials	Erosion
Delamination: corner radius delamination	Marcelled fibers	Voids
Delamination: corner/edge splitting	Mismatched parts	Warping
Delamination: edge damage	Missing plies	

Source: Heslehurst, R.B. and Scott, M., *Composite Polymers*, 3(2), 103–133, 1990.

This could be overcome by better understanding of the material properties, which may lead to a reduced factor of safety, lower costs, lighter structures, superior design, and fewer potential hazards. Unlike metals, composites are brittle, and the possible failure parameters are matrix failure, fiber failure, interface failure, and interlaminar failure. Metals generally have ductile/brittle failure associated with static or fatigue loading. The different types of defects found in composite materials are presented in Table 12.2.

12.3.1 Matrix Failure

The principal role of the matrix in FRP composites is to secure the fibers in place, to transfer the load between fibers, and also to protect the brittle fibers from damage. Matrix imperfections, such as microcracks, voids, and porosity, lead to other failure modes. Voids cannot be eliminated but can be minimized with optimized manufacturing processes.

12.3.2 Fiber Failure

The fiber is the primary load-carrying member in the composite structure. Three modes of fiber failure are tensile, compression, and shear failure. The fracture

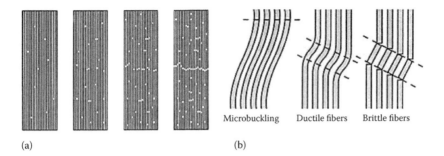

(a)   (b)

FIGURE 12.9 (a) Failure sequence in unidirectional composite with fiber-dominated strength under longitudinal tensile loading; (b) microbuckling, leading to formation of kink zones with excessive deformation or fracture planes for ductile or brittle fibers, respectively. (From Daniel, I.M. and Ishai, O., *Engineering Mechanics of Composite Materials*, Oxford University Press, 1994.)

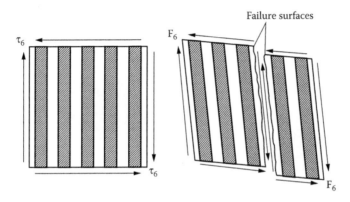

FIGURE 12.10 Failure mode of unidirectional composite under in-plane shear. (From Daniel, I.M. and Ishai, O., *Engineering Mechanics of Composite Materials*, Oxford University Press, 1994.)

behavior of E-glass (electrical-grade glass) is brittle in nature; if the tensile load/strain exceeds its critical limit, the glass fiber fracture may occur locally or globally. Fiber failure does not initiate catastrophic failure, as the load is generally distributed to neighboring fibers through the medium of the matrix. Typical longitudinal tensile and compressive failures are shown in Figure 12.9a and b. The in-plane shear failure process of unidirectional composites is shown in Figure 12.10.

Typical transverse tensile and transverse compression failures are shown in Figure 12.11a and b. Impact failure may cause fiber failure at the impacted area, depending on the size of the impact zone and the impact velocity. Typical fiber failures (tensile and compressive) and interfiber failures are shown in Figure 12.12.

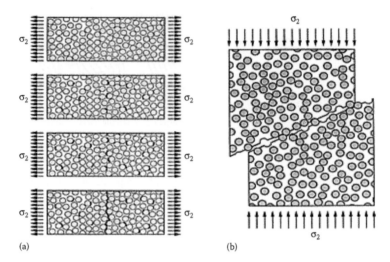

FIGURE 12.11 (a) Progressive microcracking leading to ultimate failure in unidirectional composite under transverse tension. (From Daniel, I.M. and Ishai, O., *Engineering Mechanics of Composite Materials*, Oxford University Press, 1994.) (b) Shear failure mode in unidirectional composite under transverse compression.

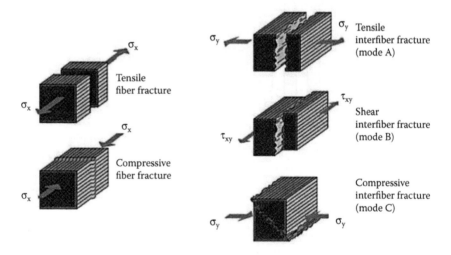

FIGURE 12.12 Fiber and interfiber failure modes. (From Maligno, A.R., Finite element investigations on the microstructures of composite materials. PhD thesis, The University of Nottingham, UK, 2007.)

12.3.3 INTERFACE FAILURE

Fiber–matrix debonding may initiate due to microscopic cracks arising in the matrix material during loading. When fiber breakage occurs, the fiber tips act as microscopic cracks and cause debonding. Good fiber–matrix interface strength controls the fiber pullout.

12.3.4 Interlaminar Failure: Delamination

Delamination is the primary failure mode in composites and reduces the strength and integrity of the structure, particularly through matrix cracking (Garg, 1988; Sridharan, 2008). Delamination is induced by interlaminar tension and shear due to inherent factors such as stiffness (elastic modulus) mismatch between the layers, structural discontinuities, free-edge effects around holes, impact by a foreign object, and through-thickness stress in curved composites. External factors such as fatigue, moisture, and temperature variation and voids may also cause delamination. Other failure modes, such as matrix cracking, can also induce delamination. The most common sources of delamination in structural composites are material and structural discontinuities, as shown in Figure 12.13.

Initial failure through delamination can be undetected, as delamination is frequently embedded between layers within the composite structure. In some cases, delamination may reduce the residual strength of the structure by as much as 60%. The problem of interlaminar tensile/compressive or shear stress was highlighted at the bonded joints and attachments of marine composite structures (Dodkins et al., 1994). Wisnom (1996; Kaczmarek et al., 1998) was one of the primary investigators to study delamination failures, using experimental, analytical, and numerical approaches to detect interlaminar failure and measure the flexural strength of a composite laminate.

The interlaminar tensile and shear stresses are the key parameters in defining the delamination strength of the composite. However, these are matrix-related properties; it is very difficult to increase the delamination strength of the specimen without external fastenings. Many methods have been suggested for arresting interlaminar failure, such as using tougher matrix polymers, interleaf layers, through-thickness

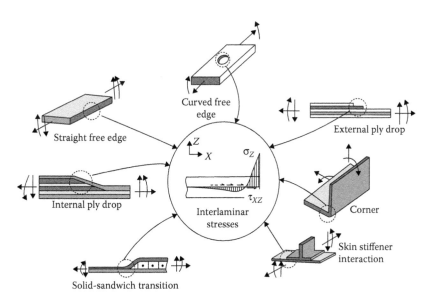

FIGURE 12.13 Delamination sources: material and geometric discontinuities. (From Sridharan, S., *Delamination Behaviour of Composites*, CRC, Boca Raton, 2008.)

reinforcement, improved fiber/matrix strength, and optimization of fabrication (Mouritz et al., 1997a, b).

12.4 NONDESTRUCTIVE EVALUATION OF COMPOSITES

Nondestructive evaluation (NDE) refers to the family of test techniques that evaluate the material integrity of a laminate (as opposed to its performance or functionality) by detecting subsurface flaws without damaging the actual test specimen. Composite structures are nonhomogeneous and anisotropic, which makes damage detection in those structures more difficult. Laminated composite materials can have a widely varying set of material properties based on the chosen fibers, matrix, and manufacturing process. This makes modeling composites complex and often nonlinear. Reliable damage detection techniques are needed to maintain the strength and integrity of the structure. Measurements from the different techniques can be interpreted to provide four possible level of diagnosis:

- Detection and progression of cracks
- Location of cracks/damage
- Damage force magnitude
- Residual strength/remaining life of the structure

Different commercially available NDE techniques have been developed. The reliability of the NDE technique is the essential issue; however, a comparison of techniques is only significant if it refers to a single task. Each NDE technique has its own set of advantages and disadvantages, and some are better suited for a particular application. The following section provides an overview of the available techniques.

12.4.1 TYPES OF NDE

Commonly available commercially NDE methods are

- Visual inspection
- Tap testing
- Ultrasound testing
- Liquid penetrant testing
- Infrared and thermal testing
- Eddy current testing
- Microwave testing
- Magnetic measurements
- Laser interferometry
- Magnetic particle testing
- Acoustic emission testing
- X-Ray radiographic testing
- Thermographic testing
- Shearographic testing
- Optical fiber sensing
- Acousto-ultrasonics

TABLE 12.3

Common NDE Techniques for Composites

Type of NDE	Damage Type	Advantages	Limitations
Visual inspection	Fatigue Cracks Delamination	Does not require sophisticated equipment Relatively inexpensive	Time consuming Limited accuracy
Ultrasonics	Fatigue Cracks BVID Corrosion Delamination	Well established and understood Sensitive to small damage Possible damage location Good depth ranging Relatively inexpensive Possible *in situ* monitoring	Point monitoring Requires coupling Often does not detect closed crack Sensitive to geometry
Eddy current	Fatigue Cracks	Detection of small cracks Possible noncontact testing Does not require coupling Possible data storage Relatively inexpensive	Used mostly for crack detection Point monitoring Requires specific monitoring skills Requires calibration Poor penetration
Acoustic emission	Fatigue Cracks BVID Corrosion Delamination	Large structures are monitored Not sensitive to geometry Possible *in situ* monitoring	Not sensitive to small damage
X-ray radiography	Fatigue Cracks BVID Corrosion Delamination	Fast monitoring Good penetration Relatively inexpensive	Not sensitive to small damage Not possible for large structures
Thermography	Fatigue Cracks BVID Delamination	Fast monitoring Large structures are monitored	Expensive Not sensitive to small damage Poor penetration
Shearography	Fatigue Cracks BVID Corrosion Delamination	Fast monitoring Large structures are monitored	Very expensive Not well developed Passive technique
Optical fiber sensing	Fatigue Cracks BVID Delamination	Damage detection in physically accessible area Possible *in situ* monitoring High signal-to-noise ratio Distributed sensing possible	Expensive Passive technique Limited sensitivity in dynamic mode Localized detection
Acousto-ultrasonics	Fatigue Cracks Delamination	Nonpassive technique	Susceptible to surface contamination Can yield false alarms Signal includes reflected and scattered modes.

Source: Staszewski et al., *Health Monitoring of Aerospace Structures: Smart Sensor Technologies and Signal Processing*, Wiley, Chichester, England, 2004; Li et al. Composite Structures, 66, 239–248, 2004; Raju et al., *Theory and Uses of Acoustic Emission*, Nova Science, New York, 2011.

NDE techniques commonly used with composites are described in Table 12.3.

Marine composite structures are generally large, and not all NDE techniques can cover such large areas. Delaminations are generally embedded in the structure; thus, penetration is required. The cost can also affect the application of NDE methods. Ultrasonic devices are sensitive to geometry, and the current thesis is based on curved composites and out-of-plane load-transfer mechanisms; thus, ultrasonics has limited applicability in the current study. Among the various forms of NDE technologies presented in Table 12.3, acoustic emission (AE) and optical-fiber sensing are well suited for composite damage analysis, as they are highly capable of *in situ* failure monitoring.

12.4.2 OPTICAL-FIBER SENSING

The different types of loads on a marine structure are quasi-static loads (wave loading, vertical and lateral combined), dynamic loads (slamming load, whipping load, and springing load), thermal effects, and the weight of the structure. These loads create a complex state of stresses in the structure. Stresses in solids cannot be measured directly; however, the strain can be directly measured, and the corresponding stress values are calculated. Acoustic emission and optical-fiber sensing are passive forms of *in situ* SHM, as no additional excitation is applied to the structure other than the operational load. Various researchers (Davis et al., 2002; Li et al., 2004) have shown that the occurrence of damage in a structure alters the local strain distribution under structural load due to the changing load path. Thus, the occurrence and extent of damage can be detected when the measured strain at a given location significantly diverges from the value expected for a healthy structure.

Strain-based damage detection is not limited by the physical properties of the target structure. This makes it suitable for a wide range of applications, from traditional metallic structures to modern fiber-reinforced composites. The major limitation of this technique is that it measures strain variations at a local scale or near locations of the installed sensors. Thus, the design of the sensor network requires knowledge of damage locations, which can be obtained from numerical simulation, experimental testing, and previous experience. For accurate strain measurement, various strain-measuring systems exist, such as semiconductor strain gages, foil gages, mercury-in-rubber strain gages, capacitive strain gages, and fiber-optic sensing. This study focuses on the application of embedded optical-fiber sensors (OFS) to measure the localized damage in glass fiber composites subjected to out-of-plane loading mechanisms.

Another NDE method suitable for damage monitoring of composite structures is optical-fiber sensing. The primary application of fiber-optic sensors (FOS) has been for data transmission in the telecommunications industry. Optical fibers can transmit light over a long distance with minimal loss, and the properties of the light inside the fiber are not affected by physical parameters outside the fiber. This implies that the fiber can be used as both the sensing element and the communication path for the signal between the sensor and the optical interrogator. These sensors are lightweight and deliver superior speed, sensitivity, and bandwidth. The key attributes of FOS are immunity to electromagnetic interference (EMI), intrinsic or extrinsic placement, water and corrosion resistance, low maintenance, ruggedness, smaller size, and low cost, and they can be multiplexed in parallel or in series. Modern FOS are suitable for the measurement of temperature, pressure, strain, angular rotation speed, acceleration, curvature, flow, refractive index, and many other parameters (Staszewski, 2004). The combination of high precision, high measurement speed, and a broad wavelength measurement range is still quite expensive;

however, in large measuring systems with many sensors per interrogator, the cost per channel can drop to a low level. Optical fibers are steadily becoming more cost effective due to recent advances in the telecommunications, optoelectronic, and micro and nano industries.

OFS offer new capabilities, provide an alternative to electronic sensors, and have a good record in terms of performance, reliability, and cost. The development of OFS focuses on niche areas; the sensors are made to match the performance of electronic sensors while retaining their inherent advantages. These advantages of OFS are (Lee and Jeong, 2002):

- Optical fiber is immune to electromagnetic fields, high voltage, high temperature (~1000°C for fused silica), and chemical reactions (corrosion, ionization). These attributes allow safe deployment of OFS in almost any environment that normally will harm electronic sensors.
- Optical fiber has a small diameter (125 μm) and is lightweight and compatible with fiberglass composites. Therefore, a large amount of FOS can be embedded arbitrarily in composites with negligible effects on the host structure.
- The performance of bare fiber sensors can be optimized by packaging techniques to meet specific requirements of a targeted application. Sensor packaging is essential, as demonstrated in the previous work on improving the performance of pressure sensitivity (Cole et al., 2002; Lin et al., 2005; Goodman et al., 2008), linearity, frequency response of either broadband (Morris et al., 2005) or resonance (Lee and Jeong, 2010), mechanical strength, and temperature stability (Horowitz and Amey, 2001).
- With a variety of multiplexing schemes available, the number of sensors per optical cable can be multiplied to reduce the cost per sensor. Among the available multiplexing schemes are time, frequency, wavelength, code, coherence and hybrid multiplexing (Cranch et al., 2003; Childs et al., 2010).

Some drawbacks also need to be addressed:

- Optical fibers have very poor elastic properties. They are very brittle and can easily break. Risk-mitigation procedures are required during sensor fabrication, measurement, and installation to avoid sensor failure.
- *In situ* measurement is cumbersome and time consuming, since measurement setup of OFS systems involves several different modules, including the light source, the coupler, and the receiver. In contrast, the well-established electronic sensor has all its system components on a single board, which is very convenient for on-site measurement.

12.4.2.1 Basic Principles

Optical fibers consist of a high–refractive index core, where broadband light travels. A low–refractive index cladding surrounds the core (mostly glass) to reflect light

Sensor parameter	Single-mode fibre (µm)	Multimode fibre (µm)	Material
Core	2–10	50–150	Silica-based
Cladding	80–120	100–250	Doped silica
Coating	250	250	Polymer
Buffer	900	900	Plastic

FIGURE 12.14 Schematic of the basic structure of an optical fiber.

back into the core, as shown in Figure 12.14. The total internal reflection guides the light inside the core. A coating (typically acrylate or polyimide) is applied to fibers to protect against environmental damage. Single-mode fiber (for distances >550 m) has a small core of 9 µm and can transmit broadband light from 1300 to 1550 nm. Multimode fibers (for shorter distances) have larger cores of 62.5 µm, capable of transmitting broadband light from 850 to 1300 nm.

Polymer optical fiber (POF) is gaining importance due to its increased strain limit of 10% (the strain limit of glass fiber is 1%, which makes it very brittle). The application of POF to advanced composite structures is limited, as the minimum diameter of POF is 1 mm, which is too thick for composites and may cause stress concentrations. Fundamental OFS used with composite structures are fiber Bragg gratings (FBG), intensity-based optical fibers, polarimetric sensors, and a range of optical-fiber interferometric sensors based on the Fabry–Perot, Mach–Zehnder, and Michelson configurations (Kuang and Cantwell, 2003).

12.4.2.2 Optical-Fiber Modulation Techniques

Commonly used modulation techniques are found in polarimetric and interferometric sensors.

1. *Polarimetric sensors*: These sensors rely on the principle of coherent interference between two light beams from a common source traveling along two different polarization axes of a common optical fiber, as shown in Figure 12.15.

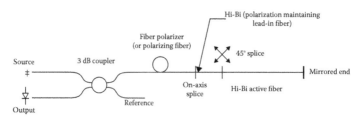

FIGURE 12.15 Schematic diagram of a polarimetric sensor configuration. (From Schwartz, M., *Smart Materials*, CRC, Boca Raton, 2009.)

2. *Interferometric sensors*: These sensors rely on the phase change between two interfering paths of light in response to external stimuli, such as strain, pressure, temperature, or chemical reaction. The different forms of interferometric sensors shown in Figure 12.16 are
   a. IFPI: intrinsic Fabry–Perot interferometric sensor
   b. EFPI: extrinsic Fabry–Perot interferometric sensor
   c. ILFE: in-line fiber etalon
   d. FBG: fiber Bragg grating
   e. LPG: long-period grating
   f. TFBG: tilted fiber Bragg grating
   g. DFBG: D-shaped fiber Bragg grating

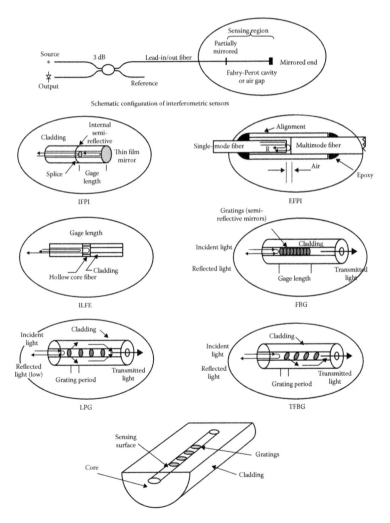

FIGURE 12.16 Schematic configurations of IFPI, EFPI, ILFE, FBG, LPG, TFBG, and DFBG sensors. (From Schwartz, M., *Smart Materials*, CRC, Boca Raton, 2009.)

12.4.2.3 Bragg Grating Sensors

Among the various types of sensors, FBG sensors are widely used in the strain monitoring of advanced composites. The basic principle of the FBG technique is shown in Figure 12.17. A periodic variation etched directly into the core of an optical fiber creates a partially reflective mirror that returns selective wavelengths in the form of a narrow spectral peak determined by the grating's period.

FBG has been demonstrated as a fundamental component in the detection of different types of measurands, including strain, temperature, and acoustic emission. The FBG is formed by introducing a periodic change of the refractive index in a continuous length of optical fiber, which can be achieved by exposure of the fiber core to an intense optical interference. Each index perturbation reflects the incoming field and constructively forms a back-reflected peak with the center wavelength determined by the grating parameters. The center or Bragg wavelength, λ_B, of an FBG is given by:

$$\lambda_B = 2n_{\mathrm{eff}}\Lambda \tag{12.1}$$

where:

n_{eff} is the effective refractive index of the optical fiber
Λ is the grating period

Figure 12.18 shows a schematic of a FBG and light interaction inside the FBG (Raju et al., 2011).

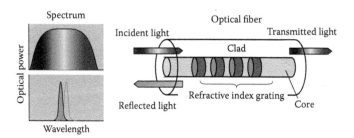

FIGURE 12.17 Schematic diagram of an FBG.

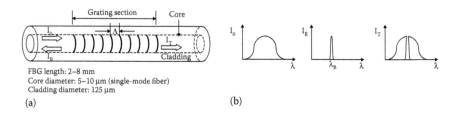

(a)   (b)

FIGURE 12.18 (a) Schematic of an FBG and (b) graphical representation of light interaction when a broadband optical field, I_0, is launched into FBG, resulting in a reflected, I_R, and a transmitted, I_T, field.

The wavelength of a FBG is affected by changes of strain and temperature. Both factors change the effective refractive index of the core as well as the index change period. The shift of Bragg wavelength due to strain and temperature change is given by Othonos and Kalli (1999):

$$\Delta\lambda_B = 2\left(\Lambda\frac{\partial n_{\text{eff}}}{\partial l} + n_{\text{eff}}\frac{\partial \Lambda}{\partial l}\right)\Delta l + 2\left(\Lambda\frac{\partial n_{\text{eff}}}{\partial T} + n_{\text{eff}}\frac{\partial \Lambda}{\partial T}\right)\Delta T \tag{12.2}$$

where:

T is the temperature
l is the grating length

The first term of Equation 12.8 represents the strain effect, which can also be written as (Meltz and Morey, 1992)

$$\Delta\lambda_B = \lambda_B\left(1 - p_e\right)\varepsilon_z \tag{12.3}$$

where ε_z is the longitudinal strain and p_e is the strain-optic constant given by

$$p_e = \frac{n_{\text{eff}}^2}{2}\left(p_{12} - \nu(p_{11} + p_{12})\right) \tag{12.4}$$

where:

ν is Poisson's ratio
p_{11} and p_{12} are the elasto-optic coefficients

For a germanosilicate optical fiber, these parameters have the following values: $\nu = 0.16$, $p_{11} = 0.113$, $p_{12} = 0.252$, and $n_{\text{eff}} = 1.482$.

The second term of Equation 12.2 gives the wavelength shift due to the temperature change, ΔT, given by (Meltz and Morey, 1992)

$$\Delta\lambda_B = \lambda_B(\alpha_\Lambda + \alpha_n)\Delta T \tag{12.5}$$

where:

α_Λ is the thermal expansion coefficient
α_n is the thermo-optic coefficient

For silica, $\alpha_\Lambda = 0.55 \times 10^{-6}$ and $\alpha_n = 8.6 \times 10^{-6}$.

With proper calibration and compensation for thermal variations, an accurate localized measurement of strain at the sensor can be obtained. The length of the sensor varies from 1 to 20 mm to achieve versatile sensing requirements, and the grating reflectivity can be approximately 100% in practice, depending on the length of the fiber (Rao, 1997).

The gratings may be written using the following techniques:

- Split beam interferometer
- Diffractive optical element phase mask technique
- Point-by-point method

To achieve a high Bragg reflectivity, the photosensitivity of the pristine fiber is enhanced, either by doping the core with elements such as Ge or Sn or by hydrogen loading the optical fiber via immersion in high-pressure hydrogen. It has been observed (Wei et al., 2002) that FBGs written in hydrogen-loaded fibers have less than 50% of the mechanical strength of FBGs that have not been hydrogen loaded. To achieve high reflectivity, this loading is unavoidable, and this strength issue must be addressed (Raju et al., 2011).

The manufacturing costs of FBGs are still relatively high; current manufacturing technology is not suitable for mass production. Although FBG manufacturing methods have improved considerably, current technology only manages to produce one FBG at a time, and most of the processes still require manual handling. In addition, running FBG facilities is expensive due to high electricity consumption, expensive ultrahigh-purity nitrogen, and the requirement for highly skilled operators.

12.4.2.4 Multiplexing

Optical fibers are well suited to multiplexed networks, as each FBG can be written at a unique wavelength, as is done in the telecommunications industry. This gives FBG sensor systems the important property of being able to simultaneously read large numbers of sensors on very few fibers, reducing the cabling requirements and simplifying installation. Three different techniques for multiplexing are

- Wavelength-division multiplexing (WDM): with WDM, the sensors in the network are initialized with their own wavelength, such that they can only respond to a particular range of wavelengths.
- Time-division multiplexing (TDM): TDM demodulates a series of discrete pulses from individual sensors, which are separated by fibers that are longer than the optical pulse.
- Spatial-division multiplexing (SDM): here a single light source is required, and the outputs from each sensor are directed to individual detectors; therefore, there is no multiplexing advantage at the detection end of the system.

12.4.2.5 FBG Interrogation

A typical FBG sensing system, shown in Figure 12.19, consists of a broadband light source, a coupler (to guide the emissive light to the specimen and the reflected light back to the optical spectrum analyzer [OSA]), an array of FBG sensors, and an OSA.

The reflected signals from the specimen (sensor) are then recorded by the OSA, in which the reflection spectrum is observed with a sharp single peak. The reflected wavelength (λ_B) is obtained using the equation in Figure 12.19, where λ_B is the wavelength at maximum reflectivity, n_{eff} is the effective mode index of

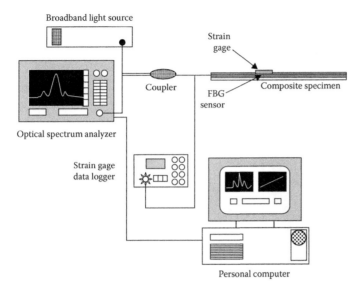

FIGURE 12.19   Data acquisition system for FBG interrogation. (From Ling, H.Y. et al., *Smart Materials and Structures*, 14, 1377–1386, 2005.)

refraction, and Λ is the grating period (Erdogan, 1997). Both the grating period and the refractive index are dependent on the temperature and mechanical strain that the FBGs experienced.

12.5   DAMAGE CHARACTERIZATION OF ADVANCED COMPOSITES USING OPTICAL-FIBER SENSORS

The sensor signal is compared with the response of a healthy structure (predetermined data), and any statistically significant variation is attributed to the presence of damage or damage progression. OFS applications have focused on concrete structures, composite structures, and metal alloys. Composites suffer various failure modes not seen in metallic alloys, such as delamination, fiber breakage, and matrix cracking. A significant number of research papers have been published concerning the use of FOS with advanced composite materials.

12.5.1   MATRIX CRACKING

- *(Ussorio et al., 2006)*: They embedded a single-mode optical fiber with a 10 mm FBG in a GFRP cross-ply laminate to monitor crack initiation and development due to matrix cracking. The spectra reflected in the case of undamaged material exhibit a single peak and a symmetrical shape. After the formation of a crack at 0.3% strain, the spectra become skewed, with a broadening of the spectra at higher wavelengths, as shown in Figure 12.20.

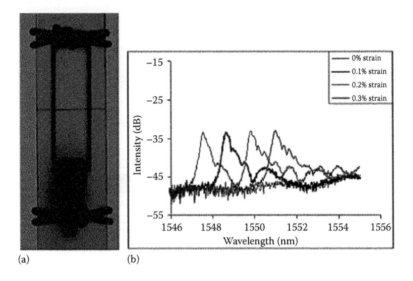

(a)　　　　　　　　(b)

FIGURE 12.20 (a) A matrix crack grown fully across the coupon; (b) reflected FBG spectra at increasing applied strains for a coupon with a crack.

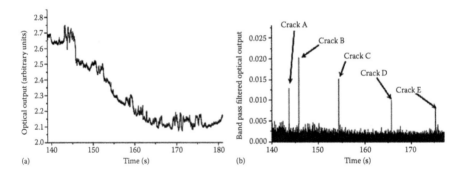

(a)　　　　　　　　(b)

FIGURE 12.21 (a) Optical output from the sensor when five cracks pass it; (b) band-pass filtered signal using fast Fourier transform (FFT) band-pass range of 100–300 Hz.

Secondary peaks in the spectrum at higher wavelengths become more prominent with increasing applied strain.

- *(Wang et al., 2005)*: A polarimetric OFS was embedded in cross-ply GFRP laminate to monitor transverse-ply crack initiation and propagation in real time. The combined observation of an extensometer, load recordings, and a video recording of the crack development enable a direct correlation between matrix crack growth past the sensor and a step-change in the sensor response, as shown in Figure 12.21.

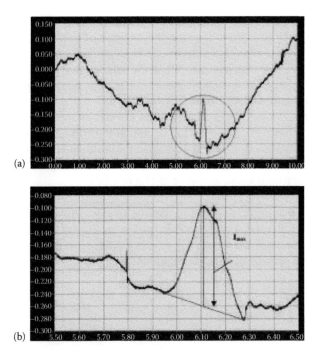

FIGURE 12.22 (a) Integral strain vs. time plot from the interferometer; (b) a detailed view of the jump.

12.5.2 Debonding

- *(Xu et al., 2005)*: Experimental and numerical studies were carried out on debonding detection based on fiber-optic interferometry using a surface-mounted optical fiber along the length of the repair material in a carbon fiber–reinforced polymer (CFRP) composite specimen. The location and extent of damage were identified using induced perturbations on the plot of the fiber integral strain vs. load position, as shown in Figure 12.22. The load was applied at different positions, and the integral strain (or total elongation) was measured with fiber-optic interferometry. The technique was simple and suitable for *in situ* damage monitoring.

12.5.3 Crack Monitoring in a Bonded Repair System

- *(McKenzie et al., 2000)*: They embedded an optical fiber with an array of FBGs for *in situ* monitoring of crack propagation beneath the bonded repair systems on cracked aluminum skins. The FBGs were multiplexed with a combination of wavelength and spatial techniques using a tunable Fabry–Perot filter to track individual gratings. The sensors were located appropriately to monitor the changing stress fields associated with the propagation of cracks beneath the bonded patch. Three-dimensional (3D) finite element analysis (FEA) was supplemented with the experimental study. During crack propagation, the strain gradient was monitored to assess the extent of damage. In addition to this test, they conducted tests on graphite epoxy doubler, bonded to both the upper and lower surfaces of the specimen, to evaluate delamination damage in both the adhesive and the composite doubler. A range of disbond and delamination lengths were considered: 0, 26, 31, 36, 41, and 46 mm from the edge of the patch.

- *(Li et al., 2006)*: They conducted experimental and numerical investigations on marine composite structural joints with various lengths of artificial disbonds. A methodology was developed for health monitoring of composite marine T-joints based on strain measurements using embedded FBG under operational loads, as shown in Figure 12.23.

They argued that the presence of disbonding significantly altered the local strain distribution, and that the damage-induced strain anomalies could be successfully detected using the embedded sensors. Using the localized strain gradient variations, the location and intensity of damage were monitored.

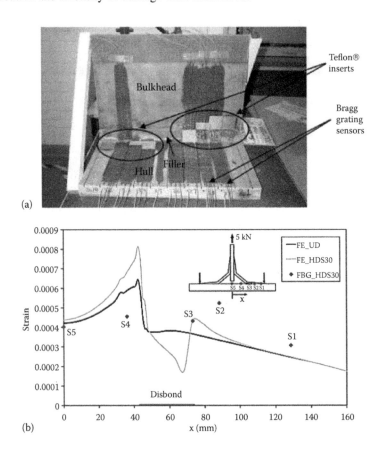

FIGURE 12.23 (a) T-joint before lamination; (b) comparison of bond-line longitudinal strain distribution from FEA, FBG, and experimental results.

12.5.4 Detection of Transverse Cracks

- *(Okabe et al., 2000)*: FBG sensors were used to detect transverse cracks, which cause nonuniform strain distribution within the gauge length in CFRP cross-ply laminate, as shown in Figure 12.24. With an increase in the crack density, the shape of the reflection spectrum was distorted. The intensity of the highest peak was reduced, some peaks appeared around it, and the spectrum broadened. As the crack density approached close to saturation, the spectrum became narrow again, and the highest peak recovered its height. The changes in the spectrum were thought to be due to the strain distribution caused by the transverse cracks in 90° ply. The occurrence of transverse cracks was detected from the changes in the reflection spectrum. One of the practical issues they encountered was that during the early stages of damage initiation, the spectrum split into two peaks due to the transverse thermal residual stresses.
- *(Takeda et al., 1999)*: Takeda is one of the key researchers in the usage of optical fibers for crack detection. In this study, the researchers tested flat coupons embedded with plastic optical fiber. It was concluded that the optical power of the fiber decreases nonlinearly after the initiation of transverse cracks when the specimen is loaded by tension or bending.

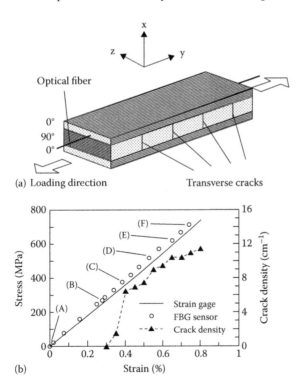

FIGURE 12.24 (a) Embedding an FBG sensor in a composite cross-ply laminate; (b) crack density and stress as a function of strain measured from a strain gage and FBG.

12.5.5 DELAMINATION

- *(Grouve et al., 2008)*: The influence of delamination on the resonant frequencies of carbon–polyetherimide flat composite specimens was studied using artificial delaminations (16 μm foil polyimide). A clear shift in the resonance frequency was observed due to the presence of delaminations. The FBGs could measure the strains up to 950 Hz. At higher frequencies, the strain variation was too small for the FBG sensors to measure. It was also observed that the standard deviation of the FBG sensor before and after manufacturing the specimen was 1 and 3.7 με, respectively, demonstrating the negative influence on the signal resolution during specimen manufacturing.

- *(Leung et al., 2005)*: An interferometric sensor was used to detect the delamination in CFRP composites with artificial Teflon inserts, as shown in Figure 12.25. A quasi-impulse load was applied at different points along the specimen length. When delamination occurred, a sudden jump in the magnitude of the integral strain was observed. It was shown that the location and extent of damage could be deduced from the points where the slope of the integral strain vs. load position curve showed a sudden change.

- *(Ling et al., 2005b)*: They conducted FEA and experimental three-point bending tests on GFRP composites with different edge delaminations in

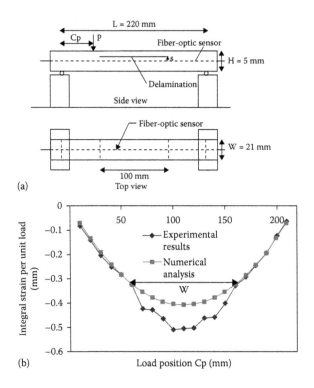

(a)

(b)

FIGURE 12.25 (a) Experimental setup and (b) integral strain vs. load position for comparison.

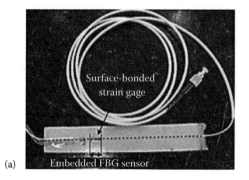

(a)

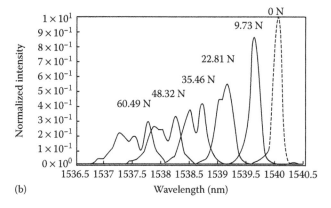

(b)

FIGURE 12.26   (a) Composite specimen with an embedded FBG sensor and (b) reflection spectra from the FBG sensor with an increasing applied load.

the through-thickness direction for eight different stacking sequences, as shown in Figure 12.26. An optical fiber with FBG sensors was embedded to characterize the delaminations in the specimen using a static-strain method. An abrupt change in the strain gradient due to the delamination caused deformation of the reflection spectra.

- *(Sorensen et al., 2007)*: Nonuniform strains resulting from the propagation of mode I delamination in double cantilever beam (DCB) specimens were measured. They embedded FBG sensors with a length of 22/35 mm in carbon fiber–polyphenylene sulfide composite specimens. The residual strains were used as the reference data. Whenever the delamination crack grew to a desired length, the FBG readings were compared with the reference data. The measured distribution of strains in the FBG sensor and numerical modeling were used in an inverse procedure to determine the actual distribution of bridging stresses during delamination.
- *(Takeda et al., 2005)*: They developed a technique for the quantitative evaluation of delamination length in CFRP laminates using Lamb wave sensing with embedded FBG sensors.

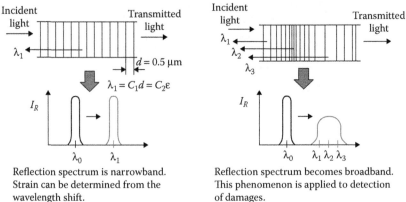

FIGURE 12.27 Response of FBG sensors to (a) uniform and (b) nonuniform strain distribution.

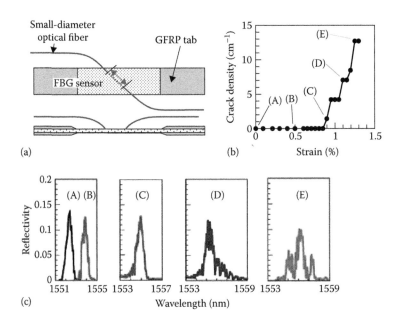

FIGURE 12.28 (a) Specimen with embedded FBG sensors, (b) transverse crack evolution, and (c) change in wavelength distribution of reflected light from the FBG sensor.

A piezo-ceramic actuator generated Lamb waves in a CFRP laminate. During wave propagation in the laminate, FBG sensors detect the transmitted waves, which correspond to the interlaminar delamination. When the Lamb waves passed through the delamination, a reduction in the peak amplitude (reflectivity) was observed, and a new wave mode emerged, as shown in Figures 12.27 and 12.28. This was confirmed with FEA

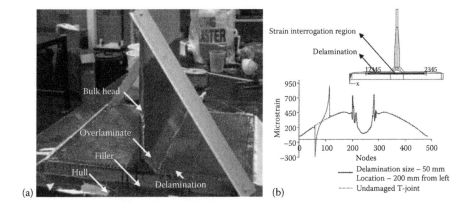

FIGURE 12.29   (a) T-joints embedded with artificial delaminations during manufacturing; and (b) strain distribution along the bond-line of a T-joint with embedded delaminations.

simulation. By analyzing the changes in the amplitude ratios and the arrival time of the new mode, they were able to evaluate the delamination length quantitatively.

* *(Kesavan et al., 2008)*: Fourteen optical-fiber sensors were embedded in GFRP composite T-joints with artificial disbonds (Teflon inserts) to detect the damage mechanism, as shown in Figure 12.29. The sensors were able to detect the presence of multiple delaminations and determine the location and extent of all delaminations present in the T-joint structure, regardless of the load (angle and magnitude) acting on the structure. Two-dimensional FEA, using MSC.NASTRAN, was conducted to study the strain behavior of T-joints with multiple delaminations.

12.5.6   Effects of Embedded Optical Fibers on Structural Integrity

It is a general assumption that the strain measured by embedded FBG sensors is the same as the strain experienced by the host material. Nevertheless, in practical applications, not all the strain is transferred to the sensor. Some energy may be damped by the protective coating, spray adhesives used, and other impurities. Successful and accurate strain measurement depends on factors such as sensor network design, laminate manufacturing process, and sensor location. The strain needs to be successfully transferred from the fiber/resin host to the core/cladding of the optical fiber through the primary resin-rich region, which acts as an interface. Past studies (Ling et al., 2005a) concluded that the strength and fracture behavior of the structure could be significantly affected by improper alignment and placement of optical fibers in the laminate. They concluded that use of embedded optical fibers for damage detection is accurate and reliable if the interaction between the optical fiber and the delamination is known. Despite their small size compared with the typical diameter of the reinforcement fibers, the diameter of optical fibers is large enough to induce stress concentrations and geometric discontinuities in the structure. Previous studies on the effect of embedding optical fiber on the mechanical properties (Jensen et al., 1992; Lee et al., 1995; Mall et al., 1996; Choi et al., 1997; Silva et al., 2009; Shivakumar and Emmanwori, 2004) concluded that structures where the optical fibers are embedded in the mid-plane of the composite laminate have the greatest effect. The severity depends on the stacking sequence, manufacturing process, and type of reinforcement (cross-plied, woven, short fibers, or unidirectional).

12.5.7 SUMMARY OF OPTICAL-FIBER SENSORS

Studies have been conducted using FOS to evaluate the damage mechanism in composites. The damage could be qualitatively analyzed by using damage detection schemes, as presented in Table 12.4.

It is evident that the failure mode of composites could be determined by analyzing the change in wavelength distribution, change in strain rate, and wavelength shift. This technique of damage detection requires a low level of signal processing and yields highly accurate results. In addition to strain variation, the FBGs could also provide information on temperature change. Although research has been carried out to determine the damage mechanism, some failure modes, such as matrix cracking, delamination, transverse cracks, and crack propagation, have not been investigated to characterize the damage based on the sensor response.

TABLE 12.4
Summary of FBG Studies

Type of Damage	Researchers
Matrix cracking	(Ussorio et al., 2006; Wang et al., 2005)
Debonding	(Xu et al., 2005)
Crack monitoring in bonded repair system	(McKenzie et al., 2000; Li et al., 2006)
Detection of transverse cracks	(Okabe et al., 2000; Takeda et al., 1999)
Impact damage	(LeBlanc, 1992; Measures et al., 1989; Martin et al., 1997; Chang and Sirkis, 1998; Pearson et al., 2007; Kitade et al., 1995)
Fatigue-induced damage	(Badcock and Fernando, 1995; Lee et al., 2001; Rodrigo et al., 2009)
Strain measurement	(Rowe et al., 1986; Everall et al., 2000; Davies et al., 1994; Froggatt and Moore, 1998; Tao et al., 2000)
Delamination	(Leung et al., 2005; Elvin and Leung, 1997; Kesavan et al., 2008; Grouve et al., 2008; Ling et al., 2005a; 2005b; Sorensen et al., 2007; Takeda et al., 2005)

12.6 EXPERIMENTAL INVESTIGATION OF TOP-HAT STIFFENER

The closed-mold manufacturing process has become increasingly popular as it provides superior strength, stiffness, high surface finish, and virtually no emissions. Hence, to compare and validate the manufacturing process, the next set of top-hat stiffeners were manufactured using vacuum-assisted resin transfer molding (VARTM).

12.6.1 VARTM Manufacturing Process

Chopped strand mat (CSM) is denser and preferable for hand layup, but continuous filament mat (CFM) is generally preferred for VARTM. However, CFM often wrinkles around bends and makes the structure unsuitable; thus, a small amount of CSM was used with VARTM in the current study. The gel time was increased to assist the resin flow throughout the mold. The basic materials used are given in Table 12.5.

Three different layups were tested, and the layup configurations are presented in Table 12.6.

TABLE 12.5
Basic Materials for VARTM

Fiber	E-glass: chopped strand mat (CSM)	451 and 225 gsm
	E-glass: double bias (DB)	611 gsm
	E-glass: unidirectional Fibers (UD)	461 gsm
	E-glass: B-axial (BE)	461 gsm
Resin	Vinylester–SPV 6036–FGI	
Hardener	Norox CHM-50: 1.5%	

TABLE 12.6
VARTM Top-Hat Stiffener Laminate Configuration

Layup	1	Layup	2	Layup	3
CSM + DB **Mold Surface**	**Thickness** **(mm)**	**CSM + DB + UD** **Mold Surface**	**Thickness** **(mm)**	**CSM + DB+UD + BE** **Mold Surface**	**Thickness** **(mm)**
DB-450	0.26	CSM-225	0.3	CSM-225	0.3
CSM-450	0.6	CSM-225	0.3	DB-450	0.26
DB-450	0.26	DB-450	0.26	CSM-225	0.3
CSM-450	0.6	CSM-225	0.3	DB-450	0.26
DB-450	0.26	DB-450	0.26	CSM-225	0.3
CSM-450	0.6	CSM-225	0.3	BE-450	0.465
DB-450	0.26	UD-461	0.26	UD-461	0.26
DB-450	0.26	CSM-225	0.3	CSM-225	0.3
		UD-461	0.26	BE-450	0.465
		CSM-225	0.3	UD-461	0.26
Total thickness	**3.1**	**Total thickness**	**2.84**	**Total thickness**	**3.17**

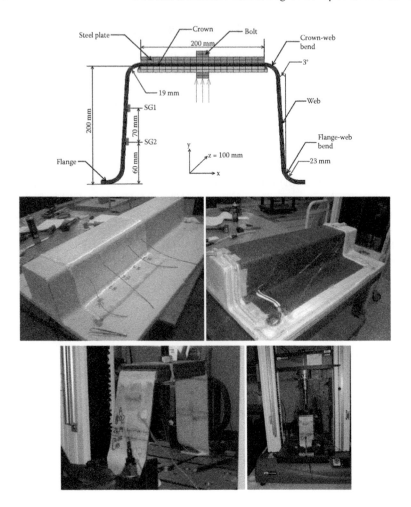

FIGURE 12.30 Top-hat stiffener manufacturing with VARTM Testing. (From Raju et al., *Ocean Engineering*, 37, 1180–1192, 2010.)

Although VARTM is expensive and requires more complex tools and equipment than the hand-layup technique, the end product has a higher fiber fraction, consistent thickness, and fewer voids, and is a superior-quality specimen. A snapshot of the top-hat stiffener specimens manufactured by VARTM is shown in Figure 12.30. The specimens are cured under the vacuum bag for 24 h at a temperature of 20°C. Later, they are sliced to the required dimension, the edges are sanded, and they are cleaned thoroughly with acetone.

12.6.2 VARTM Top-Hat Stiffener Results

Four different specimens were tested for each layup scheme under displacement control of 2 mm/min. An identical test arrangement for top-hat layup using the INSTRON machine was ensured in this series of experiments. A total of 12 specimens with three layups were tested to failure to determine the failure behavior and progression. The specimens were color coded with a permanent marker at the thickness of the laminate. During the delamination/splitting process, it was possible to identify the failure location by measuring the thicknesses on both sides of the crack, either during testing or during the analysis of various photographs taken during the test procedure. The results for each layup scheme are presented in the following subsections.

12.6.2.1   VARTM Top-Hat Stiffener Layup 1 Experimental Results

Layup 1 consisted of five DB-450 gsm and three CSM-450 gsm layers. The load–deflection behavior of this layup is shown in Figure 12.31. An initial stiffness of 1 kN/mm is seen until the first failure (Table 12.7).

Initial failure of the four specimens tested occurred around 11 mm deflection. For specimen THS VIP D02, 12 distinguishable load drops are seen on the load–deflection plot in Figure 12.31. Matrix cracking occurred at 12.48 kN and 11.85 mm deflection.

A significant load drop of 2.76 kN occurred at 13.87 mm deflection. This was a crack between the second and third layers at the radius of the left crown bend. With further application of load, the structure continued to take up the load until the right crown bend failed due to delamination and eccentric loading began on one side of the specimen at 15.66 mm (14.48 kN load). Upon further loading, both the crown bends started to disappear, and at 20 mm (18.91 kN load), the right flange bend suffered a major crack between the fourth and fifth layers. A secondary stiffness started to appear and, at 23.33 mm deflection (24.22 kN load), the left flange bend lost its stiffness, and delamination occurred between the fourth and fifth layers. Further

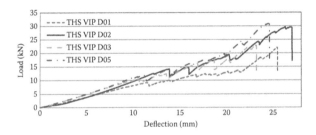

FIGURE 12.31   VARTM top-hat stiffener load deflection for Layup 1.

TABLE 12.7
VARTM-Top-Hat Stiffener Layup 1 Experimental Results

Specimen	First Failure Load (kN)	First Failure Deflection (mm)	Final Failure Load (kN)	Final Failure Deflection (mm)	Initial Stiffness (kN/mm)
D01	9.05	11.42	22.01	25.48	0.792
D02	12.48	11.85	29.57	27.04	1.053
D03	9.54	10.37	22.76	23.20	0.920
D05	10.71	10.99	30.83	24.60	0.975

FIGURE 12.32   VARTM top-hat stiffener Layup 1 damage process.

loading caused delamination propagation between other layers, and the structure finally collapsed at 27.04 mm deflection and 29.57 kN. The laminate sheared away at the left flange base below the edge of the steel plate. The damage process is shown in Figure 12.32.

12.6.2.2   VARTM Top-Hat Stiffener Layup 2 Experimental Results

Layup 2 consisted of two DB-450 gsm, four CSM-225 gsm, and two UD-461 gsm layers. The first and final failure load/deflection of all the four specimens tested are shown in Table 12.8. The first failure load and deflection were comparable for all the specimens, even though a small amount of scatter was observed.

The load–deflection plot for Layup 2 is shown in Figure 12.33. As with the hand-laid specimens, the addition of unidirectional layers created secondary stiffness in

TABLE 12.8

VARTM-Top-Hat Stiffener Layup 2 Experimental Results

Specimen	First Failure Load (kN)	First Failure Deflection (mm)	Final Failure Load (kN)	Final Failure Deflection (mm)	Initial Stiffness (kN/mm)
E01	8.02	10.01	24.26	26.30	1.248
E02	7.61	8.39	22.05	23.42	0.907
E03	8.52	12.54	19.18	30.02	0.679
E04	6.49	7.66	22.08	24.52	0.847

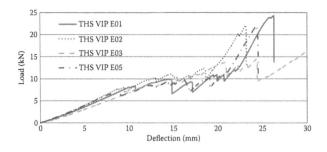

FIGURE 12.33   VARTM top-hat stiffener load deflection for Layup 2.

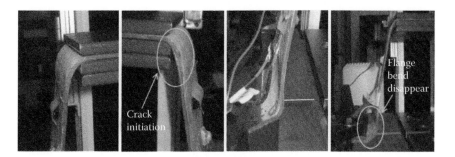

FIGURE 12.34   VARTM top-hat stiffener Layup 2 damage process.

the structure. Considerable scatter is seen at the initial failure point, but the failure mechanism and secondary stiffness process are consistent in all specimens. For specimen THS VIP E01, 10 distinguishable load drops can be seen in Figure 12.33.

The first failure was matrix cracking, with a small crack between the second and third layers at 10.01 mm deflection and 8.02 kN load, as shown in Figure 12.34. The failure process was similar to that of Layup 1—the first major crack occurred at the left crown bend between the second and third layers in the radial direction. At 20.65 mm deflection, all four bends were damaged due to delamination. Until the first failure, the stiffness was 1.248 kN/mm, and further loading caused the bends to straighten. A secondary stiffness was noticed, with a stiffness of 3.666 kN/mm. After a noticeable secondary stiffness was observed, the final failure occurred at 26.30 mm deflection at a load of 24.26 kN at the right flange bend, underneath the edge of the steel plate.

12.6.2.3   VARTM Top-Hat Stiffener Layup 3 Experimental Results

Layup 3 consisted of four CSM-225 gsm, two DB-450 gsm, two UD-461 gsm and two BE-450 gsm layers. Similarly to the other two layups, four specimens were tested till failure. The failure load, deflection, and initial stiffness for all four specimens are given in Table 12.9.

The initial failure load was around 10 kN, and the failure behavior was similar. Similarly to Layup 2, secondary stiffness was observed due to the presence of unidirectional and biaxial (BE) reinforcements. For specimen THS VIP F02, the initial

TABLE 12.9
VARTM-Top-Hat Stiffener Layup 3 Experimental Results

Specimen	First Failure Load (kN)	First Failure Deflection (mm)	Final Failure Load (kN)	Final Failure Deflection (mm)	Initial Stiffness (kN/mm)
F02	9.19	9.26	23.34	24.36	0.992
F03	10.17	13.82	18.37	29.40	0.736
F04	9.31	7.80	20.77	23.21	1.194
F05	10.32	8.04	21.52	22.82	1.284

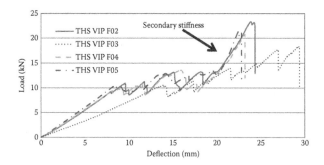

FIGURE 12.35 VARTM top-hat stiffener load deflection for Layup 3.

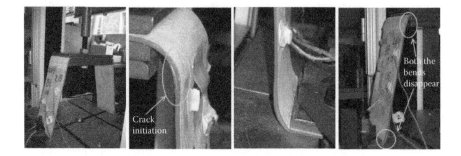

FIGURE 12.36 VARTM top-hat stiffener Layup 3 damage process.

and secondary stiffness were 0.992 and 2.48 kN/mm. Eleven load drops can be seen from the load–deflection plot (Figure 12.35).

The initial failure was matrix cracking at 7.56 mm and the major delamination occurred between the second and third layers in the right crown bend, as shown in Figure 12.36. Successive load increments caused all bends to fail due to delamination. Secondary stiffness began when the bends disappeared and the structure started to take the form of a trapezium.

12.7 FBG STRAIN SENSING

FBGs are excellent strain sensors for marine/aerospace structure applications. The manufacturing procedure for FBG as a strain sensor is explained, and the test setup process is detailed. Four FBG FOS were embedded in the curved section of the top-hat stiffener, where the maximum strain variation is expected. A minor change in FBG wavelength after the specimen manufacture process is observed due to temperature variation and residual stresses during curing. A procedure for successfully embedding the optical fibers in the curved section of the top-hat stiffener is discussed. This procedure provides satisfactory sensor performance and offers adequate protection of the optical fibers at the ingress and egress.

Top-hat stiffeners with optical fibers embedded at appropriate locations were manufactured with VARTM. Four specimens of each of three layups were tested till failure. Local strain variation during the damage process was observed, and the abnormal strain variations were captured successfully. Many researchers have applied the FBG strain sensing technique to flat specimen monitoring (Rowe et al., 1986; Davies et al., 1994; Elvin and Leung, 1997; Froggatt and Moore, 1998; Takeda et al., 1999; Everall et al., 2000; McKenzie et al., 2000; Okabe et al., 2000; Tao et al., 2000; Leung et al., 2005; Ling et al., 2005a; 2005b; Takeda et al., 2005; Wang et al., 2005; Xu et al., 2005; Li et al., 2006; Ussorio et al., 2006; Grouve et al., 2008; Sorensen et al., 2007; Kesavan et al., 2008). In this study, strain sensing is performed on the curved section of the structure, where huge strain variation could be observed.

12.7.1 FABRICATION OF FBG SENSOR

The grating was written using the phase mask technique at the University of New South Wales (Photonics and Optical Communications, School of Electrical Engineering and Telecommunications) writing facilities. As shown in Figure 12.37, the writing system consists of three main components: an ion laser, a grating fabrication enclosure, and a control system. The ion laser produces output power at 488 nm, which is then frequency doubled at 244 nm by a beta-barium borate (BBO) crystal. The standard operating power for the laser is between 200 and 300 mW at a tube current of 50 A. Standard safety procedures were implemented for working with the Class 4 laser, including wearing ultraviolet (UV) goggles and a protective coat. UV radiation exposure was minimized even when working with low power levels. The UV beam is aligned from the laser front toward the phase mask through multiple optics components, as shown in Figure 12.37.

At the turning mirror, a small portion of the beam (~1%) is passed through to a photodetector to observe beam vibration and provide feedback to the second tip/tilt mirror to compensate for the vibration. The UV beam is highly focused by a cylindrical lens so that the maximum UV power enters the fiber core. When the UV beam is incident on the phase mask, the diffracted +1 and −1 orders are maximized (each

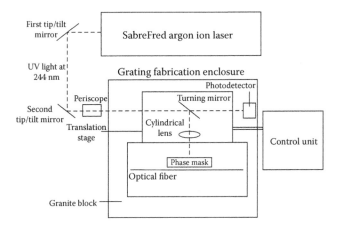

FIGURE 12.37   Setup of FBG writing system in University of New South Wales (Photonics Laboratory).

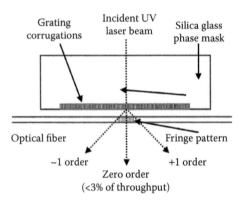

FIGURE 12.38 Phase mask technique for writing FBG.

contains approximately 35% of the total incident power), while the zero order is suppressed. This scenario is shown in Figure 12.38.

For writing a grating, a section of stripped optical fiber is placed close to the phase mask and clamped to secure its position. The UV beam is scanned along the intended FBG length by moving the translation stage at a controlled speed. A near-field fringe pattern (grating) is formed on the fiber by the plus and minus first-order diffracted beam. The period of the fringe pattern is half of the phase mask period ($\Lambda_B = \Lambda_{pm}/2$). The strength of the fringe pattern is controlled by the speed of the translation stage and the laser output power. For fabrication, several phase-mask periods were used (1056.5, 1064.7, 1076.2, and 1081.9 nm) to obtain Bragg wavelengths of 1529, 1540, 1555, and 1565 nm, respectively. Different Bragg wavelengths were attained by applying different tensions during writing. A tension of approximately 100 g produces wavelength spacing of about 1 nm. The control unit handles the entire process of writing, gives feedback to the system and, monitors the equipment status.

Once the writing process is completed, the grating position is labeled and annealed at 300°C for 5 min. Annealing is required to age the grating and to stabilize the wavelength and the spectral drifts. If the grating needs to exhibit stable long-term behavior, either a hydrogenated or a nonhydrogenated fiber is required. The final process is to bake the entire fiber in an oven at 80°C for 24 h to remove remaining hydrogen from other parts of the fiber. Then, the FBG samples are ready for experimental testing. The writing facilities and fabrication procedures yield highly reliable and repeatable processes while producing an excellent quality of FBG. Figures 12.39 through 12.41 show the main equipment and some of the processes involved in manufacturing FBGs. Figure 12.39a shows the procedure of laser alignment prior to FBG fabrication, while Figure 12.39b shows the in-process FBG manufacturing control unit.

Figure 12.40 shows a close-up view of the FBG writing equipment, while Figure 12.41 shows the annealing equipment used in this study.

The optical fiber is hydrogen loaded to increase the reflectivity of the sensor. As hydrogen loading increases the photosensitivity of the boron and germanium

FIGURE 12.39 (a) Laser alignment being performed prior to writing process to optimize lasing power and (b) the grating fabrication enclosure (left) and control unit (right) (Photonics Laboratory).

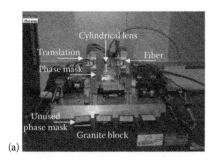

FIGURE 12.40 Inside view of the grating writing enclosure before (a) and during writing process (b) (Photonics Laboratory).

FIGURE 12.41 Equipment for annealing FBG (Photonics Laboratory).

co-doped silica fiber (Lemaire et al., 1993), hydrogen loading into single-mode fiber was performed prior to grating writing. This process is a relatively simple, cost-effective, and highly effective method of achieving high UV photosensitivity. In this process, the required length of optical fiber was mixed with hydrogen in a hydrogen cell using the standard recipe – duration of 24 h, pressure of 200 atm, and temperature of $-100°C$. The unused hydrogen-loaded fiber is kept in a refrigerator at low temperature ($-20°C$) to slow the diffusion of hydrogen from the fiber core. This retains the fiber photosensitivity for up to three days without the need for rehydrogenation.

12.7.2 LAYUP AND EMBEDDING OPTICAL FIBERS

The interlaminar stresses (tensile, compressive, or shear) are maximal around the mid-thickness of the laminate (depending on the loading mechanism) and have a near-zero value at the outer surface during structural loading. Cracks initiate between the plies, where strain anomalies within the thickness of the laminate are much greater than at the surface. Hence, embedding an OFS in the composite laminate is preferable to surface bonding for capturing strain variations due to interlaminar cracks. The FBG is well protected inside the laminate rather than being exposed on the outer surface, and embedding achieves good bonding between the host material and the sensor.

Each optical fiber consists of one FBG strain sensor. Two optical fibers were embedded at each of the two crown bends, with sensor location as shown in Figure 12.42. Tables 12.10 through 12.12 show the location of the FBG sensor at the crown bend of the top-hat stiffener in the laminate for all three different layups.

The four different wavelengths of FBGs for the four different locations were 1529, 1540, 1555, and 1565 nm. The optical fibers were color coded, as shown in Figure 12.43.

The ingress and egress points of the optical fibers must be adequately protected from excessive resin during the VARTM, and this was achieved using heat-shrink tubes. The optical fibers were inserted into 3 mm diameter heat-shrink tubes, and the tubes were shrunk with an industrial heat gun. Once the reinforcement layers were laid on the mold, FOS and the heat-shrunk tubes were placed at the appropriate positions, and the laminate was infused with resin.

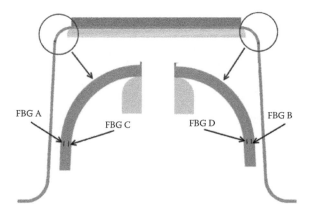

FIGURE 12.42 Location of four FBGs on a top-hat stiffener specimen.

TABLE 12.10
FBG Location in Top-Hat Stiffener Layup 1

Layer No.		Layup 1	
8	------------ DB450		
7	------------ DB450		
	-------------------------	FBG A (LEFT)/FBG B (RIGHT) – Outer	
6	------------ CSM450	(White)	(Yellow)
5	------------ DB450		
4	------------ CSM450		
3	------------ DB450		
	-------------------------	FBG C (LEFT)/FBG D (RIGHT) – Inner	
2	------------ CSM450	(Blue)	(Green)
1	------------ DB450		
	Mold Surface		

TABLE 12.11
FBG Location in Top-Hat Stiffener Layup 2

Layer No.		Layup 2	
10	------------ CSM225		
9	------------ UD450		
8	------------ CSM225		
	-------------------------	FBG A (LEFT)/FBG B (RIGHT) – Outer	
7	------------ UD450	(White)	(Yellow)
6	------------ CSM225		
5	------------ DB450		
4	------------ CSM225		
	-------------------------	FBG C (LEFT)/FBG D (RIGHT) – Inner	
3	------------ DB450	(Blue)	(Green)
2	------------ CSM225		
1	------------ CSM225		
	Mold Surface		

The heat-shrink tubes containing the optical fiber were inserted into the fiber mat (in the web section of the top-hat stiffener). An FEA analysis showed no reduction in strength of the structure due to this penetration, as the bends fail prior to the web during actual loading. A detailed discussion of the procedure for embedding the OFS in the curved bend of the top-hat stiffener is provided in Appendix G. Figure 12.44 shows the optical fibers on the mold prior to resin infusion.

Colored, protected optical fibers were again inserted into a 5 mm heat-shrink tube and the edges sealed to protect the optical fiber from resin during infusion. After curing, the top-hat stiffener was sliced to the required dimensions using a bandsaw, and the edges were sanded and cleaned with acetone, as shown in Figure 12.45.

TABLE 12.12
FBG Location in Top-Hat Stiffener Layup 3

Layer No.	Layup 3
10	----------- UD450
9	----------- BE450
8	----------- CSM225
	----------------------- FBG A (LEFT)/ FBG B (RIGHT) – Outer
7	----------- UD450 (White) (Yellow)
6	----------- BE450
5	----------- CSM225
4	----------- DB450
	----------------------- FBG C (LEFT)/ FBG D (RIGHT) – Inner
3	----------- CSM225 (Blue) (Green)
2	----------- DB450
1	----------- CSM225
	Mold Surface

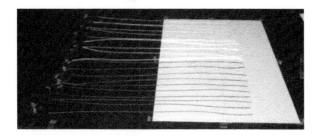

FIGURE 12.43 Optical-fiber preparation for embedding into the laminate.

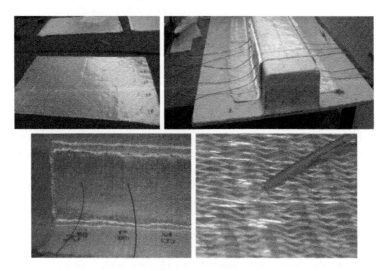

FIGURE 12.44 Positioning of optical fibers on the mold prior to resin infusion.

FIGURE 12.45   Manufacturing of top-hat stiffeners with embedded FBGs.

12.7.3   Optical Fiber Embedding Procedure with VARTM

1. Apply Mold Release coat 1 on the mold surface.
2. Plan the layup. On the mold, at least 100 mm should be left free on all sides.

3. Plan for the resin ingress and exit end space allowance, which requires at least 50 mm from each end of the mold.
4. Calculate the length and width of the specimen for cutting reinforcement fabric. Allow 25 mm on both ends and 2.5 mm per cut (including sanding/polishing).
5. Calculate the dimensions of the fabric for extra CSM layers at bottom bend—ply drop.
6. Apply Mold Release coat 2 on the mold.
7. Cut fabric to required size for laminate and for ply drop layers. Organize the fiber mats as per the layup, including ply drop layers. (Do not use tack spray at this stage.)
8. Once the layup is checked, lay the fiber mats on the mold, and remember to use ply drop layers as well. Use light tack spray between the layers to prevent fabric slippage.
9. If two layups are manufactured at a time, then add a bridging material between the layups over the mold surface to facilitate easy resin flow.
10. Cut peel ply to size and lay it over the entire laminate, 10 mm extra on all sides. Use tack spray. The resin ingress and egress ends should cover the resin inlet pipe.
11. Cut perforated film and place it on top of peel ply—use tack spray.
12. Cut flow mesh and align over the perforated film. Fix it using adhesive tape. Cut and place flow medium (50 mm wide, white honeycomb structure) at the ingress side. It should sit on top of the peel ply and a bit on the laminate and flow mesh.
13. Fix tacky-tape all around the perimeter of the mold. Cut the resin inlet hose to size and fix it at the ingress end using tacky-tape pieces over the flow medium.
14. Take a T-joint; connect a hose to the central inlet and the sides to the resin inlet pipe.
15. Cut the resin inlet hose to size and fix it at the egress end using tacky-tape pieces.
16. The resin flow needs to be restricted at the egress end. Take 600 gsm CSM, cut into pieces, apply tack spray, and fix it over the resin outlet pipe. Connect the egress pipe to the outlet pipe using T-joint.
17. Connect the ingress and egress pipe to the mold using tacky-tape.
18. Cut the nylon sheet (for vacuum bagging) to size; fix it on the mold slowly with the glued tacky-tape. Make pleats at corners for efficient circulation of resin.
19. Secure the nylon sheet, applying thumb pressure all along the length of the tacky-tape.
20. Close the inlet pipe using a clamp. Egress pipe will be in catch pot.
21. Switch on the vacuum slowly and align the nylon cover, fiber mat, and everything in the nylon bag to be secured on the mold properly. Increase the suction to 100 kPa.
22. Conduct drop test; check to see whether the vacuum pressure drops from 100 kPa, which indicates an air leak. Use acoustic emission headphones and fix all the leaks.

23. Estimate the amount of resin (Vinylester SPV 6036) and hardener (CHM 50/925H: 1.5% of resin).
24. Mix resin and hardener thoroughly in an appropriate container. The estimated gel time is around 30 min.
25. Place the container near the mold and immerse ingress pipe end in the container. Recheck the vacuum strength. Slowly release the clamp and allow resin to infuse into the laminate slowly.
26. Observe the generation of air bubbles in the ingress pipe as well as in the laminate (resin front).
27. Keep an eye on the resin container; the ingress pipe should always be immersed in the container. If there is a shortage of resin, stop the flow using the clamp, prepare new resin-hardener mix, and place it carefully under the inlet pipe.
28. Allow the whole laminate to infuse, frequently checking for any leaks, and fix the leaks by patching with tacky-tape. Disconnect the resin supply after successful infusion, and seal the ingress and egress hose.
29. Keep the mold under vacuum overnight.
30. Disconnect all the pipes, remove nylon bag, flow medium, and perforated film along with peel ply, and slowly release the laminate from the mold.

During the curing process, the OFS may be subjected to high temperature variations, and residual stresses may be induced in the laminate. This can vary the wavelength of the FBG significantly. Some optical fibers may be misaligned or broken during the installation and resin-infusion process. For this reason, the wavelengths of all the FBGs were measured again after manufacturing to determine their survival and any variation in their wavelengths. Tables 12.13 through 12.15 show the magnitude of the signal before and after embedding. FBGs marked "w" had a weak signal

TABLE 12.13
FBG Wavelength Details for Layup 1 Specimens

		Left Outer		Left Inner		Right Outer		Right Inner
FBG λ before	A1	1528.8510	C1	1556.6862	B1	1540.5393	D1	1565.1653
FBG λ after		1529.46 (w)		1557.2025		1541.1825		1565.4175 (w)
FBG λ before	A2	1528.7612	C2	1556.5934	B2	1540.4914	D2	1565.2115
FBG λ after		1529.0800		1556.7538		1541.0400		1565.4900
FBG λ before	A3	1528.8580	C3	1556.6044	B3	1540.5408	D3	1565.0314
FBG λ after		1529.6500		1557.48 (w)		1540.8800		X
FBG λ before	A4	1528.8746	C4	1556.7714	B4	1540.5485	D4	1565.1033
FBG λ after		1529.0400		1556.9970		1540.724 (w)		1565.5800
FBG λ before	A5	1528.8553	C5	1556.4556	B5	1540.5310	D5	1565.2342
FBG λ after		1529.5375		1556.8975		1541.1625		1566.2346

Note: w: FBG with weak reflection, which can be measured using tunable laser system; X: FBG either broken or drifted away from the original location.

TABLE 12.14

FBG Wavelength Details for Layup 2 Specimens

		Left Outer		Left Inner		Right Outer		Right Inner
FBG λ before	A1	1528.7818	C1	1556.8218	B1	1540.2971	D1	1564.9772
FBG λ after		X		X		1540.9800		X
FBG λ before	A2	1528.8474	C2	1556.7265	B2	1540.3359	D2	1564.6499
FBG λ after		1529.5672		X		X		X
FBG λ before	A3	1528.8322	C3	1556.8223	B3	1540.3054	D3	1565.1606
FBG λ after		X		X		X		X
FBG λ before	A4	1528.8223	C4	1556.6548	B4	1540.2222	D4	1565.1221
FBG λ after		X		X		X		X
FBG λ before	A5	1528.7583	C5	1556.5211	B5	1540.2006	D5	1565.1452
FBG λ after		X		1557.3453		1540.6996		X

TABLE 12.15

FBG Wavelength Details for Layup 3 Specimens

		Left Outer		Left Inner		Right Outer		Right Inner
FBG λ before	A1	1528.8394	C1	1556.7849	B1	1540.2433	D1	1565.1597
FBG λ after		X		X		X		X
FBG λ before	A2	1528.8473	C2	1556.8592	B2	1540.5884	D2	1565.1291
FBG λ after		1529.6375		X		1541.0650		1565.3984
FBG λ before	A3	1528.7745	C3	1556.6637	B3	1540.1857	D3	1565.2083
FBG λ after		1529.6217		1557.8308		X		X
FBG λ before	A4	1528.8197	C4	1556.5003	B4	1540.5235	D4	1565.1896
FBG λ after		1529.6000		1557.6800		X		X
FBG λ before	A5	1528.8385	C5	1556.6013	B5	1540.5528	D5	1565.0603
FBG λ after		1529.4375		X		1540.9325		X

and were unable to provide adequate results. Those marked "X" were either broken or misplaced from the original location, and were not tested during the experimental study.

12.7.4 FBG Strain Measurement Setup

The setup for the strain measurement uses the typical parts shown in Figure 12.46. An OSA is required to interrogate the FBG during the experimental test procedure.

With all four FBG strain sensors deployed in a parallel configuration, three 50:50 couplers were required to combine the reflected light from the four sensors toward the Fabry–Perot tunable filter (FPTF). The reading from the reference FBG ensured compensation for temperature-induced wavelength variation. Since the light from the FBG sensors experienced high loss after propagating through two coupler stages,

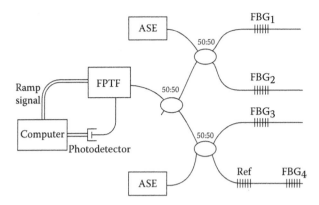

FIGURE 12.46 System setup for strain measurement. All bare FBG sensors, except the reference, were embedded in laminate. (From Prusty, B.G. and Raju, *Ships and Offshore Structures*, 9(2), 189–198, 2014.)

two C-band amplified spontaneous emission (ASE) sources were used to ensure that the FBG signal at the photodetector was distinguishable from the noise floor. The computer was equipped with a National Instrument PC1-6251 data acquisition card to digitally sample the analogue voltage at the photodetector and continuously provide the FPTF with a configured ramp signal for linear-wavelength scanning. The sampled data was processed and stored in the computer by the Labview 8.0 program. The spectral position detection was optimized by the centroid detection algorithm (CDA) (Binfeng et al., 2006).

Most of the FBG gratings designed for practical applications are not uniform— the needs of different applications vary. The need for high performance and tailored spectral and dispersive characteristics of FBGs imposes restrictions on the fabrication of special structures. These structures can be achieved by varying numerous physical parameters, such as an induced index change or periodicity along the grating, length, and fringe tilt. In this way, specialized gratings (anodized, chirped, phase shifted, and sampled) can be fabricated rather than the standard uniform Bragg gratings (de Melo, 2006).

The FBG sensors were fabricated of hydrogenated GF1 fiber 4 mm in length. A short cavity length was preferred to minimize the effect of strain variation distributed vertically, which can produce a chirped grating. Chirped gratings can be produced accurately, simply by increasing the amount of fiber translation each time the fiber is irradiated. Cosine anodization was applied to suppress noise during measurement due to side lobes. With the cosine anodization profile, the effective grating length is actually half of the total length. This requirement increases the writing time and the FBG bandwidth. The experimental setup for FBG strain measurement is shown in Figure 12.47.

Figures 12.48 and 12.49 show the reflectivity spectra for each FBG sensor and the multiplexed FBG sensor, respectively.

From these measurements, the average FBG parameters are: 3 dB bandwidth ~0.4 nm; side-mode suppression ratio >30 dB; peak reflectivity ~99%; and wavelength

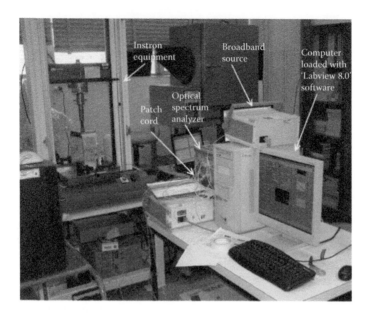

FIGURE 12.47　FBG interrogation unit used in the test procedure.

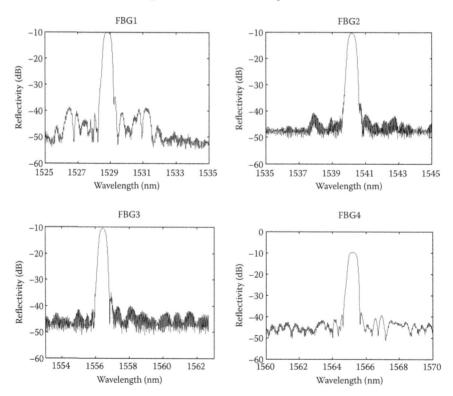

FIGURE 12.48　Measured reflectivity of FBG sensors from each wavelength.

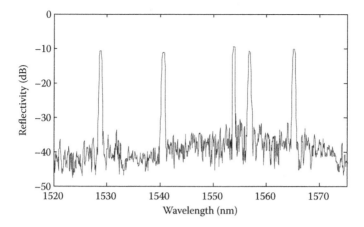

FIGURE 12.49 Measured reflection spectrum of all sensors multiplexed together: (from left to right) FBG_1, FBG_2, reference FBG, FBG_3, and FBG_4.

spacing between sensors >10 nm. However, the inclusion of the reference FBG reduced the dynamic range of FBG_2 and FBG_3. Evidently, the positioning of the FBG was crucial for avoiding wavelength crosstalk, which can cause spectral detection errors. To avoid this problem, the two FBGs with highest wavelengths (i.e., 1556 and 1565 nm) were embedded in an inner layer of the top-hat stiffener. The tension in the inner layer increases the wavelength. The two FBGs with the lowest wavelengths were embedded in the outer layer, which is subjected to compressive strains during loading. This decreases the wavelength, avoiding collision with the higher wavelengths.

12.7.4.1 Test Procedure

Four top-hat stiffener specimens for each of three different layups were manufactured by the VARTM method. Four sensors were embedded in each specimen. The experimental procedure was discussed in depth in Section 12.6. Analysis of FBG strain sensor measurement for failure initiation and progression is discussed in the next section.

12.7.5 FBG Results

When the top-hat stiffener is subjected to the designed loading condition (keel bolt pulled up), the inner and outer layers of the laminate at the bend are subjected to in-plane compression and tension, respectively. Thus, the inner FBGs (groups C and D) are in compression, and the outer FBGs (groups A and B) are in tension.

12.7.5.1 FBG Survival after Specimen Manufacture

After specimen manufacturing, some FBGs were lost due to weak reflection, optical fiber breakage, or misalignment. Summaries of the test specimens are presented in Tables 12.16 through 12.18.

The order of specimen manufacturing was Layup 2, then Layup 6, then Layup 1. Most of the manufacturing and handling mishaps were minimized by the last layup. As a result, as shown in the tables, FBG survival rate was higher for Layup

TABLE 12.16
Layup 1 FBG Survival Summary

Specimen No.	AE-Piezo	Strain-FBG
A1	Tested: OK	3 FBG tested: good
A2	Tested: OK	3 FBG tested: not good/lost spectral ref
A3	Tested: OK	2 FBG tested: OK/Spectral split
A4	Tested: OK	2 FBG tested: not good
A5	Tested: OK	4 FBG tested: good

TABLE 12.17
Layup 2 FBG Survival Summary

Specimen No.	AE-Piezo	Strain-FBG
B1	Tested: OK	1 FBG tested: good
B2	Tested: OK	1 FBG tested: good
B3	No FBG survival	
B4	No FBG survival	
B5	Tested: OK	1 FBG tested: good

TABLE 12.18
Layup 3 FBG Survival Summary

Specimen No.	AE-Piezo	Strain-FBG
C1	No FBG survival	
C2	Tested: OK	3 FBG tested: good
C3	No FBG survival	
C4	Tested: OK	2 FBG tested: partially Good
C5	Tested: problem	Tested: not OK/lost spectral reference

1 compared with the other two. To avoid repetition, the discussion of Layup 1 specimens is presented, and the failure plots and tabulation of individual failures for Layups 2 and 3 are presented in Appendix H.

12.7.5.2 VARTM Top-Hat Stiffener Layup 1 Results

Specimen THS VARTM D05 from Layup 1 had all four FBG sensors in good working condition, and the discussion of results is focused on this specimen. The 4 mm long FBG sensor yielded satisfactory strain measurements at the top two bends of the top-hat stiffeners. This was the expected failure location in the actual yacht due to excessive loading on the keel bolt.

Two different methods can be used for failure recognition based on strain measurements. First, the absolute changes in strain can be compared with those of the reference location. The overall strain distribution shows the reduction or increase in strain values at the failure points corresponding to the load drops. However, failure discrimination and determining the magnitude of the failure are not possible by analyzing the strain distribution alone. The second method is based on the detection of abnormal strain gradients. The second plot in each figure is the strain rate (μs/s) and load vs. time, showing the strain gradients with respect to time, where the damage-induced strain variation can be clearly seen. This plot shows sensor output where the sensor is in tension or compression, depending on the failure.

The cumulative strain (μs) and load vs. time are presented in Figures 12.50 through 12.53 for the left outer, right outer, left inner, and right inner FBG sensors, respectively.

A crude approach to demonstrating failure detection is made by observing the abnormal strain gradients for the four FBG sensors. No statistical analysis of the data is undertaken in this study. These abnormal strain gradients (strain rate) of the FBG sensors are presented in Table 12.19. The four FBG strain sensors are referred to as per their location in the top-hat stiffener: LO (left outer), RO (right outer), LI (left inner), and RI (right inner).

The damage characterization of top-hat stiffeners is performed by monitoring the abnormal strain rate gradients. These strain gradients have specific values (and can be positive or negative) at any given point in time, irrespective of the initial or reference values. The positive and negative values show whether the laminate is displaced up or down at that particular instant in time.

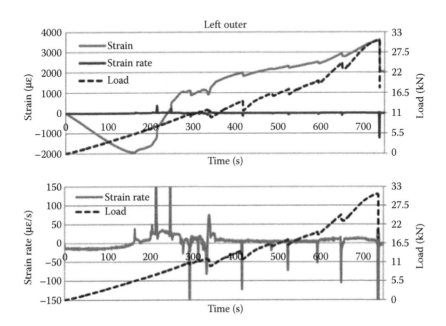

FIGURE 12.50 Top-hat stiffener Layup 1 FBG strain-sensor measurement: left outer.

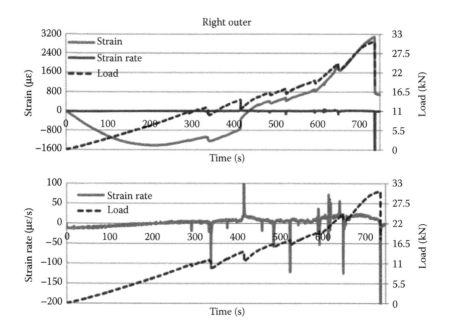

FIGURE 12.51 Top-hat stiffener Layup 1 FBG strain-sensor measurement: right outer.

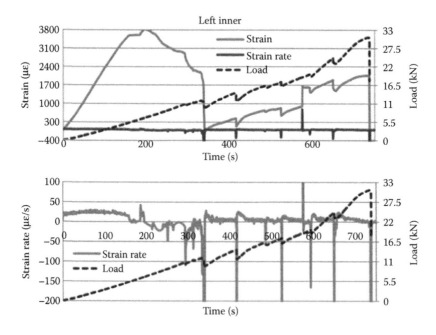

FIGURE 12.52 Top-hat stiffener Layup 1 FBG strain-sensor measurement: left inner. (From Prusty, B.G. and Raju, *Ships and Offshore Structures*, 9(2), 189–198, 2014.)

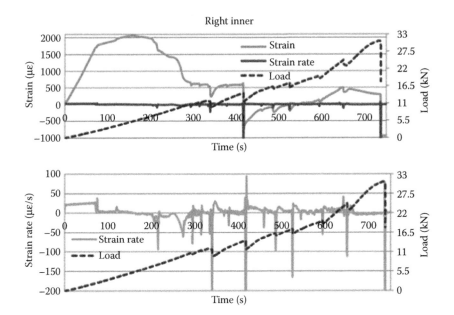

FIGURE 12.53   Top-hat stiffener Layup 1 FBG strain-sensor measurements: right inner.

TABLE 12.19
Layup 1 FBG Failure Characterization

Failure No.	Time (S)	Load (N)	Load Drop (N)	LO FBG A (µs/s)	RO FBG B (µs/s)	LI FBG C (µs/s)	RI FBG D (µs/s)	Failure Mode
1	224.66	7,571	7	373.1	0.6	−5.8	−81.7	Matrix cracking
2	299.83	10,818	484	−167.9	−18.2	−108.7	−78.55	Delamination
3	320.23	11,404	223	−32.7	4.8	−11.2	−55.51	Crack progression
4	345.35	11,254	437	60.7	−67.4	−498.7	−228.9	Delamination
5	383.36	12,337	72	6.3	−9.6	5.2	−32.2	Crack progression
6	422.50	14,200	1575	−121.9	353.6	−297.3	−306.1	Delamination
7	486.50	16,150	93	7.2	−57.9	−44.1	−83.0	Crack progression
8	530.77	17,456	631	−67	−122.1	−279.2	−165.4	Crack progression
9	592.52	19,690	257	−51.3	−53.31	789.6	−55.3	Crack progression
10	655.54	24,699	1,122	−99.1	−124.8	−310.5	−103.2	Delamination
11	743.00	30,345	12,457	−343.0	−2291.9	−2400.6	−1181.5	Fiber failure

The first failure occurred as matrix cracking between the second and third lay-ers on the left crown bend at 224.66 s, as discussed in Section 12.6.2.1. At this point, the LO FBG sensor reported a 373.1 µ/s jump in the strain rate. This failure was assumed to be matrix cracking, as a load drop of only 7 N was observed. The inner FBG sensors were placed between the second and third layers. Matrix crack-ing appeared to have initiated between these layers. The LI sensor might have been glued to the second layer, and as a result, a strain gradient of only 5.8 µ/s is observed. At this point, compared with the left crown bend, the right crown bend had a limited strain gradient (−81.7 and 0.6 µ/s for RI and RO, respectively). A major delamination occurred at the same location at 299.83 s, with strain gradients of −167.9 and −108.7 µ/s at LO and LI, respectively. For the minor crack propagation at 320.23 s, LO and RI recorded values of −32.7 and −55.51 µ/s, respectively. The next major delamination occurred at RI, with a value of −228.9 µ/s. A large delamination triggered the strain rates of −121.9, 353.6, −297.3, and −306.1 µ/s for LO, RO, LI, and RI, respectively. The next three failure points were crack propagations, whereby the failure at 592.52 s triggered a large deformation in the structure, with the FBG strain rate reading of 789.6 µ/s at LI. Comparison of the experimental results and photographic evidence of the structural damage indicates that this particular failure was caused by the disappearance of the crown bend (straightening) at the end of the steel plates. Final collapse occurred at 743 s, with readings of −343.0, −2291.9, −2400.6, and −1181.5 µ/s.

Changes in the level of intensity and bandwidth of the spectrum and a shift of Bragg wavelength are observed during the study. The shift in Bragg wavelength may be due to the alteration of the average strain distribution along the width of the grating. The abrupt changes in the strain gradient from delamination and crack propagation caused the deformation of the reflection spectra at various stages of testing. Snapshots of the FBG spectrum footprints at 221, 348, and 625 s are shown in Figure 12.54.

Similarly, the strain measurements for Layup 2 and Layup 3 are presented in Figures 12.55 through 12.60.

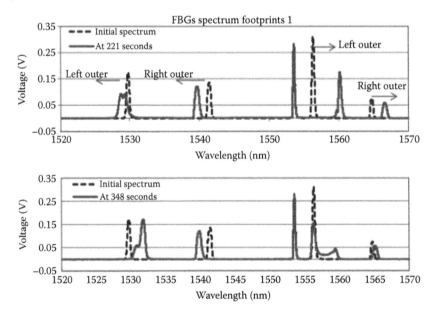

FIGURE 12.54 VARTM top-hat stiffener Layup 1 Bragg wavelength shift during different phases of the testing.

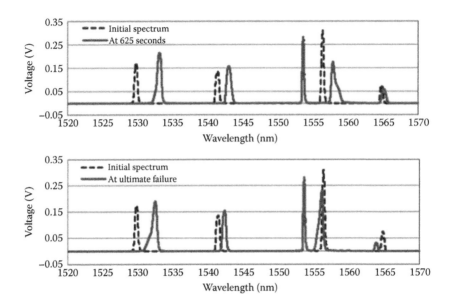

FIGURE 12.54 (Continued)   VARTM top-hat stiffener Layup 1 Bragg wavelength shift during different phases of the testing.

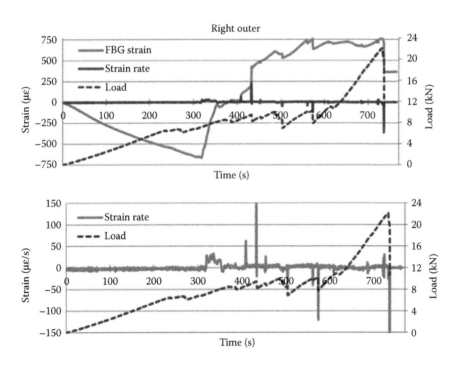

FIGURE 12.55   Top-hat stiffener Layup 2 FBG strain-sensor measurement: right outer. (From Raju and Azmi, A.I., *International Journal of Science Research*, 1(4), 531–538, 2013.)

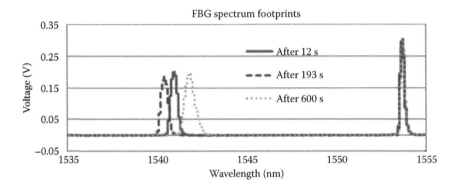

FIGURE 12.56 VARTM top-hat stiffener Layup 2 Bragg wavelength shift during different phases of the testing. (From Raju and Azmi, A.I., *International Journal of Science Research*, 1(4), 531–538, 2013.)

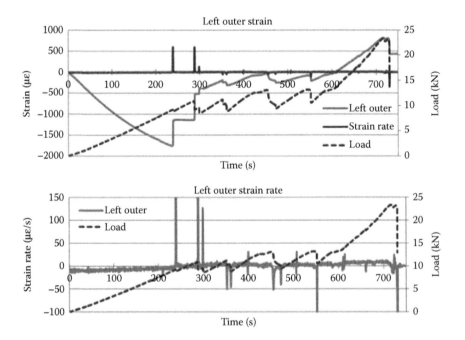

FIGURE 12.57 Top-hat stiffener Layup 3 FBG strain-sensor measurement: left outer.

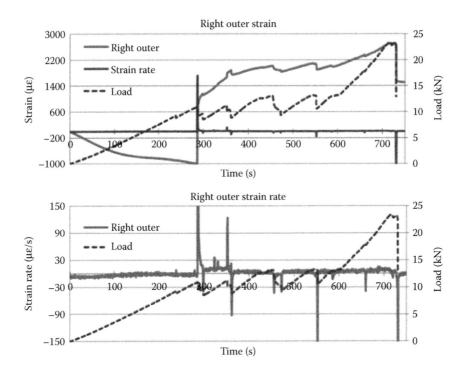

FIGURE 12.58   Top-hat stiffener Layup 3 FBG strain-sensor measurement: right outer.

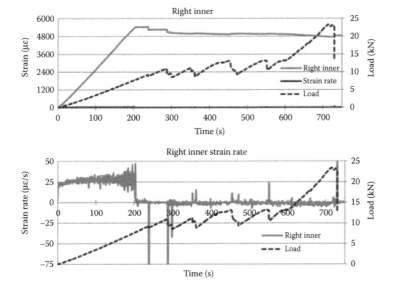

FIGURE 12.59   Top-hat stiffener Layup 3 FBG strain-sensor measurement: right inner.

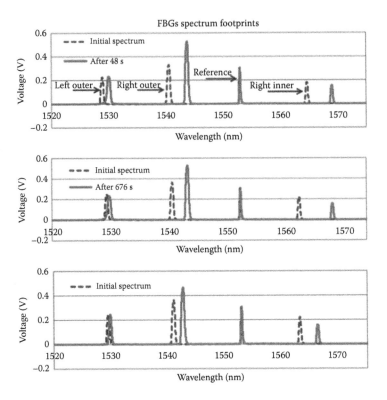

FIGURE 12.60 VARTM top-hat stiffener Layup 2 Bragg wavelength shift during different phases of the testing.

12.8 CONCLUSIONS

A methodology for damage monitoring of composite structures based on strain measurements, using embedded FBG sensors, is presented. FBG sensors were embedded at various locations of the top-hat stiffener. A state-of-the-art technique of embedding FOS using heat-shrink tubes is presented.

The study shows that the proposed technique is able to identify the occurrence and progression, size, location, and intensity of damage in a complex composite structure. It has been successfully demonstrated that the occurrence of interlaminar cracks within the laminate significantly alters the local strain distribution, and that these strain anomalies can be detected and measured using embedded FBGs. The sensors successfully detected failures from the change in the local strain distribution. Failure detection is possible by monitoring either irregular strain variations or abnormal large strain gradients—the latter method was adopted for the current study. Understanding the failure behavior of top-hat stiffeners is complex, as they are subjected to out-of-plane load transfers. The four bends of the top-hat stiffener add more complexity to the load transfer mechanism and the involved damage processes. Embedded FBGs successfully detected the occurrence and progression of delamination, demonstrating that FBGs can be successfully employed for damage monitoring in curved composite structures.

The major limitation of the proposed technique is the localized nature of damage-induced strain variations. This makes damage detection in the structure only possible on a local scale, or near the locations of the installed sensors, which requires *a priori* knowledge of failure locations. Prediction of damage location could be obtained from experience, prior knowledge of the structure, history of the in-service structures, experimental tests, and conducting numerical simulations (such as FEA). The potential of failure detection in real structures would be to implement the technique where visual inspection and other evaluation techniques may not be practical due to visibility, space, weight, or cost constraints.

REFERENCES

American Bureau of Shipping (ABS). (2000). *Guide for Building and Classing Motor Pleasure Yachts.* American Bureau of Shipping (ABS), Houston, TX.

Badcock R.A., Fernando G.F. (1995). An intensity-based optical fiber sensor for fatigue damage detection in advanced fiber-reinforced composites. *Smart Materials and Structures* 4: 223–230.

Binfeng Y., Wang Y., Li A., Cui Y. 2006. Tunable fiber laser based fiber Bragg grating strain sensor demodulation system with enhanced resolution by digital signal processing. *Microwave and Optical Technology Letters* 48(7): 1391–1393.

Chang C.C., Sirkis J.S. (1998). Design of fiber optic sensor systems for low velocity impact detection. *Smart Materials and Structures* 7: 166–177.

Childs P., Wong A.C.L., Yan B., Li M., Peng G.D. (2010). A review of spectrally coded multiplexing techniques for fiber grating sensor systems. *Measurement Science and Technology* 21(9): 1–7.

Choi C., Seo D.C., Lee J.J., Kim J.K. (1997). Effect of embedded optical fibers on the interlaminar fracture toughness of composite laminates, *Composite Interfaces* 5: 225–240.

Cole J.H., Sunderaman C., Tveten A.B., Kirkendall C., Dandridge A. (2002). Preliminary investigation of air-included polymer coatings for enhanced sensitivity of fiber: Optic acoustic sensors. *15th Optical Fiber Sensors Technical Digest* 1: 317–320.

Cranch G.A, Nash P.J., Kirkendall C.K. (2003). Large-scale remotely interrogated arrays of fiber-optic interferometric sensors for underwater acoustic applications. *IEEE Sensors Journal* 3(1): 19–30.

Daniel I.M., Ishai O. (1994). *Engineering Mechanics of Composite Materials.* Oxford University Press, Oxford.

Davies M.A., Bellemore D.G., Kersey A.D. (1994). Structural strain mapping using a wavelength/time division addressed fiber Bragg grating array. *SPIE* 2361: 342–345.

Davis C., Baker W., Moss S.D., Galea S.C., Jones R. (2002). *In situ* health monitoring of bonded composite repairs using a novel fibre Bragg grating sensing arrangement. *Smart Materials II, Proceedings of a SPIE* 4934: 140–149.

de Melo M.A.R. (2006). Implementation and development of a system for the fabrication of Bragg gratings with special characteristics. ME thesis. University of Porto, Portugal.

Dodkins A.R., Shenoi R.A., Hawkins G.L. (1994). Design of joints and attachments in FRP ship structures. *Marine Structures* 7: 364–398.

Elvin N., Leung C. (1997). Feasibility study of delamination detection with embedded optical fibers. *Journal of Intelligent Material Systems and Structures* 8: 824–828.

Encyclopædia Britannica. (2010). Keel. Retrieved September 14, 2010, from Encyclopaedia Britannica Online: http://www.britannica.com/EBchecked/topic/314095/keel.

Erdogan T. (1997). Fiber grating spectra. *Journal of Lightwave Technology* 15: 1277–1294.

Everall L., Gallon A., Roberts D. (2000). Optical fiber strain sensing for practical structural load monitoring. *Sensor Review* 20: 113–119.

Froggatt M., Moore J. (1998). Distributed measurement of static strain in an optical fiber with multiple Bragg gratings at nominally equal wavelengths. *Applied Optics* 37(10): 1741–1746.

Garg A.C. (1988). Delamination: A damage mode in composite structures. *Engineering Fracture Mechanics* 29(5): 5557–5584.

Goodman S., Tikhomirov A., Foster S. (2008). Pressure compensated distributed feedback fibre laser hydrophone. *Proceedings of SPIE* 7004 *19th International Conference on Optical Fibre Sensors*: 700426. doi:10.1117/12.785937. Perth, Australia.

Grouve W.J.B., Warnet L., de Boer A., Akkerman R., Vlekken J. (2008). Delamination detection with fibre Bragg gratings based on dynamic behaviour. *Composites Science and Technology* 68: 2418–2424.

Heslehurst R.B., Scott M. (1990). Review of defects and damage pertaining to composite aircraft components. *Composite Polymers* 3(2): 103–133.

Hinton M.J., Kaddour A.S., Soden P.D. (2004). *Failure Criteria in Fibre Reinforced Polymer Composites. The World-Wide Failure Exercise*. Elsevier Science Publishers, Amsterdam.

Horowitz S.J., Amey D.I. (2001). Ceramics meet next-generation fiber optic packaging requirements. *Photonics Spectra* 35(11): 123–126.

Jensen D., Pascual J., August A. (1992). Performance of graphite/bismaleimid laminates with embedded optical fibre optics, part I: Uniaxial tension. *Smart Materials and Structures* 1: 24–30.

Junhou P., Shenoi R.A. (1996). Examination of key aspects defining the performance characteristics of out-of-plane joints in FRP marine structures. *Composites: Part A* 27A: 89–103.

Kaczmarek K., Wisnom M.R., Jones M.I. (1998). Edge delamination in curved (04/±456)s glass-fibre/epoxy beams loaded in bending. *Composite Science and Technology* 58: 155–161.

Kedward K.T., Wilson R.S., Mclean S.K. (1989). Flexure of simply curved composite shapes. *Composites* 20(6): 527–536.

Kesavan A., John S., Herszberg I. (2008). Strain-based structural health monitoring of complex composite structures. *Structural Health Monitoring* 7(3): 203–213.

Kitade S., Fukuda T., Osaka K. (1995). Fiber optic method for detection of impact induced damage in composite laminates. *Journal of the Society of Materials Science* 44: 1196–1200.

Kuang K.S.C., Cantwell W.J. (2003). Use of conventional optical fibers and fiber Bragg gratings for damage detection in advanced composite structures: A review. *Applied Mechanical Review* 56(5): 493–513.

Larsson L., Eliasson R.E. (1994) *Principles of Yacht Design*. Adlard Coles Nautical, London.

LeBlanc M., Measures R.M. (1992). Impact damage assessment in composite materials with embedded fiber-optic sensors. *Composites Engineering* 2(5–7): 573–596.

Lee B., Jeong Y. (2002). Interrogation techniques for fiber grating sensors and the theory of fiber gratings. In *Fiber Optic Sensors*, F.T.S. Yu, S. Yin, eds., Marcel Dekker, New York.

Lee D.C., Lee J.J., Kwon I.B., Seo D.C. (2001). Monitoring of fatigue damage of composite structures by using embedded intensity-based optical fiber sensors. *Smart Materials and Structures* 10: 285–292.

Lee D.C., Lee J.J., Yun S.J. (1995). The mechanical characteristics of smart composite structures with embedded optical fibre sensors. *Composite Structures* 32: 39–50.

Lee J. R., Jeong H. M. (2010). Design of resonant acoustic sensors using fiber Bragg gratings. *Measurement Science and Technology* 21(5): 1–6.

Lemaire P.J., Atkins R.M., Mizrahi V., Reed W.A. (1993). High pressure HZ loading as a technique for achieving ultrahigh photosensitivity and thermal sensitivity in GeO$_2$ doped optical fibres. *Electronics Letters* 29(13): 1191–1193.

Leung C.K.Y., Yang Z., Ying X., Tong P., Lee S.K.L. (2005). Delamination detection in laminate composites with an embedded fiber optical interferometric sensor. *Sensors and Actuators. A, Physical* 119: 336–344.

Li H.C.H., Herszberg I., Davis C.E., Mouritz A.P, Galea S.C. (2006). Health monitoring of marine composite structural joints using fibre optic sensors. *Composite Structures* 75(1–4): 321–327.

Li H.C.H., Herszberg I., Mouritz A.P., Davis C.E., Galea S.C. (2004). Sensitivity of embedded fibre optic Bragg grating sensors to disbonds in bonded composite ship joints. *Composite Structures* 66: 239–248.

Lin Y.B., Chang K.C., Chern J.C., Wang L. A. (2005). Packaging methods of fiber-Bragg grating sensors in civil structure applications. *IEEE Sensors Journal* 5(3): 419–423.

Ling H.Y., Lau K.T., Cheng L., Jin W. (2005b). Utilization of embedded optical fibre sensors for delamination characterization in composite laminates using a static strain method. *Smart Materials and Structures* 14: 1377–1386.

Ling H.Y., Lau K.T., Lam C. (2005a). Effects of embedded optical fibre on mode II fracture behaviours of woven composite laminates. *Composites Part B* 36(6–7): 534–543.

Maligno A.R. (2007). Finite element investigations on the microstructures of composite materials. PhD thesis, The University of Nottingham, Nottingham.

Mall S. Dosedel S.B., Holl M.W. (1996). The performance of graphite-epoxy composite with embedded optical fibres under compression. *Smart Materials and Structures* 5: 209–215.

Martin A.R., Fernando G.F., and Hale K.F. (1997). Impact damage detection in filament wound tubes using embedded optical fiber sensors. *Smart Materials and Structures* 6: 470–476.

McKenzie I., Jones R., Marshall I.H., Galea S. (2000). Optical fiber sensors for health-monitoring of bonded repair systems. *Composite Structures* 50: 405–416.

Measures R.M., Glossop N.D.W., Lymer J., LeBlanc M., Dubois S., Tsaw W., Tennyson R.C. (1989). Structurally integrated fiber optic damage assessment system for composite materials. *Applied Optics* 28: 2626–2633.

Meltz G., Morey W.W. (1992). Bragg grating formation and germanosilicate fiber photosensitivity. *Proceedings of SPIE* 1516: 185–199.

Morris P., Beard P., Hurrell A. (2005). Development of a 50 MHz optical fibre hydrophone for the characterisation of medical ultrasound fields. *IEEE Ultrasonics Symposium*, Rotterdam, the Netherlands: 1747–1750.

Mouritz A.P., Gallagher J., Goodwin A.A. (1997b). Flexural strength and interlaminar shear strength of stitched GRP laminates following repeated impact. *Composites Science and Technology* 87: 509–522.

Mouritz A.P., Leong K.M., Herszberg I. (1997a). A review of the effect of stitching on the in-plane mechanical properties of fibre-reinforced polymer composites. *Composites Part A* 28: 979–991.

Okabe Y., Yashiro S., Kosaka T., Takeda N. (2000). Detection of transverse cracks in CFRP composites using embedded fiber Bragg grating sensors. *Smart Materials and Structures* 9: 832–838.

Othonos A., Kalli K. (1999). *Fiber Bragg Gratings: Fundamentals and Applications in Telecommunications and Sensing.* Artech House Optoelectronics Library, Norwood, MA.

Pearson J.D., Zikry M.A., Prabhugoud M., Peters K. (2007). Global-local assessment of low-velocity impact damage in woven composites. *Journal of Composite Materials* 41(23): 2759–2783.

Prusty B.G., Raju (2014). Failure investigation of top-hat composite stiffened panels. *Ships and Offshore Structures* 9(2): 189–198.

Raju, Azmi A.I. (2013). Structural Health Monitoring of advanced composites using acoustic emission and strain gradients with fibre optic sensors. *International Journal of Science Research* 1(4): 531–538.

Raju, Azmi, A.I., Prusty B.G. (2011). AE failure characterisation using piezo-electric sensors and fibre optic sensors for composite laminates. In *Theory and Uses of Acoustic Emission*, Burnett J.K., ed. Nova Science, New York.

Raju, Prusty B.G., Kelly D.W. (2013). Delamination failure of composite top-hat stiffeners using finite element analysis. *Proceedings of the Institution of Mechanical Engineers, Part M Journal of Engineering for the Maritime Environment* 227(1): 61–80.

Raju, Prusty B.G., Kelly D.W., Lyons D., Peng G.D. (2010). Top hat stiffeners: A study on keel failures. *Ocean Engineering* 37: 1180–1192.

Rao Y.J. (1997). In-fibre Bragg grating sensors. *Measurement Science and Technology* 8: 355–375.

Rodrigo A.S., Roberto L.A. (2009). Structural health monitoring of marine composite structural joints using embedded fiber Bragg grating strain sensors. *Composite Structures* 89(2): 224–234.

Rowe W.J., Rausch E.O., Dean P.D. (1986). Embedded optical fiber strain sensor for composite structure applications. *SPIE* 718: 266–273.

Schwartz M. (2009). *Smart Materials*. CRC, Boca Raton, FL.

Shenoi R.A., Hawkins G.L. (1995). An investigation into the performance characteristics of top-hat-stiffener to shell plating joints. *Composite Structures* 30: 109–121.

Shenoi R.A., Wellicome J.F.. (1993). *Composite Materials in Maritime Structures: Volume 2 – Practical Considerations*, 1st edition, Cambridge University Press, New York.

Shivakumar K., Emmanwori L. (2004). Mechanics of failure of composite laminates with an embedded fiber optic sensor. *Journal of Composite Materials* 38(8): 669–680.

Silva-Muñoz R.A., Lopez-Anido R.A. (2009). Structural health monitoring of marine composite structural joints using embedded fiber Bragg grating strain sensors. *Composite Structures* 89: 224–234.

Smith C.S. (1990). *Design of Marine Structures in Composite Materials*. Elsevier Applied Science, London.

Sorensen L., Botsis J., Gmür T., Cugnoni J. (2007). Delamination detection and characterisation of bridging tractions using long FBG optical sensors. *Composites Part A* 38(10): 2087–2096.

Sridharan S. (2008). *Delamination Behaviour of Composites*. CRC, Boca Raton, FL.

Staszewski W., Boller C., Tomlinson G. (2004). *Health Monitoring of Aerospace Structures: Smart Sensor Technologies and Signal Processing*. Wiley, Chichester, England.

Takeda N., Kosaka T., Ichiyama T. (1999). Detection of transverse cracks by embedded plastic optical fiber in FRP laminates. *SPIE* 3670: 248–255.

Takeda N., Okabe Y., Kuwahara J., Kojima S., Ogisu T. (2005). Development of smart composite structures with small-diameter fiber Bragg grating sensors for damage detection: Quantitative evaluation of delamination length in CFRP laminates using Lamb wave sensing. *Composites Science and Technology* 65(15–16): 2575–2587.

Tao X., Tang L., Du W., Choy C. (2000). Internal strain measurement by fiber Bragg grating sensors in textile composites. *Composites Science and Technology* 60: 657–669.

Ussorio M., Wang H., Ogin S.L., Thorne A.M., Reed G.T., Tjin S.C., Suresh R. (2006). Modifications to FBG sensor spectra due to matrix cracking in a GFRP composite. *Construction and Building Materials* 20(1–2): 111–118.

Wang H., Ogin S.L., Thorne A.M., Reed G.T. (2005). Interaction between optical fibre sensors and matrix cracks in cross-ply GFRP laminates. Part 2: Crack detection. *Composites Science and Technology* 66(13): 2367–2378.

Wei C.Y., Ye C.C., James S.W., Tatam R.P., Irving P.E. (2002). The influence of hydrogen loading and the fabrication process on the mechanical strength of optical fibre Bragg gratings. *Optical Materials* 20(4): 241–251.

Wisnom M.R. (1996). 3-D Finite element analysis of curved beams in bending. *Journal of Composite Material* 30(11): 1178–1190.

Xu Y., Leung C.K.Y., Tong P., Yi J., Lee S.K.L. (2005). Interfacial debonding detection in bonded repair with a fiber optical interferometric sensor. *Composites Science and Technology* 65(9): 1428–1435.

Index

A

ABAQUS, 168, 186, 187
ABS, *see* Acrylonitrile butadiene styrene (ABS)
Acoustic emission (AE) method, 26–28
Acoustic source localization, 26
Acrylonitrile butadiene styrene (ABS), 374
Active demodulation techniques, 81
Active system, SHM, 22
ADC, *see* Analog-to-digital converter (ADC)
Additive manufacturing process, 12
ADFP, *see* Automated dry fiber placement (ADFP)
AE, *see* Acoustic emission (AE) method
Aerospace composite structures, smart, 339–367
    advance, 339–344
        damage, 342–344
    fiber-optics in, 352–366, 345–351
        ANN-based damage detection, 361
        ANN model, 365–366
        damage detection technologies, 352–353
        damage prediction, 356
        decoding distorted FBG sensor spectra, 356–359
        development of ANN-based signal processing techniques, 360–361
        FBG sensors, 345–349
        neural networks, 361–364
        postprocessing of FBG data using ANN, 365
        PPP-BOTDA sensors, 349–351
        signal processing technologies for, sensors, 353–356
    and materials, 351–352
AF ESPI, *see* Amplitude fluctuation (AF) ESPI method
Aircraft, composite materials in, 14–16
Amplified spontaneous emission (ASE), 463
Amplitude fluctuation (AF) ESPI method, 31
Analog-to-digital converter (ADC), 110, 115, 126
Anchorage system
    distributed-based smart anchor rod, 397–400
        manufacturing and operational principle of self-sensing, 397–399
        mechanical and sensing characteristics of, 399–400
    FBG-based, 394–400
        manufacturing, 394–395
        mechanical and sensing characterization, 396–397

Anchor rod, 394
    distributed-based, 397–400
        manufacturing and operational principle of self-sensing, 397–399
        mechanical and sensing characteristics of, 399–400
    experimental application of, 408–409
ANN, *see* Artificial neural networks (ANN)
Artificial neural networks (ANN), 31, 353, 362, 363; *see also* Neural networks
    -based damage detection, 361
    development of, based signal processing techniques, 360–361
    model, 365–366
    postprocessing of FBG data using, 365
ASE, *see* Amplified spontaneous emission (ASE)
Atlas weave, 302
ATP/AFP method, 11–12
Automated dry fiber placement (ADFP), 245, 247
Auxiliary interferometer, 113
Axial strain variations, 396–397, 399–400

B

Back-propagation (BP) algorithms, 363
BDG, *see* Brillouin dynamic grating (BDG)
BFS, *see* Brillouin frequency shift (BFS)
Birefringence, 93, 94
Blackbody cavity types, 57
BOCDA, *see* Brillouin optical correlation domain analysis (BOCDA)
Boron fibers, 3
BOTDA, *see* Brillouin optical time domain analysis (BOTDA)
BOTDA/R instrument, 389
BOTDR, *see* Brillouin optical time domain reflectometry (BOTDR)
BP, *see* Back-propagation (BP) algorithms
Bragg, William Lawrence, 345
Bragg grating, 345–346
    embedded, 160–161
    sensors, 54, 434–436
Bragg mode, 86
Bragg wavelength, 71, 75, 81–83
Braiding process, 307
Bridge cables system, and suspender, 384–390
    concept and fabrication of, 385–386
    existing, 390

mechanical and sensing performance of, 389–390
operating principles of smart FRP-based, 386–388
temperature self-compensation of, 388–389
Brillouin dynamic grating (BDG), 128
Brillouin frequency shift (BFS), 122, 124, 128, 350
Brillouin optical correlation domain analysis (BOCDA), 128
Brillouin optical time domain analysis (BOTDA), 112, 124, 125
Brillouin optical time domain reflectometry (BOTDR), 118, 123, 124, 125, 133, 136, 350
Brillouin scattering, 59, 60
-based distributed temperature/strain sensing, 122–125
-based fiber-optic measuring technique, 349, 350–351
Brillouin Stokes, and anti-Stokes, 122
"Bucket and brush" approach, 8

C

Cable force monitoring, using FBG-based stress ring sensors, 288–294
analysis of long-term, 290–295
deployment of, 289–290
fundamentals of, 288–289
Canadian Communications Research Centre (CRC), 345, 346
Cantilever beam, 304
Carbon fiber–composite aircraft tailcone assembly, 244–264
development of longitudinal strain, 253–256
development of transverse strain, 250–253
infusion monitoring, 248–250
postfabrication test monitoring, 260–264
preform and sensor layup, 245–248
strain monitoring, 256–260
Carbon fiber-reinforced plastic (CFRP), 131, 133
Carbon fiber–reinforced polymer (CFRP), 30, 166, 323, 439
Carbon fiber–reinforced preform (CFRP), 244
Carbon fibers, 3, 33
Carbon nanotube (CNT), 33
CCD, *see* Charge-coupled device (CCD)
CDA, *see* Centroid detection algorithm (CDA)
Centroid detection algorithm (CDA), 463
Ceramic fibers, 3
CFM, *see* Continuous filament mat (CFM)
CFRP, *see* Carbon fiber-reinforced plastic (CFRP); Carbon fiber–reinforced polymer (CFRP); Carbon fiber-reinforced preform (CFRP)
CFRP-based prestressed plate, 380–382

manufacturing, 380
sensing performance evaluation, 380–382
CFRP bridge cable, 390
CFRP composites, 442
CFRP cross-ply laminate, 441
CFRP plate, 380, 381
Charge-coupled device (CCD), 309
Chirp, 341
Chopped strand mat (CSM), 447
Cladding mode resonances, 78, 224–225
CNT, *see* Carbon nanotube (CNT)
Cohesive zone elements (CZE), 344
Compaction model, 346
Comparative vacuum monitoring (CVM), 34
Composite arch bridge
cable force monitoring using FBG-based stress ring sensors, 288–294
analysis of long-term, 290–295
deployment of, 289–290
fundamentals of, 288–289
structural health monitoring system of, 271–274
description of FBG-based, 273–274
design and implementation of, 272–273
features of, 271–272
structural stress monitoring using FBG-based strain sensors, 274–288
analysis of long-term, 278–288
deployment of, 275–278
fundamentals of, 274–275
Composite failure, 423–428
fiber, 424–426
interface, 426
interlaminar, 427–248
matrix, 424
Composite materials, 1, 351
issues in practical implementation, 115–118
manufacturing methods and technologies, 6–13
ATP/AFP method, 11–12
design of, 12–13
filament winding method, 9–10
prepreg layup method, 6–7
pultrusion method, 10–11
resin transfer molding (RTM) method, 11
spray layup method, 8–9
structural reaction injection molding (SRIM) method, 11
wet layup method, 7–8
structural health monitoring for, 21–35
acoustic emission, 26–28
application to, 118–120
demand for, 14–16
elastic waves–based, 22–26
electromechanical impedance, 28–30
other, 33–34
smart structures and, 13–14

system for, 16–17
using distributed fiber-optic sensors,
129–148
using fiber Bragg gratings, 69–95
vibration-based, 30–33
structures and types, 2–6
matrix, 4–6
reinforcements, 3–4
Composite production, optical fiber sensors in,
202–264
challenges and future requirements, 264
industrial manufacturing
carbon fiber–composite aircraft tailcone
assembly, 244–264
laboratory manufacturing, 225–244
process temperature monitoring, 233–238
refractive index of resin, 238–242
resin infusion and flow monitoring,
225–233
strain development, 243
manufacturing processes and process
parameters, 203–210
automated fabric placement, 204
curing process, 208–209
experimental monitoring techniques, 210
postproduction monitoring, 209–210
process temperature, 207–208
resin infusion, 206–207
through-thickness reinforcement,
204–206
principles of, 211–225
optical fiber Bragg grating (FBG),
217–222
optical fiber Fresnel sensor, 211–215
tapered optical fiber sensor, 215–216
tilted fiber Bragg grating sensor (TFBG),
222–225
Compression failure, 424, 425
Continuous filament mat (CFM), 447
Core, and cladding, 310, 311
Crack, propagation of, 342
CRC, *see* Canadian Communications Research
Centre (CRC)
Cross-ply laminate, 172–174, 438
CRP, *see* Carbon-reinforced fiber plastics (CRP)
CSM, *see* Chopped strand mat (CSM)
Cure monitoring
of fiber-reinforced preform, 238–244
refractive index of resin, 238–242
strain development, 243–244
optical-fiber sensors in smart composites,
315–322
Curing process, 208–209
Curved composites, 417–418
CVM, *see* Comparative vacuum monitoring
(CVM)
CZE, *see* Cohesive zone elements (CZE)

D

Dalian University of Technology, 410
Damage localization map-creating algorithm, 30
DAQ, *see* Data acquisition card (DAQ)
DAS, *see* Distributed acoustic sensing (DAS)
Data acquisition card (DAQ), 126
Data collection, and transmission subsystem,
272–273
Data processing, and control subsystem, 273–274
DCB, *see* Double cantilever beam (DCB)
Degree, of cure/conversion, 238, 239, 240, 242
Delamination, 343, 344, 427–428, 442–445, 470
Demodulation techniques, 81–83
based on wavelength detection, 81–82
reflectometric, 82–83
DGS, *see* Dynamic grating spectrum (DGS)
Differential scanning calorimetry (DSC),
239, 240
Digital signal processor (DSP), 110
Distortion index, 354
Distributed acoustic sensing (DAS), 129
Distributed fiber-optic sensing, 120–129
measurands and technologies, 109–115
and performance parameters, 106–108
strain
Brillouin-based distributed temperature/,
122–125
development of distributed temperature
and, 127–128
FBG-based distributed temperature,
125–127
Rayleigh-based distributed temperature/,
125–127
temperature
Raman-based distributed, 120–122
Distributed fiber-optic sensors, 59–61, 129–148
issues in practical implementation, 115–118
structural health monitoring of composite
materials using, 129–148
strain sensing, applications of, 134–148
temperature sensing, applications of,
131–134
Distributed pressure sensing (DPS), 113
Distributed strain sensing (DSS), applications
of, 112, 114, 120, 123, 127, 128, 129,
134–148
composite materials and structures,
monitoring of, 136–139
deformation/load identification based on
measurement data, 144–148
steep strain distributions, monitoring in,
140–144
Distributed temperature sensing (DTS),
applications of, 110, 112, 114, 120,
123, 126, 129, 131–134
composite structures, monitoring of, 131–132

manufacturing processes, monitoring of, 133
    NDT methods, 133–134
Double cantilever beam (DCB), 443
DPS, *see* Distributed pressure sensing (DPS)
DSC, *see* Differential scanning
    calorimetry (DSC)
DSP, *see* Digital signal processor (DSP)
DSS, *see* Distributed strain sensing (DSS)
DTS, *see* Distributed temperature sensing
    (DTS)
Dynamic grating spectrum (DGS), 128
Dynamic loads, 430

E

Elastic waves–based method, 22–26
Elasto-optic effect, 73, 74
Electrical resistance, 34
Electromagnetic interference (EMI), 430
Electromechanical impedance (EMI) method,
    28–30
Electronic speckle pattern interferometry
    (ESPI), 31
Embedded Bragg gratings, 160–161
EMI, *see* Electromagnetic interference (EMI);
    Electromechanical impedance (EMI)
    method
End tip sensor, 57
Epoxy resin, 4, 85, 319, 394, 395
EP2400 resin, 242
Erbian Dadu River Arch Bridge, suspenders in,
    404–406
ESPI, *see* Electronic speckle pattern
    interferometry (ESPI)
Etalon, 52
EU FP7 "SmartFiber" project, 183
Evanescent field–based fiber sensor, 47–49

F

Fabry–Perot (F-P) fiber-optic sensor, 308
Fabry–Perot interferometers (FPIs), 52, 53
Fabry–Perot tunable filter (FPTF), 462, 463
Failure theories, 343
Far-field strains, 158
Fast Fourier transform (FFT), 113, 126, 127
Fatigue accumulative damage index, 408
FBG, *see* Fiber Bragg grating (FBG)
FE, *see* Finite element (FE) model
FEA, *see* Finite element analysis (FEA)
FEM, *see* Finite element method (FEM)
FFFDS, *see* Fixed FBG filter decoding system
    (FFFDS)
FFT, *see* Fast Fourier transform (FFT)
Fiber Bragg grating (FBG), 54, 55–56, 63, 127,
    128, 133, 134, 142, 143, 145, 147, 160,
    161, 176, 315, 325, 327, 373, 374, 385

-based distributed temperature/strain sensing,
    125–127
-based strain sensors, structural stress
    monitoring using, 273, 274–288, 289
    deployment of, 275–278
    fundamentals of, 274–275
    separating temperature effect using
        wavelet multiresolution analysis, 273,
        282–288
    strain monitoring data from, 278–282
-based stress ring sensors, cable force
    monitoring using, 288–294
    analysis of long-term, 290–295
    deployment of, 289–290
    fundamentals of, 288–289
-based temperature sensors, 273
demodulation techniques used to retrieve
    Bragg wavelength shift, 81–83
    based on wavelength detection, 81–82
    reflectometric, 82–83
fundamentals on, 70–80
    axial strain sensitivity of uniform, 73–75
    other kinds of, 78–80
    pressure sensitivity of uniform, 76
    principle of uniform, 70–71
    temperature sensitivity of uniform, 72–73
    transverse strain sensitivity of uniform,
        77–78
interrogation, 383, 389, 436–437
optical, 217–222
residual strain sensing during manufacturing
    process, 88–95
strain sensing, 452–474
    fabrication of, 453–456
    layup and embedding optical fibers,
        456–459
    measurement setup, 462–465
    optical fiber embedding procedure with
        VARTM, 459–462
    results, 465–474
temperature and strain effects, 83–88
    comparison of performances, 87–88
    pair comprising type I and type IA, 86
    pairs embedded into glass capillary,
        85–86
    weakly tilted, 86–87
Fiber Bragg grating (FBG) sensors, 25, 33, 245,
    260, 307, 319, 320, 321, 323, 324, 326,
    341, 354, 360, 361, 362, 434–436, 442,
    456, 461, 462, 465, 466
    embedding, in FRP structures, 348–349
    evolution of, 346–348
    filter system, 356, 357
    response spectra, 352, 353, 355
Fiber failure, 424–426
Fiber–matrix debonding, 426
Fiber-optic

scattering-based distributed fiber sensors,
    59–61
structural health monitoring systems for
        marine composite structures, 415–475
    curved, 417–418
    damage characterization of, 437–446
    experimental investigation of top-hat
        stiffener, 447–452
    failure, 423–428
    FBG strain sensing, 452–474
    nondestructive evaluation of, 428–437
    out-of-plane joints, 418–423
Fiber-optics, in smart aerospace composite
        structures, 352–366
    ANN
        model, 365–366
        postprocessing of FBG data using, 365
    damage
        ANN-based, detection, 361
        detection technologies, 352–353
        prediction, 356
    decoding distorted FBG sensor spectra,
        356–359
    neural networks, 361–364
    signal processing
        development of ANN-based, techniques,
            360–361
        technologies for, sensors, 353–356
Fiber-optic distributed sensors, *see* Distributed
        fiber-optic sensors
Fiber-optic sensors (FOS), 14, 16, 345–351,
        430, 456; *see also* Optical fiber (OF)
        sensors
    FBG, 33, 345–349
        embedding, in FRP structures, 348–349
        evolution of, 346–348
    PPP-BOTDA, 349–351
        Brillouin scattering-based, measuring
            technique, 350–351
Fiber-reinforced polymer (FRP), 30, 118, 339,
        340, 341, 372, 373, 374
    -based rebars/bars, 375–379
        manufacturing, 375–376
        properties of, 376–379
    -based smart components and structures,
        371–412
        applications of, 400–409
        existing, 375–400
        principles of, 373–375
    -based tube for concrete confinement,
        382–384
        manufacturing of, 383
        mechanical and sensing performance
            evaluation, 383–384
    composites, 345, 348
    embedding FBG sensors in, structures,
        348–349

operating principles of, -based cable, 386–388
    prestressed strand based on, 390–394
    rebars, 385, 386
    Young modulus of, rod, 396
Fiber-reinforced preform, cure monitoring of
    refractive index of resin, 238–242
    strain development, 243
Fiber roving, 8
FiberSensing FS2200 interrogator, 85
Filament winding method, 9–10, 383
Finite element analysis (FEA), 136, 348, 442
Finite element (FE) model, 144, 168–169, 176,
        179, 186–188, 190
Finite element method (FEM), 26, 29, 344
First ply failure (FPF), 344
Fixed FBG filter decoding system (FFFDS), 361
Fixed keels, 420
Fluorescent-based fiber sensors, 57–58
FOS, *see* Fiber-optic sensors (FOS)
FPF, *see* First ply failure (FPF)
F-P fiber-optic sensor, *see* Fabry–Perot (F-P)
        fiber-optic sensor
FPIs, *see* Fabry–Perot interferometers (FPIs)
FPTF, *see* Fabry–Perot tunable filter (FPTF)
Fresnel reflection, 319, 320, 321, 322
Fresnel sensors, 228, 238–239, 240, 242, 249, 254
FRP, *see* Fiber-reinforced polymer (FRP)
FRP-OFBG bars, 376, 385, 386
F-16 aircraft, 32
Full width at half maximum (FWHM), 354, 355
FWHM, *see* Full width at half maximum
        (FWHM)

G

GFRP, *see* Glass fiber–reinforced plastic (GFRP);
        Glass fiber–reinforced polymer
        (GFRP)
GFRP composites, 442, 445
Ghost mode, 86
Glass fiber, 3
Glass fiber–reinforced plastic (GFRP), 14, 131,
        416
Glass fiber–reinforced polymer (GFRP), 29
Graded index fiber, 50
Graphite fiber, 3

H

Harbin Institute of Technology (HIT), 375,
        376, 385
HBFs, *see* High birefringent fibers (HBFs)
HF, *see* High frequency (HF) ultrasonic wave
HiBi FBG sensor, 243, 245, 250, 251, 254, 256,
        259, 260, 263, 264
High birefringent fibers (HBFs), 54
High frequency (HF) ultrasonic wave, 34

"Hill gratings", 346
HIT, *see* Harbin Institute of Technology (HIT)
H-II Orbiting Plane-Experimental (HOPE-X),
    118, 133
Hull–bulkhead joint, 418
Human brain processes, 362
Hypodermic tube, 247, 248

I

IACC, *see* International America's Cup Class
    (IACC)
INSTRON machine, 448
Intelligent structures, 374
Intensity-modulated sensors, 45–50
    evanescent field–based fiber, 47–49
    macrobend, 50
    microbending, 49–50
Interactive theories, 343
Interface failure, 426
Interfacial debonding, 400
Interfacial shear stress, 397
Interferometric sensors, *see* Phase-modulated
    fiber-optic sensors
Interlaminar failure, 427–248
International America's Cup Class (IACC),
    118, 136
Interstate 40 Bridge, 31
Intrinsic, and extrinsic sensors, 44
Inverse, of measurement time, 115
Inverse distance weighting interpolation, 30
Inverse machining process, 12

J

Jahn, Bernhard, 16
Jili Group Inc., 389
Jiubao Bridge, Hangzhou, 271, 272, 275, 278

K

Keels, 420–423
    –hull connection, 421–423
    types of, 420–421
        canting, 421
        fin, 421
        full, 421
        swing, 421
Kernel principal component analysis, 30
K-matrices, 234, 236

L

LabView™ program, 226, 240, 463
Lamb, Horace, 22, 23
Lamb waves, 22–23, 25, 28, 443
Laminated composites, 2–3

Laser Doppler vibrometer (LDV), 24–25
Laser nonlinear wave modulation spectroscopy
    (LNWMS), 26
Laser ultrasonic (LUS) device, 25
Laser vibrometer, *see* Laser Doppler vibrometer
    (LDV)
LDV, *see* Laser Doppler vibrometer (LDV)
Lead zirconate titanate sensors, *see* Piezoelectric
    sensors
LEFM, *see* Linear elastic fracture mechanics
    (LEFM)
LF, *see* Low-frequency (LF) excitation
Limit/noninteractive theories, 343
Linear elastic fracture mechanics
    (LEFM), 344
Liquid transfer molding, *see* Resin transfer
    molding (RTM) method
LNWMS, *see* Laser nonlinear wave modulation
    spectroscopy (LNWMS)
LO, *see* Local oscillator (LO)
Localized SPR, 61
Local oscillator (LO), 112, 113
Long period fiber gratings (LPFGs), 51, 56–57,
    79, 80
Love waves, 22
Low-frequency (LF) excitation, 34
LPFGs, *see* Long period fiber gratings (LPFGs)
LUS, *see* Laser ultrasonic (LUS) device

M

Mach–Zehnder interferometric sensor (MZI),
    51, 52
Macrobending, 310–311
Macrobend sensors, 49, 50
Macro-fiber composite (MFC) rectangular
    patches, 26
Marine composite structures, fiber-optic
    structural health monitoring systems
    for, 415–475
    composite failure, 423–428
        fiber, 424–426
        interface, 426
        interlaminar, 427–248
        matrix, 424
    curved composites, 417–418
    damage characterization of advanced
        composites using optical-fiber sensors,
        437–446
        crack monitoring in bonded repair system,
            439–440
        debonding, 439
        delamination, 442–445
        detection of transverse cracks, 441
        effects of embedded optical fibers on
            structural integrity, 445–446
        matrix cracking, 437–439

experimental investigation of top-hat
  stiffener, 447–452
  VARTM manufacturing process, 447–448
  VARTM top-hat stiffener results,
    448–452
FBG strain sensing, 452–474
  fabrication of FBG sensor, 453–456
  layup and embedding optical fibers,
    456–459
  measurement setup, 462–465
  optical fiber embedding procedure with
    VARTM, 459–462
  results, 465–474
nondestructive evaluation of composites,
  428–437
  optical-fiber sensing, 430–437
  types of NDE, 428–430
out-of-plane joints, 418–423
  keels, 420–423
  T-joints, 418–419
  top-hat stiffener, 419–420
MATLAB, 187
Matrix, 4–6
  cracking, 21, 343
  failure, 424
MFC, *see* Macro-fiber composite (MFC)
  rectangular patches
M55J/M18 CFRP composite, 166–167, 189, 190,
  192
Michelson interferometers (MIs), 51–52
Microbending, 49–50, 311
Microcracking, 142, 342
Micron Optics Inc., 383, 407
Microstructured optical fibers (MOF), 61–62, 175
MIs, *see* Michelson interferometers (MIs)
MLP, *see* Multilayer perceptron (MLP)
MMF, *see* Multimode fibers (MMF)
Modal power distribution (MPD), 308, 309, 310
MOF, *see* Microstructured optical fibers (MOF)
MOI. Inc., *see* Micron Optics Inc.
MPD, *see* Modal power distribution (MPD)
MSC.NASTRAN, 445
MTS Systems Corporation, 383
Multiaxial strain sensing, strain transfer effects
  for embedded, 157–196
  coordinate systems for composite materials
    and optical fiber sensors, 161–162
  optical fiber coating optimization, 179–188
    embedded, 181–183
    examples, 192–195
    limits of applicability, 188–192
    transfer matrix tool, 184–188
  research, 159–161
    embedded Bragg gratings, 160–161
    surface-mounted Bragg gratings, 159–160
  transfer coefficient matrix, 162–179
    analytical approach, 163–167

experimental approach, 176–179
  numerical approach, 168–176
Multilayer perceptron (MLP), 365
Multimode fibers (MMF), 118, 122, 131,
  132, 133
Multiresolution analysis, 282–288
Multivariate statistical procedure, 32
MZI, *see* Mach–Zehnder interferometric sensor
  (MZI)

N

Nantong ZhiXing Laboratory, 380
NASA Dryden Flight Research Center, 144
National Instrument PC1-6251, 463
NDE, *see* Nondestructive evaluation (NDE)
NDT, *see* Nondestructive test (NDT) methods
Neural networks, 361–364
Newton–Cotes rule integration, 409
NI PXI, 226, 227
Noncontact measurement methods, 31, 32
Nondestructive evaluation (NDE), of composites,
  25, 428–437
  optical-fiber sensing, 430–431
    basic principles, 431–432
    Bragg grating sensors, 434–436
    FBG interrogation, 436–437
    modulation techniques, 432–433
    multiplexing, 436
  types of, 428–430
Nondestructive test (NDT) methods, 21, 22,
  133–134, 340
Nonfixed keels, 420
Nonlinear vibro-acoustic method, 34

O

OF, *see* Optical-fiber (OF) sensors
OFBG-FRP bars, 376, 385, 386
  smart CFRP bridge cable based on, 390
  smart steel wire bridge cable based on, 390
OFDR, *see* Optical frequency domain
  reflectometer (OFDR)
OFS, Optical-fiber sensing (OFS)
OLCR, *see* Optical low coherence reflectometer
  (OLCR)
OPD, *see* Optical path difference (OPD)
Optical fiber, 299–334
  application of, sensors in smart composites
    cure monitoring, 315–322
    monitoring composite mechanical
      properties, 323–333
  bending loss of, integrated in textile
    preforms, 310–313
    critical bend radius, 312–313
    macrobending, 310–311
    microbending, 311

embedded into textile preform for composite
applications, 300–310
3D braid preforms, 307–310
woven preforms, 300–306
sensing, 430–431, 461
basic principles, 431–432
Bragg grating sensors, 434–436
FBG interrogation, 436–437
modulation techniques, 432–433
multiplexing, 436
side-emitting properties, 313–315
type, 175–176
Optical fiber coating optimization, 179–188
embedded, 181–183
examples, 192–195
combined axial and transverse load,
193–194
global optimum, 194–195
pure axial/transverse load, 192–193
limits of applicability, 188–192
composite layup, 190–191
eccentricity, 191–192
part size, 189–190
transfer matrix tool, 184–188
determination, 185
FE implementation, 186–188
model description, 184–185
Optical fiber (OF) sensors, 41, 70, 373, 374, 376,
378, 381
in composite production, 202–264
challenges and future requirements, 264
industrial manufacturing, 244–264
laboratory manufacturing, 225–244
manufacturing processes and process
parameters, 203–210
principles of, 211–225
concept, 42–43
damage characterization of advanced
composites using, 437–446
crack monitoring in bonded repair system,
439–440
debonding, 439
delamination, 442–445
detection of transverse cracks, 441
effects of embedded, on structural
integrity, 445–446
matrix cracking, 437–439
Fresnel, 211–215
industry applications and market, 63–65
instrumentation, 62–63
measurement capabilities and advantages of,
43–62
fiber-optic scattering-based distributed
fiber sensors, 59–61
modulation schemes, 45–59
optical microfiber sensors, 62
photonic crystal fiber sensors, 61–62

surface plasmon resonance sensors, 61
types of, 44–45
Optical frequency domain reflectometer (OFDR),
82, 83, 112, 113, 114, 127, 128, 231, 305
Optical low coherence reflectometer (OLCR), 93
Optical microfiber sensors, 62
Optical path difference (OPD), 51, 53
Optical spectrum analyzer (OSA), 436
Optical switch/coupler, 389
Optical time domain reflectometer (OTDR), 82,
110, 111, 112, 114–115, 118, 120, 317,
318, 329, 330
Origin 8.0 software, 380, 402
Ormocer-coated optical fiber, 189, 190
Orthogonal Chebyshev polynomials, 26
OSA, *see* Optical spectrum analyzer (OSA)
OTDR, *see* Optical time domain reflectometer
(OTDR)
Out-of-plane joints, 418–423
keels, 420–423
–hull connection, 421–423
types of, 420–421
T-joints, 418–419
top-hat stiffener, 419–420
Oxford Instruments, 260

P

Partially interactive/failure mode-based
theories, 343
Passive demodulation techniques, 81
Passive system, SHM, 22
PBS, *see* Polarization beam splitter (PBS)
Phased array method, 24
Phase mask technique, 71, 453
Phase-modulated fiber-optic sensors, 50–54,
433, 442
Photonic crystal fiber sensors, 61–62
Piezoelectric sensors, 23, 24, 28, 29, 30, 31
Pitch-catch method, 24
Plain weave, 302
PMF, *see* Polarization-maintaining fiber (PMF)
POF, *see* Polymer optical fiber (POF)
Point sensor, and distributed sensor, 44–45
Polarimetric sensors, 432
fiber, 58–59
fiber-optic, 308
Polarization beam splitter (PBS), 125, 126
Polarization-maintaining fiber (PMF), 54,
128, 175
Polymer-matrix materials, 4
Polymer optical fiber (POF), 300, 301, 313, 314,
317, 318, 319, 328–329, 330–333, 432
Polyvinyl chloride (PVC), 31
Postfabrication
mechanical testing, 256–260
test monitoring, 260–264

PPP-BOTDA, *see* Pulse-prepump Brillouin
    optical time domain analysis
    (PPP-BOTDA)
Preform, 11
Pre-impregnated (prepreg) layup method, 6–7
Prepreg layup method, *see* Pre-impregnated
    (prepreg) layup method
Prestressed strand, based on FRP, 390–394
    monitoring principle of, 392–394
    sensing properties characterization of, 394
Prestressing, 390
Prestress loss monitoring, of beam members,
    400–408
    description of beam structure and experiment
        setup, 401
    experimental results, 402–404
        with multi-strand, 403–404
        with single smart strand, 402
    health, of bridge cable in Yanghe River, 404
    of suspenders in Erbian Dadu River Arch
        Bridge, 404–406
    of Tianjin Yanghe Bridge, 406–408
Process-induced strains, 253
Process temperature, 207–208, 233–238
PTFE tubing, 248
Pulling tests, 396
Pulse-echo method, 24
Pulse-prepump Brillouin optical time domain
    analysis (PPP-BOTDA), 345, 349–351
Pultrusion method, 10–11, 410
PVC, *see* Polyvinyl chloride (PVC)
PZT sensors, *see* Piezoelectric sensors

Q

Quasi-static loads, 430

R

Rainflow method, 408
Raman-based distributed temperature sensing,
    120–122
Raman OTDR (ROTDR), 114, 118, 133
Raman scattering, 59, 60–61, 110
Raman Stokes, and anti-Stokes, 121, 122
Rayleigh-based distributed temperature/strain
    sensing, 125–127
Rayleigh scattering, 59–60
Rayleigh waves, 22
RC, *see* Reinforced concrete (RC)
Readout resolution, *see* Sample spacing
Refractive index (RI), 54, 57, 71, 72, 73, 238–242
Refractometer instrument system, 226
Reinforced concrete (RC), 30, 372, 378, 400, 401
Reinforcements, 3–4
Residual strain, 88–95, 326
Resin

curing of, 238, 239
    infusion, 206–207
        and flow monitoring, 225–233
    refractive index of, 238–242
Resin transfer molding (RTM) method, 11, 322
Resistance foil strain gauges (RFSG), 263, 264
Resistive ladder sensors, 34
Response delay, in length, 107
RFSG, *see* Resistance foil strain gauges (RFSG)
RFSG rosette sensor, 256, 259
RI, *see* Refractive index (RI)
Righting moment (RM), 421
Root-mean-square deviation (RMSD), 30
ROTDR, *see* Raman OTDR (ROTDR)
RTM, *see* Resin transfer molding (RTM) method
RTM6 resin system, 226, 234, 236, 242, 243,
    249, 253

S

Sagnac interferometric (SI) sensors, 53–54
Sample spacing, 107
Sampling rate, 115
Sampling resolution, *see* Sample spacing
SAP, *see* Stress-applying parts (SAP)
SBS, *see* Stimulated Brillouin scattering (SBS)
Scale waveform dimension analysis, 32
Scanning laser Doppler vibrometer (SLDV), 25,
    31, 32
SDM, *see* Spatial-division multiplexing (SDM)
Self-sensing anchor, 397–399
Sensor position, effect of, 174–175
SH, *see* Shear horizontal (SH) waves
Shape memory alloys (SMA), 374
Shear failure, 424, 425
Shear horizontal (SH) waves, 22
Shear stresses, 344
Shear vertical (SV) waves, 22
SHM, *see* Structural health monitoring (SHM)
SI, *see* Sagnac interferometric (SI) sensors
Side-emitting properties, optical fiber, 313–315
Signal attenuation, 318, 319
Silica optical fibers (SOF), 300, 301, 313
Single-mode fiber (SMF), 50, 118, 122, 131
Single-phase materials, 351
SLDV, *see* Scanning laser Doppler vibrometer
    (SLDV)
SMA, *see* Shape memory alloys (SMA)
Smart components, and structures, 299, 351
    FRP-based, 371–412
        applications of, 400–409
        existing, 375–400
        principles of, 373–375
    and structural health monitoring, 13–14
SMARTEC, 61
SMF, *see* Single-mode fiber (SMF)
SM FBG sensors, 245, 250, 253, 254, 256, 259

SOF, *see* Silica optical fibers (SOF)
Spatial-division multiplexing (SDM), 436
Spatial modulation, 308
Spectral finite element method, 26
Spontaneous Brillouin scattering, 60
SPPs, *see* Surface plasmon polaritons (SPPs)
SPR, *see* Surface plasmon resonance (SPR)
    sensors
Spray layup method, 8–9
SPWs, *see* Surface plasmon waves (SPWs)
SRIM, *see* Structural reaction injection molding
    (SRIM) method
Stanford Research Systems, 226
Steel wire bridge cable, 390
Stimulated Brillouin amplification, 60
Stimulated Brillouin scattering (SBS), 124
Strain, 305
    -based damage detection, 430
    development, 243
    distribution, 140–144, 467, 470
    gauge, 32, 33
    mode shapes, 32
    monitoring, 256–260
    transfer effects for embedded multiaxial
        strain sensing and optical fiber,
        157–196
        coating optimization, 179–188
        coordinate systems for composite
            materials and, sensors, 161–162
        examples, 192–195
        limits of applicability, 188–192
        research, 159–161
        transfer coefficient matrix, 162–179
Stress-applying parts (SAP), 175
Structural health monitoring (SHM), 65, 340,
        346, 348, 350, 352, 354, 360, 373, 374
    application to, of composite materials,
        118–120
    of composite arch bridge, 271–274
        description of FBG-based, 273–274
        design and implementation of, 272–273
        features of, 271–272
    of composite materials, 129–148
        strain sensing, applications of, 134–148
        temperature sensing, applications of,
            131–134
        using fiber Bragg gratings, 69–95
    in composite structures
        demand for, 14–16
        system for, 16–17
    fiber-optic, for marine composite structures,
        415–475
        curved, 417–418
        damage characterization of, 437–446
        experimental investigation of top-hat
            stiffener, 447–452

failure, 423–428
    FBG strain sensing, 452–474
    nondestructive evaluation of, 428–437
    out-of-plane joints, 418–423
    issues in practical implementation, 115–118
    methods for composite materials, 21–35
        acoustic emission, 26–28
        elastic waves–based, 22–26
        electromechanical impedance, 28–30
        other, 33–34
        vibration-based, 30–33
    smart structures and, 13–14
Structural reaction injection molding (SRIM)
        method, 11
Structural stress monitoring, using FBG-based
        strain sensors, 274–288
    deployment of, 275–278
    fundamentals of, 274–275
    separating temperature effect using wavelet
        multiresolution analysis, 282–288
    strain monitoring data from, 278–282
Superconducting magnet coil, 260–264
Surface-bonded RFSG sensors, 259
Surface-mounted Bragg gratings, 159–160
Surface plasmon polaritons (SPPs), 61
Surface plasmon resonance (SPR) sensors, 61
Surface plasmon waves (SPWs), 61
SV, *see* Shear vertical (SV) waves

T

Tapered optical fiber sensor, 215–216
TC, *see* Transfer coefficient (TC) matrix
TDM, *see* Time-division multiplexing (TDM)
Temperature-induced stress, 284–285
Tensile failure, 424, 425
Textile preform, for composite applications
    optical-fiber embedded into
        bending loss of, 310–313
        3D braid preforms, 307–310
        woven preforms, 300–306
    optical-fiber sensors in smart, 315–333
        cure monitoring, 315–322
        monitoring mechanical properties,
            323–333
    optical fiber side-emitting properties, 313–315
TFBG, *see* Tilted fiber Bragg grating sensor
        (TFBG)
Thermal effects, 430
Thermoplastics, 4
Thermoset matrices, 4
3D carbon fiber cantilever beam, 303
3D orthogonal woven preforms, 301
3D woven preform, 317
Three-point bending tests, 442
Tianjin Yanghe Bridge, monitoring of, 406–408

Tilted fiber Bragg grating sensor (TFBG), 78, 79, 86, 222–225, 226, 227, 228, 234, 236, 237
Time-division multiplexing (TDM), 436
T-joints, 418–419, 445
TLS, *see* Tunable laser source (TLS)
Top-hat stiffener, 419–420
   experimental investigation of, 447–452
   VARTM
      manufacturing process, 447–448
      results, 448–452
Transfer coefficient (TC) matrix, 162–179, 183, 189, 190, 191–192, 193
   analytical approach, 163–167
      application to embedded sensor in M18/ M55J, 166–167
   experimental approach, 176–179
      results, 179
      setup, 176–179
   numerical approach, 168–176
      finite element model and loading conditions, 168–169
      parametric study, 169–176
   tool, 184–188
      determination, 185
      FE implementation, 186–188
      model description, 184–185
Transform-domain processing, 282
Transverse strain, 77, 93
Transverse waves, 22
Tunable laser source (TLS), 112, 113, 126
2D woven preform, 300, 311
3D braid preforms, 307–310
Two-fiber-type reflection sensors, 46
Type I, and IA FBGs, 86

U

UD, *see* Unidirectional (UD) laminate
Unidirectional (UD) laminate, 170–172, 192
University of New South Wales, 453
Unsaturated polyester
   /Durez resin blend, 315
   /Plyophen resin blend, 315

V

Vacuum-assisted resin transfer molding (VARTM), 304, 456

manufacturing process, 447–448
optical fiber embedding procedure with, 459–462
results, 448–452
   top-hat stiffener layup 1 experimental, 449–450
   top-hat stiffener layup 2 experimental, 450–451
   top-hat stiffener layup 3 experimental, 451–452
   top-hat stiffener layup 1 FBG strain-sensor measurement, 466–474
Vacuum bagging system, 8, 207, 226, 231, 234, 242, 322
VARTM, *see* Vacuum-assisted resin transfer molding (VARTM)
VBM, *see* Vibration-based method (VBM)
VCCT, *see* Virtual crack-closure technique (VCCT)
Very-high-order Lobatto polynomials, 26
Vibration-based method (VBM), 30–33
Virtual crack-closure technique (VCCT), 344
Virtual crack extension technique, 344

W

Wavelength-division multiplexing (WDM), 55, 81, 82, 251, 254, 260, 436
Wavelength-modulated sensors, 54–57
Wavelet-based nonparametric approach, 282–288
WDM, *see* Wavelength-division multiplexing (WDM)
Weakly tilted FBG, 86–87
Weight, of structure, 430
Wet layup method, 7–8
Witten, Elmar, 16
Woven preforms, 300–306
Wrapping cured concrete method, 383

Y

Yanghe River, bridge cable in, 404
Young modulus, 396

Z

ZhiXing FRP Reinforcement Nantong Co., Ltd, 380, 397
Zhou, 410

Printed and bound by CPI Group (UK) Ltd, Croydon, CR0 4YY

24/10/2024

01778306-0009